W0225830

Geology of the Cayman Islands

Brian Jones

Geology of the Cayman Islands

Evolution of Complex Carbonate Successions on Isolated Oceanic Islands

Brian Jones
Department of Earth and Atmospheric Sciences
University of Alberta
Edmonton, AB, Canada

ISBN 978-3-031-08229-0 ISBN 978-3-031-08230-6 (eBook)
https://doi.org/10.1007/978-3-031-08230-6

This Springer imprint is published by the registered company Springer Nature Switzerland AG
The registered company address is: Gewerbestrasse 11, 6330 Cham, Switzerland

Preface

Located in the Caribbean Sea, the Cayman Islands (Grand Cayman, Little Cayman, Cayman Brac) are famous for the spectacular reefs that attract tourists and scuba divers from around the world. The carbonate sediments generated in these crystal-clear tropical waters develop through complex arrays of intertwined biogenic and abiogenic processes. These islands, surrounded by deep oceanic waters, are, in reality, the summits of submarine mountains that rise from the depths of the Caribbean Sea. Isolated from any other land mass, they lie in a complex tectonic setting with each island being located on separate fault blocks that are part of the east-west Cayman Ridge that forms the northern boundary of the Cayman Trench. This is a tectonically active zone with a spreading center located off the southwest corner of Grand Cayman and the Oriente Transform Fault that forms the boundary between the Cayman Ridge and the Cayman Trench. Against this background, the carbonate succession on each island developed into response to temporal sea level changes that were dictated by the interplay between eustatic sea level changes and tectonic isostasy.

With the passage of time, a variety of diagenetic processes transformed the shallow-water carbonate sediments into well-lithified rocks through a wide range of different, complex diagenetic processes. Pervasive dolomitization of the Oligocene to Pliocene successions produced thick successions of "island dolostones" that are similar to those found on other islands throughout the oceans of the world. Collectively, these diagenetic processes add another layer of complexity for deciphering the geological evolution of these islands. Although our knowledge of the processes that control the genesis of these sediments and their subsequent diagenetic changes has greatly advanced over the last 50 years or so, much remains to be discovered.

This book provides an up-to-date overview of the vast amount of data, assembled over the last 40 years of research, that offers critical insights into the geological evolution of the islands. Although some significant advances have been made with respect to our understanding of the bedrock geology and the modern depositional environments of the islands, this research has also highlighted other enigmatic problems that still need to be resolved. This research, driven primarily by academic curiosity, has also produced some significant practical applications. The information has, for example, been extensively used in the development of the reverse osmosis plants that now provide much of the drinking water on the islands. The lessons learnt from these three small islands carry implications for deciphering the geological evolution of isolated islands throughout the oceans of the world.

Edmonton, Canada — Brian Jones

Acknowledgements

This book is based on papers that have been published in various geological journals and my own research program that has focused on the Cayman Islands since 1981. The research that I and my graduate students undertook on the Cayman Islands would have been impossible without the help of many individuals and organizations who have facilitated much of the research. I am indebted to various companies on the islands who kindly gave me permission to collect samples from the quarries on Grand Cayman, Little Cayman, and Cayman Brac, and landowners who allowed us to collect samples and, in some cases, to drill wells on their land. Many of the wells used during this research were drilled by Industrial Services and Equipment Ltd., Grand Cayman and Scott Development Co. Ltd. on Cayman Brac. Innumerable graduate students, who elected to complete their M.Sc. and/or Ph.D. theses on various aspects of the islands' geology, have greatly contributed to our current understanding of the islands. Similarly, the broad spectrum of analytical techniques used to decipher these rocks has involved many highly skilled individuals at the University of Alberta and other analytical laboratories in Canada. Funding for the research came from the Natural Sciences and Engineering Council of Canada and the University of Alberta. All of the research was conducted with all the necessary permits being in place.

I am deeply indebted to all of the people and organizations who made research on the Cayman Islands possible. Logistical support on the Cayman Islands was commonly provided by the Water Authority and the Department of Environment. In particular, I would like to thank Dr. Gelia Frederick-van Genderen, Director, Water Authority who readily supported the research that I undertook on the Cayman Islands. My sincere thanks are also extended to my great friend Hendrik van Genderen (Water Authority) who helped resolve many of the logistical issues that I encountered as well as the countless days that he spent in the field while drilling was being done and the assistance he provided as we explored the islands. I am also indebted to numerous individuals from the Water Authority who helped during the drilling operations. I must also fully acknowledge the assistance that Gina Ebanks-Petrie, Director, Department of Environment provided and the assistance of numerous individuals from that department who helped in obtaining the underwater cores and many of the sediments that were used during the course of this research. Additional help and data have been provided by APEC Consulting Engineers Ltd and DART Cayman Islands. The research embodied in this book would have been impossible without the assistance from my Caymanian friends.

Contents

Introduction

1

1.1 Introduction

The Cayman Islands, comprising Grand Cayman, Little Cayman, and Cayman Brac, are located in the central part of the Caribbean Sea where they are separated by deep oceanic waters from Cuba to the north and Jamaica to the southeast (Fig. 1.1). Little Cayman and Cayman Brac, discovered by Christopher Columbus on May 10, 1503, were originally named the "Las Tortugas"—a name derived from the numerous turtles that he saw on the shores of the islands. A map of the Caribbean Sea produced by Juan Vespucci in 1526 showed three islands that he named "*Caymanos*" and a map presented by Pierre Deceliers in 1545 recognized "Cayman Grande" and the two smaller "Caymanes" (Craton 2003). In 1556, Guillaume Le Testu produced a map of the Caribbean that included the "Caimannes" that showed the islands more accurately located and indicated North Sound on Grand Cayman for the first time. Originally, the Lessor Caymans (Little Cayman and Cayman Brac) were claimed by Spain based on their sighting by Columbus in 1503. England's claim to the islands, as early as 1662, was formally acknowledged by Spain in the Treaty of Madrid that was signed in 1670 (Craton 2003).

The first map of Grand Cayman, which also showed the locations of shipwrecks and turtle pens, was produced in 1773 by George Gould who spent about 10 days around the island (Craton 2003; Lawson 2004). At that time, the population of the island was about 450 people. By 1802 the population had increased to 933 people (census by Edward Corbet) with the largest settlement being at Bodden Town. The first maps of the sister islands of Little Cayman and Cayman Brac that showed land elevations and water depths around the islands were produced by Richard Owen, who was a crew member on the HMS Blossom that visited the islands in 1831. Additional maps were produced by the crew of the HMS Sparrowhawk who charted the islands in 1880 and 1881. A census in 1921 recorded a population of 5,253 people on all three islands (Matley 1926a). Although the population in 1970 was only 9,144 people, it has since risen to ~70,000 people (early results from 2021 census, Cayman Islands).

Early information on the geology of the Cayman Islands was provided by English (1912) who stated that "Grand Cayman is just a flat coral covered summit of a the submarine ridge which extends the southern mountain range of Cuba to the westward, so that its surface geology is of the simplest, and its fauna and flora are those of an "oceanic" island". He noted the hard surface rock on the islands is underlain by "marl".

Hydrick (1917) noted that "…the islands are of coral formation…" and recognized that on Cayman Brac that there was an older central part and that the "…lowland sections were added afterwards by the lifting of the ocean-bed". Although von Humboldt (von Humboldt and Bonpland 1889, p. 199) never landed on the islands, he described Cayman Brac as having "…a rocky wall, bare and steep towards the south and south-east" and commented that "From its whiteness and its proximity to Cuba I suppose it to be of Jura Limestone".

In the early 1880s, phosphate deposits were discovered near George Town on Grand Cayman (Hirst 1910). Operations by the "Grand Cayman Phosphate Company" began in May 1884. Additional phosphate deposits were subsequently found on the northwest part of Grand Cayman and the west end of Little Cayman. Mining of these deposits continued for five years before vast phosphate deposits were found on the west coast of Florida that proved to be more economical to mine.

The first geological survey of the Cayman Islands was undertaken by Charles Matley in 1926 (Matley 1926a). While the Government Geologist for Jamaica, Matley visited the Cayman Islands in 1924 and spent four days on Cayman Brac, two days on Little Cayman, and five days on Grand Cayman. Although the main results of his visit were summarized in his 1926 paper (Matley 1926a), he also produced some short articles that were published in various venues, including newspapers (Matley 1923, 1924a, b c, 1925). At that time, access to many parts of Grand Cayman, including

B. Jones, *Geology of the Cayman Islands*,
https://doi.org/10.1007/978-3-031-08230-6_1

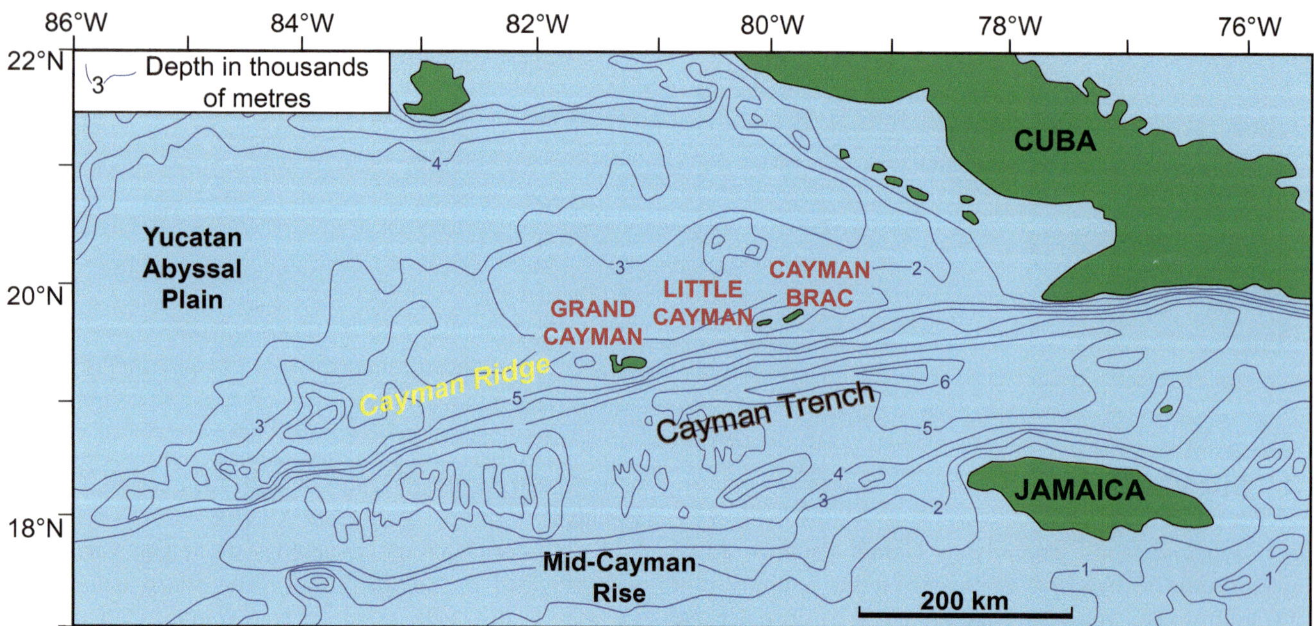

Fig. 1.1 Map showing locations of Grand Cayman, Little Cayman, and Cayman Brac on the Cayman Ridge that forms the northern boundary of the Cayman Trench. Modified from Jones (1994) with map based on information from Perfit and Heezen (1978, their Fig. 2) and MacDonald and Holcombe (1978, their Fig. 1)

the eastern part, was limited. Accordingly, his investigation of the island was based on (1) examination of the western peninsula and (2) a circuit of the island that involved land travel to Gorling Bluff (southeast corner of Grand Cayman) and then by boat along the east and north coasts before sailing across North Sound from Rum Point to George Town. Given the time that Matley spent on each island, the maps that he produced for Cayman Brac (his Fig. 2), Little Cayman (his Fig. 5), and Grand Cayman (his Fig. 8) are remarkably similar to the maps that are currently in use. His recognition and naming of the Bluff Limestone and the Ironshore Formation has stood the test of time as this simple scheme still forms the backbone of that used today. Matley (1926b) also provided a summary of the phosphate deposits that are found on the islands. Some of the foraminifera collected from limestones on the east end of Cayman Brac were sent to Vaughan (1926) for dating so that the temporal setting of the carbonate succession could be established.

The fact that the Cayman Islands are geographically isolated and difficult to reach is reflected by the dearth of geological publications in the 20 to 25 years following Matley's papers. Passing mention of the geology of the islands is made in Douglas (1940) who noted the uplifted core of Cayman Brac and Billymer (1946) who recognized that the cores of the islands are formed of the Bluff Limestone whereas the coastal areas were formed of the Ironshore Formation that contained numerous molluscs. Warthin (1959) focused on the Ironshore Formation and described the limestones around George Town that were cut by two sets of vertical master joints, ~90° to each other and spaced about 30 m apart with lesser joints 6–10 m apart. He also made note of the "solution notch" in the face that is commonly 2 to 2.5 ft high and centered ~4 in above mean tide.

Based on the land snails found on the Cayman Islands, Richards (1955) debated if the Cayman Islands had once been connected to Cuba or Jamaica. Although Schuchert (1910) suggested that there had once been a land bridge between the Cayman Islands and the Sierra Maestra (Cuba), Pilsbry (1931a, b) argued against this by noting that the mollusca on Grand Cayman were more similar to those on Jamaica and suggested that the land mollusca had been blown from Jamaica to Cayman during hurricanes. Bond (1948) suggested that migration of the snails between the islands was associated with other animals that moved between the islands.

Doran (1954) provided an overview of the general setting of the Cayman Islands and focused on the main geomorphic features of Grand Cayman. He drew attention to the "ironshore", the beach ridges formed of large cobbles (his Fig. 8), beachrock on the west coast of the island, the high peripheral rims found around the east end of the island, the sinkholes, and vertical solution-widened fractures, and the linear ridges and swales in the interior of the eastern end of the island. The various ideas that he proposed for the formation of many of these features have since given way to more realistic theories that were based on more rigorous analyses of the geology of the island. Rehder (1962) identified 36 species of molluscs that came from the Ironshore Formation at a locality about half way across the Grand Cayman between Little Bluff and Half Moon Bay.

As travel to the Cayman Islands became easier in the early 1970's, the number of geological studies of the islands increased. Important papers published in the early 1970's included descriptions of the facies and mineralogy of the sediments in North Sound (Roberts 1971a, b), examination of beachrock genesis at various localities around Grand Cayman (Moore 1973), the description of phytokarst from Hell, Grand Cayman (Folk et al. 1973), facies in the Ironshore Formation on Grand Cayman (Brunt et al. 1973), and hydrogeological examination of three major freshwater lenses on Grand Cayman that are housed in the Bluff Formation (Bugg and Lloyd 1976).

During the 1970's and early 1980's, field trips associated with various geology conferences focused largely on the modern sediments and reefs found in the crystal-clear waters around Grand Cayman. Guidebooks from those field trips provide excellent descriptions of the modern carbonate depositional settings and commonly pinpointed ideal dive locations for exploring these vibrant settings. Such publications included those by Roberts (1977) who focused on the west coast of Grand Cayman, North Sound, and the Ironshore Formation, and Roberts and Sneider (1982) who led a field trip to Grand Cayman that focused largely on the modern carbonate environments and the Pleistocene carbonates of the Ironshore Formation. Roberts (1983) used a side-Scan Sonar, sub-bottom profiler, and a linear chart recording fathometer to assess the location of sediment sinks and the areas of maximum sediment input to the shelf and if the morphology of the reefs influenced that process. He showed that (1) much of the sediment found on the southwest shelf came from sediment flushed out of South Sound by the strong reef overwash that affected that lagoon, and (2) wide gaps in the offshore reefs provided avenues by which sediment could be moved offshore and into deeper water.

In 1980, Emery and Milliman (1980) noted that some of the dredged samples from the walls of the Cayman Trough were formed of shallow-water limestones. They also drew attention to two wells drilled on Grand Cayman by K.E. Merren and D.P. Hamilton in 1956 to depths of 401 m (well #1) and 159 m (well #2) to test the oil potential of the island (see Emery and Milliman 1980, their Fig. 1). Although no indication of oil was found, the wells showed that the carbonate succession was at least 401 m thick with well cuttings revealing the presence of crystalline limestones that contained Oligocene foraminifera. Efforts to find the samples collected from those wells have been unsuccessful.

Emery (1981) showed that the shorelines around Grand Cayman were characterized by muddy peats associated with mangroves, rocky sea cliffs and platforms, sandy beaches, and beachrock. He noted the presence of submarine terraces at depths of 8 m and 15 m and shelf-breaks at 20–25 m and 73–75 m. He also reported limestones dredged from a depth of 3000 m and argued that the island had been subsiding at a rate of 10 cm/10^3 years. Based on data from wells 1 and 2 drilled in 1956, he argued that the subsidence rate was 0.1 to 1.3 cm/10^3 years.

In the early 1980's, graduate students under the guidance of Dr. David Stoddart focused their attention on the sediments found in the mangrove swamps on the western part of Grand Cayman and weathering of exposed carbonate rocks. Woodroffe et al. (1980), Woodroffe (1981, 1982) and Woodroffe et al. (1983) focused attention on the sand, mud, and peat deposits associated with the mangrove swamps on the western part of Grand Cayman. Spencer (1981, 1985a, b) and Spencer et al. (1984) examined the rates of subaerial erosion associated with reef carbonates and calcarenites exposed on the western part of Grand Cayman. Stoddart (1980) provided an overview of the geology and geomorphology of Little Cayman.

Acker and Risk (1985), Rose and Risk (1985), and Murphy et al. (2016) demonstrated the role that natural borers, such as *Cliona*, play in the production of sand-sized carbonate grains in the waters around Grand Cayman. Much of this sediment was produced from borings into corals. Johns and Moore (1988) demonstrated that *Halimeda* forms a significant part of the sediment around Grand Cayman and showed that most of this sediment originates in the deep fore reef and is swept onto the reef where it becomes an integral part of the reef framework.

Surface and subsurface karst is common on all of the Cayman Islands. Gilleland (1998) described the numerous caves found on Cayman Brac with some being later described by Tarhule-Lips (1993, 1999) in her M.Sc. and Ph. D. theses and various papers that emanated from that work (Tarhule-Lips and Ford 1996, 1998a, b, 1999).

Terrestrial deposits on the Cayman Islands commonly contain an abundant, diverse biota that includes the remains of insectivores, rodents, and crocodiles (Morgan and Patton 1979; Morgan and Albury 2013; Morgan et al. 2019).

1.2 Physiography of the Cayman Islands

The fact that the physiography of Grand Cayman, Little Cayman, and Cayman Brac are different has a significant impact on bedrock exposures and hence the methods that have to be employed to characterize the geological framework of each island. In this respect, the following points are important.

- Grand Cayman, which is 35 km long with an average width of 6.5 km, covers an area of 90.7 km^2. Most of the island is low-lying with the maximum elevation on the "Mountain" that is ~17 m asl. The dominant feature is North Sound, on the west end of the island, which covers an area of ~91 km^2 (Fig. 1.2). Most of the western peninsula and the area along its southern and eastern

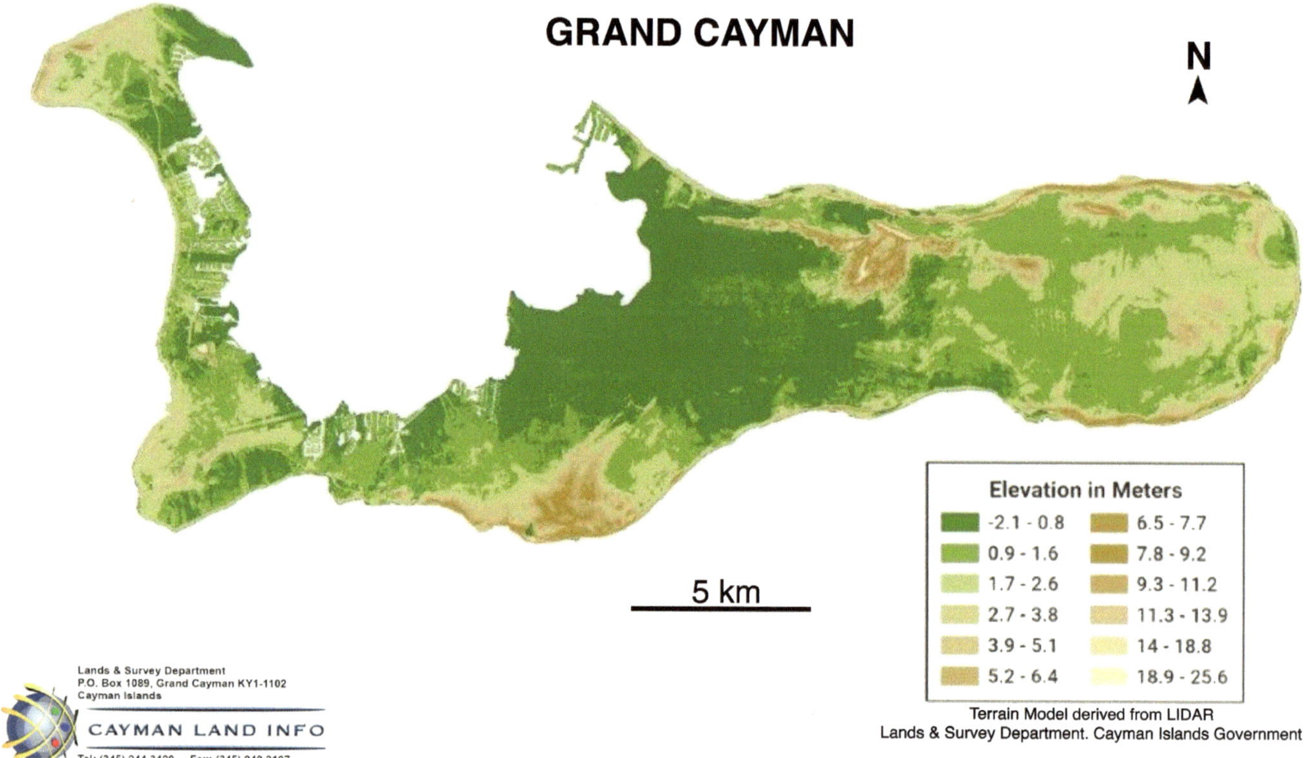

Fig. 1.2 Topographic map of Grand Cayman. Note the highest elevations are around the Mountain (light to dark brown) on the north-central part of the island and in the ridge that follows the north, east, and part of the south coastlines. Map based on data from the Cayman Islands Lands and Survey Department and is used herein by permission (Lands and Survey Department © Cayman Islands Government; all rights reserved. *Source* www.caymanlandinfo.ky)

margins are generally ~2 m asl. For the central and eastern parts of the island, the highest elevations are typically associated with the elevated ridge that parallels many sections of the coastline (Fig. 1.2).

- Little Cayman, which is 16 km long with an average width of 1.6 km, covers an area of 26 km^2. Most of the island is low-lying with the maximum elevation being ~15 m asl on the west end of the island (Fig. 1.3a). Along the northern coast there a ridge with many parts being >10 m asl.
- Cayman Brac, which is 19.3 km long with an average width of 2.4 km, covers an area of 39 km^2. The island rises from sea-level at its west end to ~44 m asl at its east end (Fig. 1.3b). A relatively flat platform of variable width is found along the south, west, and north coasts of the island. In contrast, there is no fringing platform on the east end of the island and the vertical to overhanging cliffs plunge directly into the ocean.

On Grand Cayman (Fig. 1.2), bedrock outcrops are widely scattered and generally of limited aerial extent. On the western part of the island, there are scattered outcrops along the western margin of North Sound and along the southwest and northwest coastline. Most of the western peninsula is devoid of outcrops because of its low-lying nature and the extensive development that has occurred there. In some areas, material dredged from canals that cut through the mangrove swamps provides evidence of the bedrock. On the central and eastern part of the island, scattered exposures are found along road-cuts and coastal areas. The interior of the island is difficult to access because of the rugged karst terrain and thick tropical vegetation. Any rock that is exposed is typically weathered and discoloured. Assessment of the bedrock is generally restricted to quarries, including Pedro Castle Quarry on the south coast and High Rock Quarry in the central east part of the island. Both quarries are no longer active and exposures in High Rock Quarry are now limited to scattered, low outcrop areas around its margin. Overall, the exposures on Grand Cayman provide a very limited view of the stratigraphic succession. Accordingly, most of the information on the bedrock succession on Grand Cayman has come from the well cuttings and cores that have been obtained from numerous wells that have been drilled on the island.

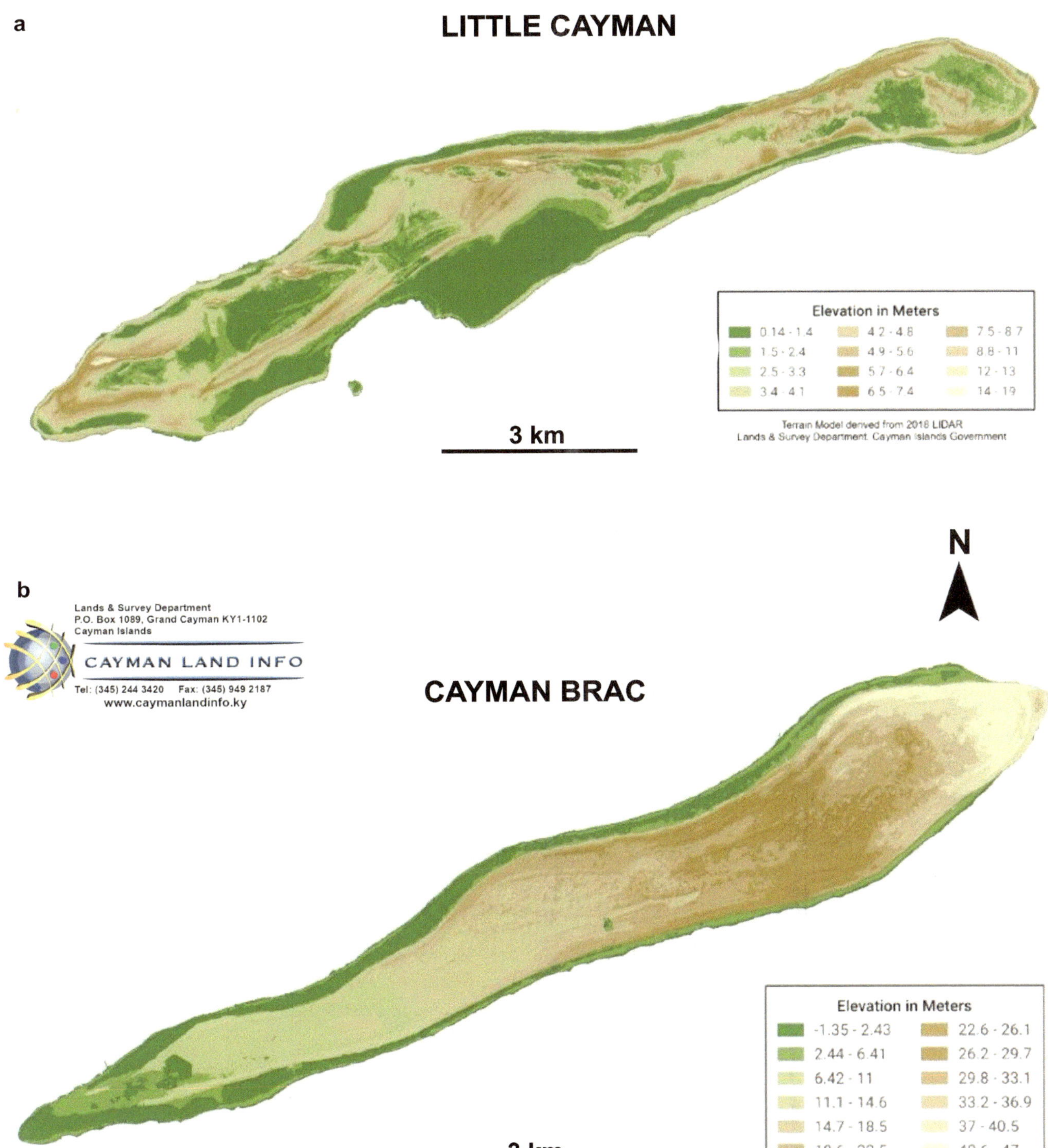

Fig. 1.3 **a** Topographic map of Little Cayman showing highest land at east end of island and various areas on the ridge that parallels the north coast. **b** Topographic map of Cayman Brac showing central core that rises from the west to the east and flanked by a narrow, low-lying platform along the south, west, and north coasts of the island. East end of island characterized by vertical to overhanging cliffs. Maps based on data from the Cayman Islands Lands and Survey Department and is used herein by permission (Lands and Survey Department © Cayman Islands Government; all rights reserved. *Source* www.caymanlandinfo.ky)

On Little Cayman (Fig. 1.3a), exposures are limited to the area around Salt Rocks on the west coast, the small quarry in the west-central part of the island, and scattered outcrops along the northeast corner of the island and along some of the roads. Most of the island is inaccessible because of the dense vegetation and rugged phytokarst. Given that no wells have been drilled and sampled on Little Cayman, information on the bedrock succession of the island is very limited.

Cayman Brac, which rises to ~44 asl at its eastern end, is characterized by an uplifted core with well-exposed, vertical to overhanging cliff faces that are generally inaccessible (Fig. 1.3b). The upper surface of the island, as on Grand Cayman and Little Cayman, is characterized by well-developed phytokarst and dense vegetation. Information on the bedrock geology comes from sections beside roads and paths that cross the island, Jennifer Bay Quarry and Scotts Quarry, and wells that have been drilled and sampled at various locations on the island.

1.3 Cayman Research—University of Alberta

In the early 1980's, many graduating students from the University of Alberta found employment in various oil companies that were based in Calgary, Alberta. A critical part of their geological education was a good working knowledge of carbonate sedimentology as many of the oil and gas reserves in western Canada are located in carbonate strata of various ages. In order to improve their educational experience, I started to run annual field trips to the Cayman Islands so that they could visit the spectacular reefs, carbonate sediments, and the older lithified carbonates found on land. It quickly became apparent that these islands offered superb natural laboratories for research on many different aspects of carbonate sedimentology and diagenesis. As a result, small projects that led to the publication of various papers quickly expanded as graduate students elected to complete their M.Sc. and Ph.D. theses on various aspects of the islands' geology. Such research has now been ongoing for 40 years as critical research questions continued to arise. During that period, 10 Ph.D. theses and 24 M.Sc. theses that dealt with various aspects of Cayman geology were completed at the University of Alberta—as follows:

Ph.D. theses (listed by year of completion).

Tongpenyai, B., 1989. An assessment of the use of image analysis in carbonate sedimentology.

Ng, K.C., 1990. Diagenesis of the Oligocene-Miocene Bluff Fomation of the Cayman Islands—a petrographic and hydrogeochemical approach.

Hunter, I.G., 1994. Modern and ancient coral associations of the Cayman Islands.

Blanchon, P.A., 1995. Controls on modern reef development around Grand Cayman.

Li, C., 1997. Foraminifera: their distribution and utility in the interpretation of carbonate sedimentary processes around Grand Cayman, British West Indies.

Zhao, H., 2013. Origin of island dolostones: case study based on Tertiary dolostones from Cayman Brac, British West Indies.

Li, R., 2014. The Pleistocene Ironshore Formation, Grand Cayman: diagenetic response to sea level change.

Liang, T., 2015. Evolution of eorsional unconformities in the Cenozoic succession of the Cayman Islands.

Ren, M., 2017. Origin of island dolostone: case study of Cayman Formation (Miocene), Grand Cayman, British West Indies.

Booker, S.D., 2020. Variation in climatic conditions from the Cayman Islands through stable isotope and element analyses fom coral and sediment cores.

M.Sc. theses (listed by year of completion).

Lockhart, E.B., 1986. Nature and genesis of caymanite in the Oligocene-Miocene Bluff Formation of Grand Cayman, British West Indies.

Pleydell, S.M., 1987. Aspects of diagenesis and ichnology in the Oligocene-Miocene Bluff Formation of Grand Cayman Island, British West Indies.

Smith, D.S., 1987. The genesis of speleothemic calcite deposits on Grand Cayman, British West Indies.

Squair, C.A., 1988. Surface karst on Grand Cayman Island, British West Indies.

Cerridwen, S.A., 1989. Paleoecology of Pleistocene mollusca from the Ironshore Formation, Grand Cayman, B. W. I.

Rehman, J., 1992. Diagenetic alteration of *Strombus gigas, Siderastrea siderea,* and *Montastrea annularis* from the Pleistocene Ironshore Formation of Grand Cayman.

Shourie, A., 1993. Depositional architecture of the Late Pleistocene Ironshore Formation, Grand Cayman, British West Indies.

Kalbfleisch, W.B.C., 1995. Sedimentology of Frank Sound and Pease Bay, two modern shallow-water hurricance-affected lagoons, Grand Cayman, British West Indies.

Wignall, B.D., 1995. Sedimentology and diagenesis of the Cayman (Miocene) and Pedro Castle (Pliocene) Formations at Safe Haven, Grand Cayman, British West Indies.

Hills, D., 1997. Rhodolite development in the modern and Pleistocene of Grand Cayman.

Vézina, J.L., 1997. Stratigraphy and sedimentology of the Pleistocene Ironshore Formation at Rogers Wreck Point, Grand Cayman.

Montpetit, J.C., 1998. Sedimentology, depositional architecture, and diagenesis of the Cayman Formation at Tarpon Springs Estates, Grand Cayman, British West Indies.

Willson, E.A., 1998. Depositional and diagenetic features of the Middle Miocene Cayman Formation, Roger's Wreck Point, Grand Cayman, British West Indies.

Arts, A.E., 2000. Sedimentology and stratigraphy of the Pedro Castle Formation, SW Grand Cayman, B.W.I.

Beanish, J.M.R., 2000. Sedimentology of a current-dominated lagoon: case study of South Sound, Grand Cayman, B.W.I.

MacKinnon, L., 2000. Sedimentology of North Sound, Grand Cayman, British West Indies.

MacNeil, A.J., 2001. Sedimentology, diagenesis, and dolomitization of the Pedro Castle Formation on Cayman Brac, British West Indies.

Coyne, M.K., 2003. Transgressive-regressive cycles in the Ironshore Formation, Grand Cayman, British West Indies.

Etherington, S.S., 2004. Sedimentology and depositional architecture of the Cayman (Miocene) and Pedro Castle (Pliocene) Formations on western Grand Cayman, British West Indies.

Corlett, H.J., 2005. Epiphyte growth and community structure on *Thalassia testudinum*: a case study from Grand Cayman.

Uzelman, B.C., 2009. Sedimentology, diagenesis, and dolomitization of the Brac Formation (Lower Oligocene), Cayman Brac, British West Indies.

Der, A., 2012. Deposition and sea level fluctuations during Miocene times, Grand Cayman, British West Indies.

Zheng, E., 2017. Environmental controls on alternating aragonite-calcite laminations in notch-speleothems from Cayman Brac, Cayman Islands, British West Indies.

McCormick, C.A., 2019. Eustatic and tectonic controls on the development of the stratigraphic architecture of the Cayman Islands, British West Indies.

This research program on the geology of the Cayman Islands has resulted in 112 papers that have been published in various international journals and conference proceedings. Some of these papers were based on the theses listed above, whereas other papers resulted from and projects that I undertook independent of the student thesis work. These papers are referenced in the appropriate chapters throughout this book.

1.4 Methods

Research on the Cayman Islands has been rooted in extensive field work so that the settings of the modern sediments and the lithified, on-land strata could be completely and accurately described and sampled with the resultant interpretations being based on extensive databases. This research has involved the examination of all accessible outcrops on each of the islands and collection of samples from numerous wells that have been drilled on Grand Cayman and Cayman Brac. As yet, no wells have been drilled on Little Cayman. The fieldwork and drilling provided the framework for assessing all of the detailed laboratory analyses that have been completed. Laboratory analyses have involved thin section petrography, X-ray diffraction analyses, scanning electron microscope analyses, electron microprobe analyses, age-dating through the use of C^{14} and U/Th systems, stable isotope analyses, radiogenic isotope analyses, and geochemical analyses for determination of trace elements and rare-earth elements. Each type of analysis has been based on stringent analytical methods and techniques that are outlined below.

1.4.1 Field Work and Drilling

Field work provided the in-situ framework for assessing the stratigraphic and sedimentological framework of the carbonate successions found on each of the Cayman Islands. Where possible, exposed sections were measured and systematically sampled so that the sequence of formations, their constituent facies, and the key stratigraphic boundaries could be established.

The low-lying topographies on Grand Cayman and Little Cayman limit vertical sections to about 12 m on Grand Cayman and even less on Little Cayman where the scattered outcrops provide vertical successions that are <5 m thick. On Cayman Brac, the situation is different because the exposed core of the island rises to ~40 m above sea level at its east end. In general, however, it is impossible to scale the vertical to overhanging cliff faces and establishing the stratigraphic boundaries and the measurement of sections is restricted to a few locations where roads and footpaths have been established. Given the limitations imposed by the topographical nature of the islands, most of the information on the constituent formations had to be obtained from cores and well cuttings obtained from wells drilled at various locations on Grand Cayman and Cayman Brac. As yet, no wells have been drilled on Little Cayman. On each island, access to the various active and dormant quarries have provided critical, clean exposures of the bedrock.

Well drilling on the islands has been undertaken using a variety of systems. During 1981 and 1982, The Water Authority drilled 20 shallow wells around Lower Valley and Pedro Castle with cores being taken to assess the porosity and permeability of the rocks (Ng and Beswick 1996). Starting in 1991, my research group started to use a JKS Boyles Winkie Portable Drill that was capable of reaching depths of 122 m (400 ft). Between 1991 and 1998, ~50 wells were drilled and cored to various depths (usually determined by the nature of the subsurface rocks that are commonly friable and caused drilling to cease) on Grand Cayman with the main areas of attention being the western peninsula, the area south of North Sound, and around Rogers Wreck Point on the northeast corner of the island. The deepest cored well managed with this system was 100 m at Rogers Wreck Point. This phase of drilling ceased in 1997 once the drilling equipment failed beyond repair.

With the ever-increasing need for drinking water in the early 1990's, the Water Authority on the Cayman Islands had to develop reverse osmosis plants in order to meet the demand. This process involved the drilling of deep wells by Industrial Services and Equipment Ltd., Grand Cayman with both well cuttings and cores being obtained from the wells. The samples from such wells provided further information on the subsurface strata of Grand Cayman. In subsequent years, additional wells cuttings and cores were obtained from wells that Industrial Services and Equipment Ltd drilled at various locations in the central and eastern part of Grand Cayman for research purposes. On Cayman Brac, well drilling was undertaken by Scott Development Co. Ltd. Their drilling rig was limited to the production of well cuttings to a maximum depth of 61 m (200 ft).

1.4.2 Thin Sections

Small (4.7 × 2.7 cm) and large (7.5 × 5.0 cm) thin sections of rocks that had been impregnated with blue epoxy provided the basic information of the field and well samples. Analysis of these thin sections allowed identification of the allochems, matrices, cements, and porosity as well as the distribution of calcite and dolomite throughout the samples. Fluorescent and cathodoluminescent microscopy was also used for the analysis of some samples in order to examine the porosity styles and the compositional changes in some of the cements. The thin sections were routinely stained with Alizarin Red S solution in order to clearly separate the calcite (stained red) from the dolomite (no stain).

1.4.3 Scanning Electron Microscopy

Numerous samples were examined on various models of Scanning Electron Microscopes (SEM) at the University of Alberta in order to assess the microscopic aspects of the samples, including detection and imaging of the main components of the samples, diagenetic fabrics, and the morphology of microorganisms (e.g., bacteria, fungi, algae). The elemental content of specific components of the samples were determined using the Energy-Dispersive X-ray systems that were attached to the SEMs.

In most cases, fractured samples (typically ~1–2 cm cubes) were used for the SEM analyses. Examination of the morphology and internal structures of the dolomite crystals, however, required a different approach that involved either (1) using thin sections that were 50 μm thick rather than the standard 30 μm thickness, or (2) thin, polished wafers of dolomite. Both types of samples were etched in concentrated HCl for 20–30 secs before being rinsed, dried, and mounted on SEM stubs and then coated with gold or carbon. The SEM analyses of these samples allowed imaging and analysis of the internal structures of the dolomite crystals that had been highlightd by the acid-etching process. For the thin section samples, subsequent thinning to 30 μm then provided standard petrographic analyses and images that could be compared with the SEM images of the etched dolomite.

1.4.4 X-Ray Diffraction Analysis

The powders used for X-ray diffraction (XRD) analyses were prepared using an agate mortar and pestle in order to avoid any possible contamination of the samples. Here, it is important to note that large samples were ground up so that the same sample could also be used for other geochemical analyses (e.g., trace element, REE).

The XRD analyses were used to (1) obtain a basic mineralogical analysis of samples, and/or (2) determine the %Ca ((molar Ca/(Ca + Mg)) × 100) of the dolomite. Most of the XRD analyses were acquired using a Rigaku Geigerflex sealed-tube X-ray generator with a Co tube, run at 40 kV and 35 mA. Each backpacked sample container held ~1 g of sample.

Jones et al. (2001) developed a XRD procedure for determining the %Ca of dolomite that involved adding a quartz standard to the powdered dolomite sample and performing a slow XRD scan from 29° to 38° 2Θ. The resultant X-ray diffractograms were analyzed using a peak-fitting technique. Based on these X-ray diffraction analyses, low-calcian dolomite (LCD) is defined as dolomite that contains <55% mol %$CaCO_3$ (hereafter referred to as %Ca), whereas high-calcian dolomite HCD contains >55%Ca (Jones et al. 2001). This method can be applied to any type of sample irrespective of the crystal size of the dolomite.

1.4.5 Electron Microprobe

Polished thins sections were analyzed on a JEOL 8900R electron microprobe at the University of Alberta that was operated with a 15 kV accelerating voltage, 5 nA beam current measured at a Faraday cup, 100 s count time on Ca, Mg, Sr, and Fe peaks, and 50 s count time on the backgrounds. Backgrounds were measured for each analytical point. Although a beam diameter of 5 or 10 μm was generally used, a beam diameter of 3 to 4 μm had to be used on some of the smaller crystals. The resultant data were reduced using the ZAF program provided by JOEL, and CO_2 was calculated by stoichiometry.

1.4.6 Stable Isotope Analyses

Most of the stable isotope analyses presented in this document were determined by Isotope Tracer Technologies Inc. (Waterloo, Ontario, Canada) who used a Finnigan MAT DeltaPlus XL spectrometer. During these analyses, the powdered sample was held at a constant temperature of 50°C following evacuation with ultrapure helium. Samples were treated with 100% phosphoric acid for two hours. Two international standards (NBS-18 and NBS-19) and an in-house standard were measured repeatedly in order to obtain reliable results. The isotope values are reported using the δ notation relative to the Vienna Pee Dee Belemnite (VPDB). Results are specified relative to VPDB standard normalized to NBS-10 and NBS-18 standards. The analytical uncertainties are ± 0.2‰ for $\delta^{18}O$ and $\delta^{13}C$.

1.4.7 $^{87}Sr/^{86}Sr$ Analyses

The $^{87}Sr/^{86}Sr$ analyses of the carbonate samples from the Cayman Islands were completed in the radiogenic isotope laboratory of Dr. Robert Creaser at the University of Alberta. The values are corrected for mass fractionation relative to 0.1194 for $^{87}Sr/^{86}Sr$ and normalized to a value of 0.710245 for SRM-987 standard reference material. Based on replicate analyses of the samples and the standard, the analytical uncertainty is ±0.000010.

1.4.8 Elemental and Rare Earth Elements

Powdered samples were initially ingested in 8 ml HF and 2 ml HNO_3 and left to dry on a 130 °C hotplate overnight until completely dry. Then 5 ml HCl and 2 ml HNO_3 were added and heated at 130 °C for two hours before being transferred to a 15 ml centrifuge tube. Deionized water was added to produce 15 ml final solution: 0.1 ml HNO_3, 0.1 ml of internal standards (In, Bi, Sc), and 8.8 ml deionized water were added to 1 ml of this solution. This solution was analyzed on an Quadrupole Inductively Coupled Plasma Mass Spectrometer (Elan 6000 ICP-MA, Perkin Elmer) in order to determine the concentrations of 36 elements (Li, Be, B, Na, Mg, Al, P, K, Ca, Ti, V, Cr, Fe, Mn, Co, Ni, Cu, An, Ga, Ge, As, Se, Rb, Sr, Zr, Nb, Mo, Ru, Pd, Ag, Cd, Sn, Te, Cs, and Ba). For the Cayman carbonates, most of these elements are below detection limits.

Rare earth elements (REE) and yttrium (Y) were also determined from the same diluted samples. Appropriate oxide interference corrections were applied by running a standard solution (CeCO.Ce < 3%) before calibration. The detection limits were between 0.02 and 0.06 ppm. The REE concentrations are presented after normalization against a standard such as the Post-Archean Australian Shale (PAAS) or Chondrite standards (suffixes 'PN', 'CN', respectively) and plotted against their atomic number (cf., McLennan 1989). Y is inserted between Dy and Ho based on its ionic radius (cf., Bau and Dulski 1996).

1.4.9 U/Th Dating

U/Th dates were completed at the Radiochronology Laboratory at the GEOTOP-UQAM-McGill Research Center, Canada. Each powdered sample was dissolved in nitric acid in a Teflon beaker before a calibrated mixed spike ^{233}U-^{236}U-^{229Th} was added and then allowed to slowly evaporate to dryness. This was then dissolved in $HNO_3$7N and ~10 mg of iron carrier was added and the solution left overnight for spike-sample equilibrium. U and Th were co-precipitated with $Fe(OH)_3$ by adding ammonium hydroxide drop by drop until a pH of 8–9 was attained. The precipitate obtained by centrifuge was washed twice with deionized water before being dissolved in 6 N HCl. The U–Th separation was done in a 2 ml A1X8 anionic resin with the thorium fraction being recovered through elution with 6 N HCl, and the U and Fe fraction, with water. The U fraction was purified on a 0.2 m U-Teva resin volume. Fe was eluted with 3 N HNO_3 and the U fraction with 0.02 N HNO_3. After drying, thorium purification was done in a 2 mL AG1X8 resin in 7 N HNO3 and eluted with 6 N HCl. U-Th measurements were performed using a multi-collector inductively coupled plasma mass spectrometry Nu instrument TM. ^{236}U-^{235}U-^{234} U-^{233} U and ^{232}Th-^{230}Th-^{229}Th were measured on the ion counter (IC0) in peak switching mode for the uranium and thorium isotopes, respectively. ^{238}U was calculated from $^{235}U/^{236}U$ ratios, assuming a constant $^{238}U/^{235}U$ mass ratio (137.88). Knowing the $^{236}U/^{233}U$ of the spike, the mass bias corrections in atomic mass unit (amu1) were calculated and used to correct measured ratios of uranium and thorium isotopes.

References

Acker KL, Risk MJ (1985) Substrate destruction and sediment production by the boring sponge *Cliona caribbea* on Grand Cayman Island. J Sediment Petrol 55:705–711

Bau M, Dulski P (1996) Distribution of yttrium and rare-earth elements in the Penge and Kuruman iron-formations, Transvaal Supergroup, South Africa. Precambr Res 79:37–55

Billymer JHS (1946) The Cayman Islands. Geogr Rev 36:29–43

Bond J (1948) Origin of the bird fauna of the West Indies. The Wilson Bull 60:207–229

Brunt MA, Giglioli MEC, Mather JD, Piper DJW, Richards HG (1973) The Pleistocene rocks of the Cayman Islands. Geol Mag 110: 209–221

Bugg SF, Lloyd JW (1976) A study of freshwater lens configuration in the Cayman Islands using resistivity methods. Q J Eng Geol 9: 291–302

Craton M (2003) Founded upon the seas: a history of the Cayman Islands and their people. Ian Randle Publishers, Kingston, Jamaica, p 532

Doran E (1954) Landforms of Grand Cayman Island, British West Indies. Tex J Sci 6:360–377

Douglas AJA (1940) The Cayman Islands. Geogr J 95:126–131

Emery KO, Milliman JD (1980) Shallow-water limestones from slope off Grand Cayman Island. J Geol 88:483–488

English TMS (1912) Some notes on the natural history of Grand Cayman, Handbook of Jamaica

Folk RL, Roberts HH, Moore CH (1973) Black phytokarst from Hell, Cayman Islands, British West Indies. Geol Soc Am Bull 84:2351–2360

Gilleland T (1998) The caves of Cayman Brac. Nat Speleol Soc News 56:202–207

Hirst GSS (1910) Notes on the History of the Cayman Islands. P.A. Benjamin Manf. Co., Kingston, Jamaica, p 412

Hydrick JL (1917) Report on Hookworm survey of the Cayman Islands from April 18, 1917 to June 20, 1917 (The Rockefellar Foundation)

Johns HD, Moore CH (1988) Reef to basin sediment transport using *Halimeda* as a sediment tracer, Grand Cayman. Coral Reefs 6: 187–193

Jones B (1994) Geology of the Cayman Islands. In: Brunt MA, Davies JE (eds) The Cayman Islands: natural history and biogeography. Kluwer, Dordrecht, The Netherlands, pp 13–49

Jones B, Luth RW, MacNeil AJ (2001) Powder X-ray analysis of homogeneous and heterogeneous dolostones. J Sediment Res 71:791–800

Lawson W (2004) Shipwrecks of the Cayman Islands. Aqua Press, Essex, England

MacDonald KC, Holcombe TL (1978) Inversion of magnetic anomalies and sea-floor spreading in the Cayman Trough. Earth Planet Sci Lett 40:407–414

Matley CA (1923) Report of a reconnaissance geological survey of the Cayman Islands. Geol Survey Cayman Islands, pp 41–45

Matley CA (1924a) Geological Survey of the Cayman Islands. Annual General Report, Jamaica, pp 41–45

Matley CA (1924b) Reconnaissance geological survey of Cayman Islands, B.W.I. The Pan-American Geologist XLII, pp 313–315

Matley CA (1924c) Report of a reconnaissance geological survey of the Cayman Islands. Supplement to the Jamaica Gazette, 13 June 47, pp 69–73

Matley CA (1925) A reconnaissance geological survey of the Cayman Islands, British West Indies. In: British Association for the advancement of science, Report of the 92nd Meeting. Toronto, pp 392–393

Matley CA (1926a) The geology of the Cayman Islands (British West Indies) and their relation to the Bartlett Trough. Q J Geol Soc Lond 82:352–387

Matley CA (1926b) Phosphate in the Cayman Islands. In: 14th International geological congress, Madrid, pp 777–779

McLennan SM (1989) Rare earth elements in sedimentary rocks: influence of provenance and sedimentary processes. Rev Mineral Geochem 21:169–200

Moore CH (1973) Intertidal carbonate cementation, Grand Cayman, West Indies. J Sediment Petrol 43:591–602

Morgan GS, Albury NA (2013) The Cuban crocodile (Crocodylus rhombifer) from Late Quaternary fossil deposits in the Bahamas and Cayman Islands. Fla Museum Nat Hist Bull 52:161–236

Morgan GS, Patton TH (1979) On the occurrence of Crocodylus (Reptilia, Crocodilidae) in the Cayman Islands, British West Indies. J Herpetol 13:289–292

Morgan GS, MacPee RDE, Woods R, Turvey ST (2019) Late Quaternary fossil mammals from the Cayman Islands, West Indies. Bull Am Mus Nat Hist 428:1–19

Murphy GN, Perry CT, Chin P, McCoy C (2016) New approaches to quantifying bioerosion by endolithic sponge populations: applications to the coral reefs of Grand Cayman. Coral Reefs 35:1109–1121

Ng K-C, Beswick RGB (1996) Management of ground water resources on Grand Cayman: a methodology for developing small fresh water lenses. Apett J 30:51–59

Perfit MR, Heezen BC (1978) The geology and evolution of the Cayman Trench. Geol Soc Am Bull 89:1155–1174

Pilsbry HA (1931a) Results of the Pinchot South Sea Expedition-II Land Mollusks of the Canal Zone, the Republic of Panama and the Cayman Islands. Proc Acad Nat Sci Philadelphia 82:339–354

Pilsbry HA (1931b) Results of the Pinochot South Sea expedition I. Land mollusks of the Caribbean Islands, Grand Cayman, Swan, Old Providence and St. Andrews. Proc Nat Acad Sci Philadephia 82:221–261

Rehder HA (1962) The Pleistocene mollusks of Grand Cayman Island, with notes on the goelogy of the island. J Paleontol 36:583–585

Richards HG (1955) The geological history of the Cayman Islands. Notulae Naturae 284:1–11

Roberts HH (1971a) Environments and organic communities of North Sound, Grand Cayman Island, B.W.I. Carib J Sci 2:67–79

Roberts HH (1971b) Mineralogical variation in lagoonal carbonates from North Sound, Grand Cayman Islands, B.W.I. Sed Geol 6:201–213

Roberts HH (1983) Shelf Margin Reef Morphology: A clue to major off-shore sediment transport routes, Grand Cayman Island, West Indies. Atoll Res Bull 263:1–11

Roberts HH, Sneider RM (1982) Reefs and associated sediments of Grand Cayman Island, B.W.I.-Recent Carbonate Sedimentation, Field Trip Guide Book. Geol Soc Am, 50p

Roberts HH (1977) Field guidebook to the reefs and geology of Grand Cayman Island, B.W.I. In: Third International symposium on coral reefs, 41p

Rose SR, Risk JM (1985) Increase in *Cliona delitrix* infestation of *Montastrea cavernosa* heads on the organically polluted portion of the Grand Cayman fringing reef. Mar Ecol 6:345–363

Schuchert C (1910) Paleogeography of North America. Bull Geol Soc Am 20:427–606

Spencer T (1981) Micro-topographic change on calcarenites, Grand Cayman Island, East Indies. Earth Surf Proc Land 6:85–94

Spencer T (1985a) Marine erosion rates and coastal morphology of reef limestones on Grand Cayman Island, West Indies. Coral Reefs 4:59–70

Spencer T (1985b) Weathering rates on a Caribbean reef limestone: results and implications. Mar Geol 69:195–201

Spencer T, Woodroffe CD, Stoddart DR (1984) Calcareous crusts and contemporary weathering on raised reef limestones, Grand Cayman Island, West Indies. In: Advances in Reef Science, Miami, Abstracts of Papers, 117p

Stoddart DR (1980) Geology and geomorphology of Little Cayman. Atoll Res Bull 241:11–16

Tarhule-Lips RFA (1993) Speleogenesis on Cayman Brac, Cayman Islands, British West Indies. M.Sc. thesis, McMaster University

Tarhule-Lips RFA, Ford DC (1996) Timing and causes of speleothem dissolution on Cayman Brac, Cayman Islands, BWI. In: Climate change: the Karst Record, vol 2. Karst Waters Institute Special Publication, Bergen, Norway, pp 165–166

Tarhule-Lips RFA, Ford DC (1998a) Condensation corrosion in caves on Cayman Brac and Isla de Mona. J Cave Karst Stud 60:84–95

Tarhule-Lips RFJ, Ford DC (1998b) Morphometric studies of bell hole development on Cayman Brac. Cave Karst Sci 25:119–130

Tarhule-Lips RFA (1999) Karst processes on Cayman Brac, a small oceanic island. Ph.D. thesis, McMaster University, Hamilton

Vaughan TW (1926) Species of *Lepidocyclina* and *Carpentaria* from the Cayman Islands and their geological significance. Geol Mag 82:388–400

von Humboldt A, Bonpland A (1889) Equinoctial regions of America during the years 1799–1804, vol 3. Ross T (Trans, Eds). George Bell and Sons, London, 409p

Warthin AS (1959) Ironshore in some West Indian islands. Trans NY Acad Sci Series 2(21):649–652

Woodroffe CD (1981) Mangove swamp stratigraphy and Holocene transgression, Grand Cayman Island, West Indies. Mar Geol 41:271–294

Woodroffe CD (1982) Geomorphology and development of mangrove swamps, Grand Cayman, West Indies. Bull Mar Sci 32:381–398

Woodroffe CD, Stoddart DR, Giglioli MEC (1980) Pleistocene patch reefs and Holocene swamp morphology, Grand Cayman, West Indies. J Biogeogr 7:103–113

Woodroffe CD, Stoddart DR, Harmon RS, Spencer T (1983) Coastal morphology and Late Quaternary history, Cayman Islands, West Indies. Quatern Res 19:64–84

2 Tectonic Setting of the Cayman Islands

Abstract

Grand Cayman, Cayman Brac, and Little Cayman are each located on separate fault blocks that are part of the Cayman Ridge that, on its south side, is separated from the depths of the Cayman Trench by the Oriente Transform Fault. The Mid-Cayman Rise, which is an active spreading center, is located off the southwest corner of Grand Cayman. Grand Cayman commonly experiences earthquakes that are triggered by activity along the Oriente Transform Fault or the Mid-Cayman Rise. Given their locations on separate fault blocks, each of the Cayman Islands have tectonic histories that are independent of each other. This is clearly evident from the fact that Cayman Brac experienced uplift and tilting, probably during the Pliocene, whereas the other islands have not undergone any tectonic tilting. Although fractures, joints, and photolineaments are evident on Grand Cayman and Cayman Brac, there is no evidence that the bedrock of each island experienced any faulting that resulted in offset of their strata.

2.1 Introduction

Grand Cayman, Cayman Brac, and Little Cayman are situated on the Cayman Ridge that extends from the Gulf of Honduras to the southern part of Cuba (Fig. 2.1). The ridge defines the northern margin of the Cayman Trough (Fig. 2.2) that is 100 to 150 km wide with depths up to 7,686 m (Uchupi 1975; Rigby and Roberts 1976; MacDonald and Holcombe 1978). It formed during the Cretaceous period as the southward subduction of oceanic crust produced an island arc along the North American–Caribbean plate boundary (Perfit and Heezen 1978; Iturralde-Vinent 2006; Boschman et al. 2014). Relative to the North American Plate, the Caribbean plate changed from northerly to easterly movement during the Eocene as it collided with the Bahamas (Pindell and Kennan 2009; Boschman et al. 2014). Activity along these transform faults produced the left-lateral motion of the North American Plate relative to the Caribbean Plate (Holcombe et al. 1973; Emery and Milliman 1980; Durham 1985; Rosencrantz et al. 1988).

Located off the southwest corner of Grand Cayman is the north–south trending Mid-Cayman Rise (also known as the Mid-Cayman Spreading Center–Tinker 1990), which is a spreading centre (Hayman et al. 2011; Peirce et al. 2019) that became active ~49.4 Ma (MacDonald and Holcombe 1978; Leroy et al. 2000). MacDonald and Holcombe (1978) suggested that the spreading rates were 4 cm/yr between 2 to 6 Ma, but only 0.4 cm/yr over the last two million years. From the south end of the spreading centre, the Swan Islands Transform Fault extends westward, whereas at the north end, the Oriente Transform Fault extends to east and parallels the Cayman Ridge (Fig. 2.2). For a distance of 50 km from the spreading center the earth's crust is only 2 to 3 km thick (ten Brink et al. 2002). On the Mid-Cayman Rise, the Von Damm Vent Field is located at a depth of 2,300 m, whereas the Beebe Vent Field is at a depth of 4,960 m (Connelly et al. 2012). Copper-enriched fluids vent from the Beeb Vent Field with the buoyant plume rising 1,100 m. Plouviez et al. (2015) described the unusual faunal assemblages associated with the Mid-Cayman Spreading Center. McDermott et al. (2018) described the geochemistry of the fluids emitted from the Beebe Vents and Beebe Woods black smokers that are located in the Piccard hydrothermal field, where waters ejected at a temperature of 398°C are rich in Cl, SiO_2, Ca, Br, Fe, Cu, and Mn.

Seismic activity along the Oriente Transform Fault is responsible for the earthquakes that frequently impact Grand Cayman with at least 7 earthquakes affecting the island in 2020 (USGS Earthquake Center). Although most are low magnitude earthquakes, an earthquake of magnitude 7.7 occurred on January 28, 2020 (Tadapansawut et al. 2020). Some of those earthquakes caused minor damage on Grand Cayman.

B. Jones, *Geology of the Cayman Islands*,
https://doi.org/10.1007/978-3-031-08230-6_2

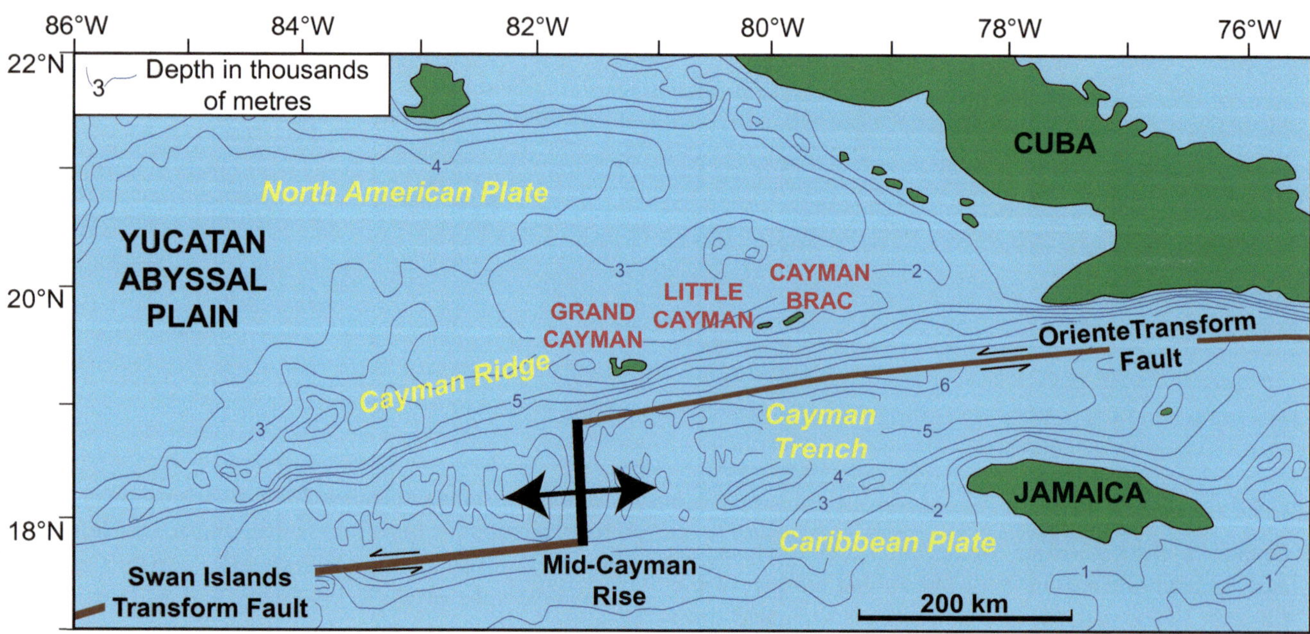

Fig. 2.1 Map of central Caribbean Sea showing location of the Cayman Islands on the Cayman Ridge and proximal to the Oriente Transform Fault, the Swan Islands Transform Fault, and the Mid-Cayman Rise. Based on map in Jones (1994, his Fig. 2.1) that was developed from maps in Perfit and Heezen (1978, their Fig. 2) and MacDonald and Holcombe (1978, their Fig. 1)

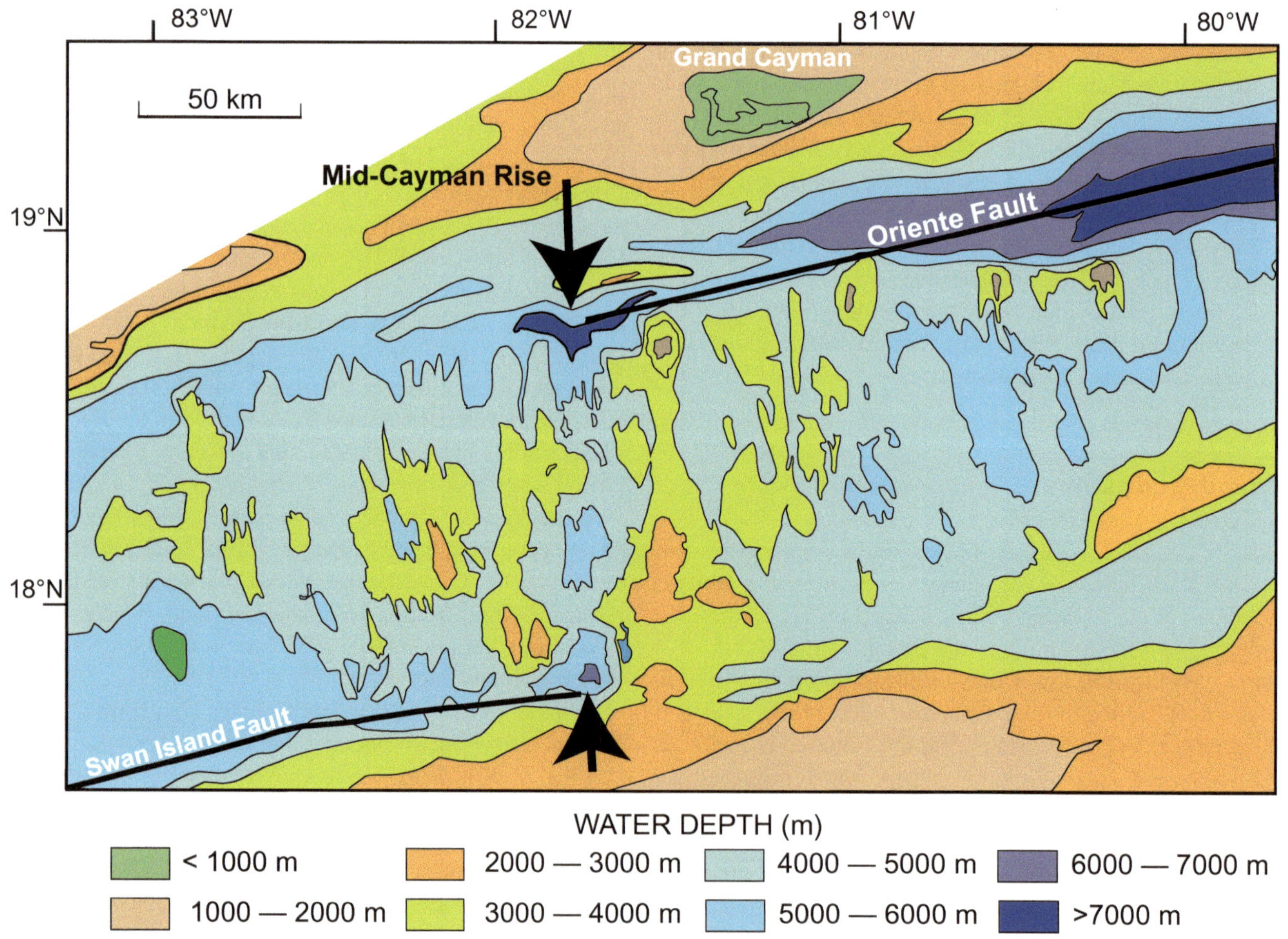

Fig. 2.2 Map showing water depths and locations of the Mid-Cayman Rise, the Oriente Fault and the Swan Islands Fault in the Cayman Trench (modified from ten Brink et al. 2002, their Fig. 2). Note difference in fabric of the seafloor with north–south oriented trends dominant within the zones north of the Swan Island Fault and south of the Oriente Fault that contrast with the ENE-WSW trends associated with the Cayman Ridge

Seismic data indicates that the Cayman Ridge, which rises 1,500 to 2,000 m above the surrounding seafloor, is an uplifted fault block (Fahlquist and Davies 1971; Dillon et al. 1972; Malin and Dillon 1973; Perfit and Heezen 1978). According to Bowen (1968) and Fahlquist and Davies (1971), this fault block is delimited by faults that dip at 30°. The Cayman Ridge was thought to have been a shallow carbonate bank until the Miocene when it started to subside at a rate of 6 to 10 cm per 100 years (Perfit and Heezen 1978; Emery and Milliman 1980). After middle Miocene times, localized uplift elevated central America, the Swan Islands, the Cayman Islands, Jamaica, and large parts of southern Cuba above sea level (Perfit and Heezen 1978). Each of the Cayman Islands are probably located on separate fault blocks that have operated independently (Stoddart 1980; Woodroffe et al. 1983; Woodroffe 1988)—a posit that is readily appreciable when the physiography of Cayman Brac is compared to that of Little Cayman. Nevertheless, the fact that the wave-cut notch that formed ~125,000 years ago when sediments that now form Unit D of the Ironshore Formation were being deposited is today at the same elevation on Grand Cayman and Cayman Brac indicates that there has been no differential vertical movement of these islands since that time.

Grand Cayman, Cayman Brac, and Little Cayman are, in reality, the tops of pinnacles that rise from the depths of the ocean. Perfit and Heezen (1978) and Iturralde-Vinent and MacPhee (1999) suggested that the Cayman Ridge was segmented into blocks during the early Miocene when faults perpendicular to the Oriente Transform Fault developed. It seems that each island is founded on a granodiorite base that is capped by basalt and then Tertiary carbonates. Off the south coast of Grand Cayman, diorite is evident at depths of 600 to 3,400 m (Emery and Milliman 1980). The exact thickness of the carbonate succession is unknown because the base had not been reached by drilling on the islands. On Grand Cayman, for example, an exploratory well drilled on the south coast near Bodden Town was still penetrating Oligocene carbonates when it was terminated at a depth of 401 m (Emery and Milliman 1980).

Ituralde-Vinent and MacPhee (1999), based on information obtained from ODP Hole 998 that was drilled on the Cayman Rise, showed that the Eocene–Oligocene transition and the late Middle/early Late Miocene were characterized by very low sedimentation rates that correlate with uplifts and hiatuses documented on the Cayman Islands and elsewhere in the Caribbean.

2.2 Earthquakes and Tsunamis

The location of the Cayman Islands on the Cayman Ridge alongside the Orient Transform Fault and close to the Mid-Cayman Rise (Fig. 2.1) means that they are in a tectonically active area. As a result, earthquakes are common and minor tsunamis have been generated by some of the stronger earthquakes. The epicentres for most of these earthquakes are located along the Oriente Transform Fault or the Mid-Cayman Rise. Novalo-Casanova and Suárez (2010) noted, however, that there are no historical records of any major event over the last 300 years.

Since the beginning of 2020, the islands have experienced numerous earthquakes with magnitudes generally less than 5, but with some that are much stronger (Table 2.1). Although Grand Cayman is impacted by all of these earthquakes, Little Cayman and Cayman Brac have generally experience little or no effects from them. Novalo-Casanova and Suárez (2010) made note of an earthquake with magnitude of 6.8 that affected Grand Cayman on December 14, 2004. On January 28, 2020, Grand Cayman experienced an earthquake with a magnitude of 7.7, which is probably one of the most severe earthquakes to have affected the Cayman Islands. That earthquake was followed by eight aftershocks that had magnitudes of up to 6.5 (Novelo-Casanova and Suárez 2010).

Earthquake activity has the potential of activating movement of faults/fractures on the islands and/or generating tsunamis. Although there is anecdotal evidence of fractures appearing in shoreline exposures of the Ironshore Formation, there is no direct proof that this has happened. Following the 7.7 earthquake at the end of 2004, there were documented examples of small sinkholes (Xu et al. 2022, their Fig. 6) that developed in some of the sandy areas on Grand Cayman, and some roads were damaged.

Although tsunamis are commonly generated by major earthquakes, this has not been the case on the Cayman Islands. Following the 7.7 earthquake at the end of 2004, a small tsunami resulted in an increase in sea level of ~0.5 m

Table 2.1 List of earthquakes—Grand Cayman 2020 to end of 2021 (from United States Geological Survey earthquake list; VD = Volcano Discovery (https://www.volcanodiscovery.com/region/86144/earthquakes/cayman-islands/archive/2020.html)

Date	Magnitude	Epicentre	
2021-12-26	4.2	152 km S of George Town	USGS
2021-11-25	4.4	52 km S of George Town	USGS
2021-10-16	5.4	73 km SSW of George Town	USGS
2021-01-28	6.1	55 km SE of George Town	USGS
2020-11-25	5.1	51 km S of George Town	VD
2020-09-08	4.5	102 km ESE of East End	USGS
2020-04-28	4.1	37 km S of George Town	USGS
2020-03-22	4.6	51 km SE of East End	USGS
2020-04-28	4.1	38 km S of George Town	VD
2020-03-23	4.3	66 km SE of USGS	VD
2020-03-22	4.5	49 km SE of George Town	USGS
2020-02-26	3.4	66 km S of Bodden Town	USGS
2020-02-19	4.4	47 km SSE of Bodden Town	USGS
2020-02-01	4.2	63 km SE of George Town	VD
2020-01-30	4.7	59 km SE of George Town	VD
2020-01-29	4.0	51 km SE of George Town	VD
2020-01-29	4.2	51 km SE of East End	USGS
2020-01-28	7.7	58 km SE of East End	USGS
2020-01-28	5.2	44 km S of Bodden Town	USGS
1999-12-01	6.3	209 km SSW of George Town	USGS

in George Town Harbour (Novelo-Casanova and Suárez 2010). Xu et al. (2022), based on a detailed analysis of the data derived from the earthquake and tsunami, argued that the tsunami was actually caused by a submarine landslide that had been triggered by the earthquake.

2.3 Tectonic Elements on the Cayman Islands

Tectonic features are hard to detect on the Cayman Islands because of the well-developed phytokarst that exists over large areas and the dense, commonly impenetrable tropical vegetation that covers large tracts of land. On Grand Cayman, for example, much of the northern part of the western peninsula and the central area east of North Sound are covered with mangrove swamps that effectively cover the underlying bedrock. Emery (1981) demonstrated that there are six marine terraces on Grand Cayman that are above present-day sea level. From these, he showed that the terrace at 2 m above sea level was probably associated with deposition of the Ironshore Formation during the Sangamon highstand ~125,000 years ago. He was unable to provide dates for the other terraces because they were extensively weathered and attempts to date them, using various radiometric techniques failed because the fossils had been altered by weathering. He did argue, however, that the island had not experienced any uplift over the last 125,000 years.

Rigby and Roberts (1976) examined the orientation of joints that are apparent in the Cayman Formation at various locations on the eastern part of Grand Cayman and produced a series of rose diagrams that show their orientations (Fig. 2.3a). Examination of air photographs shows the presence of photolineaments at various locations around the eastern end of Grand Cayman and Cayman Brac (Liang and Jones 2015, their Figs. 3 and 6). On Grand Cayman, there are three sets of photolineaments: Set I that trends N-S, Set II that trends ENE-WSW, and Set III that trends NNW-SSE (Fig. 2.3a). Attempts to locate these lineaments in the field have failed because of the rugged phytokarst and the dense vegetation that covers the area. Liang and Jones (2015) used digital elevation models (DEM) for the area to the west of High Rock Quarry (HRQ) to show that the photolineaments track a series of valleys (20 to 100 m wide) that are separated from each other by symmetrical ridges that are up to 2 m high with slopes of up to 5°. The orientations of the photolineaments are consistent with the joint orientations that Rigby and Roberts (1976, their Fig. 30) produced from various areas on Grand Cayman (Fig. 2.3a). On Cayman Brac, photolineaments are limited to the uplifted core that is formed of the Cayman Formation, are divided into sets IV, V, and VI (Fig. 2.3b). Set IV is similar to set II on Grand

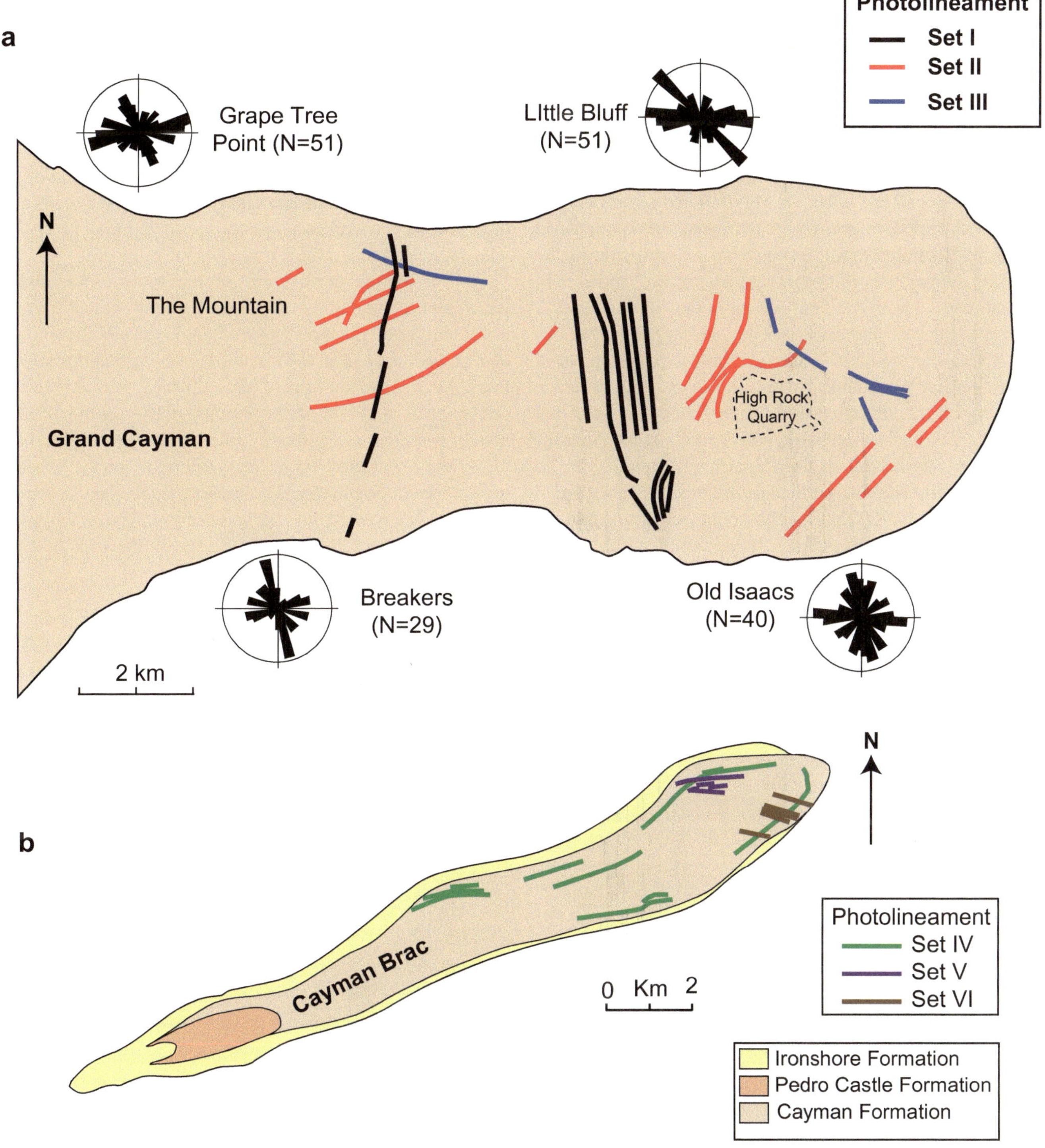

Fig. 2.3 Maps showing main structural features of Grand Cayman and Cayman Brac. **a** Map of eastern part of Grand Cayman showing joint orientations (from Rigby and Roberts 1976, their Fig. 30) and Sets I, II, and III of photolineaments (modified from Liang and Jones 2015, their Fig. 4). **b** Map of Cayman Brac showing Sets IV, V, and VI of photolineaments evident in the uplifted core of the island that is formed of dolostones of the Cayman Formation (modified from Liang and Jones 2015, their Fig. 6)

Cayman. The orientations of the photolineaments in Sets V, and VI, however, are different from the joint orientations on Grand Cayman. As for Grand Cayman, Digital Elevation Models show that the ridges (up to 5 m high), which define the photolineaments, define the margins of valleys that are up to 150 m wide. Roed (2006) noted the presence of northeast and northwest trending lineaments that are evident on the surface of the uplifted core of Cayman Brac. Liang and Jones (2015) suggested that the orientations of the photolineaments seem to parallel the faults that define the fault block on which Cayman Brac is located.

Roed (2006), based on air photographs taken in 2004, mapped various structures on Grand Cayman and Cayman Brac. For the area just to the west of the East End Quarry (also known as High Rock Quarry), he noted a series of N–NW trending ridges that butted up against a series of NE trending ridges (Roed 2006, his Fig. 2.5) and suggested that this was a fault line that could be traced for 7 km. Roed (2006, p. 60) noted that the ridges on the west side of the fault "…are distinctly folded or bent adjacent…" and argued that the west side of the fault had been moved to the south relative to the east side. Based on this, he suggested that this was a "transcurrent fault" with a substantial displacement of at least several kilometers. There is, however, little evidence to support the presence of such a fault. Numerous wells that have been drilled in that area reveal thick successions of dolostones that belong to the Cayman Formation with no differences between those drilled on either side of the "fault" proposed by Roed (2006). It is also important to note that the notion that this may be a "transcurrent fault" (more commonly known as a strike-slip fault) is questionable because it goes against the dominant east–west motions that are associated with the Oriente Transform Fault, which is the strike-slip fault located off the south coast of Grand Cayman. There is no evidence of offset on the Oriente Transform Fault that would be expected if movement on the fault postulated by Roed (2006) had taken place.

Fractures and joints are evident in many exposures of the Cayman Formation. The possibility that faulting has affected the strata on Cayman Brac was raised by Brunt et al. (1973) who suggested that a fault was evident in the cliff face near Spot Bay on the north coast of the island. The precise location of the fault noted by Brunt et al. (1973) is unknown, but at least two "fractures" are evident in the cliff faces to the east of Spot Bay (Fig. 2.4). Although those features stretch from sea level to the top of the cliff face, it is difficult to determine if they are faults because there are no distinct marker beds that can provide evidence of vertical off-sets of the strata. Vegetation cover and weathering of the exposed rocks further complicates the situation. Equivalent structures have not been noted on the south coast, where the strata appear to be laterally continuous. Given the lack of clear

Fig. 2.4 Views of northeast coast of Cayman Brac showing fractures (arrows) that cut through the Brac Formation and overlying Cayman Formation from just above sea level to the top of the cliff that is ~40 m above sea level. There is no evidence of strata offset across these discontinuities

Fig. 2.5 Filled fractures on the coastlines **a** to east of Pedro Castle, and **b** east end of island, south end of High Rock Drive. **a** North–south joint cutting through dolostones of the Cayman Formation, up to ~2 m wide at its maximum (right margin indicated by arrows), filled with assorted lithoclasts (up to 50 cm long) of caymanite and dolostones, derived from surrounding bedrock, that are cemented by calcite. **b** East–west trending filled joint in dolostones of the Cayman Formation (CF), locally lined with calcite flowstone and filled with lithoclasts derived from the surrounding bedrock. Note modern corals and other debris that is now collecting in this joint.

Fig. 2.6 View of south wall in Pedro Castle Quarry showing vertical fractures (white arrows) cutting through the Cayman Formation (CF), the overlying Pedro Castle Formation (PCF), and the Cayman Unconformity that separates the two formations. Note that there is no offset of the formational boundary associated with the fractures

evidence, these structures are herein treated as fractures rather than faults. Careful examination of the cliff faces along the south and north coasts of Cayman Brac also failed to identify any other fractures/faults like those evident near Spot Bay.

On Grand Cayman, joints are evident in coastal exposures near Pedro Castle and in the coastal area south of the south end of High Rock Drive. In Pedro Castle Quarry, fractures evident in the quarry walls cut through the Cayman Formation and the overlying Pedro Castle Formation (Fig. 2.6). There is no evidence of the strata on each side of the fractures being offset relative to each other and the unconformity that separates the two formations is not offset by the fractures. The fact that these fractures are filled with terra rossa suggests that they were not formed by blasting operations when the quarry was still active. Fractures evident in the Cayman Formation and Pedro Castle Formation along the coastal platform to the south of the quarry and east of Pedro Castle itself are up to 2 m wide and typically filled with various combinations of terra rossa, assorted lithoclasts, and/or flowstone (Fig. 2.5a). It is evident that such fractures must have opened up and possibly been widened by weathering before being filled. Although the evidence is not conclusive, it is possible that the widest joints may have evolved through repeated cycles of widening and filling. On the eastern end of Grand Cayman, at the south end of High Rock Drive and west of Blowholes, the Cayman Formation is exposed in a raised platform (Jones and Hunter 1992) that is cut by numerous fractures/joints that are generally <1 m wide (Fig. 2.5b). Most of these fractures/joints are filled with flowstone, various types of lithoclasts, and/or terra rossa (Fig. 2.5b). Modern corals that must have been washed onshore during storms are also collecting in some of these fractures (Fig. 2.5b).

2.4 Conclusions

The following elements regarding the tectonic setting of the Cayman Islands are important because they have dictated many aspects of the geological development of each island.

- The Cayman Islands are located on the Cayman Ridge, which forms the northern boundary of the Cayman Trench.
- The Oriente Transform Fault forms the boundary between the Cayman Trench and the Cayman Ridge.
- The Mid-Cayman Rise, which is an active spreading centre, is located off the southwest corner of Grand Cayman.
- Grand Cayman, Little Cayman, and Cayman Brac are each located on separate fault blocks that, from a tectonic perspective, appear to have acted independently of each other. This is well illustrated by the uplift and westward tilting of Cayman Brac that may have occurred during the Pliocene.
- Earthquakes, which are a common occurrence on Grand Cayman, are triggered by movement on the Oriente Transform Fault or tectonic activity associated with the Mid-Cayman Rise.
- There is no clear evidence that faulting affected the bedrock of each island.
- Fractures and photolineaments evident on each island indicate that tectonic activity may have had some minor effects on the bedrock of each island.

References

Boschman LM, Van Hinsbergen DJJ, Torsvik TH, Spakman W, Pindell JL (2014) Kinematic reconstruction of the Caribbean region since the early Jurassic. Earth Sci Rev 138:102–136

Brunt MA, Giglioli MEC, Mather JD, Piper DJW, Richards HG (1973) The Pleistocene rocks of the Cayman Islands. Geol Mag 110: 209–221

Connelly DP, Copley JT, Murton BJ, Stansfield K, Tyler PA, German CR, Van Dover CL, Amon D, Furlong M, Grindlay N, Hayman N, Hühnerbach V, Judge M, Le Bas T, Mcphail S, Meier A, Nakamura K, Nye V, Pebody M, Pedersen RB, Plouviez S, Sands C, Searle RC, Stevenson P, Tawa S, Wilcox S (2012) Hydrothermal vent fields and chemosynthetic biota on the world's deepest seafloor spreading centre. Nature Communications, 1–9

Dillon WP, Vedder JG, Graf RJ (1972) Structural profile of the northwestern Caribbean. Earth Planet Sci Lett 17:175–180

Durham JW (1985) Movement of the Caribbean Plate and its importance for biogeography in the Caribbean. Geology 13:123–125

Emery KO (1981) Low marine terraces of Grand Cayman Island. Estuarine, Coast Shelf Sci 12:569–578

Emery KO, Milliman JD (1980) Shallow-water limestones from slope off Grand Cayman Island. J Geol 88:483–488

Fahlquist DA, Davies DK (1971) Fault-block origin of the western Cayman Ridge, Caribbean Sea. Deep-Sea Res 18:243–253

Hayman NW, Grindlay NR, Perfit MR, Mann P, Leroy S, Mercier De Lépinay B (2011) Oceanic core complex at the ultraslow spreading mid-Cayman spreading center. Geochem Geophys Geosyst 12:1–21

Holcombe PRV, Vigt PR, Matthews JE, Murchison RR (1973) Evidence for sea-floor spreading in the Cayman Trough. Earth Planet Sci Lett 20:357–371

Iturralde-Vinent MA (2006) Meso-Cenozoic Caribbean paleogeography: implications for the historical biogeography of the region. Int Geol Rev 48:791–827

Iturralde-Vinent MA, Macphee RDE (1999) Paleoeogeography of the Caribbean region: implications for Cenozoic biogeography. Bull Am Mus Nat Hist 238:1–95

Jones B (1994) Geology of the Cayman Islands. In: Brunt MA, Davies JE (eds) The Cayman Islands: natural history and biogeography. Kluwer, Dordrecht, The Netherlands, pp 13–49

Jones B, Hunter IG (1992) Very large boulders on the coast of Grand Cayman: the effects of giant waves on rocky coastlines. J Coastal Res 8:763–774

Leroy S, Masuffret A, Patriat P, Mercier De Lépinay B (2000) An alternative interpretation of the Cayman Trough evolution from a reidentification of magnetic anomalies. Geophys J Int 141:539–557

Liang T, Jones B (2015) Ongoing, long-term evolution of an unconformity that originated as a karstic surface in the Late Miocene: a case study from the Cayman Islands, British West Indies. Sed Geol 322:1–18

Macdonald KC, Holcombe TL (1978) Inversion of magnetic anomalies and sea-floor spreading in the Cayman Trough. Earth Planet Sci Lett 40:407–414

Malin PE, Dillon WP (1973) Geophysical reconnaissance of the western Cayman Ridge. J Geophys Res 78:7769–7775

Mcdermott JM, Sylva SP, Ono S, German CR, Seewald JS (2018) Geochemistry of fluids from Earth's deepeast ridge-crest hot-springs: Piccard Hydrothermal Field, mid-Cayman Rise. Geochim Cosmochim Acta 228:95–118

Novelo-Casanova DA, Suárez G (2010) Natural and man-made hazards in the Cayman Islands. Nat Hazards 55:441–466

Peirce C, Robinson AH, Campbell AM, Funnell MJ, Grevemeyer I, Hayman NW, Van Avendonk HJA, Castiello G (2019) Seismic investigation of an active ocean-continent transform margin: the interaction between the Swan Islands fault zone and the ultraslow-spreading mid-Cayman spreading centre. Geophys J Int 219:159–184

Perfit MR, Heezen BC (1978) The geology and evolution of the Cayman Trench. Geol Soc Am Bull 89:1155–1174

Pindell JL, Kennan L (2009) Tectonic evolution of the Gulf of Mexico, Caribbean and Northern South America in the mantle reference frame: an update. In: KH James MA Lorente, JL Pindell (eds) The origin and evolution of the Caribbean Plate. Geological Society Of London Special Publication London, pp 1–55

Plouviez S, Jacobson A, Wu M, Van Dover CL (2015) Characterization of vent fauna at the mid-Cayman Spreading Center. Deep-Sea Res I 97:124–133

Rigby JK, Roberts HH (1976) Grand Cayman Island: geology, sediments and marine communities 4. Brigham Young University, Geology Studies, Special Publication No. 4, pp 97

Roed MA (2006) Islands from the sea. Cayman Free Press, Grand Cayman, Geologic Stories of Cayman

Rosencrantz E, Ross MI, Sclater JG (1988) Age and spreading history of the Cayman Trough as determined from depth. J Geophys Res 93:2141–2157

Stoddart DR (1980) Geology And Geomorphology of Little Cayman. Atoll Res Bull 241:11–16

Tadapansawut I, Okuwaki R, Yagi Y, Yamashita S (2020) Rupture process of the 2020 Caribbean earthquake along the Oriente Transform Fault, involving supershear rupture and geometric complexity of fault. Geophy Res Lett 1–9

Ten Brink US, Coleman DF, Dillon WP (2002) The nature of the crust under the Cayman Tough from gravity. Marine Petrol Geol 19:971–987

Tinker MN (1990) Seismic reflection data analysis of the Orient and Swan Island fracture zones bounding the Cayman Trough. In: Geuer RA (ed) CRC handbook of Geophysical Exploration at Sea. Taylor And Francis Group (Crc Press), U.S.A., pp 191–226

Uchupi E (1975) Physiography of the Gulf of Mexico and Caribbean Sea. In: Nairn AEM, Stehli FG (eds) The ocean Basins and margins, 3. Plenum Publications, New York, The Gulf Of Mexico And The Caribbean, pp 1–64

Woodroffe CD (1988) Vertical movement of isolated oceanic Islands at plate margins. Z Geomorphol 69:17–37

Woodroffe CD, Stoddart DR, Harmon RS, Spencer T (1983) Coastal morphology and Late Quaternary history, Cayman Islands, West Indies. Quatern Res 19:64–84

Xu Z, Sun L, Radhman MNA, Liang S, Shi J, Li H (2022) Insights on the small tsunami from January 28, 2020, Caribbean Sea M_w Earthquake by numerical simulation and spectral analysis. Natural Hazards, 1–17

3 Bluff Group—the Brac Formation

Abstract

The Late Oligocene Brac Formation, the lowermost formation in the Bluff Group, is known only from the cliff faces on the east of Cayman Brac and some of the wells on Cayman Brac and Grand Cayman. The foraminifera and $^{87}Sr/^{86}Sr$ ratios indicate that the limestones in the upper part of this formation represent sedimentation that took place during the Late Oligocene. The formation, which is at least 50 m thick, is capped by a topographically uneven unconformity that separates it from the overlying Cayman Formation. The basal boundary of the formation has not yet been located. The limestones, found on the north coast of Cayman Brac, are characterized by numerous large benthic foraminifera (mainly *Lepidocyclina*), encrusting foraminifera, molluscs, red algae, echinoid plates, and rare corals in wackestone to packstone matrices. On the south coast of Cayman Brac, the formation is formed largely of sucrosic dolostones (zoned euhedral crystals up to 1.5 mm long) that are characterized by numerous different diagenetic stages that included the development of limpid dolomite cements, calcite cements, and complex vadose cements formed of calcite and dolomite. The Brac Formation on Grand Cayman is poorly understood because it is not exposed and is known only from well cuttings obtained from some of the deep wells. The original sediments of the Brac Formation formed and accumulated on open banks in shallow water that was probably no more than 25 m deep.

3.1 Introduction

The Brac Formation was formally designated by Jones et al. (1994b) for strata exposed in the vertical to over-hanging cliff faces on the east end of Cayman Brac (Figs. 3.1, 3.2 and 3.3). They designated section LCB, located on the northeast coast of Cayman Brac (Fig. 3.1) as the type section of the Brac Formation and section SCD, located on the southeast coast (Fig. 3.1), as the reference section because that area includes extensive dolostones that are absent on the north coast.

When first defined, information about the Brac Formation was limited because no drilling had been done on Cayman Brac and the sequence had not been identified on Little Cayman or Grand Cayman. Subsequent drilling on the eastern part of Cayman Brac in 2002 (wells EOR#1—43 m deep, CRQ#1—61 m deep) and 2003 (wells KEL#1—53 m deep, APL#1—61 m deep) provided additional data on the subsurface strata (Fig. 3.1). It should be noted that the drilling equipment available on Cayman Brac limited drilling to a maximum depth of 61 m with only chip samples being obtained. Data from KEL#1 was supplemented by examination of the Cayman Formation that was exposed in the cliff face ($\sim$12 m high) adjacent to the drill site. CRQ#1 was drilled from the floor of a quarry (Jennifer Bay Quarry) and sampling of the quarry walls to a height of $\sim$15 m above sea level provided additional information on the overlying Cayman Formation. The vertical cliff-face beside well APL#1 could not be measured and sampled. For the EOR#1 well, located beside the road, there are no nearby surface exposures that could be examined. Although data from these wells provided additional information about the formation, the fact that only chip samples could be obtained meant that only limited information on the textures of the rocks could be ascertained and no porosity or permeability data could be derived.

Jones (1994) and Jones et al. (1994b) treated the Brac Formation as a single unit with no members being defined. Uzelman (2009), in her unpublished M.Sc. thesis, which was based on data from sections SCD and LCB and information from wells APL#1, KEL#1, CRQ#1, and EOR#1 (Fig. 3.1), argued that the sucrosic dolostones exposed in and around section SCD should be assigned to the Pollard Bay Member as a way of recognizing the fact that dolostones and limestones were present in different geographic areas of the

B. Jones, *Geology of the Cayman Islands*,
https://doi.org/10.1007/978-3-031-08230-6_3

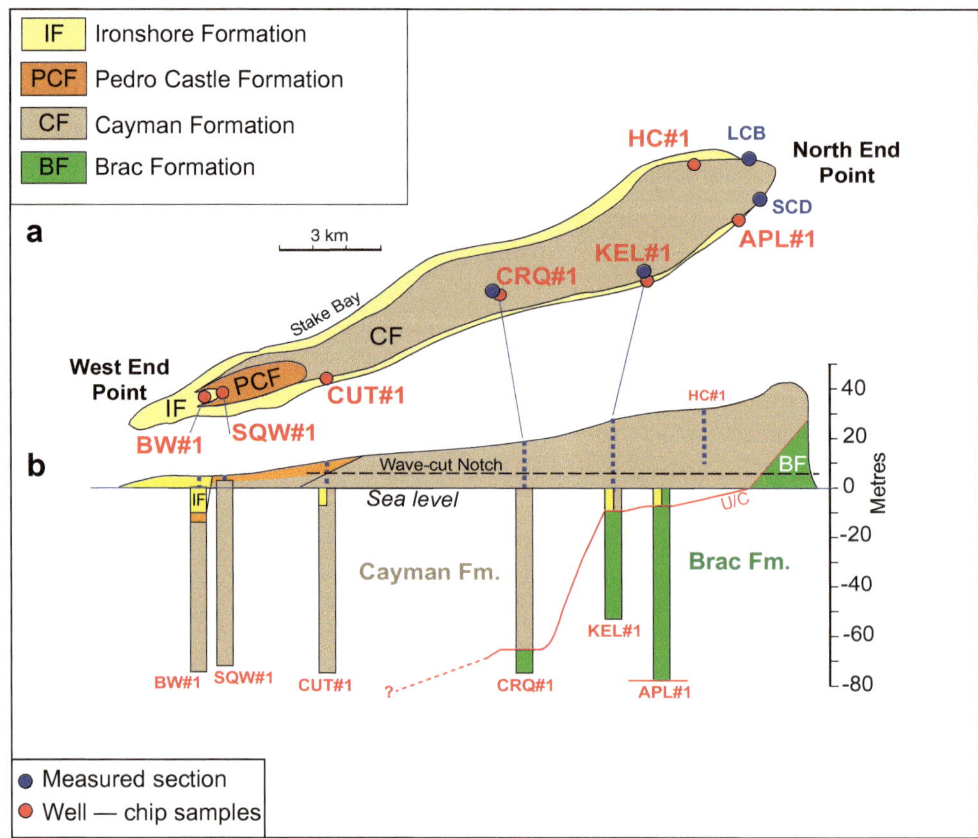

Fig. 3.1 Locations of measured sections and wells where the Brac Formation was identified and sampled. **a** Locations of measured sections SCD and LCB and wells EOR#1. APL#1, KEL#1, and CRQ#1 were samples where collected from the Brac Formation. **b** Cross-section along the length of Cayman Brac, from North end Point to West End Point, showing the distribution of the Brac Formation, Cayman Formation, Pedro Castle Formation, and Ironshore Formation. The positions and depths of wells APL#1, KEL#1, and CRQ#1, drilled on the south coast are shown along with measured sections KEL#1 and CRQ#1. Well EOR#1 was drilled on the north coast close to measured section LCB

formation. That suggestion is not followed here because with the presently available information it is still impossible to precisely define the extent of the sucrosic dolostones and data from the north coast is limited to that from section LCB and EOR#1 as no other wells have been drilled on the north coast between Spot Bay and the cross-island road.

3.2 Distribution

The Brac Formation is only exposed in the vertical cliff faces on the east end of Cayman Brac (Figs. 3.2 and 3.3). It has also been recognized in the subsurface on Grand Cayman, especially on the western part of the island. It is possible that the deeper strata in well NSC#1/3 (located in central-eastern part of Grand Cayman), which was drilled to a depth of 244 m, includes a thick succession of strata that belong to the Brac Formation. The succession in that well, however, remains problematic because only chip samples could be obtained from the strata deeper than 125 m. Nevertheless, there are no obvious formation boundaries and no "jumps" in the $^{87}Sr/^{86}Sr$ ratios that commonly denote the boundary between Brac Formation and Cayman Formation as in other wells on the western part of the island.

The Brac Formation has not been found on Little Cayman because the island is low-lying, exposures are limited, and no wells have been drilled on the island.

3.3 Lower Boundary

The lower boundary of the Brac Formation remains unknown because it is not exposed on any of the islands and drilling on Cayman Brac has yet to reveal any lithological changes at depth that might denote a change to a different formation.

3.4 Upper Boundary

The upper boundary, clearly evident in the cliff faces on the east end of Cayman Brac (Figs. 3.2 and 3.3), is the unconformity that separates the Brac Formation from the overlying Cayman Formation. Many of the caves, which are common in the Brac Formation on the southeast coast, have flat roofs that are defined by the unconformity (Fig. 3.3). For ease of reference, this unconformity has been designated as the Brac Unconformity.

Fig. 3.2 View of vertical cliff face on north coast of Cayman Brac, just to east of section LCB, showing position of the unconformity that separates the Brac Formation from the overlying Cayman formation. Cliff face is approximately 30 m high

Synthesis of data from outcrop and wells drilled on the east part of Cayman Brac shows that there is at least 25 m of relief on the Brac Unconformity with the surface sloping westward from the highest part at the northeast corner of the island (Fig. 3.4). In the cliff faces on Cayman Brac the unconformity always appears planar with little evidence of topographic variations that could be attributed to karst development (Figs. 3.2 and 3.3).

3.5 Thickness

The thickness of the Brac Formation is unknown because its lower boundary has yet to be found and defined on any of the islands. Estimates based on the cliff faces at the east end of the island below the lighthouse and the wells drilled on the south coast of Cayman Brac suggests that the formation is at least 50 m thick.

3.6 Lithology

3.6.1 Cayman Brac

In the type section (LCD) on the northeast coast of Cayman Brac (Fig. 3.5), the formation is formed of wackestones to packstones that are characterized by numerous foraminifera (Fig. 3.6). Included in this biota, are well-preserved disc-shaped *Lepidocyclina* that are up to 35 mm diameter (Fig. 3.6a–c). Other foraminifera species are scattered among the *Lepidocyclina* (Fig. 3.6d) and some epiphytic foraminifera, characterized by a flat base, are also present (Fig. 3.6e, f). Scattered red algae and gastropods are associated with these foraminifera. One bed, near the top of the formation, is characterized by gastropods and large, articulated, smooth-shelled bivalves. Dolomite is restricted to scattered rhombs and small pods (<1 m long, < 0.5 m thick) near the top of the formation.

Fig. 3.3 View of vertical cliff face on south coast of Cayman Brac, just west of section SCD, showing the position of the unconformity that separates the Brac Formation from the overlying Cayman Formation. Note caves with flat roofs that are defined by the unconformity that separates the two formations. Top of the cliff face is ~40 m above sea level

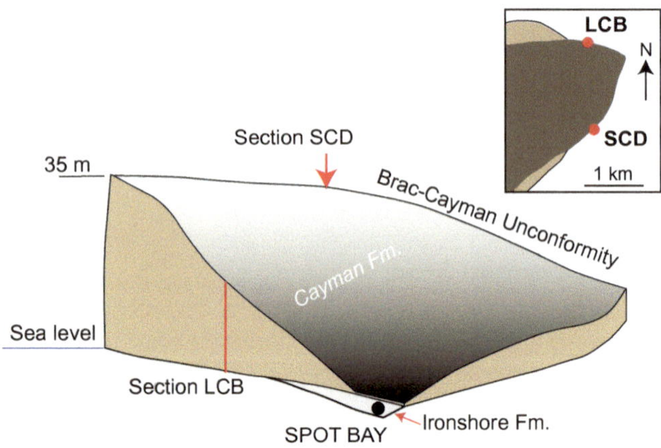

Fig. 3.4 Simplified schematic diagram showing surface of the unconformity that separates the Cayman formation from the underlying Brac formation. Modified from Jones (1994)

In section SCD-1, located on the southeast coast approximately 750 m south of section LCD (Fig. 3.1), the formation is formed largely of massive, coarsely crystalline sucrosic dolostone with localized pods of limestone (Fig. 3.5). There, an old spring vent is located at the boundary between the Brac Formation and the overlying Cayman Formation (Fig. 3.7). The surface below the spring vent is coated with calcite that was precipitated from the flowing spring waters (Fig. 3.7). Fossils are generally rare in the dolostones found on the southeast coast. Although no fossils have been found in the lower 10 m of section SCD, moulds of articulated bivalves and gastropods are present in the upper 7 m of the section (Fig. 3.8a). Scattered *Lepidocyclina*, like those found in section LCD, are also present (Fig. 3.8b). When the sea is calm, it is possible to access the lower part of the cliff faces to the east of section SCD. There, sucrosic dolostones that dominate the succession are characterized by leached fossils that appear to be mostly large *Lepidocyclina* (Fig. 3.9c). These fossils and those evident higher in the section clearly indicate that the biota is equivalent to the biota found in the limestones on the north coast.

Lenses of limestone, up to 10 m long and 2 m thick, are located in the upper part of the formation on the south coast. These limestones, like those found on the north coast, contain large *Lepidocyclina*, articulated bivalves, gastropods, red algae, echinoid plates, and other foraminifera (Fig. 3.9). Dissolution has destroyed the aragonite shells of the gastropods and bivalves so that they are now represented by moulds (Fig. 3.8a). The dolostones are formed largely of subhedral to euhedral dolomite crystals, up to 1 mm long, that have a dark core encased by a clear outer rim (Fig. 3.9c, d). Evident in the cores of these crystals are ghost structures

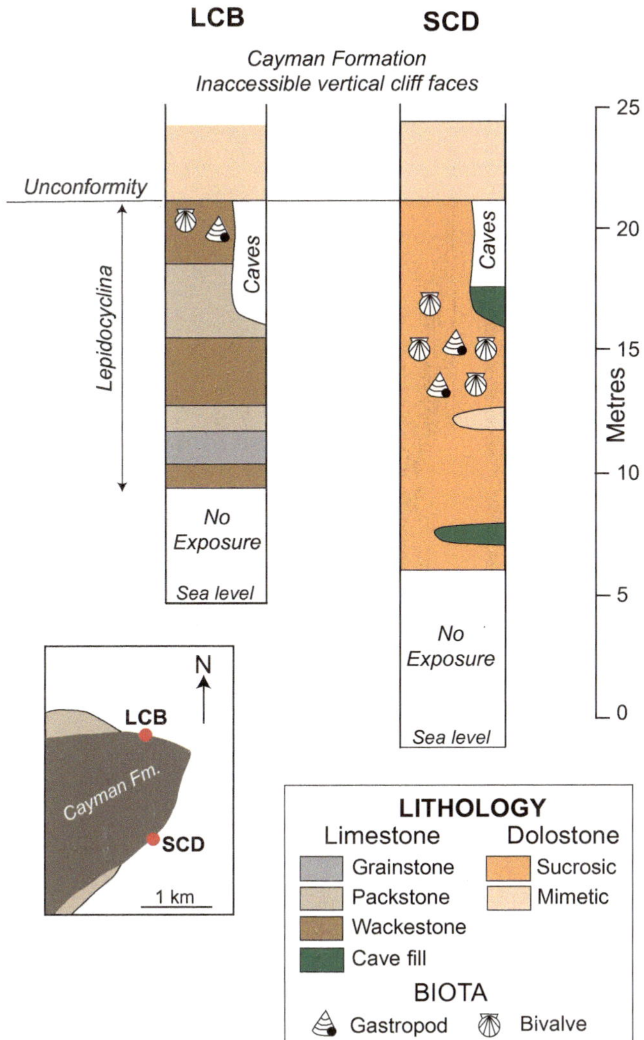

Fig. 3.5 Correlation of facies in the upper part of the Brac formation exposed in sections SCD and LCB. Modified from Jones et al. (1994b, their Figs. 6 and 7)

of red algae and foraminifera (Fig. 3.9f). Microcrystalline dolostone is found only in the upper 5 m of the formation.

The Brac Formation exposed on Cayman Brac is characterized by numerous large benthic foraminifera, bivalves, red algae, and echinoid plates, but only very rare corals. Small rotaliids and *Carpenteria* are associated with the dominant large benthic foraminifera (*Lepidocyclina*). In section LCB, fragments of *Porites* are present in the uppermost 2 m of the formation.

Uzelman (2009, her Table 2.1) based on data from sections LCB and SCD and wells APL#1, KEL#1, CRQ#1, and EOR#1 defined six facies in the Brac Formation (Fig. 3.10). The three facies recognized in the limestones and dolomitic limestones (>50% calcite) are the *Lepidocyclina*, Mollusc, and Foraminifera facies, whereas those in the dolostones and calcareous dolostone (>50% dolomite) are the Fabric Retentive Finely Crystalline Dolomite, Fabric Destructive Finely Crystalline, and Sucrosic Dolomite facies.

Lepidocyclina Facies

The facies is characterized by whole and fragmented Lepidocyclinid foraminifera, up to 4 cm in diameter, along with miliolids, rotalids, and small encrusting foraminifera, *Carpenteria*, scattered fragmentary red algae, echnoid plates, and *Porites* that are held in a wackestone to grainstone matrix.

Mollusc Facies

The facies is similar to the *Lepidocyclina* facies but also contains bivalves and gastropods, most of which are leached so that only fossil mouldic vugs remain.

Foraminifera Facies

Benthic and encrusting foraminifera along with lesser numbers of planktonic foraminifera are held in a wackestone to grainstone matrix. The chambers in some tests are lined with isopachous dolomite cement, whereas other tests have been completely replaced by fabric retentive dolomite. *Porites* and *Siderastrea,* which are locally common, are typically surrounded by grainstones. *Halimeda* plates are leached, echinoid spines are rare, and leached bivalves are scattered.

Fabric Retentive Dolomite Facies

Many of the dolomitic wackestones to packstones are characterized by fabric retentive fabrics. The dolomite, which is finely crystalline (crystals average 50 μm long), has replaced the matrix and the fossils of the original limestones. Skeletal allochems evident in these dolostones, include coralline red algae, rotalids, miliolids, encrusting foraminifera, and echinoid spines. Skeletal grains derived from *Porites,* bivalves, ostracods, and *Halimeda* were leached. No *Lepidocyclina* are found in this facies. Porosity, which is mainly fossil mouldic, is 10 to 30%.

Fabric Destructive Finely Crystalline Dolomite Facies

This dolostone is formed of dolomite crystals that are 20 to 100 μm long and limpid dolomite crystals (<300 μm long) that fill some of the fractures and voids. Little or no porosity is evident. The original depositional fabrics of this facies were obliterated by the dolomitization.

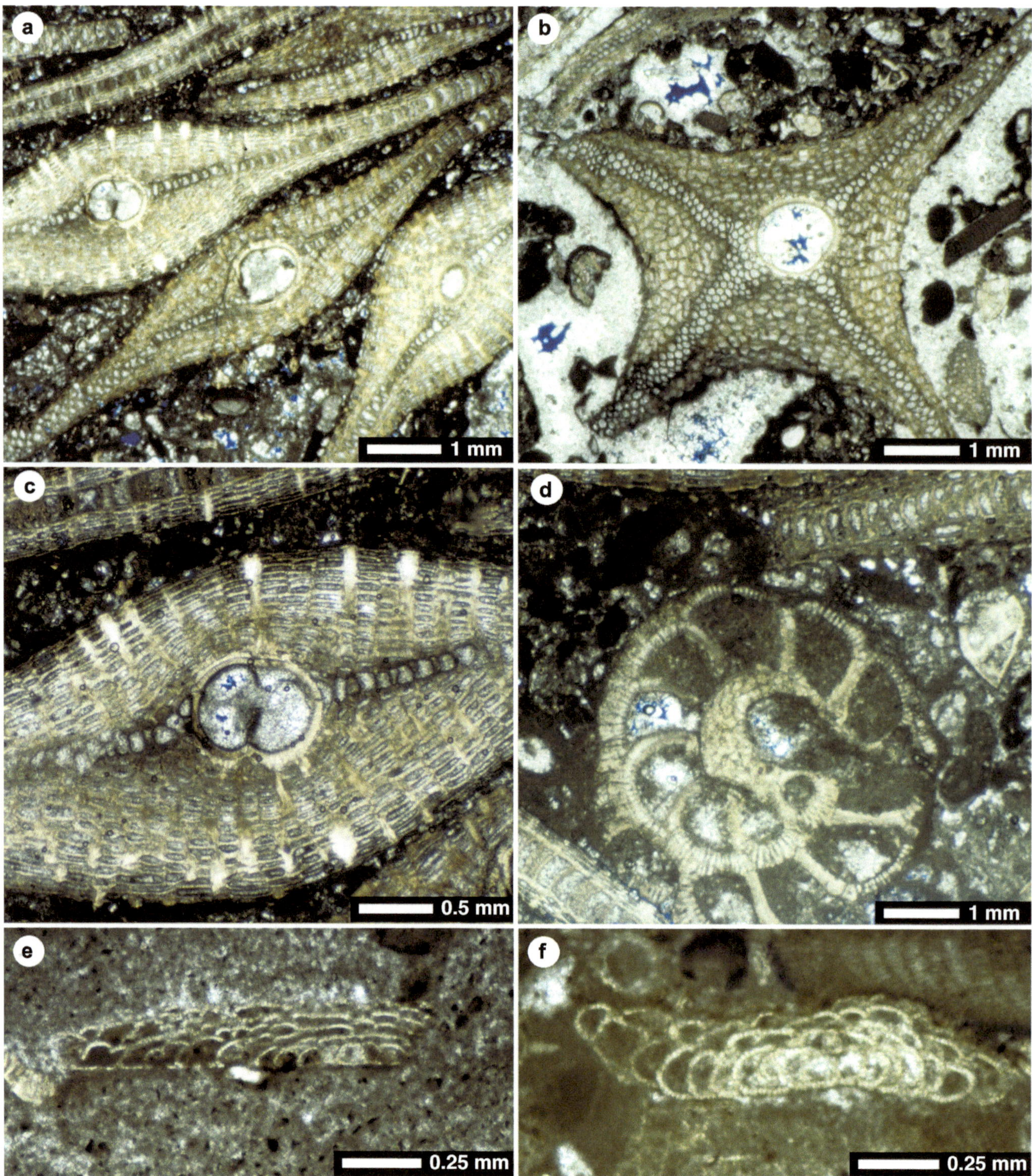

Fig. 3.6 Thin section photomicrographs of well-preserved foraminifera in limestone, lower part of section LCB. **a** Vertical cross-section through *Lepidocyclina*. **b** Transverse section through *Lepidocyclina*. **c** Enlarged view of core region of *Lepidocyclina*. **d** Other, less common foraminifera associated with *Lepidocyclina*. **e**, **f** Foraminifera with flat bases indicating that they were encrusting forms that may have been attached to seagrasses

Fig. 3.7 Unconformity near the top of section SCD, showing inactive spring orifice at unconformity between the Brac Formation (BF) and the overlying Cayman Formation (CF). Calcite precipitated from the spring waters coats the surface below the spring vent

Sucrosic Dolomite Facies

Although the original components and depositional fabrics were largely destroyed by dolomitization, textures indicative of *Lepidocyclina* packstone to grainstone are locally apparent. The euhedral dolomite crystals, up to 1.5 mm long, are characterized by dirty, inclusion-rich cores that are encased by clear dolomite rims. Hollow dolomite crystals are rare. Ghost structures of various allochems, including *Lepidocyclina,* are common. Leached bivalves and gastropods are common. Some of the larger vugs are lined with microstalactitic cements that are formed of alternating calcite and dolomite laminae.

Uzelman (2009) noted that most of the limestone facies are found on the north side of the island whereas the dolostone facies are found on the south side, with most facies being restricted to a single locality. She also suggested that most of the facies in the Brac Formation were deposited in shallow water (<10 m), moderate energy conditions, and that there was no evidence of reef development.

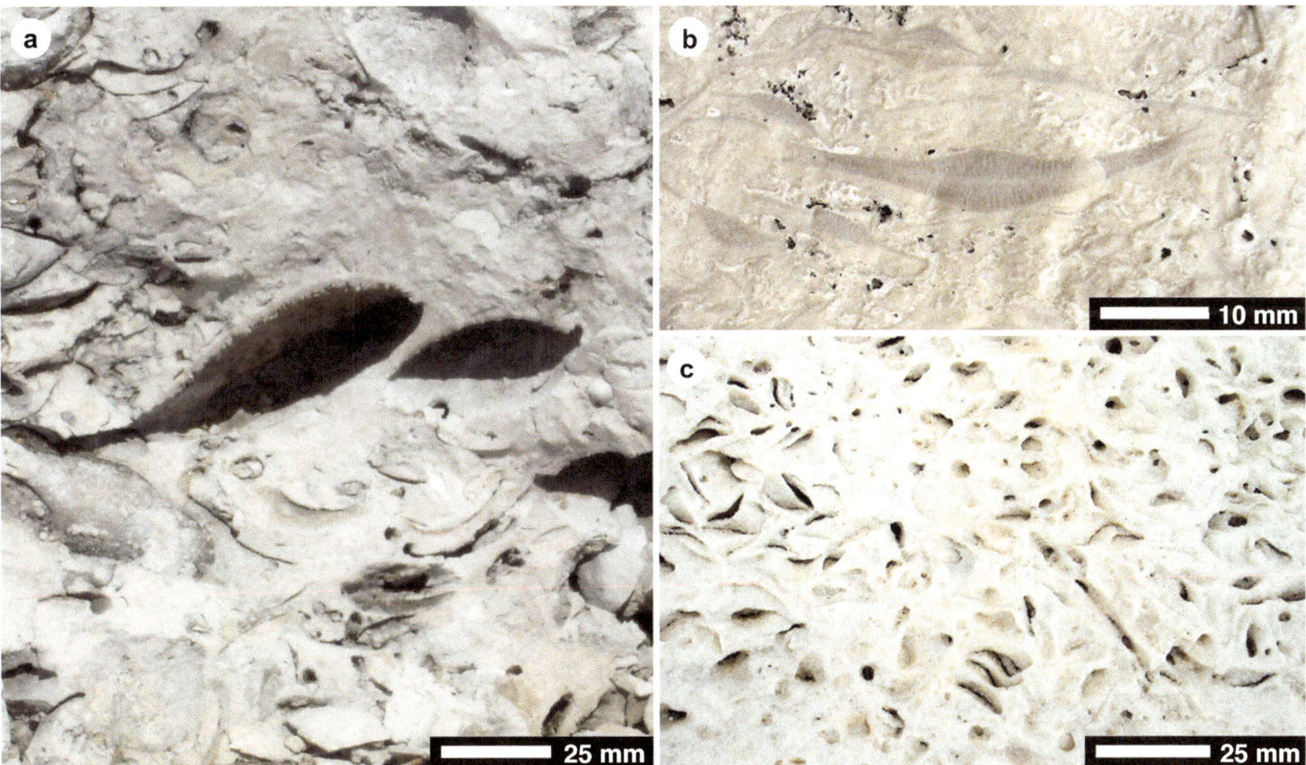

Fig. 3.8 Field photographs of main rock types in Brac Formation on Cayman Brac. **a** Limestone pod with mouldic vugs formed by leaching of bivalve shells, just below unconformity at top of formation. Section SCD. **b** *Lepidocyclina* in limestone, section LCB. **c** Leached foraminifera (possibly small *Lepidocyclina*) in sucrosic dolostone, section SCD, near sea level

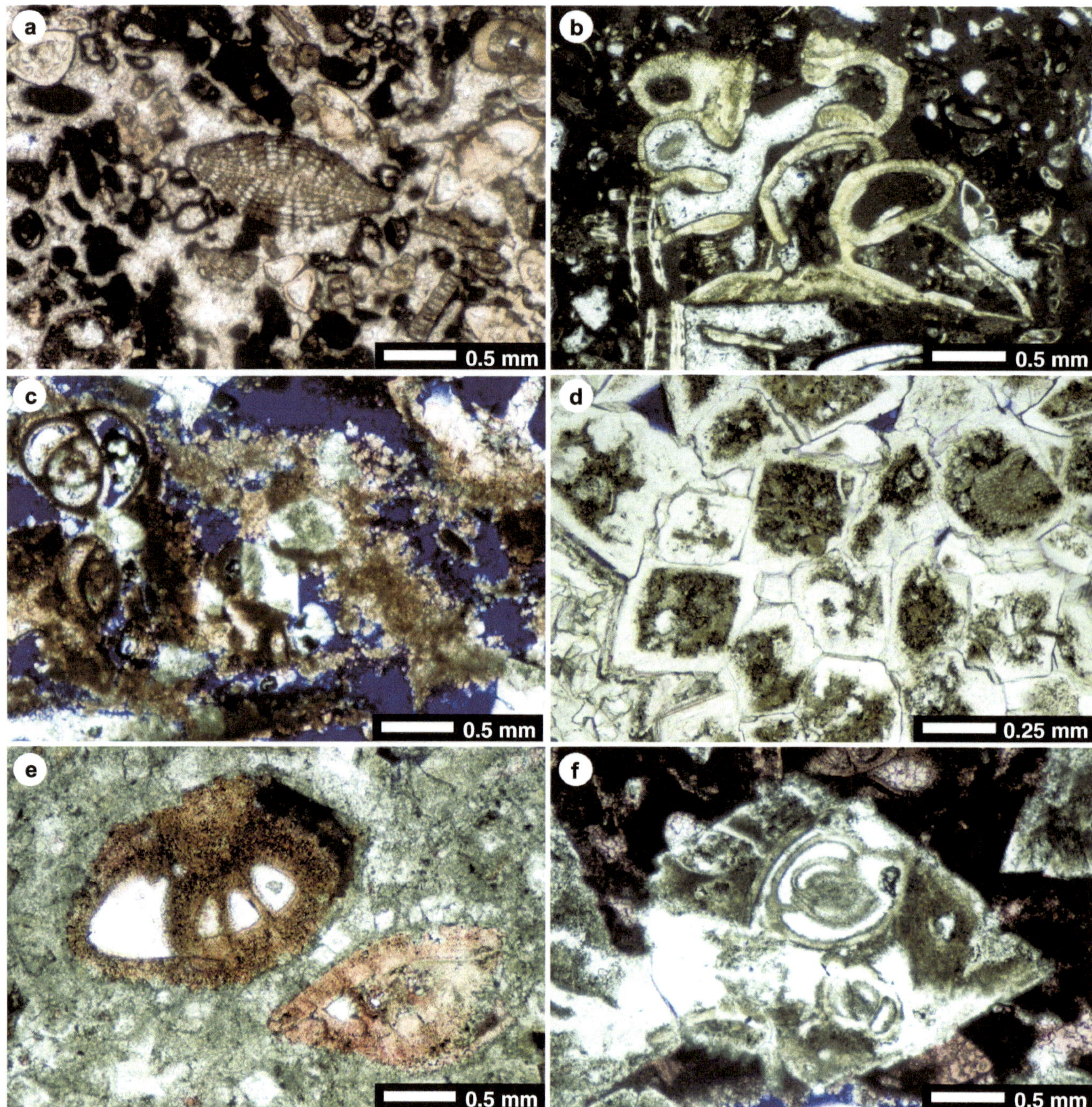

Fig. 3.9 Thin section photomicrographs showing common microfabrics in upper part of Brac Formation. All thin sections stained with Alizarin Red S solution (red = calcite; no stain = dolomite). **a** Limestone with foraminifera and other fossils embedded in calcite cement. Unit SCE-9. **b** Encrusting foraminifera (*Carpenteria*?) in micrite matrix with patches of calcite cement (white). Unit LCB-2. **c** Porous fine-grained limestone with scattered euhedral dolomite crystals (white) that include ghost structures inherited from the original limestone. Section SCC. **d** Interlocking euhedral dolomite crystals, each with a dirty core surrounded by a clear rim. Note ghost structures of various fossils in some of the dirty cores. Unit SCE-10. **e** Calcitic foraminifera in finely crystalline dolomite matrix. Note partial replacement of the peripheral areas of the foraminifera. Unit SCD-1. **f** Large euhedral dolomite crystal with ghost structures of foraminifera and red algae inherited from the original limestone. Unit SCD-2

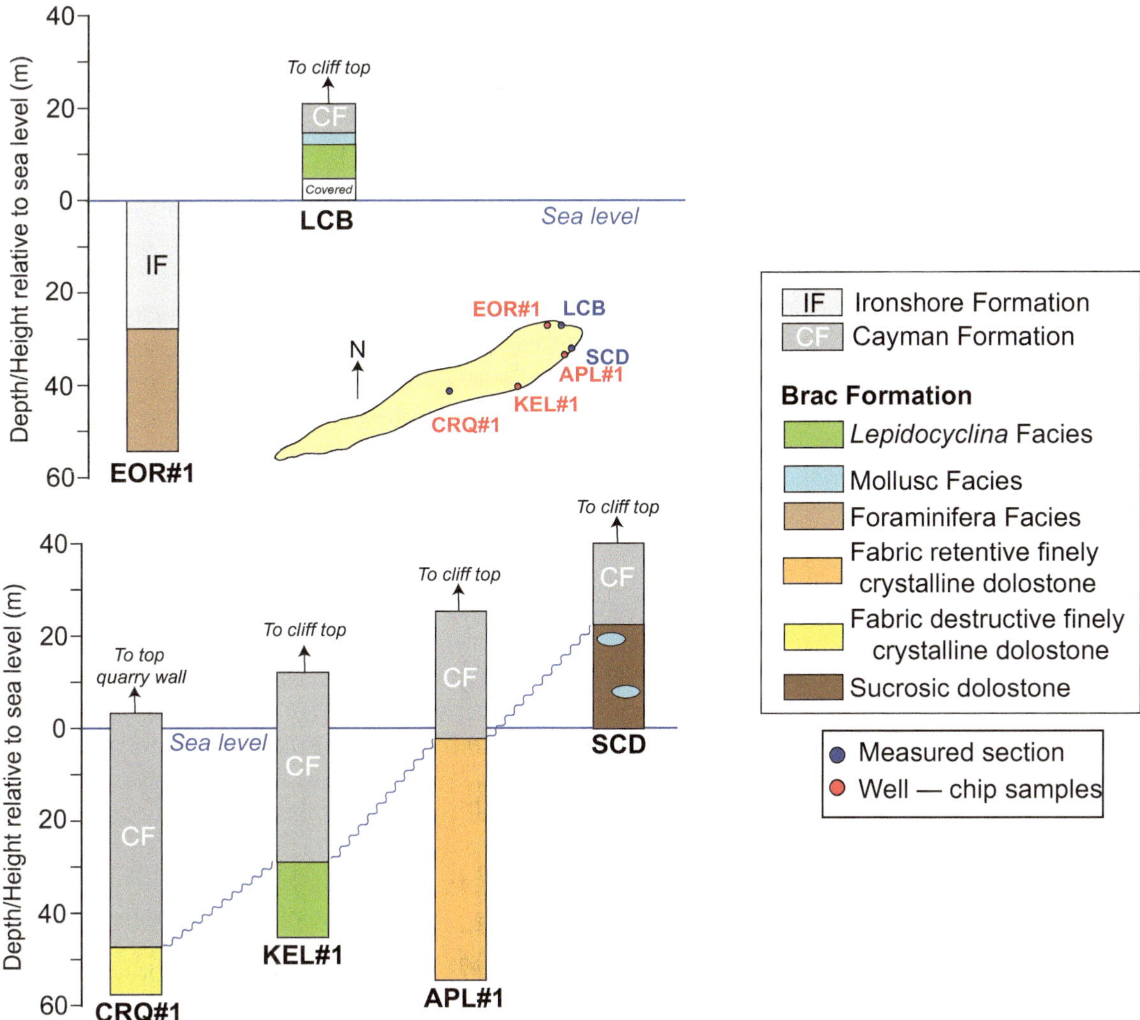

Fig. 3.10 Distribution of facies in the Brac formation (modified from Uzelman, 2009, her Fig. 2.6)

3.6.2 Grand Cayman

Little is known about the lithology of the Brac Formation on Grand Cayman because wells cuttings have only been obtained from some of the deeper wells on the island. As yet, no cores have been obtained from this formation. Available samples indicate that the Brac Formation on Grand Cayman is formed largely of dolostone that is similar to that in the overlying Cayman Formation. Little is known about the biota of this formation on Grand Cayman and *Lepidocyclina* has not yet been found in any of the samples.

3.7 Porosity

Based on digital analyses of thin sections of the dolostones from the Brac Formation in sections SCD, WOJ#3, WOJ#3, WOJ#7, and CRQ#1, Zhao and Jones (2012, their Table 1) showed that the porosity of the dolostones ranged from 0 to 13%. These should probably be regarded as minimum values because the large vugs that are common in these dolostones are not evident in the thin sections and therefore cannot be captured into the porosity calculations.

3.8 Diagenesis

The limestones found in the Brac Formation on the north coast of Cayman Brac are characterized by minor amounts of calcite cement that are found in small, irregular shaped pores in the groundmass and body cavities in the fossils (Fig. 3.9a–b). Dolomite is generally absent apart from scattered crystals found in the upper part of the formation just beneath the Brac Unconformity. The scattered limestones found along the south coast, like those on the north coast, contain calcite cements that filled the intergranular pores. These limestones generally lack dolomite or contain only scattered dolomite crystals.

The dolostones found in the Brac Formation on the south coast are characterized by complex and variable diagenetic fabrics that involved replacive dolomite and various combinations of calcite and dolomite cements (Figs. 3.9c–f, 3.11). The percentage of replacive dolomite varies from sample to sample. Some limestones contain only randomly scattered subhedral to euhedral crystals that are up to 1.5 mm long, and commonly include ghost structures of the original components of the limestone (Fig. 3.9d, f). The replacive dolomite crystals in the pervasively dolomitized limestones, are up 1.5 mm long and typically have a dirty core encased by a clear rim (Fig. 3.9d). The styles of dolomitization are variable. In some samples, the dolomite has preferentially replaced the calcitic matrix while leaving the foraminifera intact or with minor dolomite replacement around their edges (Fig. 3.9e). In contrast, the dolomite in other samples has replaced both the allochems and the original calcite cement (Fig. 3.9f). These dolomite crystals commonly display ghost structures of the foraminifera and red algae that they replaced (Fig. 3.9f).

The cements in the limestones and dolostones of the Brac Formation range from simple to complex. In some of the dolostones, it is evident that calcite cement postdated some of the dolomite cement that lined some of the cavities in the rock (Fig. 3.11a). Thus, cavities lined with euhedral crystals that have a dirty core and clear rims commonly had the remaining spaces filled with calcite cement (Fig. 3.11a). In contrast, other cavities in other dolostones were filled with clear dolomite crystals (Fig. 3.11b). Some of the larger cavities in the dolostones of the Brac Formation were filled with complex, multi-generational successions of calcite and dolomite cements and some calcite sediment (Figs. 3.3 and 3.11c, d). Some cavities are filled with up to 10 isopachous layers of dolomite and calcite cements, and in some cases, thin layers of calcitic sediment (Fig. 3.11c). In other cavities, microstalactitic cements, formed of calcite and dolomite are evident (Fig. 3.11d). These cements, which commonly seem to be formed of radiating, needle- to columnar-shaped crystals are commonly formed of alternating bands of dolomite and calcite (Fig. 3.11d). The isopachous cements probably formed in the phreatic zone, whereas the microstalactitic cements probably formed in the vadose zone.

Based on image analysis of 16 thin sections that included various types of dolostones from the Brac Formation, Zhao and Jones (2012) showed that the dolostones are formed of 35–83% (average 62%) replacive dolomite, 17–47% dolomite (average 32%) cement, and 0–14% (average 5%) calcite. In the fine to medium crystalline dolostones, there is 18–28% dolomite cement whereas in the coarsely crystalline dolostones, the dolomite cement ranges from 26 to 54%.

3.9 Dolomite Stoichiometry

Dolomite stoichiometry, expressed by the percentage of $CaCO_3$ in the dolomite (hereafter abbreviated to %Ca) was determined using the X-ray diffraction method of Jones et al. (2001). As proposed by Jones and Luth (2002), dolomite with less than 55%Ca is termed "low calcium dolomite" (LCD), whereas dolomite with more than 55%Ca is termed "high calcium dolomite" (HCD).

Uzelman (2009), based on samples from sections CRQ#1, EOR#1, KEL#1, APL#1, and WOJ-2/7, showed that the dolostones in the Brac Formation are non-stoichiometric with %Ca ranging from 51.70%Ca to 60.7%Ca with an average of 57.1%Ca (Fig. 3.12). In contrast, Zhao and Jones (2012), based on 41 samples from sections SCD, WOJ#7, WOJ#3, CRQ#1 and KEL#1 showed that the %Ca ranges from 55.0 to 57.5%Ca with an average of 56.7%Ca. The differences between the two sets of analyses relate to the dolomite from EOR#1 that contain 57.6–60.7%Ca and dolomites from APL#1 with 52.2–60.7% Ca (Fig. 3.12). If the outliers in those two sets of samples are excluded, then the %Ca of remaining samples are largely consistent with the %Ca values from the other samples (Fig. 3.12). Irrespective, of these nuances, the dolostones from the Brac Formation are typically formed of HCD (cf. Jones et al. 2001).

Uzelman (2009) suggested that the variance in the %Ca might be related to the type of dolomite given that the samples in EOR#1 and APL#1, which yielded the highest variance, are characterized by a variety of dolomite types, including finely crystalline dolomite, fabric retentive dolomite, and limpid dolomite cement. In EOR#1, most of the dolomite is a spar, pore-lining cement. Although this is recognized as a potential problem, it is impossible to segregate the different types of dolomite for XRD analysis.

3.10 Stable Isotopes

Uzelman (2009) and Zhao and Jones (2012) used $\delta^{18}O$ and $\delta^{13}C$ isotopes as part of their investigation of the limestones and dolostones of the Brac Formation on Cayman Brac. Uzelman (2009) considered the stable isotopes from the limestones and dolostones whereas Zhao and Jones (2012) focused largely on the stable isotopes from the dolostones. In both studies, the $\delta^{18}O$ and $\delta^{13}C$ values are reported relative to the PeeDee Belemnite standard normalized to NBS-18 using the per mil (%) notation.

Uzelman (2009), based on 138 samples of limestones from 6 different wells and measured sections (Fig. 3.13a), showed that $\delta^{18}O$ values ranged from −6.11 to +1.24‰ (average −1.82‰) and the $\delta^{13}C$ ranged from −10.68 to

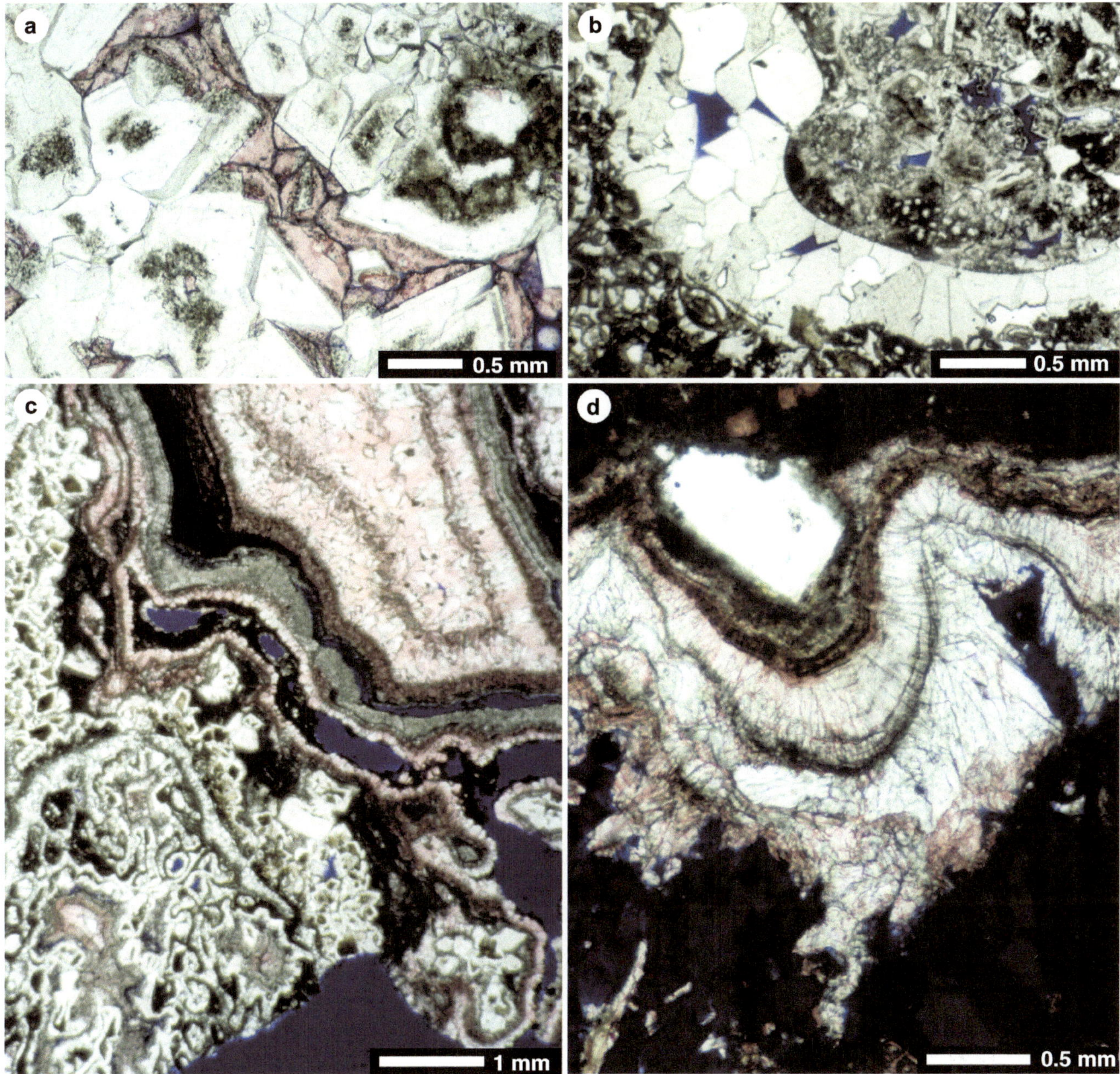

Fig. 3.11 Cements in dolostones of the Brac Formation. All thin sections stained with Alizarin Red S solution (red = calcite; no stain = dolomite). **a** Calcite cement filling cavity in dolostone formed of zoned dolomite crystals. Unit SCD-1. **b** Coarsely crystalline dolomite cement filling cavity created by leaching of a bivalve shell. Unit SCD-5. **c** Cavity in dolostone from section SCD with complex succession of dolomite cements, calcite cements, and calcitic sediment that probably originated in the phreatic zone. **d** Microstalactitic cement formed of calcite and dolomite. Unit SCD-4

+1.16‰ (average –2.34‰). In contrast, Uzelman (2009) based on the analyses of 156 samples (Fig. 3.13b), showed that the $\delta^{18}O$ values of the dolostones ranged from −4.64 to +4.73‰ (average +1.39‰) and the $\delta^{13}C$ ranged from −4.53 to +3.80‰ (average +1.36‰).

For the calcite there is a positive covariant trend between the $\delta^{18}O$ and $\delta^{13}C$ (Fig. 3.13a) with the most variability being evident in samples from section APL#1 (Fig. 3.13c). The calcite isotope values fluctuate with depth, as is apparent from the succession in well APL#1 (Fig. 3.13c).

The isotopes for the dolostones from sections WOJ-2 and WOJ-7, which together are equivalent to section SCD, have $\delta^{18}O$ values ranged from + 1.00 to +3.00‰ and $\delta^{13}C$ values from 0 to +3.00‰ (Fig. 3.13d). In contrast the dolostones from wells EOR#1 and APL#1 yielded samples that had both positive and negative $\delta^{18}O$ and $\delta^{13}C$ values

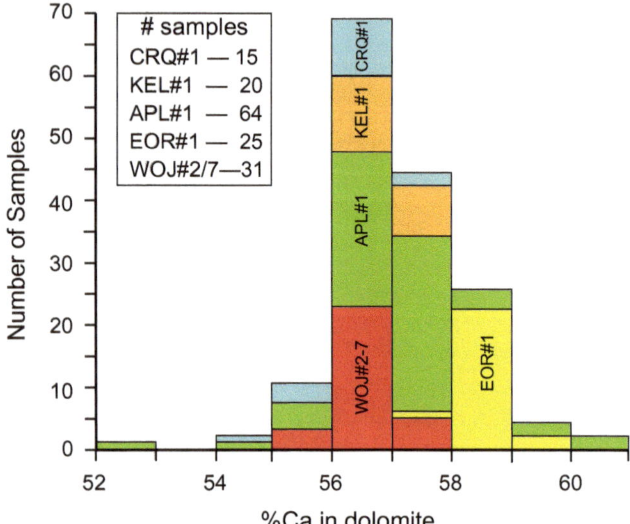

Fig. 3.12 %Ca in dolomite from the Brac formation in sections CRQ#1, KEL#1, APL#1, EOR#1, and WOJ#2/7. Modified from Uzelman (2009, her Fig. 3.5)

(Fig. 3.13e). In wells CRQ#1, KEL#1, and WOJ-2/7, the δ^{18}O and δ^{13}C values are relatively constant with depth. In contrast, the δ^{18}O and δ^{13}C values in wells APL#1 are more variable without there being any specific relationship with depth (Fig. 3.13e). Uzelman (2009) suggested that there is a poor relationship between the %Ca and the δ^{18}O of the dolomite (Fig. 3.14).

3.11 ^{87}Sr/^{86}Sr Geochronology

Jones and Luth (2003, their Fig. 9) showed that the ^{87}Sr/^{86}Sr ratios of the dolostones in the upper part of the Brac Formation are different from ^{87}Sr/^{86}Sr ratios of the coeval limestones but similar to those of the dolostones in the basal part of the overlying Cayman Formation (Fig. 3.15). With the drilling and sampling of wells APL#1, KEL#1, and CRQ#1 in 2002 and 2003, additional ^{87}Sr/^{86}Sr ratios were available for use by Uzelman (2009) who focused her attention on the Brac Formation on the eastern part of Cayman Brac (Fig. 3.16). Using these data, Uzelman (2009) showed that the ^{87}Sr/^{86}Sr ratios for the limestones from sections EOR#1, APL#1, and EOR#1 range from 0.708000 to 0.709144, with an average of 0.708569 ± 0.000022 (Fig. 3.16). In wells CRQ#1 and KEL#1, the dolostones of the Cayman Formation are characterized by relatively constant ^{87}Sr/^{86}Sr ratios of ~0.709000 with only a few samples deviating slightly from that ratio (Fig. 3.16). In contrast, the ^{87}Sr/^{86}Sr ratios from the Brac Formation in those wells and well APL#1 are more variable (Fig. 3.16). In all of the wells, the change from the Brac Formation to the Cayman Formation is clearly evident from the contrast in ^{87}Sr/^{86}Sr ratios (Fig. 3.16).

3.12 Age

Based on descriptions provided in Matley (1926), it appears that his *Lepidocyclina* Bed came from the vicinity of section LCB, where he considered the strata to be horizontal. He collected his samples from fallen blocks and suggested that they came from the uppermost part of the cliff face. In that area, where there are numerous fallen blocks, the lower part of the cliff face is commonly masked by limestone and dolostone lithoclasts that are cemented in place (Fig. 3.17). These breccias are probably equivalent to the breccias found on the south coast that Jones and Ng (1988) attributed to localized collapse of the cliff face. Careful inspection of the visible part of the cliff face on the north coast clearly shows that the *Lepidocyclina*-rich limestones are located in the lower part of the cliff face and below the unconformity that defines the top of the Brac Formation. Also, careful inspection of the exposures on the top of the island failed to locate any foraminiferal limestones.

Vaughan (1926), who received the foraminifera collection from Matley, identified the presence of *Lepidocyclina (Lepidocyclina) yurnangunensis* Cushman, *Lepidocyclina (Nephrolepidina) undosa* Cushman, *Lepidocyclina (Nephrolepidina) undosa* var. *tumida* nov. Vaughan, *Lepidocyclina gigas* Cushman, *Lepidocyclina*. sp. cf. *marginaata* (Michelotti), *Lepidocyclina* sp. indet., and *Carpenteria americana* Cushman. Matley (1926) adopted Vaughan's suggestion that this fauna indicated a Middle Oligocene (Rupelian) age.

An ^{87}Sr/^{86}Sr value of 0.70808, the average derived from five samples of limestone from the Brac Formation, plotted against the temporal ^{87}Sr/^{86}Sr trends of DePalolo and Ingram (1985) and Hess et al. (1986), yielded an age of about 28 million years (Jones et al. 1994a). This age is consistent with the Rupelian age suggested by the foraminifera fauna (cf., Vaughan, 1926). Subsequently, Jones and Luth (2003) showed that the limestones and dolostones in the Brac Formation on Cayman Brac have an average ^{87}Sr/^{86}Sr values of 0.708189 and 0.708939, respectively. In comparison, the basal dolostones in the overlying Cayman Formation yielded

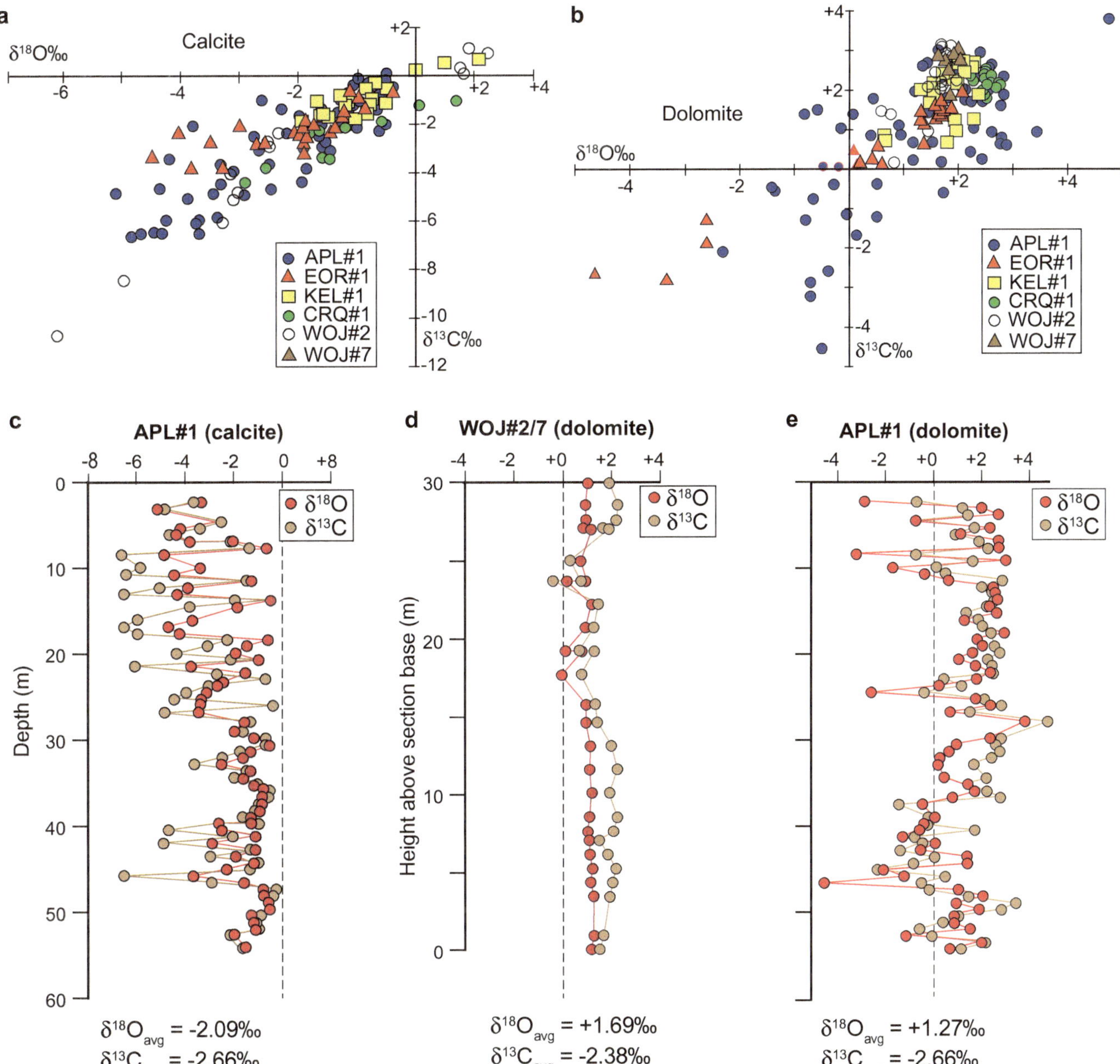

Fig. 3.13 Stable isotopes from the Brac Formation, east end of Cayman Brac. **a**, **b** Cross-plot of $\delta^{18}O$ against $\delta^{13}C$ for calcite **a** and dolomite **b** from the Brac Formation in sections APL#1, EOR#1, KEL#1, CRQ#1, WOJ#2 and WOJ#7 (latter two sections are equivalent to section SCD). Modified from Uzelman (2009, her Fig. 3.1 and 3.3a respectively). **c** ^{18}O and ^{13}C versus depth for calcite in the Brac Formation in well APL#1. **d**, **e** ^{18}O and ^{13}C versus depth for dolomite in wells WOJ#2/7, respectively. Modified from Uzelman (2009, her Figs. 3.2, and 3.4)

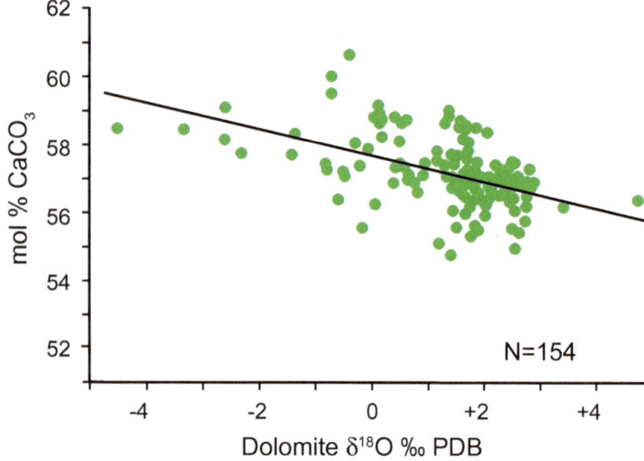

Fig. 3.14 $\delta^{18}O$ versus %Ca in dolomite from the Brac Formation. Modified from Uzelman (2009, her Fig. 3.6)

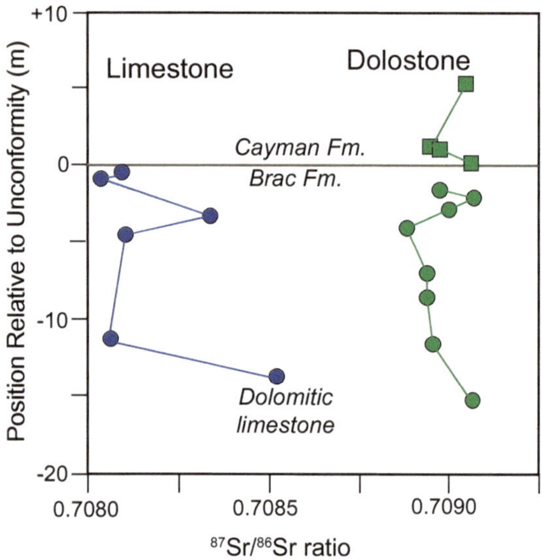

Fig. 3.15 Comparison of $^{87}Sr/^{86}Sr$ ratios in limestones and dolostones in the upper part of the Brac Formation and lower part of Cayman Formation in section SCD. Modified from Jones and Luth (2003, their Fig. 9)

average $^{87}Sr/^{86}Sr$ values of 0.708992. Plotted against the temporal $^{87}Sr/^{86}Sr$ trend of Hodell et al. (1989), Jones and Luth (2003) argued that the $^{87}Sr/^{86}Sr$ values from the limestones also yielded an age of ~28 million years, whereas the $^{87}Sr/^{86}Sr$ values of the dolostones, which indicate a late Miocene age, were reset when dolomitization took place during the Late Miocene.

3.13 Depositional Regimes

Given that the basal boundary of the Brac Formation has not yet been discovered on any of the Cayman Islands, it is impossible to determine the surface topography of the banks on which the sediments of this formation were deposited. In effect, the depositional conditions that existed during the development of the succession can only be determined for the upper part of the formation.

The fact that each of the Cayman Islands are located on separate fault blocks coupled with the lack of evidence of intertidal or supratidal deposits in the Brac Formation, indicates that sedimentation must have taken place on isolated banks under the control of sea level fluctuations and/or tectonic subsidence. Although there is no evidence of reef development around the margins of the Cayman Brac, there is the possibility that they may have been destroyed by subsequent coastal erosion as there is evidence that the central core of Cayman Brac has retreated inland due to erosion (Jones and Ng, 1988).

Potentially, the presence of *Lepidocyclina* in the Brac Formation is important because it has been recorded from many other successions throughout the world. These large benthic foraminifera are common throughout the Oligocene strata of the Caribbean (Cole, 1957, 1958; Sachs, 1959). Frost et al. (1983), based on the locations of large *L. undosa* relative to the Oligocene shelve and reefs in Puerto Rico suggested that they inhabited and accumulated in water that was ~100 m deep. Elsewhere in the world, depth estimates for *Lepidocyclina* range from 30 to 60 m in Somalia (Bosellini et al. 1987), outer ramp settings in the Tethyan region (Buxton and Pedley, 1989), back reef and reef flat environments in southern Italy (Bosellini and Russo, 1992), and shallow (<12 m) sea-grass communities in the Australian Oligocene (Chaproniere, 1975).

The rarity of corals (fragmentary *Porites*) in the Brac Formation on Cayman Brac contrasts sharply with the diverse biota in the overlying Cayman Formation that includes numerous corals. It is, however, difficult to pinpoint the reason(s) for this contrast because there are any number of factors, such as salinity, water turbidity, lack of nutrients, and/or water temperature that could be responsible (Jones and Hunter, 1994). After evaluating the available evidence, Jones and Hunter (1994) suggested that the sediments that now form the Brac Formation may have accumulated in shallow-water conditions where a muddy substrate favoured the development of sea-grass beds.

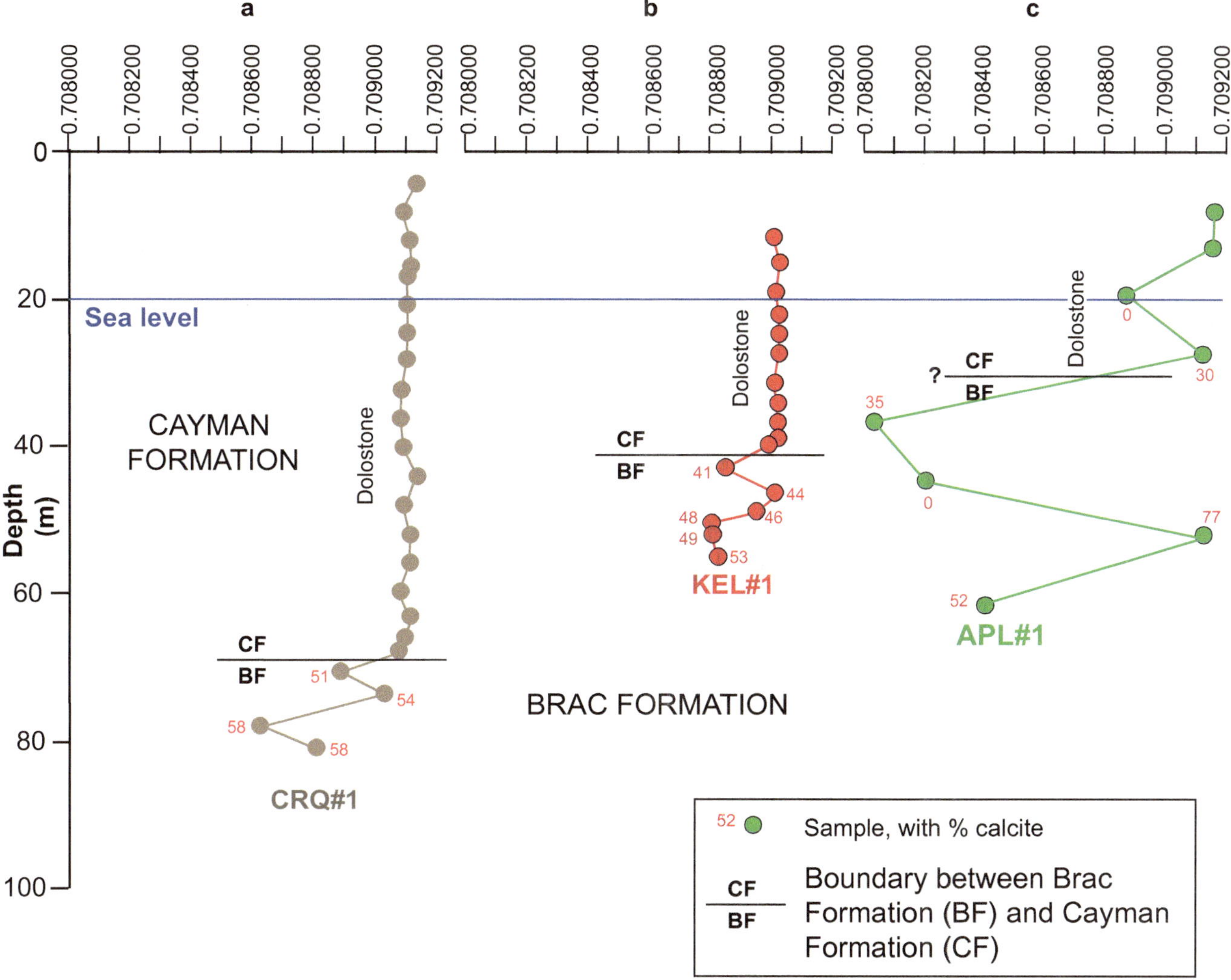

Fig. 3.16 Comparison of $^{87}Sr/^{86}Sr$ ratios from the Brac Formation and Cayman Formation in wells CRQ#1, KEL#1, and APL#1. Note fluctuating $^{87}Sr/^{86}Sr$ ratios in the Brac Formation (BF) relative to the almost constant values from the dolostones in the overlying Cayman Formation (CF). Graphs based on data in Appendix 1 of Uzelman (2009)

3.14 Conclusions

The following important conclusions are known regarding the Brac Formation.

- The Brac Formation is known only from exposures of its upper beds exposed on the east end of Cayman Brac and from deep wells drilled on Grand Cayman and Cayman Brac.
- The lower boundary of the formation has not yet been located.
- The Brac Formation is separated from the overlying Cayman Formation by an unconformity.
- The formation is at least 50 m thick.
- In the cliff faces on the east end of Cayman Brac, the fossiliferous limestones in the Brac Formation on the north coast contrast with the sucrosic dolostones and scattered pods of fossiliferous limestones found on the south coast.
- The foraminifera and $^{87}Sr/^{86}Sr$ ratios show that the upper part of the Brac Formation is Upper Oligocene in age.
- The lithologies and fossils indicate that the original sediments were deposited in relatively calm water that was <25 m deep.
- The high-calcian dolostones are formed of large dolomite crystals that have each have a "dirty" core encased by a clear rim.
- The dolostones and limestones developed through multiple phases of diagenesis.
- Dolomitization of the limestones was probably mediated by normal to slightly evaporated seawater.

Fig. 3.17 Cemented breccia at base of cliff face on northeast coast of Cayman Brac, close to the type section of the Brac Formation (section LCB). The clasts in the breccia are formed of dolostones that were probably derived from the Cayman Formation that forms the upper part of the cliff face. Hammer handle is 0.3 m long

References

Bosellini A, Russo A (1992) Stratigraphy and facies of an oligocene fringing reef (Castro Limestone, Salento Peninsula, Southern Italy). Facies 26:145–166

Bosellini A, Russo A, Arush MA, Cabdulqadir MM (1987) The oligo-miocene of Eil (NE Somalia): A prograding coral - *Lepidocyclina* system. J Afr Earth Sci 6:583–593

Buxton MWN, Pedley HM (1989) A standardized model for Tethyan Tertiary carbonate ramps. J Geol Soc Lond 146:746–748

Chaproniere GCH (1975) Palaeoecology of oligo-miocene larger foraminiferida, Australia. Alcheringa 1:37–58

Cole WS (1957) Variations in American oligocene species of *lepidocyclina*. Bull Am Paleontol 38:31–51

Cole WS (1958) Names of and variation in certain American larger foraminifera. Bull Am Paleontol 38:179–212

DePaolo DJ, Ingram BL (1985) High-resolution stratigraphy with strontium isotopes. Science 227:938–941

Frost SH, Harbour JL, Beach DK, Realini MJ, Harris PM (1983) Oligocene reef tract development southwestern Puerto Rico, Part 1, 2, 3. Sedimenta, 4. The Comparative Sedimentology Laboratory, Division of marine geology and geophysics. Rosenstiel Sch Mar Atmos Sci, Miami:141

Hess J, Bender ML, Schilling JG (1986) $^{87}Sr/^{86}Sr$ evolution from cretaceous to present - applications to paleoceanography. Science 231:979–984

Hodell DA, Mueller PA, McKenzie JA, Mead GA (1989) Strontium isotope stratigraphy and geochemistry of the late Neogene ocean. Earth Planet Sci Lett 92:165–178

Jones B (1994) Geology of the Cayman Islands. In: Brunt MA, Davies JE (eds) The Cayman Islands: Natural History and Biogeography. Kluwer, Dordrecht, The Netherlands, pp 13–49

Jones B, Hunter IG (1994) Evolution of an isolated carbonate bank during oligocene, Miocene, and Pliocene times, Cayman Brac, British West Indies. Facies 30:25–50

Jones B, Hunter IG, Kyser TK (1994a) Revised stratigraphic nomenclature for Tertiary strata of the Cayman Islands, British West Indies. Carib J Sci 30:53–68

Jones B, Hunter IG, Kyser TK (1994b) Stratigraphy of the bluff formation (miocene-pliocene) and the newly defined Brac formation (oligocene), Cayman Brac, British West Indies. Carib J Sci 30: 30–51

Jones B, Luth RW (2002) Dolostones from Grand Cayman, British West Indies. J Sediment Res 72:560–570

Jones B, Luth RW (2003) Temporal evolution of Tertiary dolostones on Grand Cayman as determined by $^{87}Sr/^{86}Sr$. J Sediment Res 73:187–205

Jones B, Luth RW, MacNeil AJ (2001) Powder X-ray analysis of homogeneous and heterogeneous dolostones. J Sediment Res 71:791–800

Jones B, Ng K-C (1988) Anatomy and diagenesis of a pleistocene carbonate breccia formed by the collapse of a seacliff, Cayman Brac, British West Indies. Bull Can Pet Geol 36:9–24

Matley CA (1926) The geology of the Cayman Islands (British West Indies) and their relation to the Bartlett Trough. Q J Geol Soc Lond 82:352–387

Sachs KN (1959) Puerto rican upper oligocene larger foraminifera. Bull Am Paleontol 39:399–417

Uzelman BC (2009) Sedimentology, diagenesis, and dolomitization of the Brac formation (lower oligocene), Cayman Brac, British West Indies. M.Sc. thesis, University of Alberta

Vaughan TW (1926) Species of *lepidocyclina* and *carpentaria* from the Cayman Islands and their geological significance. Geol Mag 82:388–400

Zhao H, Jones B (2012) Genesis of fabric-destructive dolostones: a case study of the Brac formation (oligocene), Cayman Brac, British West Indies. Sed Geol 267–268:36–54

4 Bluff Group—The Cayman Formation

Abstract

The lower and upper boundaries of the Cayman Formation, which separate it from the underlying Brac Formation and overlying Pedro Castle Formation, are unconformities with significant karstic relief. Although the Cayman Formation is at least 150 m thick, its true original thickness is unknown because of the relief on these boundaries. This formation developed during the Middle Miocene when marine sediments were deposited on an open bank in waters <25 m deep. On Cayman Brac, the western part of Grand Cayman, and eastern coastal areas of Grand Cayman, the formation is formed of fabric-retentive, porous, finely-crystalline dolostones. In the central-eastern part of Grand Cayman, however, the formation is formed of limestones and locally, limestones overlain by dolostones. These dolostones and limestones commonly contain numerous corals, bivalves, gastropods, foraminifera, red algae, *Halimeda*, and rare echinoids. There are, however, only scattered, small coral patch reefs. Fossil-mouldic porosity is common given that the aragonitic skeletons of the corals and other organisms were commonly lost during early diagenesis. The dolostones, characterized by high porosity (up to 48%), are commonly cut by complex arrays of joints and fractures. The constituent dolomite crystals, typically <50 μm long, commonly have cores of high-calcian dolomite encased by an outer zone formed of low-calcian dolomite. The complex diagenetic fabrics in the dolostones involve numerous phases of limpid dolomite, calcite, and/or internal sediments. Dolomitization, which probably occurred during the Late Miocene and Pliocene was probably mediated by normal or slightly evaporated seawater.

4.1 Introduction

The term Bluff Limestone, originally established by Matley (1926), was subsequently used in studies by Mather (1972), Brunt et al. (1973), Bugg and Lloyd (1976) Rigby and Roberts (1976), Stoddart (1980), Woodroffe et al. (1980), Emery (1981), Woodroffe (1981), Woodroffe et al. (1983), and Spencer (1985). The sequence is widely exposed on Grand Cayman and Cayman Brac (Fig. 4.1a, c), but the scattered low-lying exposures on Little Cayman may belong to either the Cayman Formation or the overlying Pedro Castle Formation (Fig. 4.1b). Although the term Bluff Limestone was used as a formation name, it was never formally described and no type section was ever designated. Use of the term "Bluff Limestone" also disguised the fact that the formation is formed largely of finely crystallized dolostones, as demonstrated by Jones et al. (1984), Pleydell (1987), and Pleydell and Jones (1988). Given this, Jones and Hunter (1989) designated the 8 m thick succession exposed in Pedro Castle Quarry as the type section of the Bluff Formation and defined the Cayman Member and the Pedro Castle Member with the two members being separated by an unconformity (Fig. 4.2a). Use of these names also removed the lithological connotation that had been inherent with usage of the term Bluff Limestone. Although the section in Pedro Castle Quarry is not an ideal type section because of its limited thickness, it was the thickest exposed succession on Grand Cayman that included the unconformity that separated the two units. At that time, few wells had been drilled and sampled on Grand Cayman.

Jones et al. (1994b) formally designated the Brac Formation and noted that the overlying Bluff Formation was formed of the Cayman Member and Pedro Castle Member.

B. Jones, *Geology of the Cayman Islands*,
https://doi.org/10.1007/978-3-031-08230-6_4

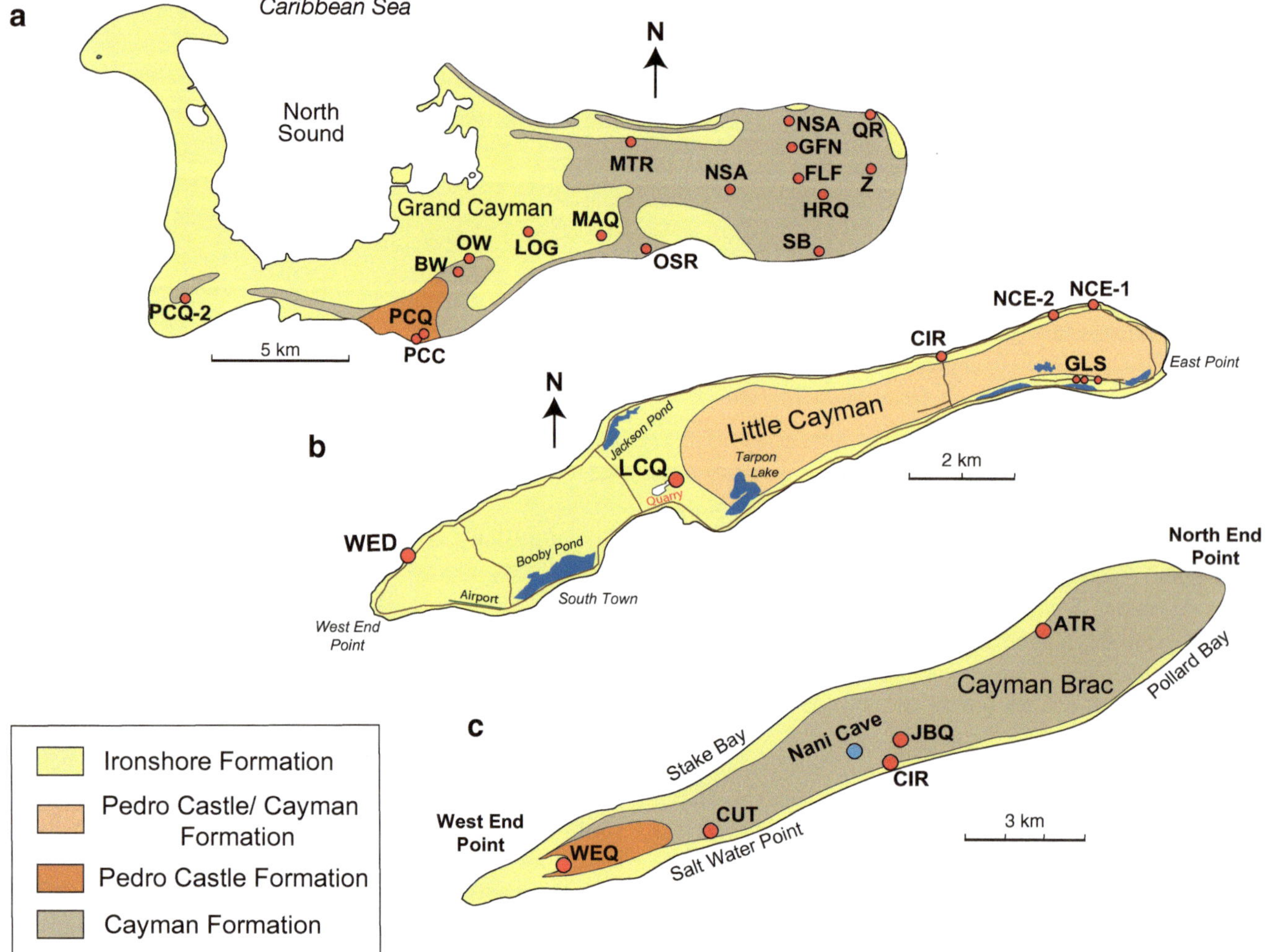

Fig. 4.1 Maps showing distribution of the Cayman Formation on each of the Cayman Islands. For Little Cayman, unit is shown as the Cayman Formation/Pedro Castle Formation because of the uncertainty regarding formation assignment for the scattered outcrops of limestone/dolostone. Important localities cited in the text are shown by red dots with locality code

They estimated that the Cayman Member, formed of finely crystalline dolostones was ~100 m thick, whereas the Pedro Castle Member, formed of limestones, partly dolomitized limestones, and dolostones was 15–20 m thick. Subsequently, Jones et al. (1994a) revised the stratigraphy so that the Bluff Group included the Brac Formation, Cayman Formation, and Pedro Castle Formation. They used a composite section on Cayman Brac to illustrate the stratigraphic setting of the Cayman Formation between the Brac Formation and Pedro Castle Formation. Jones (1994) provided a summary description of the Cayman Formation in his overview of the geology of the Cayman Islands.

The finely crystalline dolostones of the Cayman Formation commonly contain numerous fossil moulds given that the aragonitic skeletons of the corals, bivalves, and gastropods were dissolved during diagenesis. Although identifiable in outcrop and core, it is impossible to identify the presence of such fossils from well cuttings. The cuttings from the wells have primarily been used to determine the percentage of calcite and dolomite, stable isotopes, $^{87}Sr/^{86}Sr$ ratios, and rare earth elements (REE) in each sample.

4.2 Distribution

Surface exposures of the Cayman Formation are common on Grand Cayman and Cayman Brac but may be absent on Little Cayman where outcrops are limited and difficult to assess (Fig. 4.1). Assessing the thickness and facies on these islands is difficult for reasons that are specific to each island. Outcrops of the Cayman Formation in the cliff faces on Cayman Brac (Fig. 4.2) are important because they allow a detailed assessment of the constituent facies and the biota that characterizes the formation. They provide a basis for the interpretation of the cores and well cuttings that come from the wells drilled on Cayman Brac and Grand Cayman.

Fig. 4.2 Cliff face on south coast at east end of Cayman Brac showing massive, almost featureless dolostones of the Cayman Formation. Gray to black colour due to surface weathering. The vertical to overhanging cliff face is about 30 m high

Grand Cayman is a low-lying island that is topographically very different from Cayman Brac. On Grand Cayman, the "Mountain", which attains a maximum height of ~17 m above sea level (asl), is the highest area on the island. Most of the island is, however, <6 m asl with much of the western peninsula <3 m asl. Outcrops are typically <2 m high and restricted to road cuts. It is only in the quarries that thicker successions are exposed with the walls in Pedro Castle Quarry, which is now inactive, being up to 12 m high (Fig. 4.3a). Inland exposures of the formation are scattered and on the western peninsula, the formation is largely buried beneath the Pedro Castle Formation and/or the Ironshore Formation. The interior of Grand Cayman is generally inaccessible due to the rugged phytokarst and dense tropical vegetation. It was not until the Queen's Highway opened in 1983 that it was possible to drive around the island and access to the eastern interior only became easier with the opening of the Botanical Gardens in 1991. By necessity, therefore, most of the information about this formation on Grand Cayman comes from the numerous wells that have been drilled on the island (Fig. 4.4). Herein, attention is focused primarily on the wells that provide the most complete successions through the Cayman Formation from different parts of the island. Although many wells are not mentioned in this discussion it must be stressed that the information derived from those wells fully supports the data and interpretations that are derived from the main wells.

On Cayman Brac, the Cayman Formation is superbly exposed in the cliff faces that rise up to ~43 m asl at the east end of the island and define the elevated core of the island (Fig. 4.2). These vertical to overhanging cliff faces are, however, generally impossible to examine geologically except where roads or pathways run inland from the coastal areas. Prime examples of such access include the pathway that leads to the "Hurricane Caves" (also known as Peter's Cave), located near the top of the cliff face on the north side of the island close to Spot Bay, and Bluff Road that crosses the central part of the island (Fig. 4.1c). On the west end of the island, the formation is only exposed in the low cliff faces that characterize that part of the island.

On Little Cayman, exposures of the limestones and dolostones that belong to the Bluff Group (Fig. 4.1b) are typically <2 m high, widely scattered, and commonly covered with dense vegetation. Although Jones (2019) suggested that most of these exposed dolostones and calcareous dolostones probably belonged to the Pedro Castle Formation

Fig. 4.3 Pedro Castle Quarry. **a** View of type section exposed in wall at west end of Pedro Castle Quarry (PCQ—Fig. 1a) showing Pedro Castle Formation (PCF) unconformably overlying the Cayman Formation (CF). **b** *Lithophaga* boring opening at upper surface of the Cayman Formation. **c** Small open and filled cavities in Cayman Formation just below the unconformity (U/C) that separates it from the overlying Pedro Castle Formation. **d** Oblique cut through uppermost part of the Cayman Formation showing sponge borings filled with white dolomitic limestone that is identical to that in the basal bed of the overlying Pedro Castle Formation. **e** View of coastal area to east of Spotts Bay showing planar surface, which is the unconformity that separates the Cayman Formation from the Pedro Castle Formation. Houses have now been built on this site

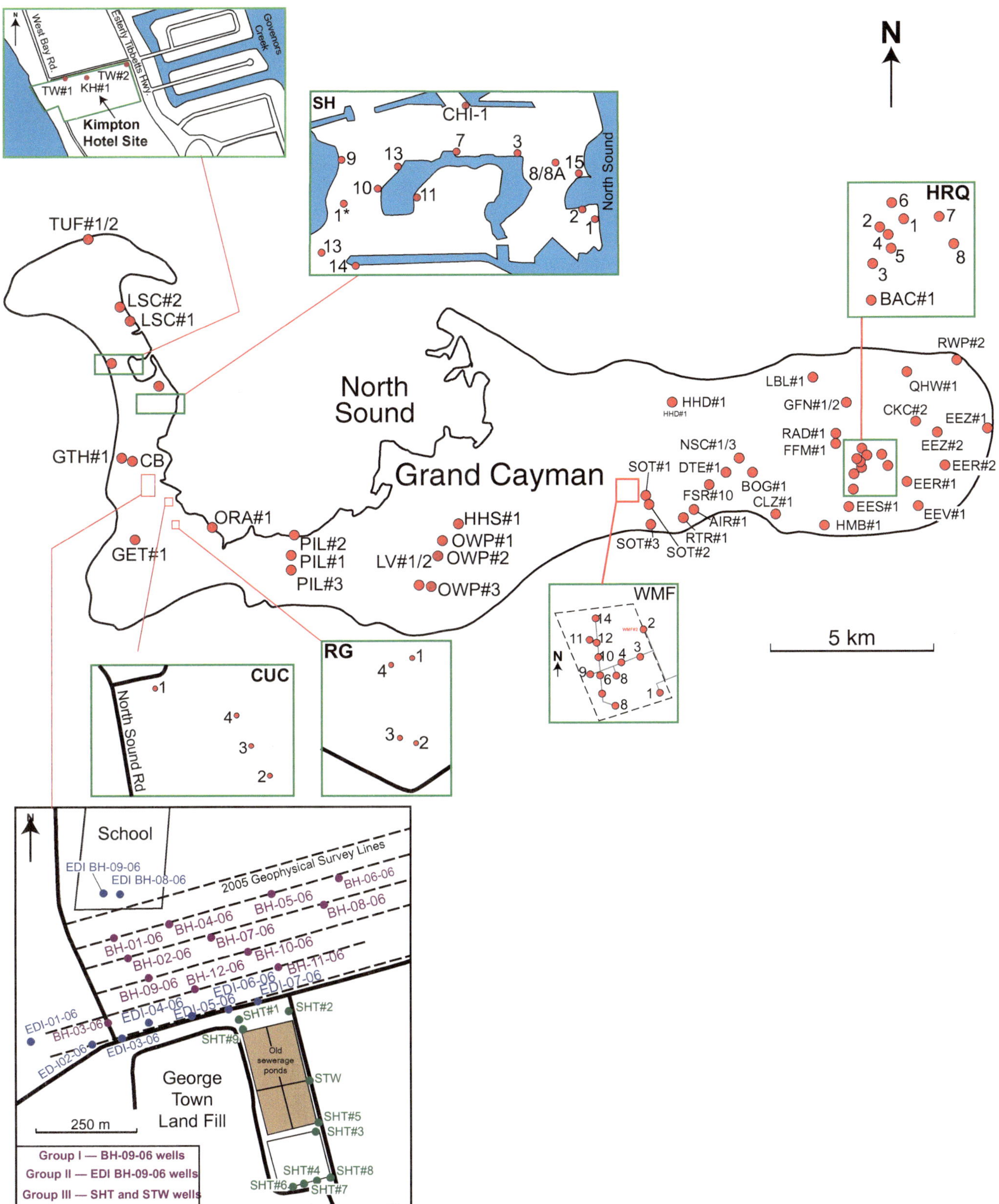

Fig. 4.4 Map showing locations of wells drilled on Grand Cayman from which data and samples have been collected as part of the investigation of the Cayman Formation

rather than the Cayman Formation, further research is required to fully resolve this issue. The fact that no wells have been drilled on this island means that the complete succession through the Bluff Group cannot be fully assessed.

On Grand Cayman, much of the information on the Cayman Formation comes from cuttings and core collected from numerous wells that have been drilled on the island (Fig. 4.4). Initially, the research group based at the University of Alberta used a portable drilling rig to drill and core numerous wells on the island. Although the depth of those wells was generally <100 m, the cores provided excellent coverage of the various formations, including the Cayman Formation. A large array of samples and information has also come from wells that have been drilled by the Water Authority as they needed to provide water supplies for the ever-growing population of the island, and the construction industry as development continues at a rampant pace. Over the last 20 years, the deepest well drilled and sampled was to a depth of ~245 m on the Water Authority site on eastern Grand Cayman. Most wells, however, are <125 m deep because of inherent limitations with the drilling rigs used on the island. Sampling from these wells range from cuttings (most common), to a mixture of well cuttings and core from selected intervals, to completely cored (rare). In all, samples and information are available from 104 wells that have been drilled on Grand Cayman (Fig. 4.4). Fewer wells have been drilled and sampled on Cayman Brac, where they are limited to a maximum depth of 61 m because of the limitations associated with the only drilling rig on the island. No coring is possible and thus, only well cuttings are available from those wells.

4.3 Lower Boundary

As defined by Zhao and Jones (2012a), the lower boundary of the Cayman Formation on Cayman Brac is denoted by changes (1) from coarsely crystalline, fabric-destructive dolomite of the Brac Formation to fabric-retentive dolostone of the Cayman Formation, and (2) from a *Lepidocyclina*–dominated fauna with rare corals and bivalves to a coral-dominated fauna with abundant *Halimeda* and bivalves. For ease of reference, this unconformity has been referred to as the Brac Unconformity. As yet, however, neither of these criteria have been evident in cores and/or well cuttings obtained from the deeper wells on Grand Cayman. Jones and Luth (2003), Liang and Jones (2014), and McCormick and Jones (2021) all argued that this boundary is also marked by paleokarst features (e.g., paleosols, erosive contacts, and penetrative dissolution) and a marked break in the vertical succession of $^{87}Sr/^{86}Sr$ ratios. For example, in well SHT#4, the Brac Unconformity, located 77 m below sea level (bsl), is highlighted by a 1 cm—thick layer of terra rossa (McCormick and Jones 2021). These criteria are used herein for locating the Brac Unconformity on Grand Cayman.

On the western peninsula of Grand Cayman, the Brac Unconformity is characterized by a subdued relief with a northward dip from 72 m bsl in well CUC#4 to 83 m bsl in well TW#2 (McCormick and Jones 2021). In contrast, the Brac Unconformity along the southern part of the island becomes deeper to the east, occurring at 123 m bsl in well LV#2 and 129 m bsl in well RTR#1. These numbers indicate that there is at least 57 m of relief on this boundary in that area.

To the east of Lower Valley, locating the Brac Unconformity has been largely unsuccessful despite the fact that some wells (e.g., NSC#1/3) have been drilled to depths of 245 m. This may be due to (1) the Brac Unconformity being located at depths that exceed even the base of well NSC#1/3, (2) there is a yet unrecognized north-south fault, located in the middle part of the island, with the eastern part of the island being on the down-throw side, or (3) it has not being recognized because of the paucity of data due to the fact that most samples from the deeper parts of the wells on the eastern part of the island are represented only by well cuttings. In well GFN#2, which was fully cored to a depth of 92 m, the entire succession is relatively monotonous with no major changes in facies and no evidence of the basal unconformity of the Cayman Formation. The geochemical parameters, including the $^{87}Sr/^{86}Sr$ ratios are also relatively constant throughout the succession. Similarly, in well NSC#1/3, there is no evidence of a significant change in lithology and there is no distinct break in the $^{87}Sr/^{86}Sr$ ratios like that seen in the wells on the western part of the island. The sparse notes that are available from the Merren Well #1, drilled to a depth of 401 m (but samples only collected to a depth of 282 m), also reports a relatively monotonous succession of sparsely fossiliferous limestones. With the information that is presently available it is difficult to explain the apparent lack of the Brac Unconformity on the eastern part of the island.

On Cayman Brac, the unconformity that defines the lower boundary of the Cayman Formation is evident in the cliff faces along the north and south coasts at the east end of the island (Figs. 3.2 and 3.3). That boundary is also detectable in well cuttings from wells APL#1, KEL#1, and CRQ#1 on Cayman Brac (Fig. 4.5). Drilling of wells to the west of CRQ#1 were terminated before the lower boundary of the Cayman Formation was reached because of limitations with the drilling rig.

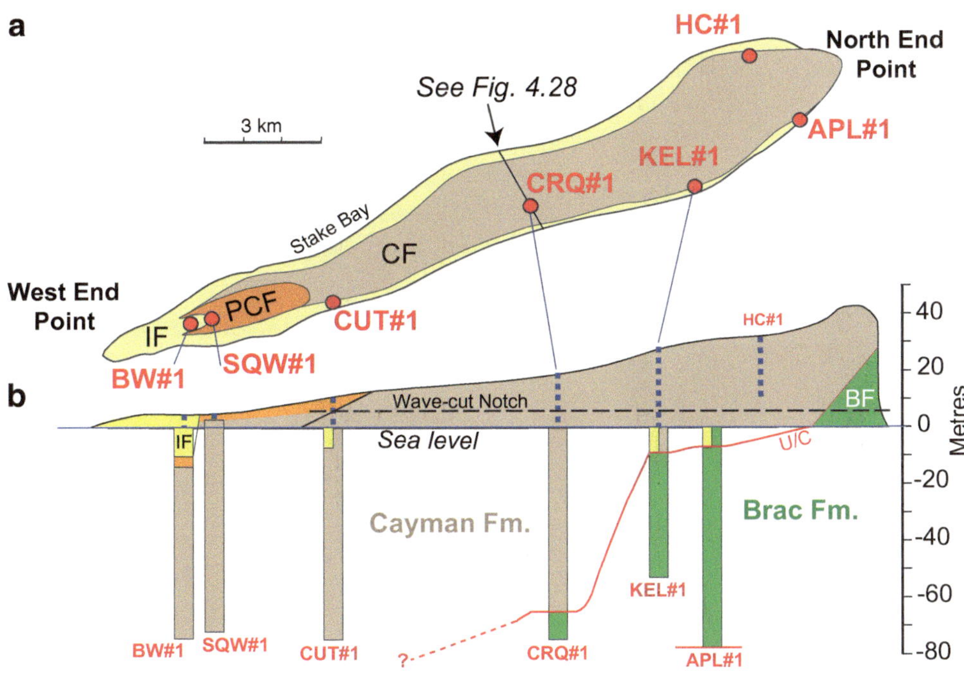

Fig. 4.5 **a** Map of Cayman Brac showing locations of measured sections and wells used to establish the relationship between the Cayman Formation and the older Brac Formation and younger Pedro Castle Formation. **b** Cross-section along length of island showing relationship between the Brac Formation, Cayman Formation, and Pedro Castle Formation

4.4 Upper Boundary

The upper boundary of the Cayman Formation, herein referred to as the Cayman Unconformity, is a complex karst surface that is still, on the eastern part of Grand Cayman, evolving today (Jones and Hunter 1994b; Liang and Jones 2015). On Grand Cayman, the unconformity is exposed in Paul Bodden Quarry (Fig. 4.6), Pedro Castle Quarry (Fig. 4.3a), and along the coastline to the east of Spotts Bay (Fig. 4.3e). In Paul Bodden Quarry (PBQ#1), the top of the Cayman Formation became exposed when the overlying limestones of the Ironshore Formation were removed by quarrying (Fig. 4.6). Today, that surface is not visible as it is covered with soil and grass and used as a soccer field. There, the friable limestones of the Ironshore Formation are still visible in the quarry walls even though they have been extensively weathered and are now partly hidden by vegetation. There is no evidence of the Pedro Castle Formation in this quarry. In Pedro Castle Quarry (PCQ), with walls up to 12 m asl, the unconformity is obvious as the darker coloured dolostones of the Pedro Castle Formation contrast sharply with the white to off-white dolostones of the Cayman Formation (Fig. 4.3a). It should be noted, however, that when freshly exposed the colours of the dolostones in both formations are very similar. Likewise, after extended periods of weathering the rocks in both formations have a dark grey surface colour and it becomes difficult to separate them on this basis.

In Pedro Castle Quarry, the Cayman Unconformity is characterized by numerous borings and solution cavities that open at that surface (Fig. 4.3b–d). Borings include those made by lithophagid bivalves (Fig. 4.3b) and sponges (Fig. 4.3c). The solution cavities, which are of variable sizes (up to 3 cm diameter on vertical face), occur up to 15 cm below the unconformity (Fig. 4.3d). Prior to 1986, the quarry wall in the southwest corner of the quarry included a vertical cross-section through a cave in the dolostones of the Cayman Formation (Fig. 4.7) that had been filled with various types of sediment, flowstone, and terra rossa (Jones 1992). Although the sediments had been dolomitized, the flowstone and terra rossa were not dolomitized. As quarrying proceeded it became evident that the west end of the cave was connected to a tunnel that led to a sinkhole that had its opening at the Cayman Unconformity (Fig. 4.7). The sediments that filled the sinkhole were also dolomitized. Collectively, the evidence from Pedro Castle Quarry indicates that the Cayman Unconformity originated as a karst surface during an extended period of subaerial exposure. The borings and caves were filled during the sedimentation that was associated with the initial phase of the transgression that led to deposition of the sediments that now form the Pedro Castle Formation.

The notion that the Cayman Unconformity is an ancient karst surface is supported by the topography that characterizes its surface. On the western part of the island, the Cayman Unconformity is generally below sea level and buried beneath the Pedro Castle Formation and/or the Ironshore Formation (Fig. 4.8). In wells SH#5 and LV#2, the Cayman Unconformity is ~30 m bsl, which is ~40 m below its level in Pedro Castle Quarry. The Mountain, with its crest ~17 m asl, is located near the north coast to the northeast of Lower Valley. As such, the Cayman Formation

Fig. 4.6 Paul Bodden Quarry, western part of Grand Cayman, showing Ironshore Formation unconformably overlying the Cayman Formation. Photograph taken in 1989—floor of old quarry now covered with soil and grass and used as a soccer pitch

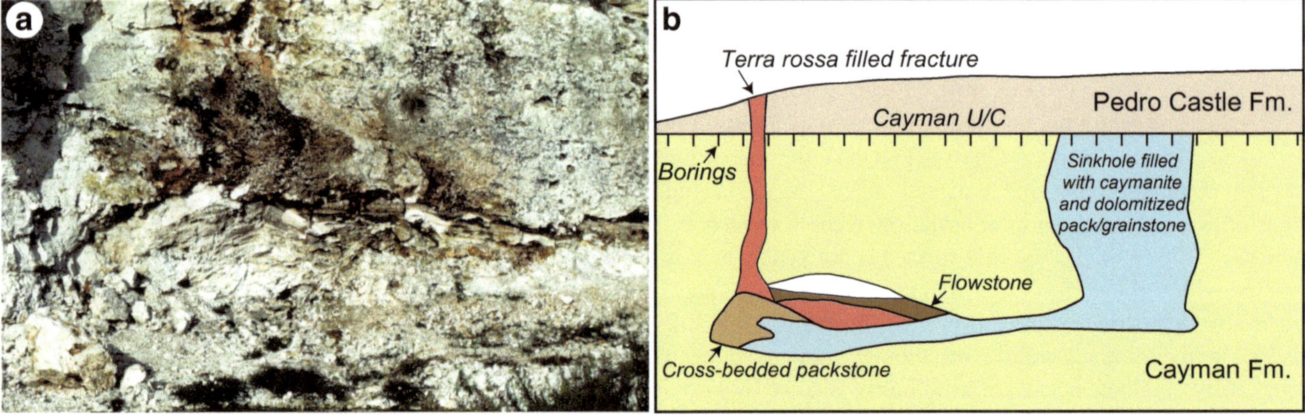

Fig. 4.7 **a** View of wall in Pedro Castle quarry (photograph taken in 1984) showing filled cave in upper part of the Cayman Formation (see text for description of cave fills). This wall was subsequently destroyed by quarrying activity. **b** Schematic cross-section of quarry wall showing the filled cave and the tunnel that extended from one end to a nearby sinkhole that was filled with dolomitized packstone and grainstone. The filled sinkhole is cross-cut by the unconformity that forms the boundary between the Cayman Formation and the overlying Pedro Castle Formation. Modified from Jones (1992, his Fig. 2)

on the top of the Mountain is ~46 m above the Cayman Unconformity in well LV#2. These examples clearly illustrate the rugged topography that developed on the Cayman Formation prior to deposition of the sediments that now form Pedro Castle Formation.

On the eastern part of the Grand Cayman, the situation is different because the top of the Cayman Formation is largely defined by the present-day land surface (Fig. 4.9). On that part of the island, one of the most significant features is the peripheral ridge, formed of the resistant dolostones of the

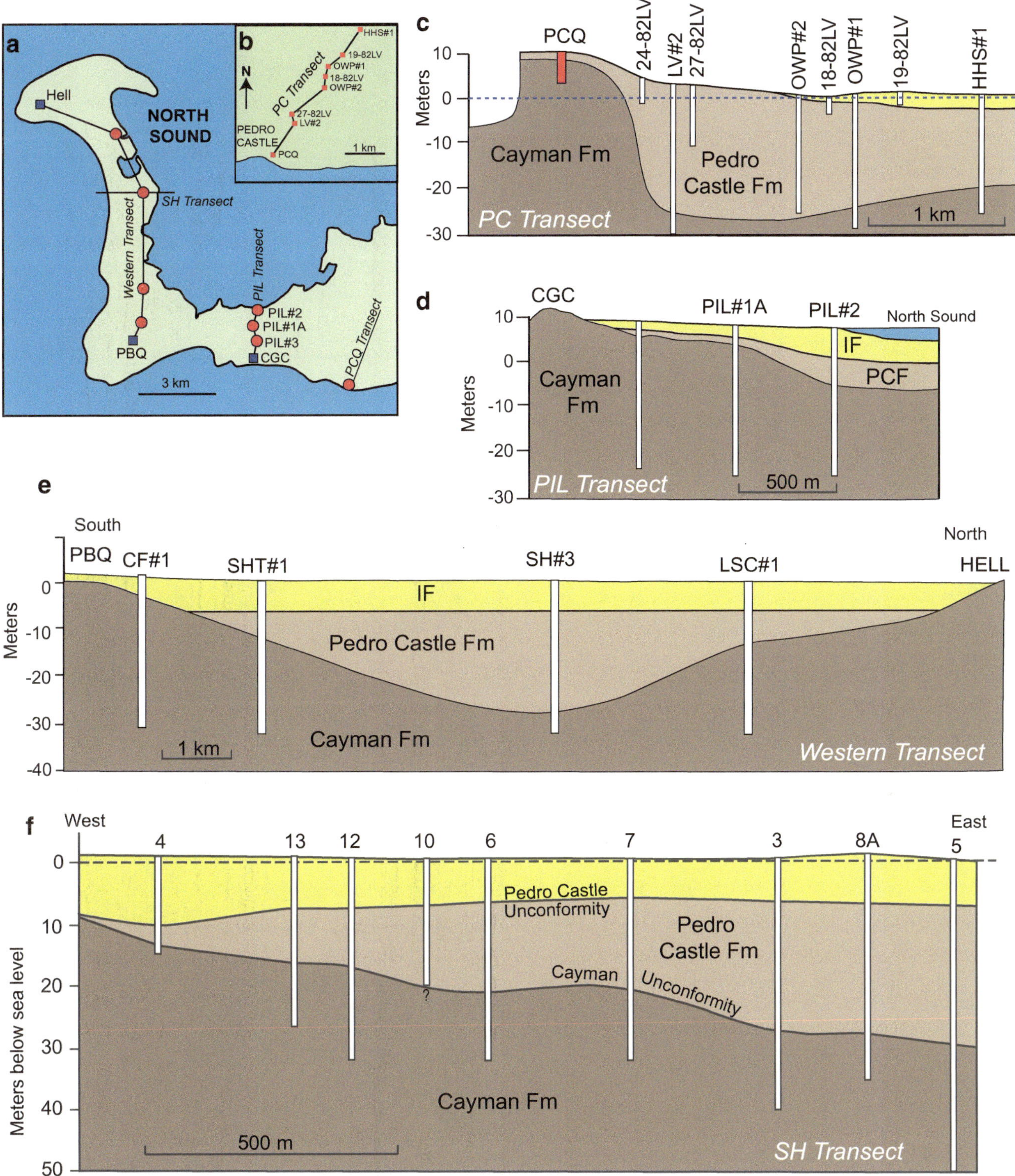

Fig. 4.8 Stratigraphic cross-sections on the western part of Grand Cayman showing relationship between the Cayman Formation, the Pedro Castle Formation, and the overlying Ironshore Formation. **a** Location of cross-sections and sampled wells. **b–f** Stratigraphic variations in the PC, PIL, Western, and SH transects, respectively. Note relief on top of the Cayman Formation

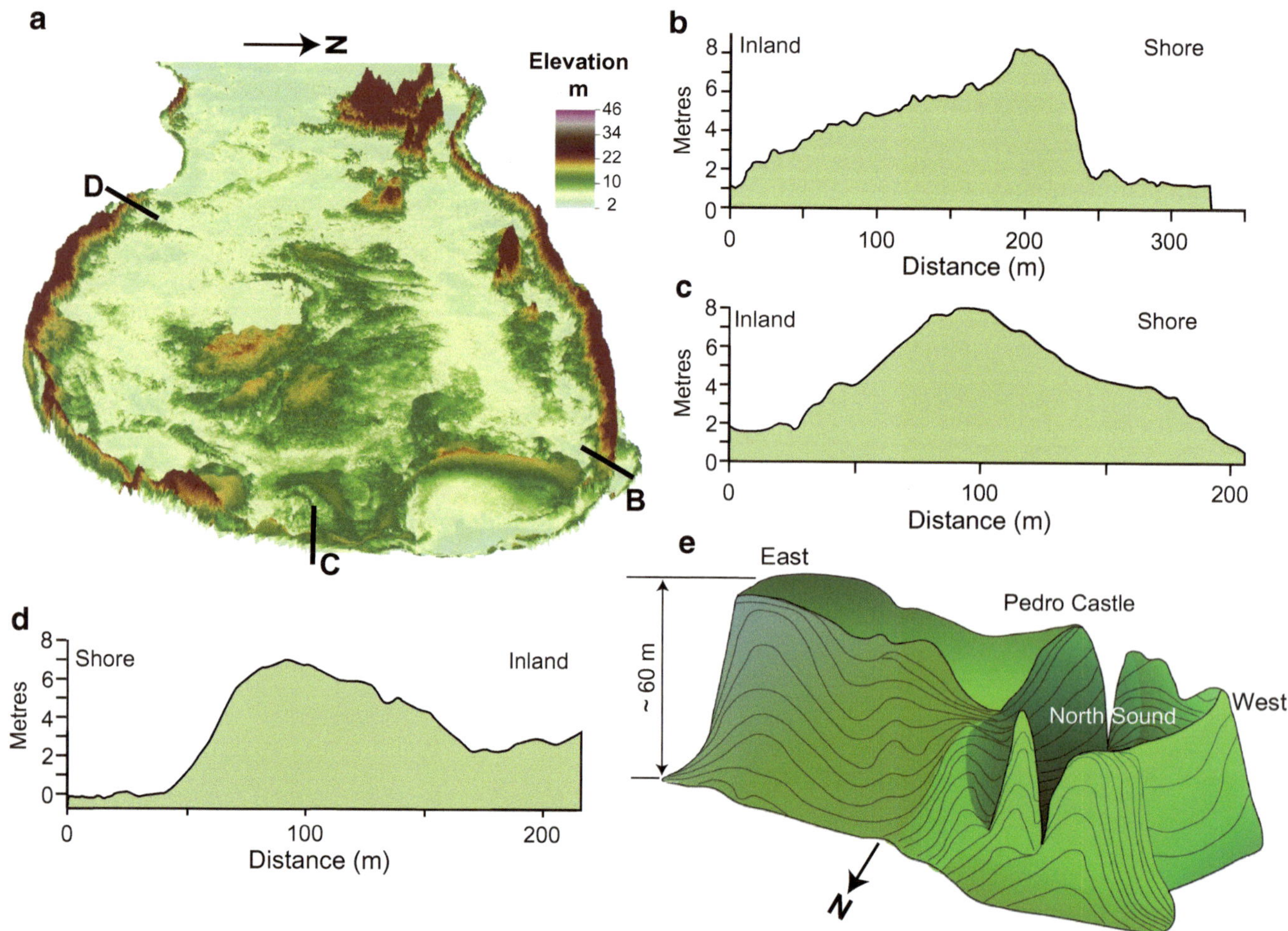

Fig. 4.9 Topography of the Cayman Unconformity that forms the upper boundary of the Cayman Formation. **a** Oblique view, from the east, showing 3-D model of topography of the eastern half of Grand Cayman with locations of cross-sections B, C, and D. Different colours indicate different elevations—note high coastal ridges. Modified from Liang and Jones (2015, their Fig. 4). **b**, **c**, **d** Cross-sections showing topography of the coastal ridge relative to the land seaward and landward of the coast ridge. Modified from Liang and Jones (2015, their Fig. 8). **e** Three-dimensional model showing topography of Cayman Unconformity on Grand Cayman. Oblique view from northwest of island. Modified from Jones and Hunter (1994b, their Fig. 12a)

Cayman Formation that parallel the coastline (Fig. 4.9). In many areas, the ridge is higher than the interior parts of the island (Fig. 4.9). In effect, the Cayman Unconformity is still developing today as weathering lowers the dolostone surface (Jones and Hunter 1994b; Liang and Jones 2015).

4.5 Thickness

The original thickness of the Cayman Formation on Grand Cayman and Cayman Brac is unknown because of the topographies that exist on its lower and upper boundaries. Jones and Hunter (1994b) showed that the upper boundary, labelled as the Cayman Unconformity by Liang and Jones (2015), represents a karst surface that is characterized by considerable relief (at least 41 m) and substantial topographic variations, especially on the western part of the island. In well LV#2, where the upper and lower boundaries of the formation have been identified, the formation is 98 m thick. It must be noted, however, that the top of the formation in that well is 25 m bsl whereas in Pedro Castle Quarry the boundary is, on average, ~10 asl (Fig. 4.2a).

In wells LV#2, CUC#1, CUC#2, CUC#3, CUC$4, SHT#4, GTH#1, RG-4, and TW-2 on Grand Cayman (Fig. 4.4), where both boundaries of the Cayman Formation are present, the thickness of the formation ranges from 45 to 97 m. But, as noted previously, there is ~46 m of relief between the upper boundary of the Cayman Formation in LV#2 and the top of the Mountain that is formed of dolostones that belong to the Cayman Formation. This means that the formation is at least 143 m thick. Although it is impossible to determine the maximum original thickness of the formation, it was probably significantly greater than 143 m given the time period over which the formation has been subject to erosion on the Mountain and over the eastern part of the island.

A cross-section along the length of Cayman Brac also shows that there is significant topography on the Brac Unconformity (Fig. 4.5). Between wells KEL#1 and CRQ#1, for example, there is a depth increase of 56 m in the position of the Brac Unconformity even though the two wells are only ~4.5 km apart (Fig. 4.5). In well BW#1, located on the east end of the island, the Cayman Unconformity is ~20 m lower than where it occurs in the exposed quarry wall beside well SQW#1, which is only 1.3 km to the ENE (Fig. 4.5). These examples clearly demonstrate the topography that exists on the upper boundary of the Cayman Formation on Cayman Brac. On the east end of Cayman Brac, the Cayman Formation is about 25 m thick with its upper surface being the modern weathered surface. Although the exact thickness of the Cayman Formation on Cayman Brac is not known with certainty, it has been estimated to be at least 100 m thick (Zhao and Jones 2012b). This means that at least 75 m of the formation has been lost to erosion on the east end of the island.

The thickest known section of ~65 m (CRQ#1), located in Jennifer Bay Quarry (Cayman Brac), was constructed from a well drilled from the floor of the quarry (close to sea level) and measurement of formation exposed in the quarry walls. Although the basal contact with the underlying Brac Formation is evident in the well, the upper contact with the overlying Pedro Castle Formation is not present in that area (Fig. 4.5). Well SQW#1, drilled in the quarry on the west end of the island, started 3 m below the Cayman Unconformity that was evident in the quarry wall beside the well. When drilling ceased at a depth of 61 m below sea level, the Brac Formation had not been reached. Based on the known positions of the formational boundaries, it is estimated that the Cayman Formation is at least 100 m thick on Cayman Brac. This estimate, however, may be low because the lower and upper boundaries are characterized by significant topographic relief and a complete sequence through the formation has not been found in any of the wells that have been drilled on Cayman Brac (Fig. 4.5).

4.6 Limestone—Dolostone Distribution

Although the Cayman Formation was originally considered to be formed entirely of dolostone (Jones et al. 1990, 1994b), information obtained from wells drilled on various parts of Grand Cayman has shown that limestone is part of the formation in some areas of the island. This is especially true in the central area of the western part of Grand Cayman, where significant thicknesses of limestone are found, as in wells NSC#1 (Fig. 4.10) and GFN#1 (Fig. 4.11). On the northeast corner of Grand Cayman, the Cayman Formation in well RWP#2 is formed entirely of dolostone to the base of the well (Willson 1998) at a depth of 97 m (Fig. 4.12). Although the Cayman Formation in well EER#1, located inland of the coast, is formed largely of dolostone, minor amounts of limestone are found in the basal 49 m and the upper 6 m of the succession (Fig. 4.13). In contrast, the Cayman Formation in wells RTR#1 (Fig. 4.14), LV#2 (Fig. 4.15), SHT#4 (Fig. 4.16), and GTH#1 (Fig. 4.17), which are located on the western part of the island (Fig. 4.4), is formed entirely of dolostones.

On Cayman Brac, the Cayman Formation is formed largely of dolostones (Figs. 4.18 and 4.19). Some of the minor amounts of calcite detected in some of the well cuttings from this formation may be late stage cements that have filled some of the pores in the dolostones.

4.7 Biota

The Cayman Formation on Grand Cayman is characterized by numerous fossils (Figs. 4.20 and 4.21) that are testimony to the diverse, abundant marine biota that flourished in the tropical seas that covered the islands during the Miocene. The biota is dominated by corals (Fig. 4.20a–d), bivalves (Fig. 4.20c), and gastropods (Figs. 4.20e and 4.21b) along with red algae, green algae (mostly *Halimeda*), benthic foraminifera, and scattered pelagic foraminifera.

Identification of the corals found in the Cayman Formation is an onerous task because dissolution of their aragonitic skeletons has destroyed many of the features needed for their precise identification. Nevertheless, Hunter (1994) identified 25 species (Fig. 4.22) that he organized into seven coral associations based on species diversity and recurring sets of species. These corals are similar to those found in Miocene strata throughout the Caribbean region (Hunter 1994). The Coral Associations found in the Cayman Formation, as designed by Hunter (1994) are as follows:

- ***Stylophora* Association**: With 11 species of corals, this association is dominated by *Stylophora* along with scattered *Porites baracoaensis* and other less common species. Found throughout the formation, *Stylophora* typically occurs in thickets and broken branches are common.
- ***Stylophora–Porites* Association**: This association, with only four species and dominated by its nominal species, has only been found in small patch reefs near Cottage Point on Grand Cayman.
- ***Porites baracoaensis* Association**: This association, with two species, is dominated by *P. baracoaensis*. It is found in small patch reefs.
- ***Montastrea limbata* Association**: This diverse association, with 12 species, comprises scattered large colonies of

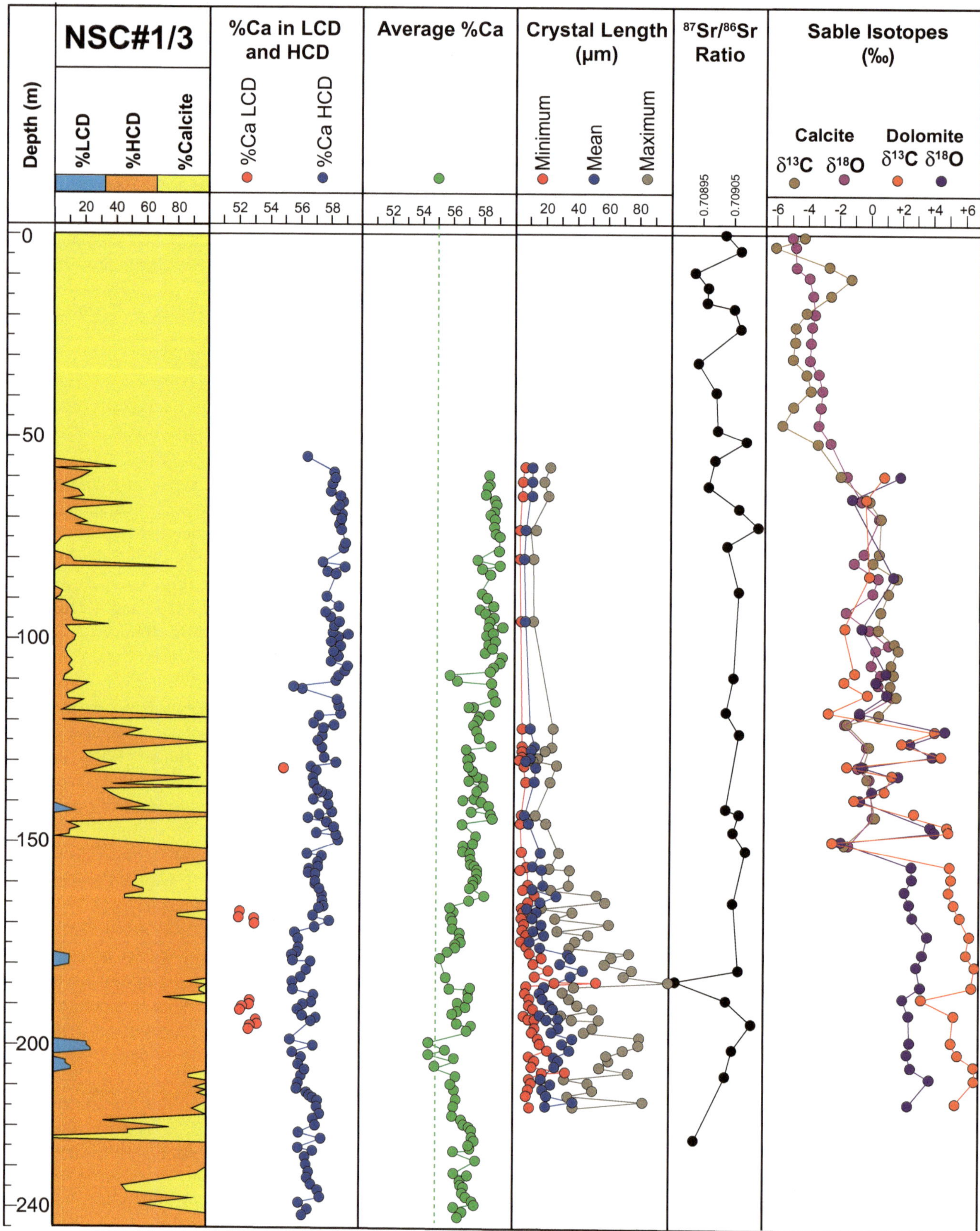

Fig. 4.10 Well NSC1/3—east central part of Grand Cayman. Cayman Formation. Lithology, %Ca in LCD and HCD, average %Ca in dolomite, minimum, mean, and maximum lengths of dolomite crystals (measured from SEM images of etched samples, each sample based on 50 measurements), $^{87}Sr/^{86}Sr$ ratios, and stable isotopes

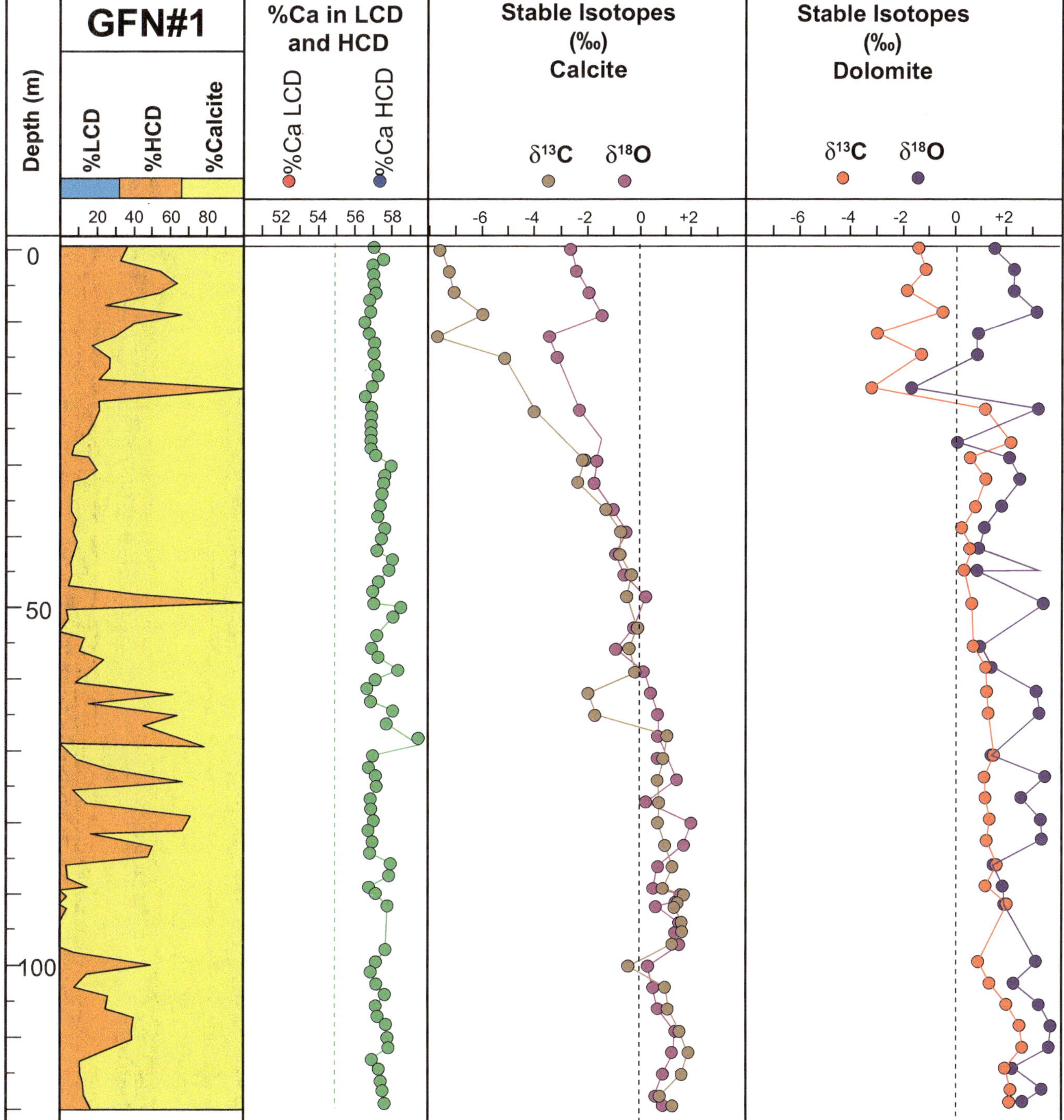

Fig. 4.11 Well GFN#1—east central part of Grand Cayman. Cayman Formation. Lithology, %Ca in dolostone, and stable isotopes

M. limbata, *M. endothecata*, *Diploria* spp., *Porites* sp. 1, *Favia* spp. and 7 other less common species. Locally, the free-living *Trachyphyllia* and *Teleiophyllia* occur with the large colonial corals. Common on Grand Cayman, these corals were extensively bored by sponges, worms, and molluscs.

- ***Goniopora hilli*** **Association**: With nine species, this association is dominated by numerous large colonies, up to 3 m diameter, of *G. hilli*, *M. limbata*, *Siderastrea* sp., *Porites* sp. 1 and other corals, but no *Stylophora*. Locally, this association forms patch reefs.

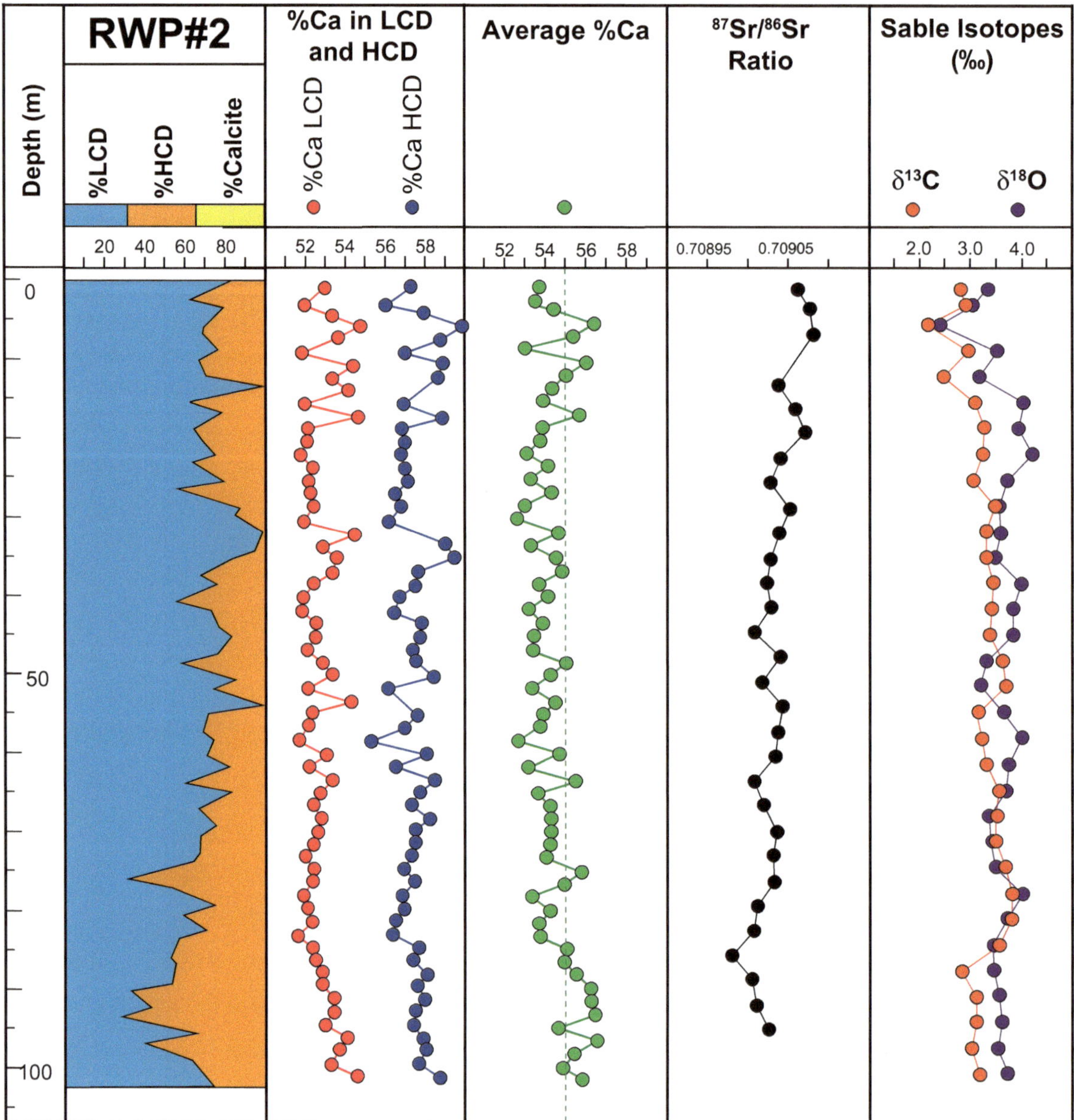

Fig. 4.12 Well RWP#2—northeast corner of Grand Cayman. Cayman Formation. Lithology, %Ca in LCD and HCD, average %Ca in dolostone, $^{87}Sr/^{86}Sr$ ratios, and stable isotopes

- ***Leptoseris* Association**: Comprising five species, this association is formed primarily of platy colonies of *Leptoseris*, *Montastrea limbata*, and *Porites* sp. 1. It has only been found in Pedro Castle Quarry on Grand Cayman.
- **Free-living coral Association**: The 15 species in this association are dominated by free-living corals that include *Trachyphyllia* spp., *Tysanus excentricus*, *Teleopphyllia grandis*, *Placocyanthus* spp., and *Antillocyathus* spp. This association has only been found at locality CIQ on Cayman Brac.

The fact that the corals in the Cayman Formation were capable of significant sediment rejection may indicate that they grew under conditions of rapid sedimentation (Hunter 1994). Given that the coral associations grade laterally into each other, Hunter (1994) suggested that substrate conditions may have been the primary control over the distribution of these coral associations along with water depth and wave energy.

Borings are common features of the corals from the Cayman Formation on Grand Cayman and Cayman Brac (Fig. 4.20d). The amount of boring varies from coral to coral

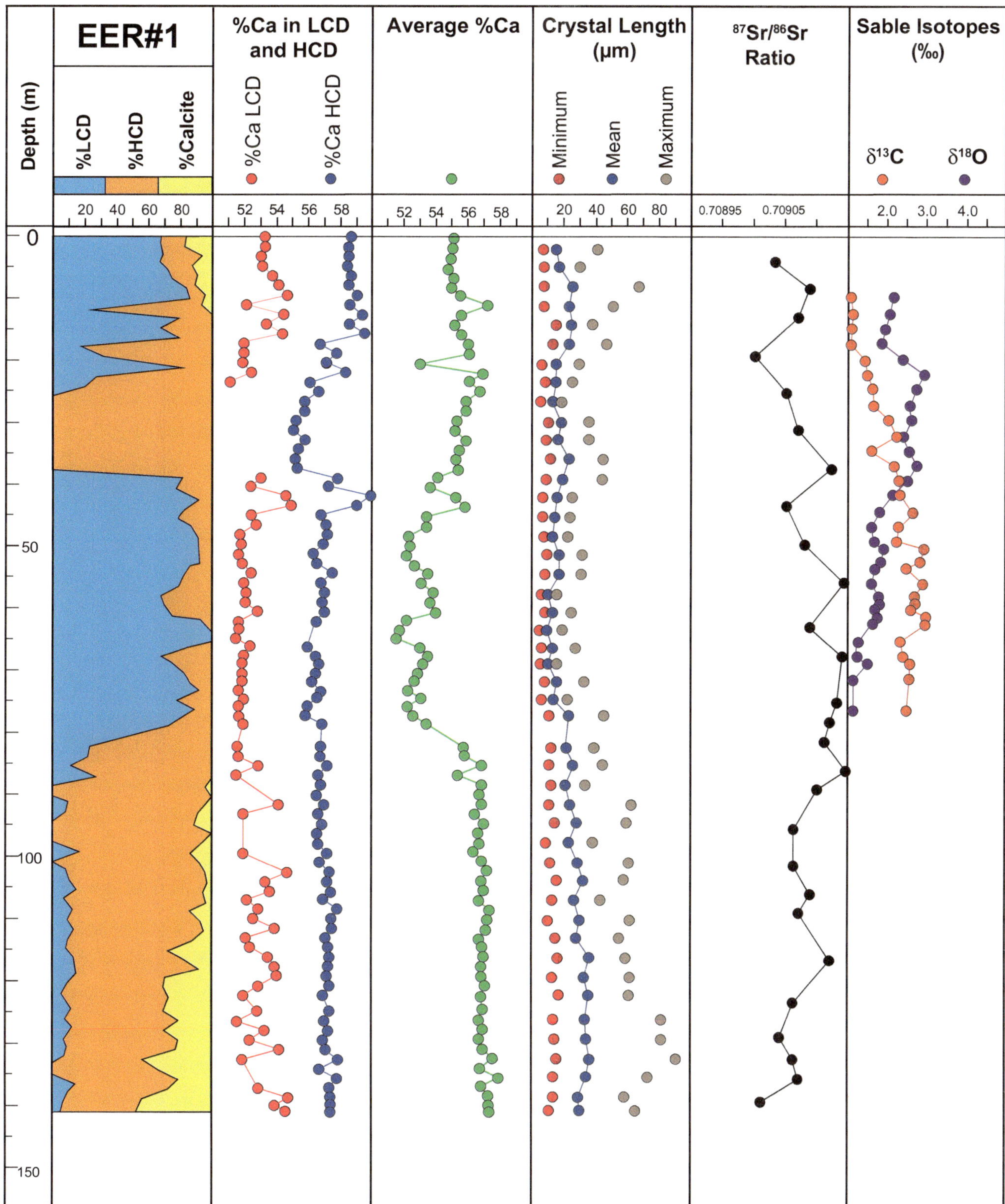

Fig. 4.13 Well EER#1—inland from coastline at East End, Grand Cayman. Cayman Formation. Lithology, %Ca in LCD and HCD, average %Ca in dolostone, minimum, mean, and maximum lengths of dolomite crystals (measured from SEM images of etched samples, each sample based on 50 measurements), $^{87}Sr/^{86}Sr$ ratios, and stable isotopes

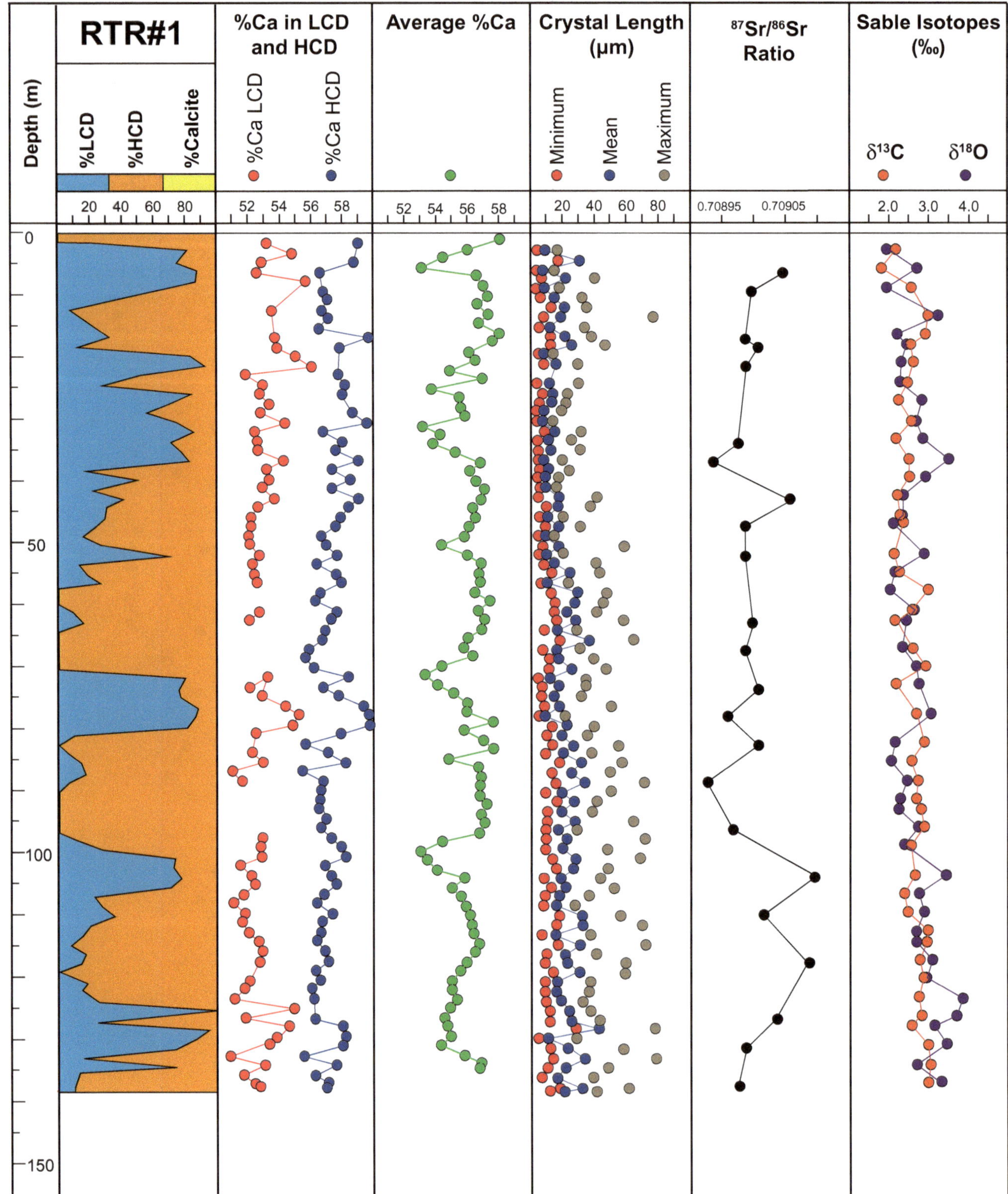

Fig. 4.14 Well RTR#1—central south coast of Grand Cayman. Cayman Formation. Lithology, %Ca in LCD and HCD, average %Ca in dolostone, minimum, mean, and maximum lengths of dolomite crystals (measured from SEM images of etched samples, each sample based on 50 measurements), $^{87}Sr/^{86}Sr$ ratios, and stable isotopes

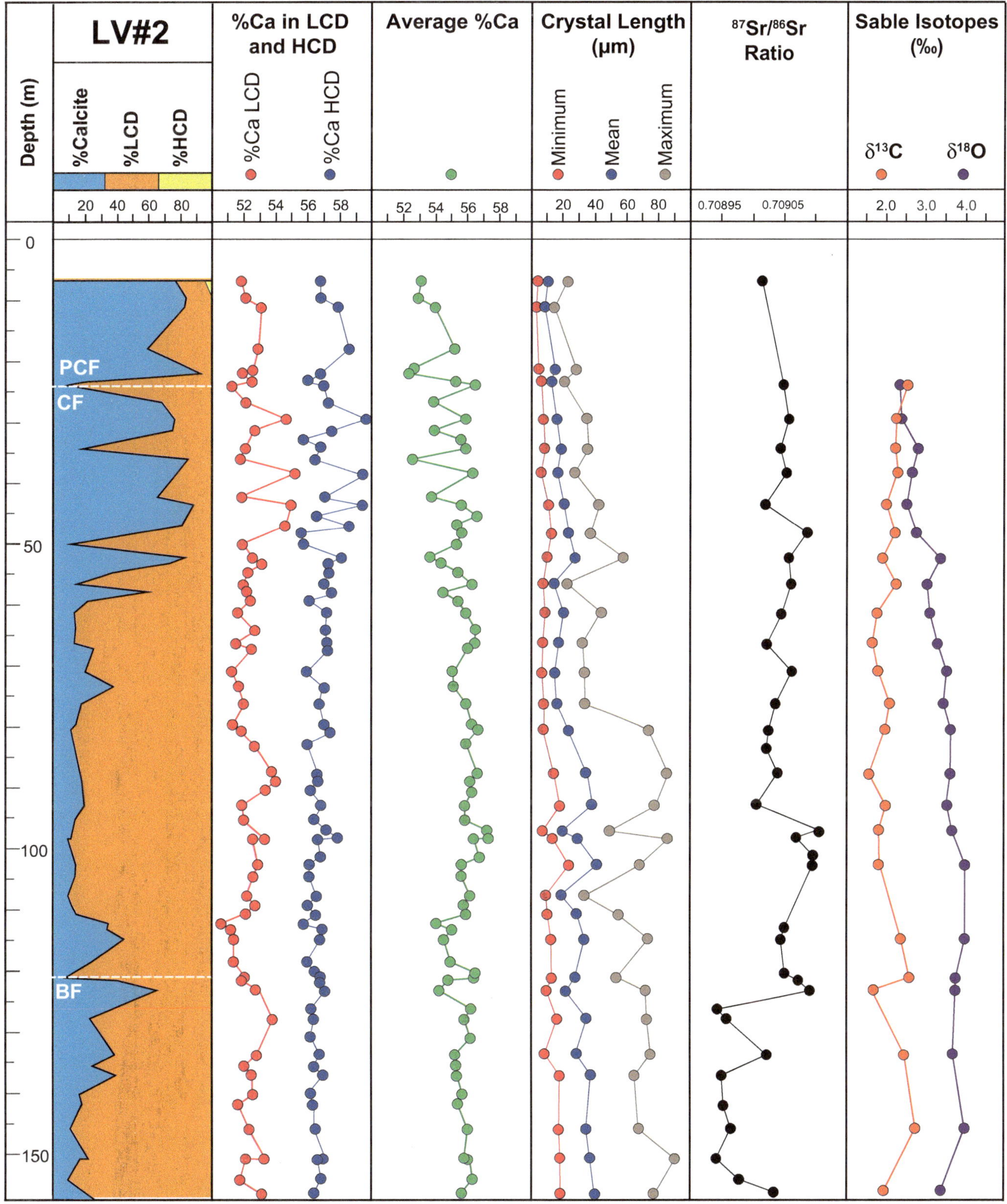

Fig. 4.15 Well LV#2—inland from Pedro Castle, west central part of Grand Cayman. Sequence through Ironshore Formation, Pedro Castle Formation, Cayman Formation, and uppermost part of the Brac Formation. Lithology, %Ca in LCD and HCD, average %Ca in dolostone, $^{87}Sr/^{86}Sr$ ratios, and stable isotopes

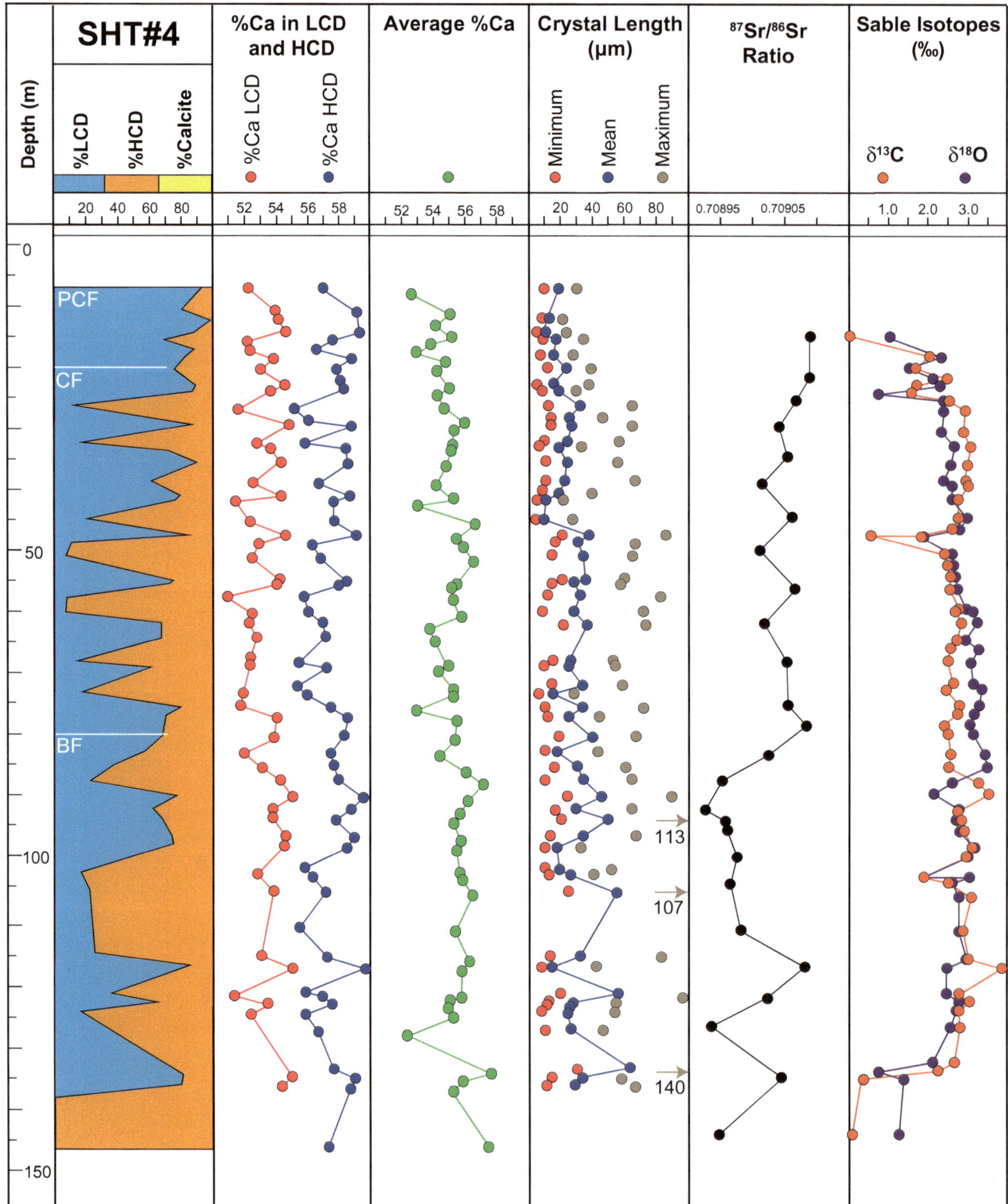

Fig. 4.16 Bluff Group in Well SHT#4—Near George Town, western part of Grand Cayman. Lithology, %Ca in LCD and HCD, average %Ca in dolostone, minimum, mean, and maximum lengths of dolomite crystals (measured from SEM images of etched samples, each sample based on 50 measurements), $^{87}Sr/^{86}Sr$ ratios, and stable isotopes

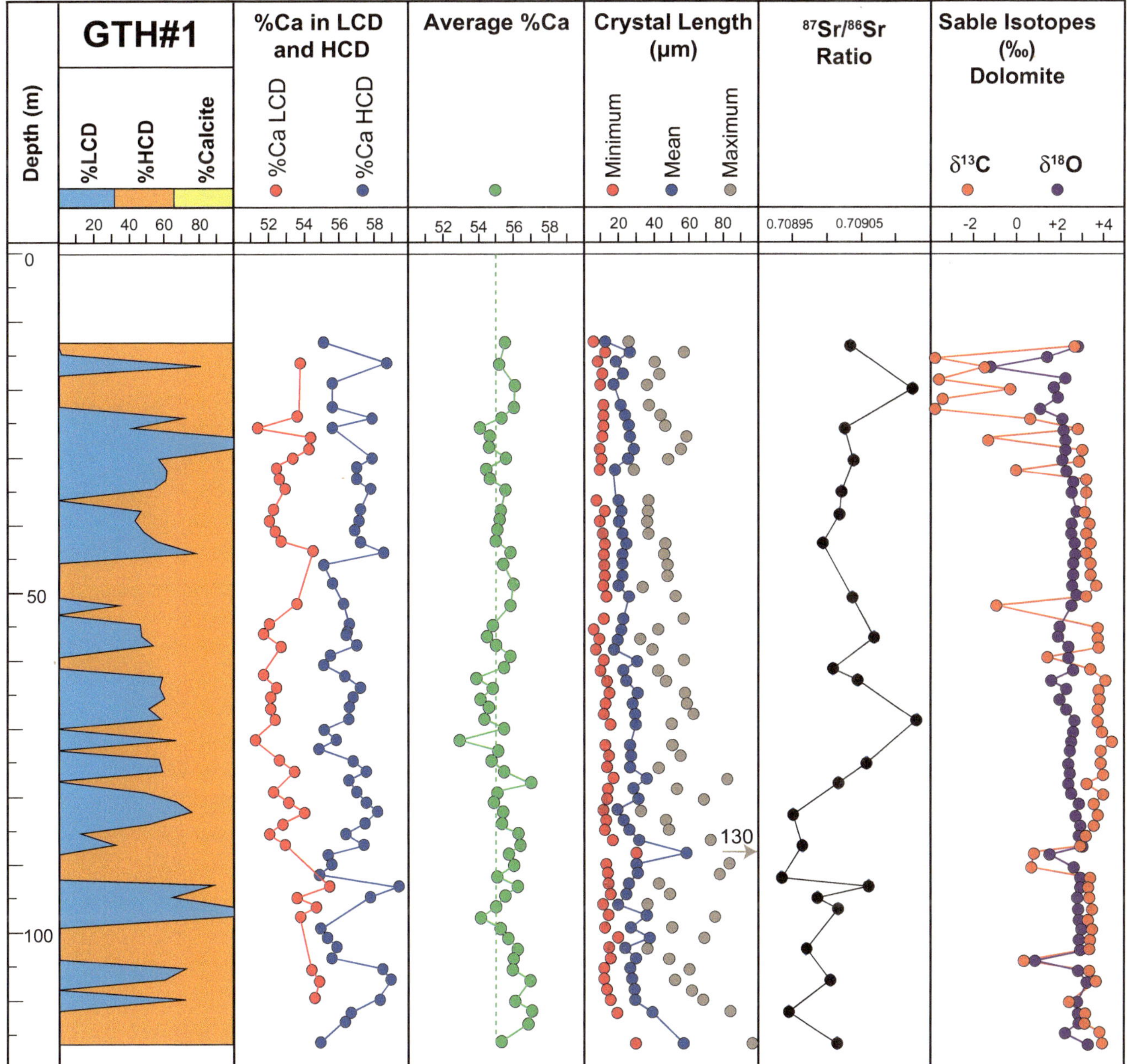

Fig. 4.17 Well GTH#1—North of George Town, western part of Grand Cayman. Cayman Formation. Lithology, %Ca in LCD and HCD, average %Ca in dolostone, minimum, mean, and maximum lengths of dolomite crystals (measured from SEM images of etched samples, each sample based on 50 measurements), $^{87}Sr/^{86}Sr$ ratios, and stable isotopes

and in some cases, from branch to branch of the same colony. Attributed to the activities of boring sponges (~75%), worms, and bivalves, these borings are similar to those found in the modern reefs around Grand Cayman. Pleydell (1987) and Pleydell and Jones (1988) documented the presence of the ichnotaxa *Entobia dendritica*, *E. cateniformis*, *E. geometrica*, *E. laquea*, *E. mammillata*, *E. megastroma*, *E. ovula*, *E. paradoxa*, *E. volzi*, *Uniglobites glomerata*, *Gastrochaenolites torpedo*, *G. tubinatus*, *Trypanites solitarius*, *T. weisei*, *Maeandropolydora* cf. *M. sulcans*, *Talpina* sp., and *Caulostrepsis* cf. *C. cretacea* in the corals.

Echinoids are rare, with specimens having only being found in Spotts Bay Quarry, Pedro Castle Quarry (one specimen from each quarry), and the eastern part of High Rock Quarry (30 specimens). Based on these specimens, Donovan et al. (2016) recorded the presence of *Arbacia*? sp., *Schizaster* sp. cf. *S. americanus*, and *Brissus* sp. cf.

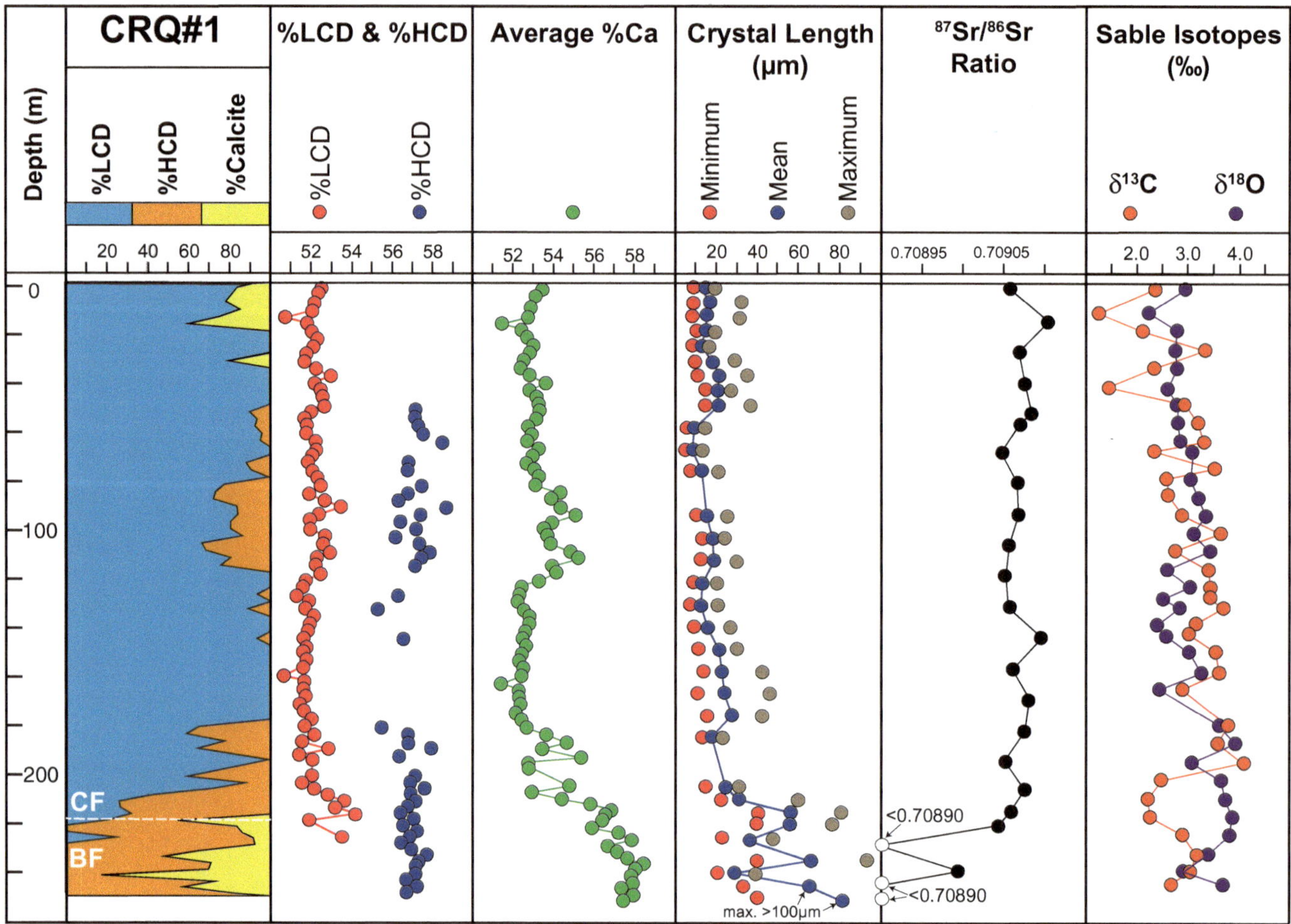

Fig. 4.18 Well CRQ#1—Jennifer Bay quarry, central part of Cayman Brac. Cayman Formation (CF) and uppermost part of Brac Formation (BF). Lithology, %Ca in LCD and HCD, average %Ca in dolostone, minimum, mean, and maximum lengths of dolomite crystals (measured from SEM images of etched samples, each sample based on 50 measurements), $^{87}Sr/^{86}Sr$ ratios, and stable isotopes

B. oblongus from East End quarry and *Clypeaster* sp. from Pedro Castle quarry (Fig. 4.20f) and Spotts Bay quarry. They suggested that these echinoids were of Miocene age.

4.8 Facies

For the Cayman Formation on Grand Cayman and Cayman Brac, facies are generally defined according to the type of samples that are available. For outcrops and core, facies have been defined based on their lithology and biota. This approach, however, is difficult to apply to samples formed entirely of well cuttings because the skeletons of the organisms that were originally formed of aragonite (e.g., corals, bivalves, gastropods) are generally impossible to detect given that their skeletons were preferentially lost during diagenesis (Figs. 4.20 and 4.21). Although facies definitions for these samples must rely largely on a lithological/mineralogical approach, some the smaller fossils (e.g., foraminifera) that were not lost to diagenesis can sometimes be identified in thin sections made from the well cuttings. Definition of the constituent facies is further complicated by the fact that on Grand Cayman, the Cayman Formation on the western part and coastal areas on the eastern part of the island is formed almost entirely of dolostones whereas the successions in the interior of the eastern part of Grand Cayman commonly includes substantial thicknesses of limestone (Der 2012; Ren and Jones 2016, 2018; Ren 2017). In contrast, no limestone has been found in the Cayman Formation on Cayman Brac. The variable facies schemes that have been developed for the Cayman Formation reflect, to a large extent, the distribution of the limestones and dolostones as well as the main focus of each study.

Jones (1994) and Jones et al. (1994a) provided overviews of the main lithologies and facies that are found in the Cayman Formation on Grand Cayman and Cayman Brac. Subsequent studies of the Cayman Formation on Grand

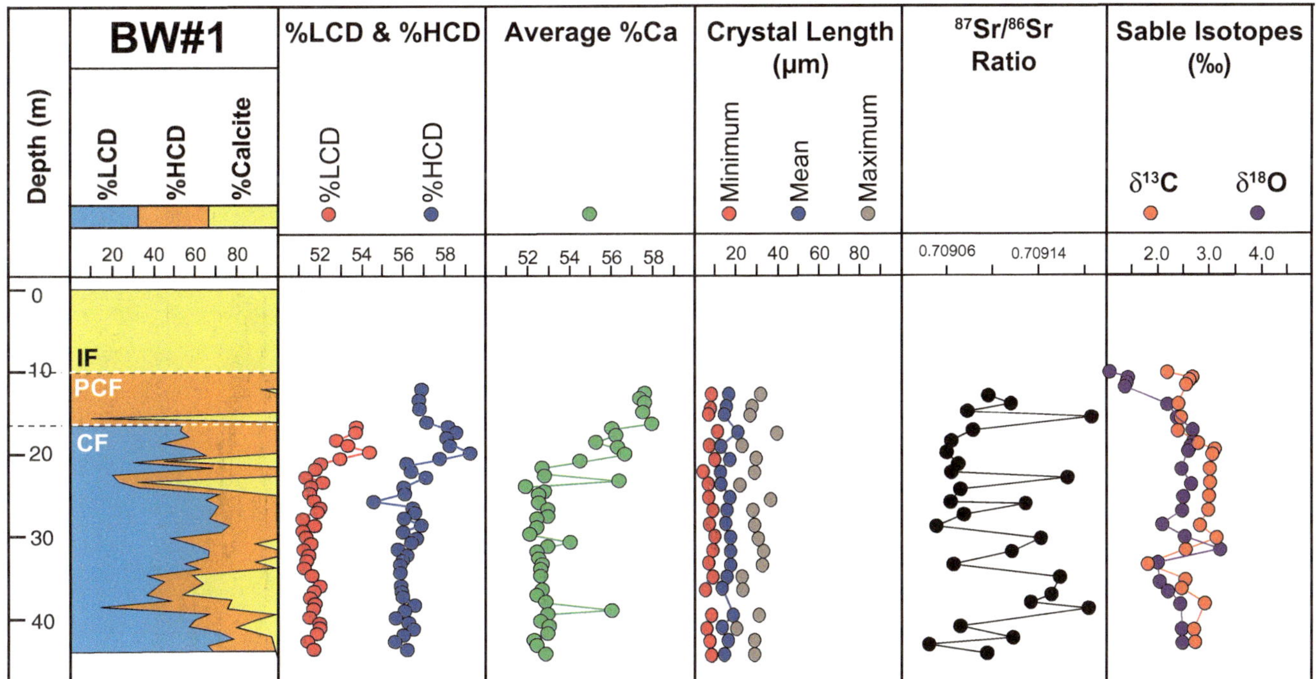

Fig. 4.19 Well BW#1 west end of Cayman Brac. Ironshore Formation (IF), Pedro Castle Formation (PCF), and Cayman Formation (CF). Lithology, %Ca in LCD and HCD, average %Ca in dolostone, minimum, mean, and maximum lengths of dolomite crystals (measured from SEM images of etched samples, each sample based on 50 measurements), ^{87}Sr/^{86}Sr ratios, and stable isotopes

Cayman and Cayman Brac allowed further refinement of the facies architecture and the depositional environments in which the facies developed.

4.8.1 Grand Cayman

Willson (1998) delineated five facies in the cores from wells RWP#2 and QHW#1 (northeast corner of Grand Cayman) that she designated as the (1) *Stylophora* floatstone, (2) rhodolite-finger coral floatstone, (3) rhodolite coral fragment rudstone-grainstone, (4) the *Porites-Stylophora* facies, and (5) the *Leptoseris-Montastrea* floatstone facies. She suggested that these facies represented deposition in a shallow water environment that experienced variable wave energy. Later, Etherington (2004), based on cores from across Grand Cayman that included those from Rogers Wreck Point, Tarpon Springs Estate, Pedro Castle Quarry, and wells ORA#1, SHT#1, SHT#2, SH#3, LSC#1, and LSC#2 on the western peninsula of Grand Cayman, identified nine facies in the Cayman Formation as the (1) *Stylophora* facies, (2) *Stylophora–Halimeda* facies, (3) *Styophoria*–bivalve facies, (4) *Stylophora–Porites* faceis, (5) Laminar coral Facies, (6) Coralline algae–*Amphistegina* facies, (7) Coralline algae–*Halimeda–Amphistegina*–bivalves facies, (8) *Porites*–intraclast–rhodolith facies, and (9) Intraclast facies.

Wells GFN#2 and RWP#2, each fully cored to depths of ~92 m, highlight the contrast in facies between the successions found in the east-central and northeast areas of the island (Fig. 4.23). The succession in GFN#2 is formed largely of limestones that include mudstones and wackestones along with some grainstones and rudstones (Fig. 4.23). Ren and Jones (2016) divided the biota in those limestones into faunal associations FA1, FA2, and FA3.

- **FA1.** Found in the lower part of the succession (53.0–92.2 m), FA1 is formed largely of skeletal floatstones and rudstones, with a biota of branching and domal corals, green algae, red algae, bivalves, gastropods, and benthic foraminifera. Planktonic foraminifera are found in the mudstones between 63.0 to 68.7 m and 80.0 to 88.0 m (Fig. 4.23).
- **FA2.** Located in the middle part of the succession (29.0–53.0 m), FA2 is formed largely of mudstones that contain planktonic foraminifera (mostly *Globigerinoides*? and *Globorotalia*?) and peloids (Fig. 4.23).
- **FA3.** Located in the upper part of the succession (6.0–29.0 m), FA3 is formed largely of grainstones that contains benthic foraminifera (mainly *Amphistegina*), and scattered bivalve and coral fragments (Fig. 4.23). Micritized grains are common.

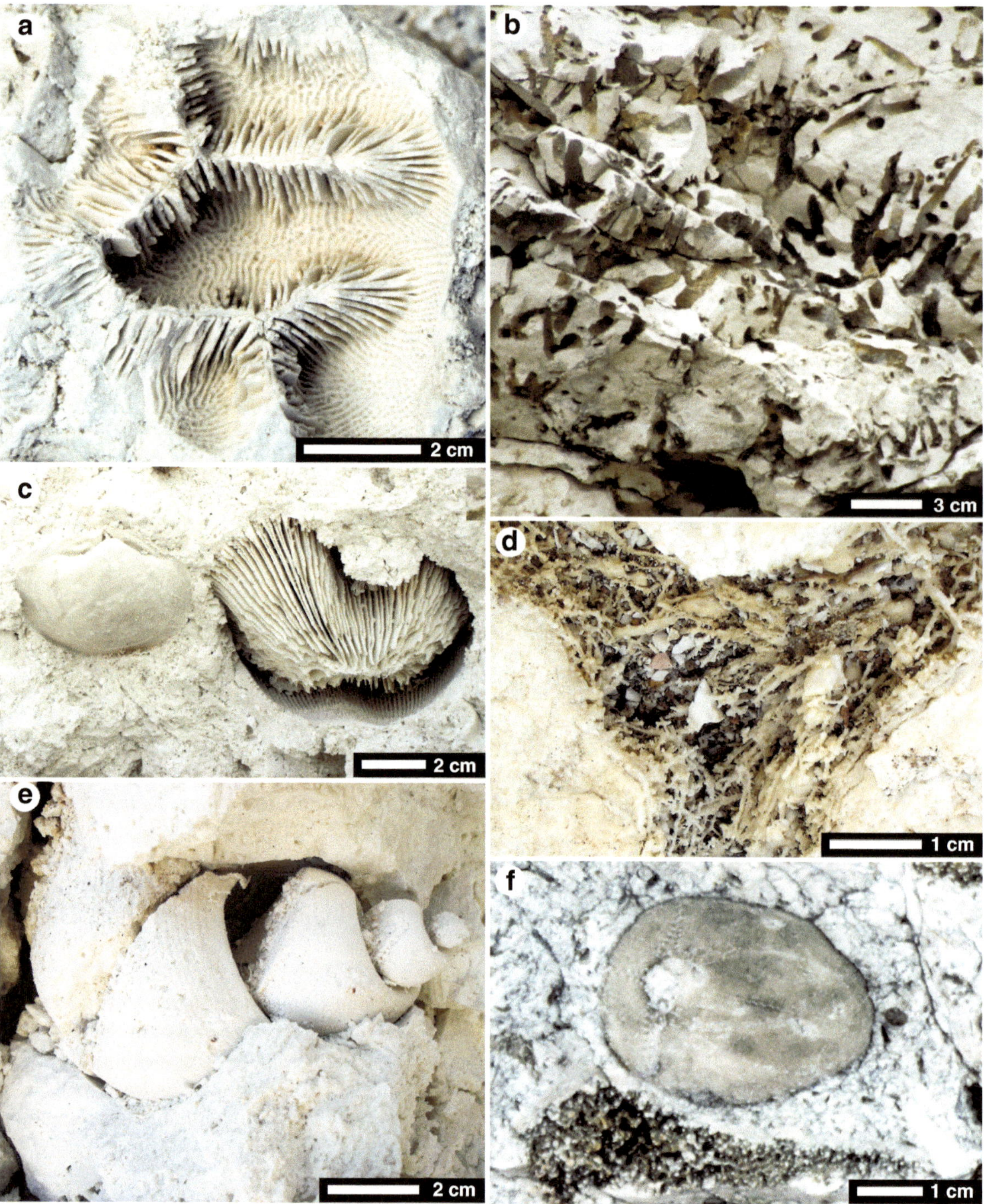

Fig. 4.20 Fossils in the Cayman Formation. **a** Leached colonial coral. Jennifer Bay Quarry, Cayman Brac. **b** Leached *Stylophora* with cavities partly filled with terra rossa. South end of Bluff Road, Cayman Brac. **c** Leached free-living coral (right) and interior mould of bivalve. Jennifer Bay Quarry, Cayman Brac. **d** Leached branching coral revealing filled sponge (?) borings. Anne Tatum Drive, Cayman Brac. **e** Interior mould of large gastropod shell. Jennifer Bay Quarry, Cayman Brac. **f** Irregular echinoid (*Brissus?* sp.). Pedro Castle Quarry, Grand Cayman

Core from well RWP#2, located ~3.5 km ENE of GFN#2, is formed entirely of dolostones that comprise alternating units of floatstone and rudstone that belong to the coral-rhodolith floatstone–rudstone facies association (FA4), as defined by Ren and Jones (2016). Corals, including *Stylophora*, *Porites*, *Montastrea*, and *Leptoseris*, rhodoliths, and broken coral fragments dominate the succession (Fig. 4.23). Willson (1998) defined five facies within this

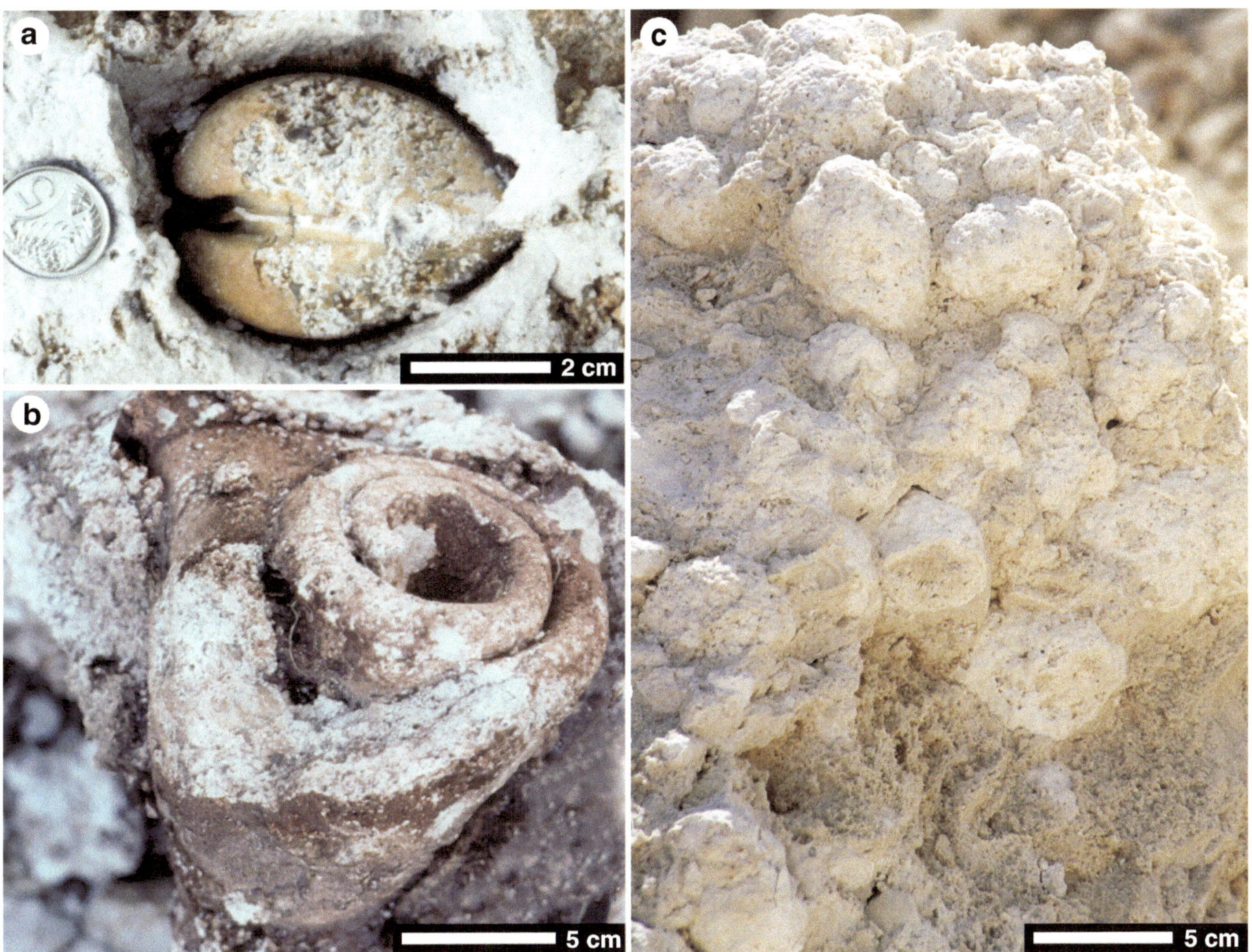

Fig. 4.21 Fossils in the Cayman Formation. **a** Interior mould of leached articulated bivalve shell. Jennifer Bay Quarry, Cayman Brac. **b** Interior mould of leached *Strombus* (conch) shell. Pedro Castle Quarry, Grand Cayman. **c** Large rhodoliths in grainstone matrix. Jennifer Bay Quarry, Cayman Brac

succession whereas Ren and Jones (2016) noted that there was no systematic pattern to the vertical distribution of the facies. The succession in RWP#2 is significantly different from that in GFN#2, both in terms of its mineralogy and facies (Fig. 4.23).

Well NSC#1/3, located ~4.5 km WSW of GFN#2, is characterized by a succession that is dominated by dolostones in the lower part and limestones in the uppermost 122 m (Fig. 4.10). Although the facies cannot be resolved as for GFN#2 because only well cuttings are available, it is evident that the succession is dominated by mudstones and wackestones that contain benthic foraminifera, red algae, bivalve, and rare coral fragments (Fig. 4.24). Pelagic foraminifera, like those in GFN#2, are also present (Fig. 4.24j). Porosity is variable with local leaching of some of the aragonite allochems. In the upper 122 m of the succession, dolomitization is patchy (Fig. 4.24e, f) and commonly replaces the matrix rather than the skeletal allochems (Fig. 4.24f). Calcite cement, common in many parts of the succession, commonly fills the leached allochems. Dolomitization is more extensive in the lower part of the succession. Der (2012), who examined the Cayman Formation in NSC#1/3 and other wells on the eastern part of Grand Cayman, delineated eight facies in the Cayman Formation: (1) *Amphistegina*–Bivalve Facies, (2) rhodolite–coral-*Amphistegina* Facies, (3) Rhodolite–branching coral—*Amphistegina* Facies, (4) Porites—*Amphistegina* Facies, (5) Branching coral–benthic foraminifera Facies, (6) Branching and domal coral—*Amphistegina* Facies, (7) Branching and platy coral-*Amphistegina*–red algae facies, and (8) *Leptoseris*– *Amphistegina* Facies.

Limestones are, based on currently available information, restricted to the interior areas of the eastern part of the island. In addition to the limestones found in wells GN#2 and NSC#1, similar limestone dominated successions are present in wells BOG#1, DTE#1 and the lower part of FFM#11.

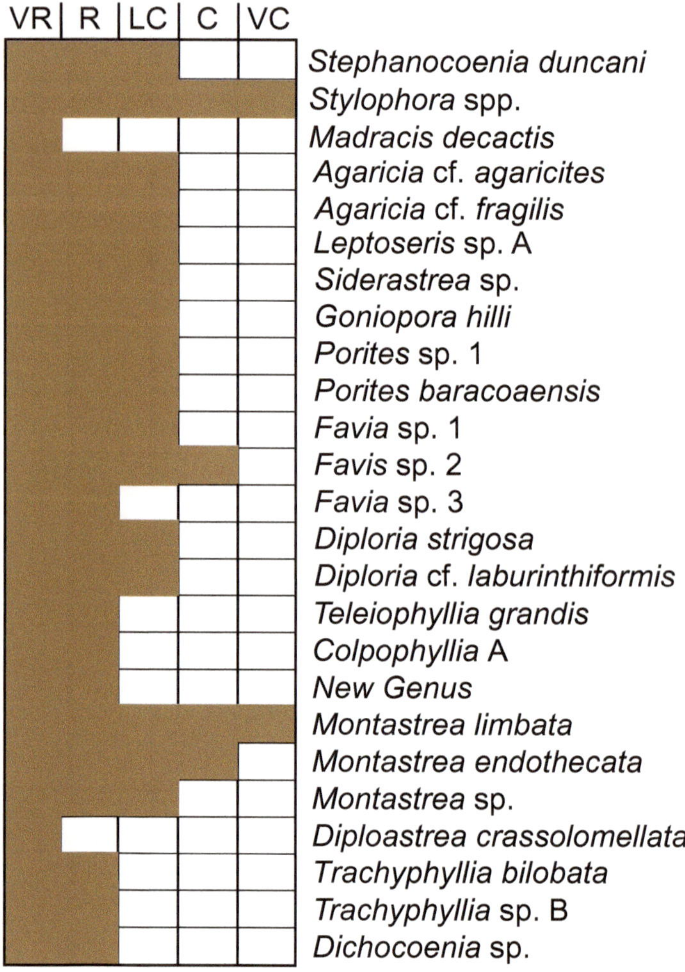

Fig. 4.22 List of corals from the Cayman Formation with estimates of overall abundance of each species. Modified from Hunter (1994, his Fig. 10.4)

Limestones are also present in the lower parts of well EER#1 (Fig. 4.13). Limestone successions like these have not been found in any of the wells drilled on the western part of the island.

The Cayman Formation in wells RTR#1 (Fig. 4.14), LV#2 (Fig. 4.15), SHT#4 (Fig. 4.16), and GTH#1 (Fig. 4.17) drilled between the south-central part of the island (RTR#1) and the west part of the island (SHT#4 and GTH#1) is formed entirely of dolomite. Although these wells are cited as examples, it is important to note that the Cayman Formation found in wells throughout the western part of Grand Cayman is always characterized by dolostone with no calcite apart from scattered calcitic cements that line or fill some of the cavities in the dolostones.

Based on the presence and relative percentages of calcite, LCD, and HCD, it is apparent that there are some important geographical variations in the Cayman Formation of Grand Cayman. The most critical aspects are (1) limestones are restricted to the interior of the eastern part of the island, and (2) the successions in the coastal areas on the north and northeastern part of the island being formed largely of LCD, whereas those on the southern and western parts of the island are formed of sub-equal amounts of LCD and HCD.

4.8.2 Cayman Brac

The Cayman Formation on Cayman Brac is formed largely of finely crystalline fossiliferous dolostones. Although bedding planes are difficult to identify in the weathered cliff faces, most of the formation appears to be formed of thickly bedded strata (Figs. 4.2 and 4.25). On freshly cut faces in Jennifer Bay Quarry, bedding planes are highlighted because of dissolution that was focused along them (Fig. 4.25). The fact that crystalline calcite cements (flowstone) are found on some of those bedding planes also offers testimony of groundwater movement along them. There, the beds are typically ~2 m thick. Although freshly exposed surfaces of the Cayman Formation are typically white to off-white, local red staining by iron washed down from the overlying terra rossa is common (Fig. 4.25). Prolonged weathering of these dolostones rapidly transforms the exposed faces to light grey

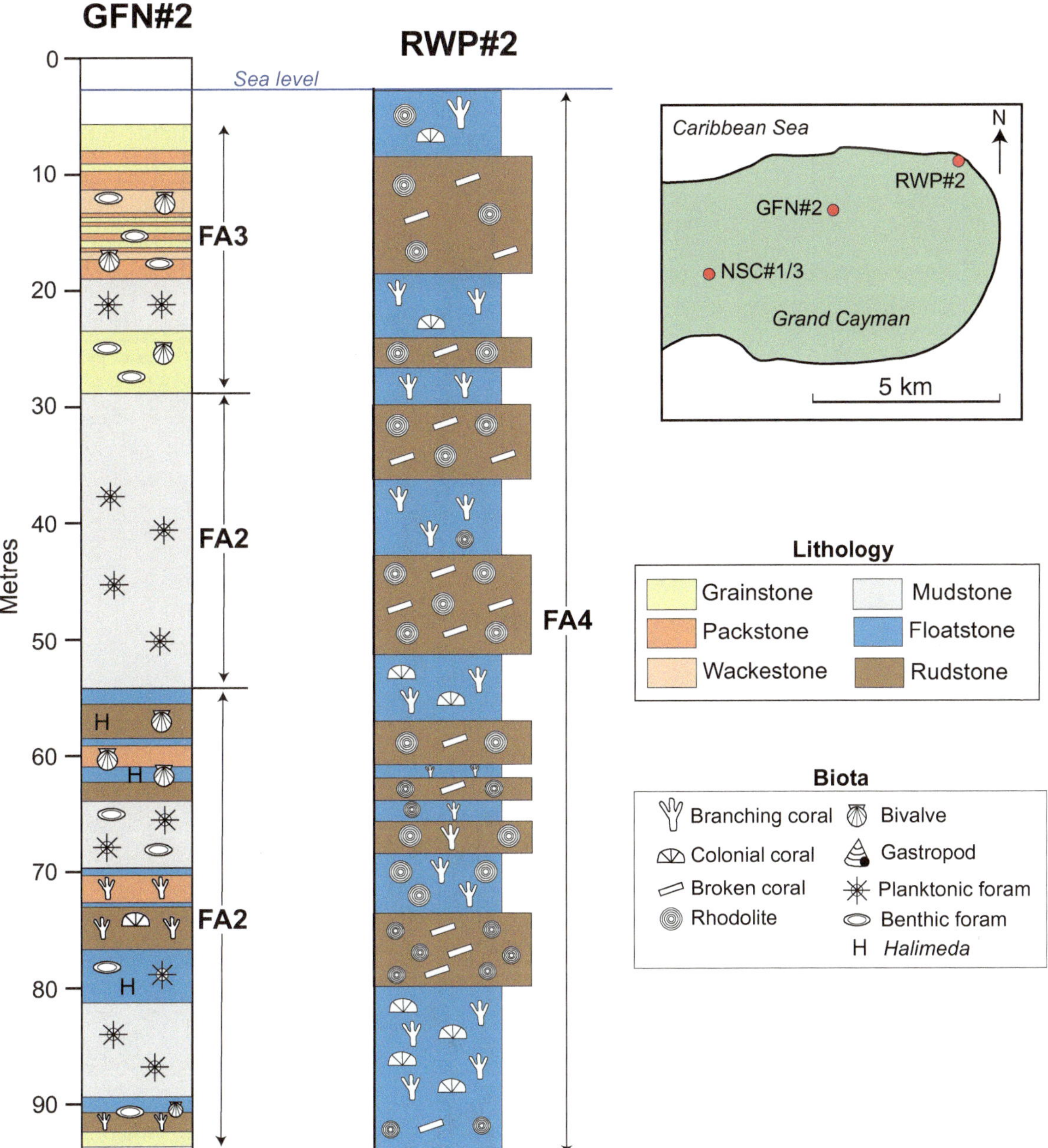

Fig. 4.23 Comparison of Cayman Formation in wells GFN#2 and RWP#2 showing variance in facies and faunal assemblages (FA1–FA4). Modified from Ren and Jones (2016, their Figs. 4 and 9)

to black (Fig. 4.2). The discolouration, due in part to biological weathering mediated by various microbes, effectively masks the rocks and detection of their components becomes difficult.

The Cayman Formation on Cayman Brac is formed of finely crystalline dolostones. As yet, no limestones have been found in this formation on Cayman Brac. For well CRQ#1, the dolomite crystal lengths for 25 samples (based on measurement of 100 crystals on SEM images of each sample) range from 4.8 to 29.4 μm with an average of 15.9 μm (Fig. 4.18) and for well BW#1, located on the west end of the island (Fig. 4.5), based on 18 samples, the crystal length ranged from 12.2 to 21.3 μm with an average of 15.5 μm (Fig. 4.19). The consistency in the size of the crystals is readily apparent when the crystal sizes are plotted against the depth of the samples in each well (Figs. 4.18 and 4.19).

Facies in the dolostones of the Cayman Formation on Cayman Brac are formed of mudstones, wackestones, packstones, and grainstones that contain variable populations of colonial corals, free-living corals, gastropods, bivalves, red algae, *Halimeda*, and foraminifera (Figs. 4.20a–e, 4.21a and 4.26a, b). Rhodolites, up to 13 cm

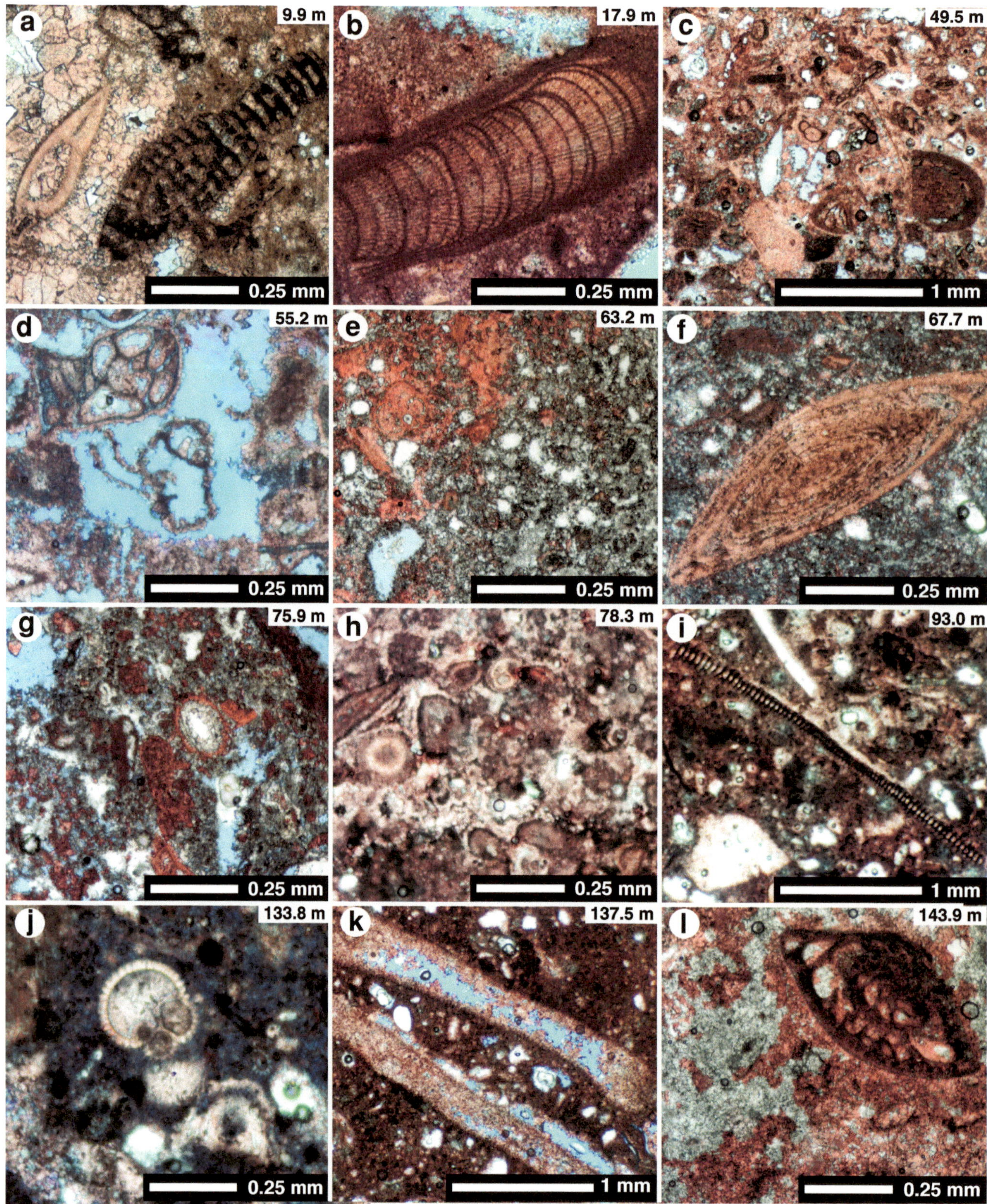

Fig. 4.24 Thin section photomicrographs of samples from well NSC1/3 to a depth of 143.9 m, impregnated with blue epoxy. Thin section stained with Alizarin Red S solution—red indicates calcite, no stain indicates dolomite. Blue indicates porosity. Note abundance of foraminifera, dolomite replacing matrix in some samples, leached fossil fragments, and pelagic foraminifera. Specified depths are from land surface

Fig. 4.25 View of dolostones of Cayman Formation exposed in south wall (estimated to be ~15 m high) of Jennifer Bay Quarry, Cayman Brac. Black arrows on right side of image indicate possible bedding planes that have been locally widened by dissolution. Note terra rossa filled crevices, fractures, and shallow hollows near the land surface. Red staining of dolostones in upper part of quarry wall originated from the terra rossa

long, are locally concentrated in distinct beds (Fig. 4.21c). Intergranular cavities in many of the dolostones are lined with limpid dolomite cement (Fig. 4.26c, d, f) and locally filled with calcite cement (Fig. 4.26d). Some of the larger cavities are partly filled with internal sediments that are overlain by calcite cement (Fig. 4.26e). The karst surface at the top of the formation is commonly characterized by various borings that are filled with dolomitic muds of the overlying Pedro Castle Formation (Fig. 4.26g). Preferential dissolution of the aragonitic coral skeletons in the dolostones of the Cayman Formation means that they are now represented by moulds (Fig. 4.29). Various borings are commonly evident in these moulds (Fig. 4.29).

As evident from sections HC and SWQ#1 (Fig. 4.27) and various outcrops along Bluff Road (Fig. 4.28), there is no consistent pattern to the lateral or vertical distribution of the facies. In many cases it is even difficult to trace facies between sections that are located close to each other—as is the case for sections CQA, CQB, CQC, and CQD in Jennifer Bay Quarry (Fig. 4.27). The facies evident in all of the sections are indicative of deposition that took place in shallow, relatively quiet waters. Periods with higher energy conditions probably led to the development of the rhodolites and deposition of some of the grainstones.

Only the *Stylophora*, *Montastrea limbata*, *Leptoseris*, and Free-living Coral associations have been recognized in the Cayman Formation on Cayman Brac Hunter (1994). Of these, the *Stylophora* and *M. limbata* associations are the most common (Hunter 1994). The Free-living Coral Association, found in the Jennifer Bay Quarry and the nearby road cut at the south end of Bluff Road, are the only localities in the Cayman Formation on any of the islands that are characterized by this association.

4.9 Depositional Regimes

The diverse array of corals found in the Cayman Formation suggests that deposition of the sediments that now form the Cayman Formation developed in normal, open water marine environments (Jones and Hunter 1994a) that were probably similar to those found around the islands today. Despite the abundance of corals, no reefs are evident apart from scattered patch reefs and local *Stylophora* thickets. The coral assemblage is similar to the *Diploria-Montastrea-Porites* assemblages that characterize modern banks in the Caribbean Sea with water depths <45 m.

Based largely on the sequences on the western peninsula of Grand Cayman, Etherington (2004) suggested that the Cayman Formation included sediments that formed in two distinct depositional regimes, which she labelled as Depositional Unit I (lower unit) and Depositional Unit II (upper unit). She argued that water depth was <30 m and commonly <20 m, and that the sediments in Depositional Unit I developed under moderate to high energy conditions, whereas Depositional Unit II formed under low energy conditions with high sedimentation rates. These two units are, however, difficult to trace across Grand Cayman and are not evident in the Cayman Formation on Cayman Brac. As noted by Etherington (2004), the low coral diversity in the

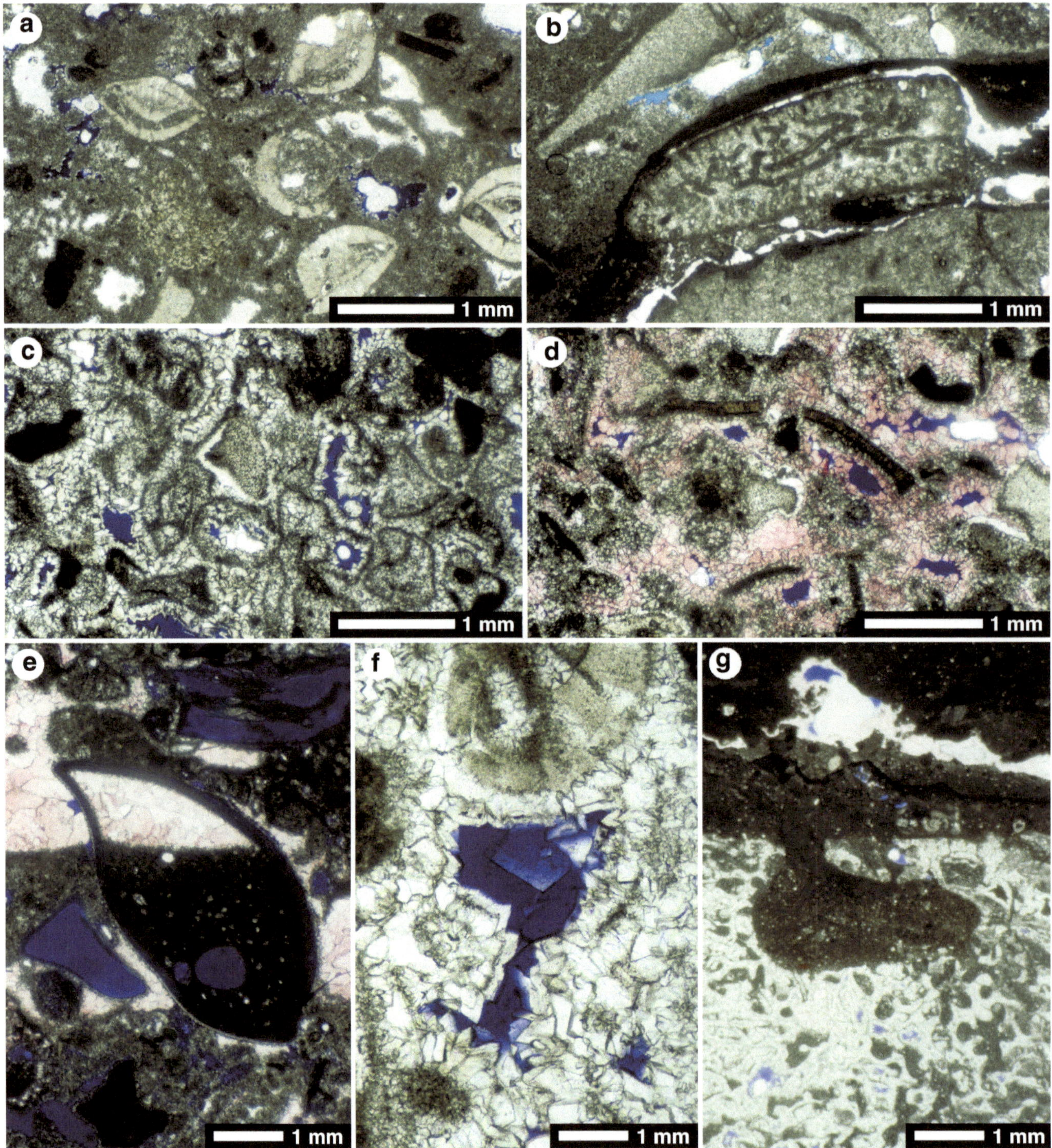

Fig. 4.26 Thin section photomicrographs of dolostones from the Cayman Formation on Cayman Brac. All thin sections stained with Alizarin Red S solution (red = calcite, no stain = dolomite). Blue = porosity. **a** Foraminifera in mudstone matrix. Jennifer Bay Quarry. **b** Dolomitized *Halimeda* plate in wackestone matrix. Jennifer Bay Quarry. **c** Grainstone with dolomitized grains with isopachous dolomite cement. Locality GMD-2. **d** Dolomitized allochems with calcite cement (red) filling most of the intergranular pores. **e** Cross-section through dolomitized bivalve shell with geopetal fill (dolomitized sediment at base, calcite cement (red) at top. Note calcite cement in cavities. Jennifer Bay Quarry. **f** Intergranular pores lined with zoned limpid dolomite cement. Locality GMD-2. **g** Contact between the Cayman Formation and Pedro Castle Formation with sponge (?) boring at unconformity. Locality ASQ-1

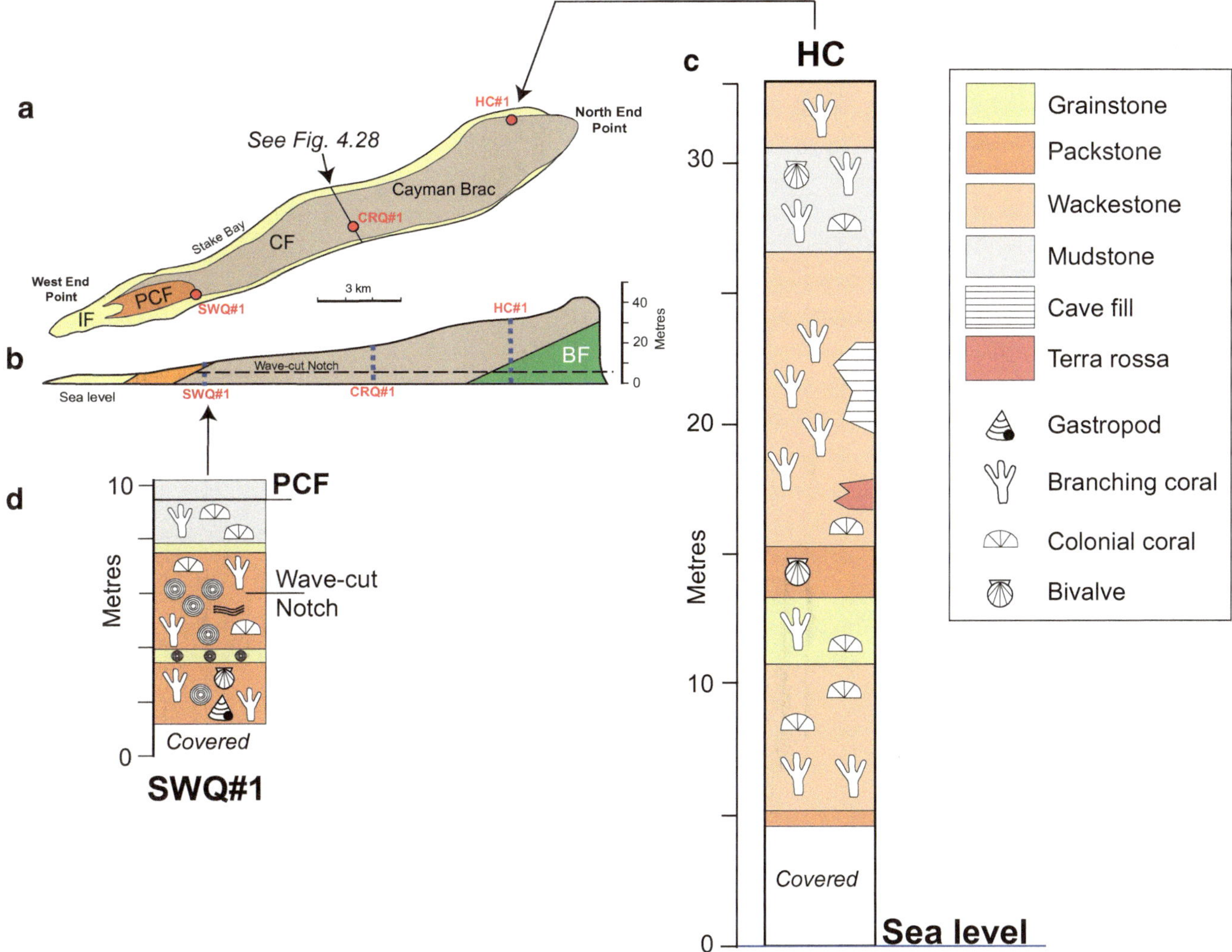

Fig. 4.27 Facies succession in Cayman Formation at localities HC and SWQ#1, Cayman Brac

Cayman Formation on the western part of Grand Cayman relative to elsewhere on Grand Cayman and Cayman Brac is probably a reflection of the muddier conditions that existed in that part of the island.

In the Cayman Formation on Grand Cayman and Cayman Brac there are no obvious patterns to the lateral and vertical facies successions. Jones and Hunter (1994a) suggested that substrate type exerted the main control over the dispersal of the corals. They noted, for example, that the massive head corals are most common in the packstone to grainstone facies, rare in the wackestones facies, and absent from the mudstone facies. Conversely, the branching corals are most common in the mudstone facies. The Free-living Coral Association, found only on Cayman Brac, is associated with wackestone facies where other corals are absent. These relationships indicate that the amount of mud in the water column may have been the primary control over the different corals. For example, the branching *Stylpohora* and *Porites* survived in muddy areas because their polyps were above the substrate and any mud that settled on their branches would easily have been shed by tentacular movement or stomadaeal uptake. In contrast, the massive head corals could only survive in areas where local currents washed away any mud that settled on them. The general paucity of fragmented fossils in the Cayman Formation suggests that energy levels were generally low. Exceptions include local occurrences of clean skeletal sands, local accumulations of rhodolites, bivalves and broken coral branches, bioclasts, and rare overturned corals (Jones and Hunter 1994a). These types of deposits may reflect local variations in energy levels or may have been generated as storms swept across the area.

For the Cayman Formation, the initial transgression onto the karstic surface at the top of the Brac Formation established high-energy conditions that led to the production of foraminiferal-red algae sands. As the water became deeper, conditions stabilized and the resident coral-dominated biota

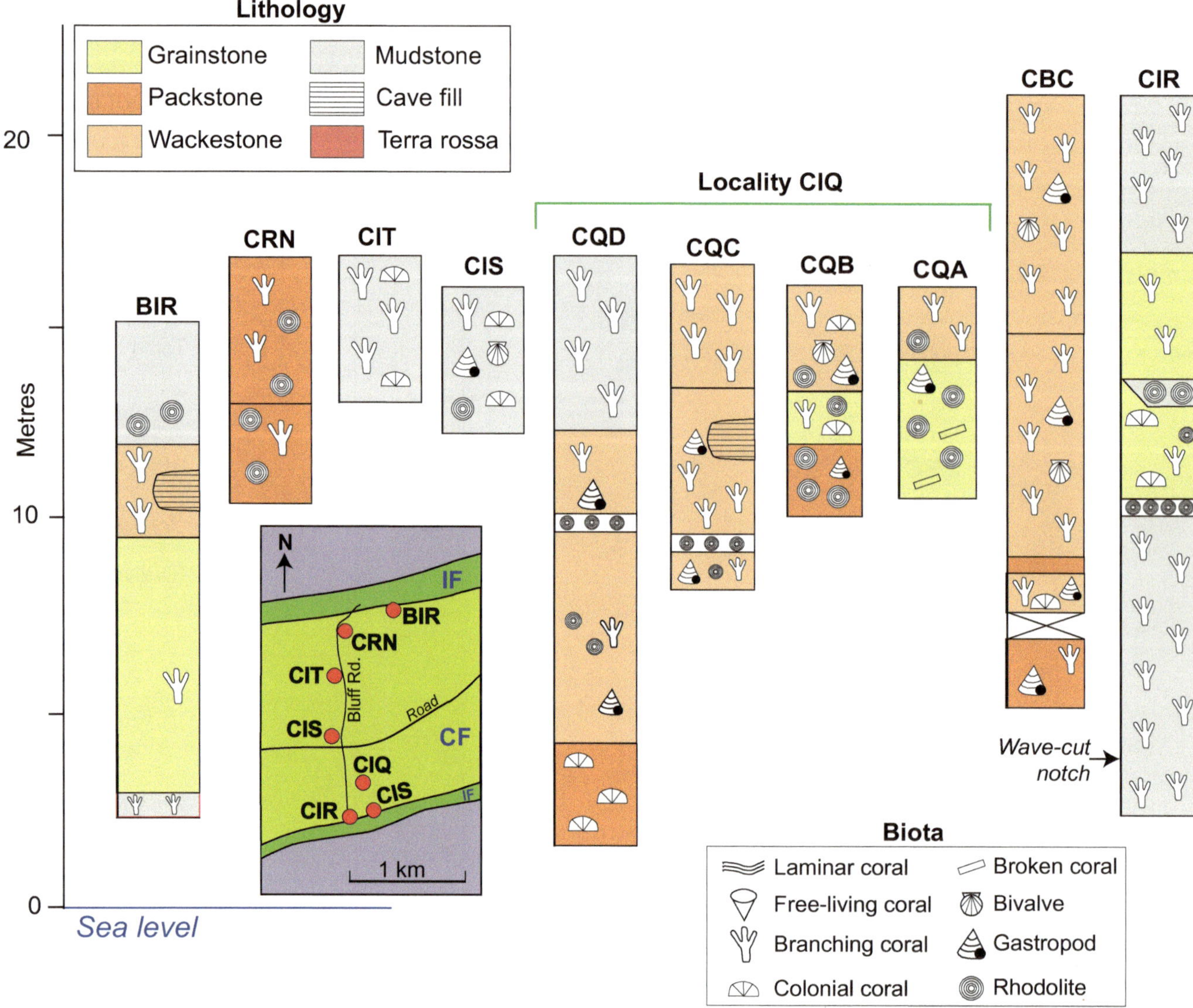

Fig. 4.28 Vertical and lateral facies changes in the Cayman Formation based on exposures along Bluff Road and in Jennifer Bay Quarry (locality CIQ) and cliff face near CIR, Cayman Brac

became established. Although the biota indicates that deposition was largely within the photic zone, establishing an accurate water depth is problematic. Based on comparisons with modern-day Caribbean banks, Jones and Hunter (1994a) suggested that the water was generally <45 m deep. This notion is supported by the presence of amphisteginid and miliolid foraminifera throughout the Cayman Formation that, in modern settings, prefer water that is <35 m deep (e.g., Wright and Murray 1972; Chaproniere 1975). In their survey of rhodolites from the eastern Caribbean, Reid and Macintyre (1988) found that rhodolites with coral branches as their nuclei were most common in water <30 m deep. Generation of some of the coarser-grained units was probably related to periodic storm/hurricane activity.

4.10 Porosity and Permeability

The porosity and permeability of the limestones and dolostones in the Cayman Formation on Grand Cayman are highly variable at all scales. Porosity ranges from microscale intercrystalline pores that are commonly <1 mm long to huge caves to open fractures and joints of all sizes. Virtually every outcrop of the Cayman Formation is characterized by cavities of various sizes and shapes that have formed through the dissolution of aragonitic skeletal material and/or localized dissolution of the matrix (Fig. 4.29). Dome-shaped cavities up to 2 m in diameter formed through the dissolution of the aragonite skeletons of large domal corals. Such cavities, however, are impossible to measure and integrate

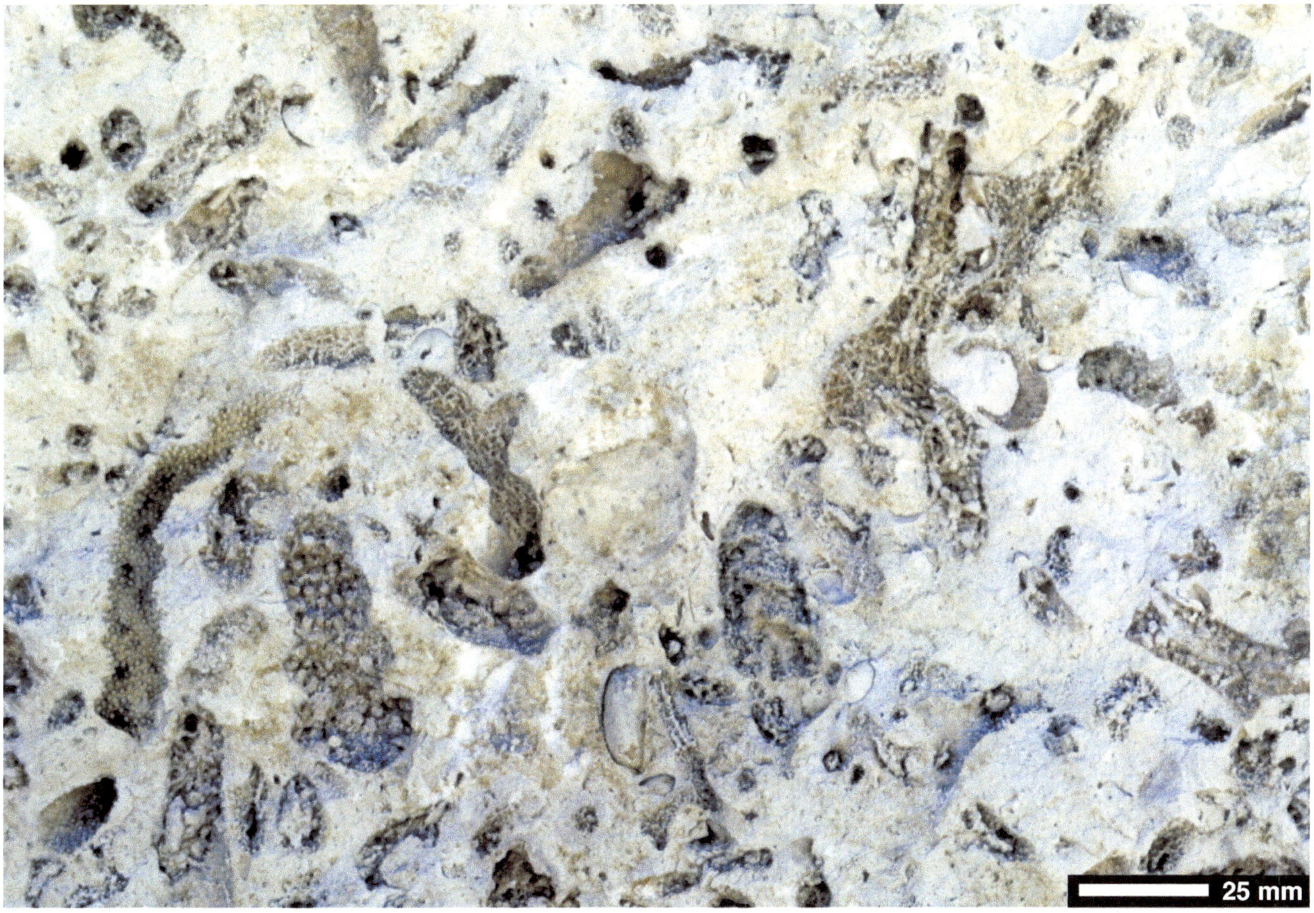

Fig. 4.29 High porosity in dolostones of the Cayman Formation, High Rock Quarry (east end, Grand Cayman) due to preferential dissolution of the aragonite skeletons of *Stylophora*. Note sponge borings in coral branches that were formed before leaching of the aragonite skeleton took place

into the overall porosity of the rocks. The situation is compounded by the fact that laboratory determination of the porosity and permeability demands samples that do not include large cavities and/or fractures that pass through the entire sample. Thus, the laboratory determination of porosity and permeability is typically based on core segments (typically ~10–12 cm long) that are lithologically homogeneous and devoid of fractures and large cavities. In effect, these samples come from the solid rock that occurs between the large caves and cavities evident in outcrop. Thus, the tested porosities are commonly the minimum values because they exclude the large cavities that are so common in the dolostones and limestones of the Cayman Formation.

The porosity of the dolostones (common) and limestones (rare) in the Cayman Formation, based on 222 tested samples from 9 different wells ranges from 2 to 49% (Fig. 4.30). In some wells, there is only a narrow range of tested porosities, whereas in other wells the porosities of individual

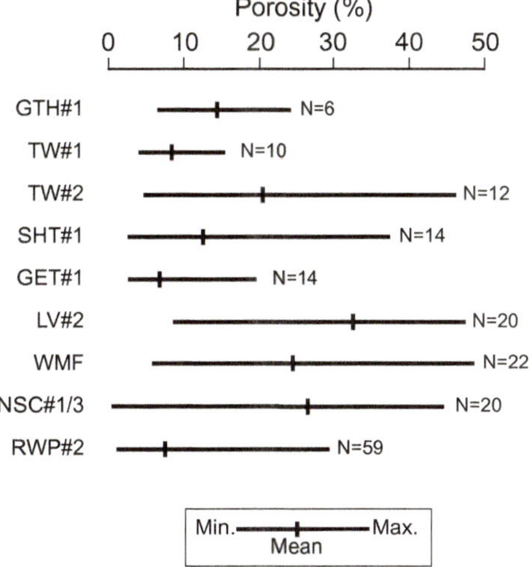

Fig. 4.30 Range of tested porosity values for Cayman Formation from various wells on Grand Cayman. N equals number of samples tested from each well

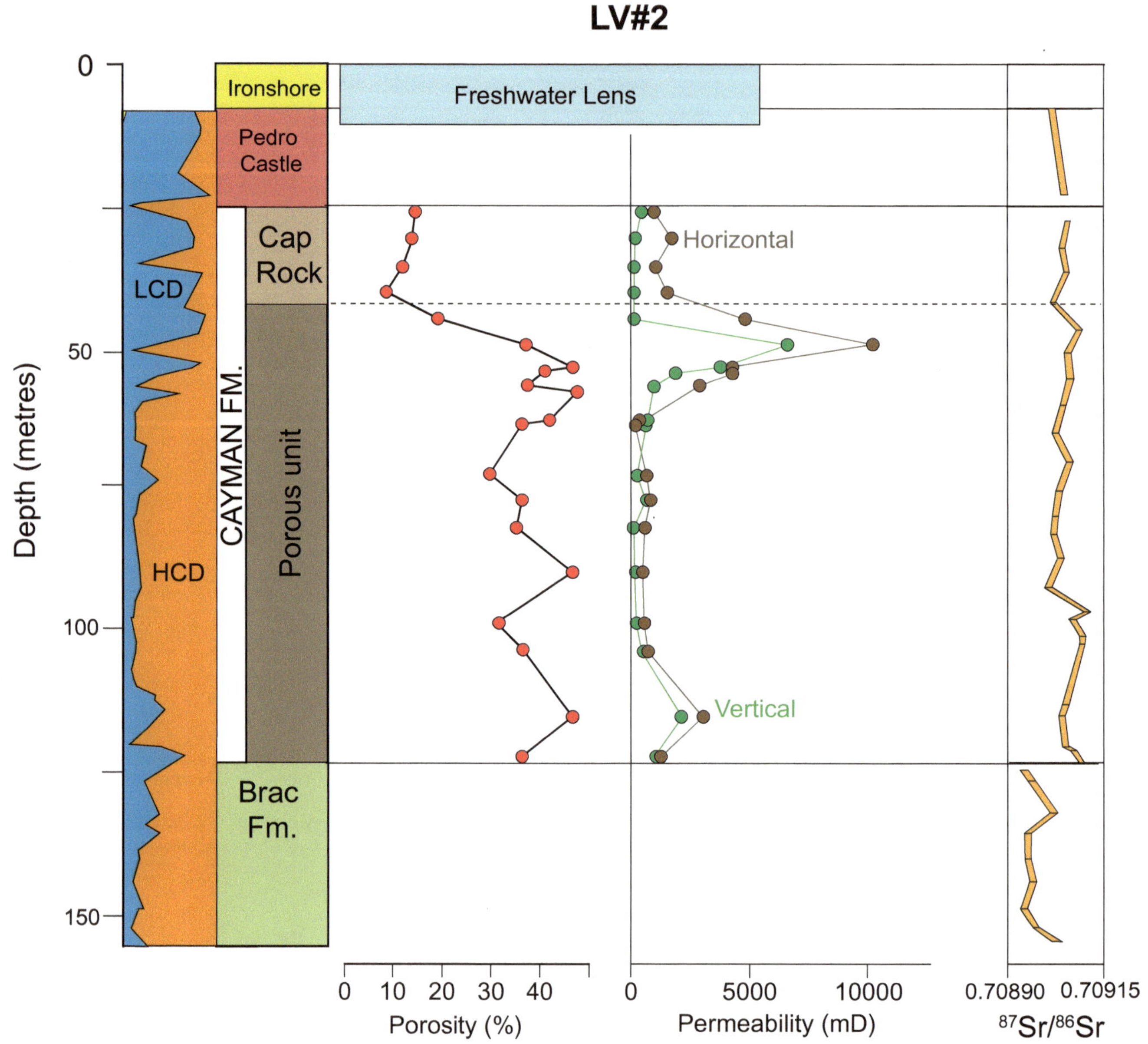

Fig. 4.31 Lithological succession in well LV#2, Grand Cayman, showing stratigraphic variation in percentage of LCD and HCD relative to the Brac Formation, Cayman Formation, and Pedro Castle Formation. The uppermost Ironshore Formation is formed entirely of limestone. The Cayman Formation is divided into the lower porous unit and upper Cap Rock based on tested porosity and permeability values. The $^{87}Sr/^{86}Sr$ curve is characterized by the distinct break at the top of the Brac Formation

samples range from 2 to 49% (Fig. 4.30). In well LV#2, the Cayman Formation can be divided into (1) the cap rock, ~13 m thick, that is defined by low porosities (<15%) and low horizontal permeability and very low vertical permeability, and (2) a lower zone, ~76 m thick that is characterized by porosities that are as high as 48% (Fig. 4.30). The upper part of the lower unit is also characterized by high lateral and vertical permeabilities (Fig. 4.31). The informally named "Cap Rock" is formed of dense, very hard finely crystalline dolostones. During drilling of many wells on the western part of the island, the transition from the "soft" dolostones in lowermost part of the Pedro Castle Formation to the "hard" dolostones that form the Cap Rock of the Cayman Formation is typically marked by a significant decrease in the drilling rates. It should be noted, however, that the successions in the Cayman Formation on the eastern part of the island are more complex and the division into the two informal units is not always apparent.

Table 4.1 Summary data for stable isotopes for calcite and dolomite from samples from the Cayman Formation in wells on Grand Cayman

	Calcite							Dolomite						
	δ18O(‰)			δ13C(‰)				δ18O(‰)			δ13C(‰)			
Well	Min	Mean	Max	Min	Mean	Max	#	Min	Mean	Max	Min	Mean	Max	#
HRQ#1	−1.50	0.20	2.15	−2.56	−0.90	0.20	4	1.04	1.90	2.46	0.32	1.29	2.33	22
CKC#1	−5.34	−4.05	−4.0	−5.34	−4.77	−4.20	2	−0.68	1.53	2.53	0.52	1.79	2.55	36
NSC#1	−5.28	−1.13	3.97	−6.36	−1.13	2.12	46	−1.18	2.94	7.93	−0.62	1.98	4.07	47
RTR#1	–	–	–	–	–	–	–	1.93	2.66	3.81	1.82	2.57	3.01	45
HMB#1	–	–	–	–	–	–	–	1.11	2.53	3.54	2.19	2.93	3.51	17
DTE#1	−6.77	−4.89	−3.06	−4.84	−3.29	−2.14	17	–	–	–	–	–	–	–
LV#2	–	–	–	–	–	–	–	2.36	3.32	4.01	1.56	2.09	2.74	23
GTH#1	–	–	–	–	–	–	–	−4.33	2.49	4.44	−1.23	2.45	3.34	65
RG#1	–	–	–	–	–	–	–	0.60	2.79	3.76	−1.39	1.88	2.87	26
SHT#4	–	–	–	–	–	–	–	0.05	2.60	3.98	−0.09	2.50	3.50	57

4.11 Stable Isotopes

On Grand Cayman, the $\delta^{18}O$ values of the dolostones from the Cayman Formation range from −4.33 to +7.93‰, whereas the $\delta^{13}C$ values range from −1.70 to +4.07‰ (Table 4.1, Fig. 4.32). If the outliers are removed, then most samples have $\delta^{18}O$ values +0.5 and +5.0‰ and $\delta^{13}C$ values between +0.5 and +3.5‰. For the limestones, the $\delta^{18}O$ values range from −7.93 to +3.97‰ and the $\delta^{13}C$ values range from −6.36 to +2.12‰ (Fig. 4.32, Table 4.1).

In individual wells, the $\delta^{18}O$ and $\delta^{13}C$ values tend to be relatively constant (Figs. 4.10, 4.11, 4.12, 4.13, 4.14, 4.15,

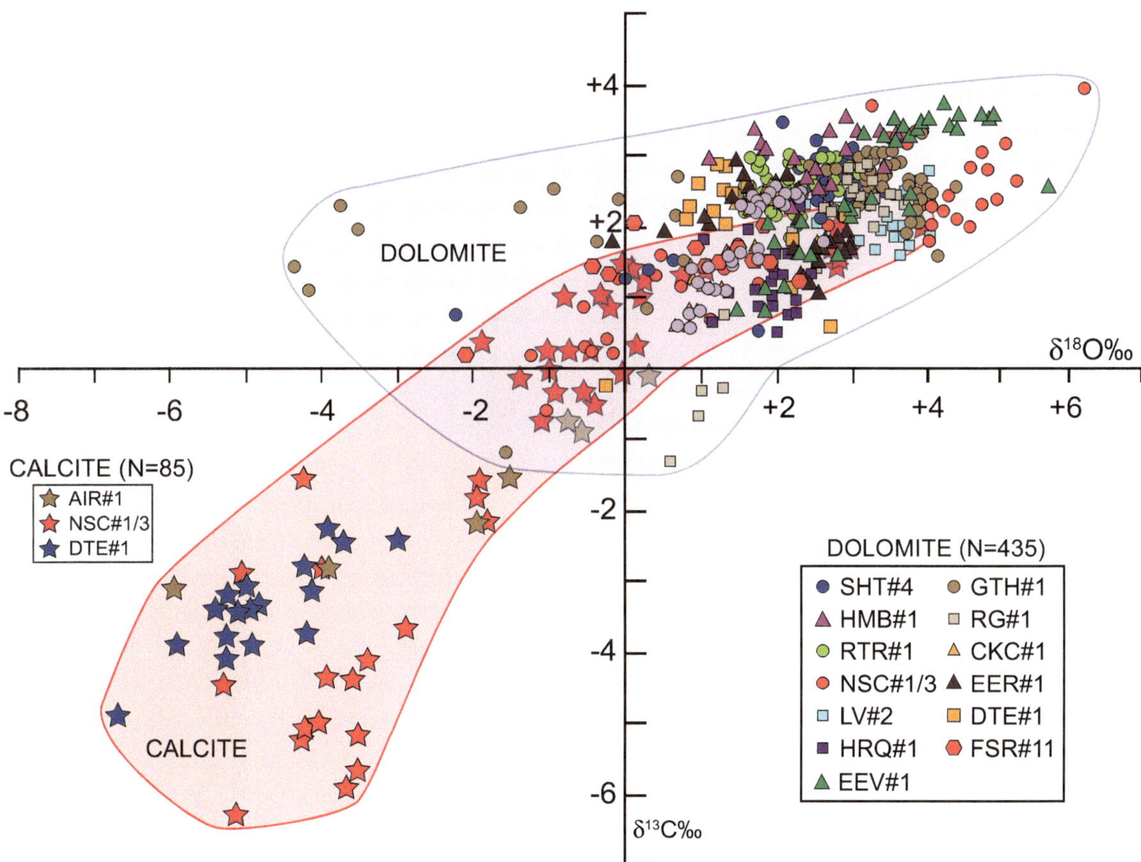

Fig. 4.32 $\delta^{18}O$ versus $\delta^{13}C$ for calcite and dolomite from various wells on Grand Cayman. Sequences with less than 5 analyses for calcite in wells HRQ#1, CKC#1, EEV#2 and FSR#11 were not included in this graph. See Table 4.1 for further details regarding these isotope analyses

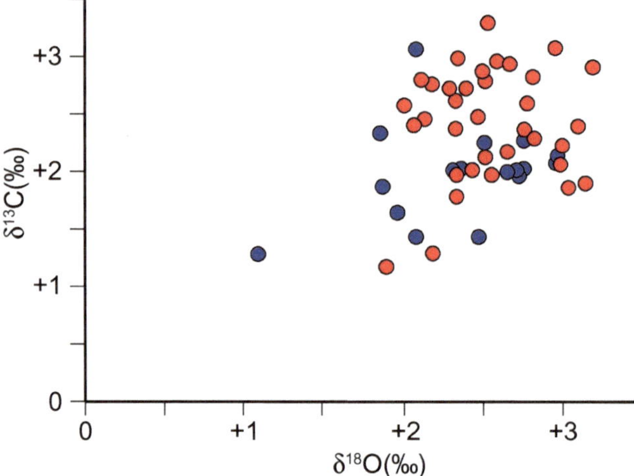

BW-1 / CRQ-1	δ^{18}O		δ^{13}C	
	BW-1	CRQ-1	BW-1	CRQ-1
Minimum	1.09	1.91	1.29	1.15
Mean	2.36	2.54	1.98	2.45
Maximum	1.96	3.19	3.06	3.33
Number	17	35	17	35

Fig. 4.33 δ^{18}O versus δ^{13}C for calcite and dolomite from wells BW-1 and CRQ#1 on Cayman Brac

4.16, 4.17, 4.18 and 4.19). Major changes in these values are only apparent in those wells (e.g., NSC#1/3, Fig. 4.10) where there is a major vertical change in the proportions of calcite and dolomite.

On Cayman Brac, the δ^{18}O from the dolostones in section BW#1, based on 17 samples, ranges from 1.09 to 1.96‰ with an average of 2.36‰, whereas the δ^{13}C ranges from 1.29 to 3.06‰ with an average of 1.98‰ (Fig. 4.33). In comparison, the δ^{13}C from the dolostones in section CRQ#1, based on 35 samples, ranges from 1.91 to 3.19‰ with an average of 2.54‰, whereas the δ^{13}C ranges from 1.15 to 3.33‰ with an average of 2.45‰ (Fig. 4.33). Although the δ^{18}O and δ^{13}C values fluctuate between their extremes, they are relatively consistent throughout the successions in sections CRQ#1 and BW#1 (Figs. 4.18 and 4.19).

4.12 ^{87}Sr/^{86}Sr Ratios

On Grand Cayman, 269 samples of dolostone from the Cayman Formation from various wells across the island yielded ^{87}Sr/^{86}Sr ratios that range from 0.708880 to 0.709240 with an average of 0.0709041 (Fig. 4.34a), whereas 71 samples of limestone yielded a range of 0.708970–0.709170 with an average of 0.709053 (Fig. 4.34b). The range of values for the dolostones is essentially the same as that for the limestones.

On Cayman Brac, the ^{87}Sr/^{86}Sr ratios from 52 dolostones from sections KEL#1, CRQ#1, and BW#1, range from 0.708890 to 0.709187 with an average of 0.709072 (N = 57) (Fig. 4.34b). These numbers reflect the fact that the ^{87}Sr/^{86}Sr ratios are relatively constant throughout the formation and similar to those obtained from the Cayman Formation on Grand Cayman (Fig. 4.34c).

4.13 Age

The fact that the Cayman Formation is sandwiched between the Brac Formation and the Pedro Castle Formation suggests that it is probably of Miocene age. Nevertheless, precise dating of the Cayman Formation is hampered by (1) the lack of well-preserved age-diagnostic fossils, (2) unconformable lower and upper boundaries with significant topographic relief, and (3) pervasive dolomitization of the strata that altered many of the geochemical characteristics. Herein, dating of the Cayman Formation relies on a combination of information derived from the (1) fossils, (2) ^{87}Sr/^{86}Sr geochronology, and (3) lowstand-highstand sea level history given that the islands were submerged while sedimentation was actively taking place. Nevertheless, it should be stressed that the age of the Cayman Formation is still open to some debate.

4.13.1 Fossils

Matley (1926) suggested that the massive white limestone found on the east end of Grand Cayman, which he called the Bluff Formation, was Miocene in age. The corals that he recorded from these strata included *Antillia walli*, *Antillia* sp., *Rotalia* sp., *Thysanus* sp., and fungid corals that T.W. Vaughan regarded as being Miocene (Matley 1926, p. 872). Matley (1926) argued that the Bluff Formation was equivalent to the White Limestone of Jamaica and that the *Lepidocyclina undosa* bed found on Cayman Brac indicated a middle Oligocene age—but here it must be remembered that the strata that contain *Lepidocyclina* are part of the Brac Formation, which is separated from the overlying, younger

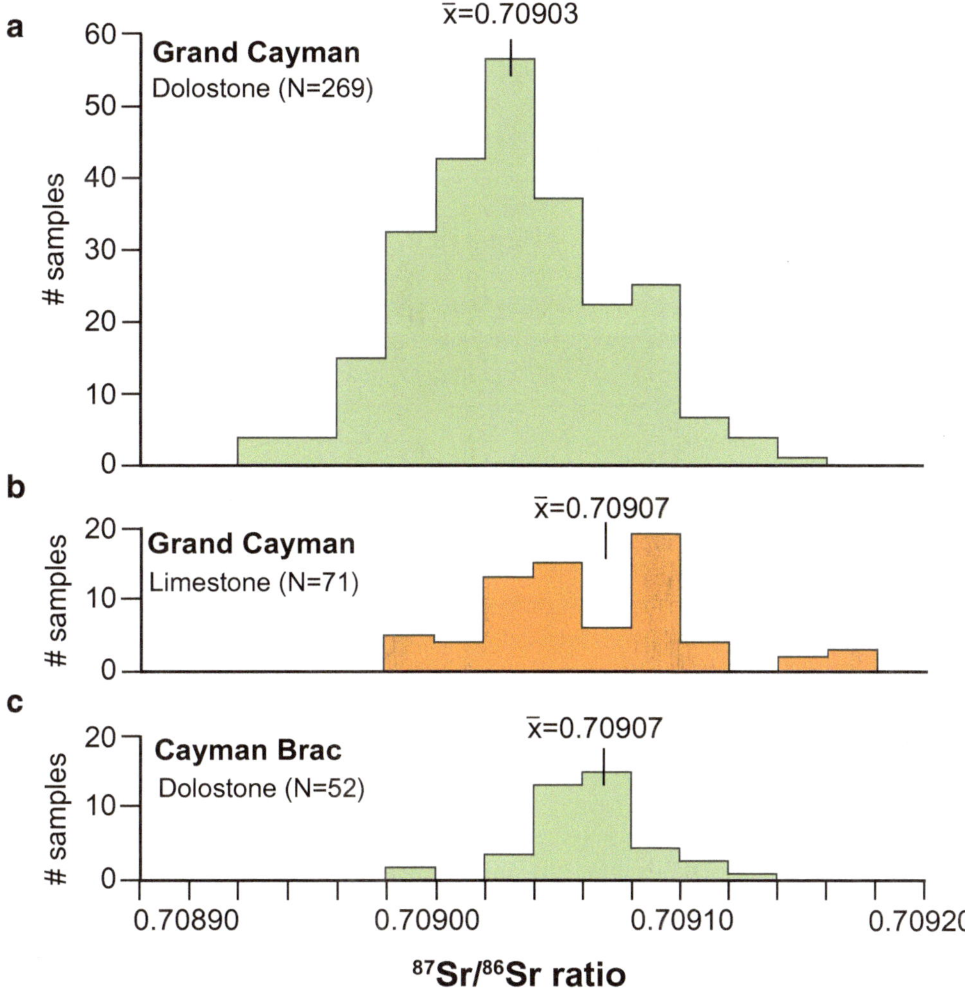

Fig. 4.34 Histogram showing distribution of the $^{87}Sr/^{86}Sr$ values from the Cayman Formation for dolostones (**a**) on Grand Cayman, (**b**) limestones on Grand Cayman, and (**c**) dolostones on Cayman Brac. X = average $^{87}Sr/^{86}Sr$ value for each set of values. N = number of samples in each histogram

strata by an unconformity. Accordingly, that age cannot be applied to the entire carbonate succession found on the island.

Hunter (1994, his Fig. 9.3) noted that most corals in the Cayman Formation are long ranging and therefore of debatable value in assessing if the strata are lower, middle, or upper Miocene in age. The age range of each coral species is also debatable because these faunas have not been widely assessed in terms of their stratigraphic ranges in the Caribbean region. *Diploastrea crassolomellata*, for example, that was considered to be indicative of the Upper Oligocene (Vaughan 1919) has since been found in the Lower Miocene of the Dominican Republic (Hunter 1994). Although the echinoids from the Cayman Formation on Grand Cayman are also indicative of the Miocene (Donovan et al. 2016) they do not allow more precise dating.

The biota evident in the Pedro Castle Formation, like that in the Cayman Formation, does not provide any clear evidence of its age. As noted by Jones (1994), the presence of *Stylophora* in the formation, indicates that it is probably no younger than Pliocene. The numerous free-living corals in the Pedro Castle Formation are consistent with the diverse communities of free-living corals found in Pliocene strata elsewhere in the Caribbean region before they abruptly became extinct 1–2 million years ago (Klaus et al. 2011).

4.13.2 $^{87}Sr/^{86}Sr$ Geochronology

Potentially, the $^{87}Sr/^{86}Sr$ values can be used for dating for limestones given that the $^{87}Sr/^{86}Sr$ ratio of seawater has steadily increased over the last 40 million years (McArthur et al. 2012, their Fig. 7.1). As noted by Wang et al. (2019, p 15) dating strata based on their $^{87}Sr/^{86}Sr$ values assumes that (1) the $^{87}Sr/^{86}Sr$ values are inherited from the seawater from which the limestones/dolostones formed, (2) for the dolomitization, none of the Sr was inherited form the precursor limestone, (3) the $^{87}Sr/^{86}Sr$ values of the seawater from which the limestones/dolostones formed was not modified in any way, and (4) the limestones/dolostones, once formed, did not experience any further diagenetic changes that could have reset the $^{87}Sr/^{86}Sr$ values. If all of these criteria are met, then the $^{87}Sr/^{86}Sr$ values should provide the oldest possible age of the limestone and the oldest possible age of dolomitization (Vahrenkamp and Swart 1988; Budd 1997). This use of $^{87}Sr/^{86}Sr$ for dating is further complicated by the fact that numerous different $^{87}Sr/^{86}Sr$—curves have been proposed, as shown in Fig. 2a of Wang et al. (2019). Thus, for some $^{87}Sr/^{86}Sr$ values, like those from the Cayman Islands, ages may differ by up to 4 Ma depending on which curve is used (Wang et al. 2019, their Fig. 2b), which is significant when dealing with potential ages that are in the 2–10 Ma range. Herein, the $^{87}Sr/^{86}Sr$ values are considered relative to the $^{87}Sr/^{86}Sr$ temporal trend of McArthur et al. (2012) that was developed from rigorous analysis of available data.

For the Cayman Formation, 269 $^{87}Sr/^{86}Sr$ ratios have been obtained from dolostones on Grand Cayman and 52 values have been obtained from Cayman Brac (Fig. 4.34). In addition, 71 $^{87}Sr/^{86}Sr$ ratios have been obtained from limestones in the Cayman Formation on Grand Cayman. Samples formed of dolomite (i.e., dolostone) range from 0.708800 to 0.709240 with an average of 0.709030 on Grand Cayman and 0.70907 on Cayman Brac. Samples formed entirely of calcite (i.e., pure limestone) have $^{87}Sr/^{86}Sr$ values from 0.708970 to 0.709170 with an average of 0.70907. The fact that the average values from the dolostones are lower than the average value from the limestones suggests that the values from the limestones reflect diagenetic factors and may not represent the time when they were originally deposited.

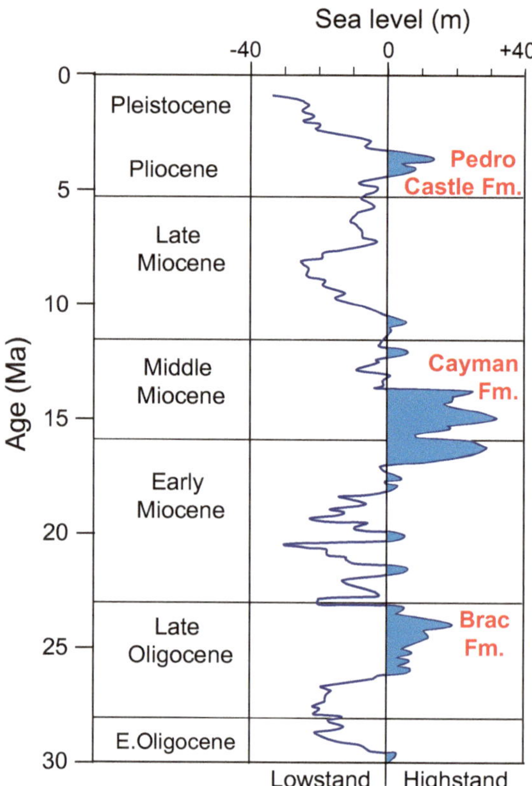

Fig. 4.35 Smoothed eustatic sea level curve from Early Oligocene to Pleistocene (Miller et al. 2020a, their Fig. 2). Blue shaded areas indicate highstands when sediments of the Brac Formation, Cayman Formation, and Pedro Castle Formation probably accumulated

4.13.3 Sea Level

Various eustatic sea level curves have been proposed for the Oligocene to Pleistocene (Vail et al. 1977; Haq et al. 1987; Miller et al. 2005, 2020a, b). In the context of sedimentation on the Cayman Islands, the critical issue is that of determining when the surfaces of each island were below sea level and in the realm of active sedimentation. This situation, however, can develop as a result of eustatic changes in sea level and/or tectonic uplift/subsidence of each island. For the Cayman Islands the situation is further complicated by the fact that each island is located on different fault blocks that may have acted independently of each other.

The smoothed eustatic sea level curve proposed by Miller et al. (2020a) and Miller et al. (2020b) shows highstand positions (i.e. sea levels above present day sea level) during the late Late Oligocene and the Middle Pliocene (Fig. 4.35). Available paleontological evidence and $^{87}Sr/^{86}Sr$ data indicates that the sediments of the Brac Formation probably accumulated during the Late Oligocene—which corresponds with the Late Oligocene highstand proposed by Miller et al. (2020a, b). Dating of the Cayman Formation and Pedro Castle Formation is difficult because (1) its diverse fauna is dominated by long-ranging organisms that do not provide precise dating, (2) most of the formation has been dolomitized and the $^{87}Sr/^{86}Sr$ ratios reflect the time of dolomitization rather than the time of sedimentation, and (3) the $^{87}Sr/^{86}Sr$ ratios of the limestones, which are the same as those of the dolostones, probably reflect the effects of diagenesis rather than the original sediments. Although there is no clear evidence for the age of the Cayman Formation, it seems probable that the original sediments were deposited during the Early Miocene and early Middle Miocene highstand that is evident on the eustatic sea level curve of Miller et al. (2020a, b). Available evidence indicates that the original sediments of the Pedro Castle Formation were probably deposited during the mid-Pliocene highstand that has been identified throughout the Caribbean region and southeast United States of America (Cronin and Dowsett 1993; Raymo et al. 2011). The $^{87}Sr/^{86}Sr$ ratios from the limestones of the Pedro Castle Formation are also consistent with this suggested age.

4.14 Conclusions

Detailed, multifaceted analyses of the Cayman Formation have provided the following conclusions.

- The Cayman Formation, which is up to 150 m thick, is bounded by unconformities that are characterized by significant karstic relief.
- Most of the Cayman Formation is formed of finely crystalline dolostones. In the central- eastern-part of Grand Cayman, however, the formation is formed of limestones or limestones that are overlain by dolostones.
- The Cayman Formation is characterized by a rich, diverse biota that includes corals, bivalves, foraminifera, red algae, *Halimeda*, and scattered echinoids. Only small, scattered patch reefs are known.
- The original sediments were deposited on an open bank in water that was less than 25 m deep.
- The finely crystalline dolostones are characterized by anhedral to euhedral dolomite crystals, typically <50 μm long, that commonly have a core formed of high-calcian dolomite that is encased by an outer zone formed of low-calcian dolomite.
- Dolomitization was probably mediated by normal or slightly evaporated seawater under near-surface conditions.
- The dolostones of the Cayman Formation are commonly cut by fractures and/or joints that provide avenues for lateral and vertical water flow.
- Porosity and permeability are highly variable from location to location. On the western part of Grand Cayman, a low-porosity-low permeability cap rock overlies a lower highly porous unit. The cap rock is not evident on the eastern part of Grand Cayman or on Cayman Brac.
- The dolostones in the Cayman Formation are commonly characterized by complex diagenetic histories that involved numerous phases of dissolution, precipitation of dolomite and calcite cements, and/or deposition of internal sediments.

References

Brunt MA, Giglioli MEC, Mather JD, Piper DJW, Richards HG (1973) The Pleistocene rocks of the Cayman Islands. Geol Mag 110: 209–221

Budd DA (1997) Cenozoic dolomites of carbonate islands: their attributes and origin. Earth Sci Rev 42:1–47

Bugg SF, Lloyd JW (1976) A study of freshwater lens configuration in the Cayman Islands using resistivity methods. Q J Eng Geol 9: 291–302

Chaproniere GCH (1975) Palaeoecology of Oligo-Miocene larger foraminiferida, Australia. Alcheringa 1:37–58

Cronin TM, Dowsett HJ (1993) PRISM warm climates of the Pliocene. Geotimes 38:17–19

Der A (2012) Deposition and sea level fluctuations during Miocene times, Grand Cayman, British West Indies. MSc thesis, University of Laberta, Edmonton.

Donovan SK, Jones B, Harper DAT (2016) Neogene echinoids from the Cayman Islands, British West Indies. Geol J 51:864–879

Emery KO (1981) Low marine terraces of Grand Cayman Island. Estuar Coast Shelf Sci 12:569–578

Etherington SS (2004) Sedimentology and depositional architecture of the Cayman (Miocene) and Pedro Castle (Pliocene) Formations on western Grand Cayman, British West Indies. MSc thesis, University of Alberta, Edmonton

Haq BV, Hardenbol J, Vail PR (1987) Chronology of fluctuating sea levels since the Triassic. Science 235:1156–1167

Hunter IG (1994) Modern and ancient coral associations of the Cayman Islands. PhD thesis, University of Alberta, Edmonton

Jones B (1992) Void-filling deposits in karst terrains of isolated oceanic islands: a case study from Tertiary carbonates of the Cayman Islands. Sedimentology 39:857–876

Jones B (1994) Geology of the Cayman Islands. In: Brunt MA, Davies JE (eds) The Cayman Islands: natural history and biogeography. Kluwer, Dordrecht, The Netherlands, pp 13–49

Jones B (2019) Diagenetic processes associated with unconformities in carbonate successions on isolated oceanic islands: case study of the Pliocene to Pleistocene sequence, Little Cayman, British West Indies. Sed Geol 386:9–30

Jones B, Hunter IG (1989) The Oligocene-Miocene Bluff Formation on Grand Cayman. Carib J Sci 25:71–85

Jones B, Hunter IG (1994a) Evolution of an isolated carbonate bank during Oligocene, Miocene, and Pliocene times, Cayman Brac, British West Indies. Facies 30:25–50

Jones B, Hunter IG (1994b) Messinian (Late Miocene) karst on Grand Cayman, British West Indies: an example of an erosional sequence boundary. J Sediment Res 64:531–541

Jones B, Hunter IG, Kyser TK (1994a) Revised stratigraphic nomenclature for Tertiary strata of the Cayman Islands, British West Indies. Carib J Sci 30:53–68

Jones B, Hunter IG, Kyser TK (1994b) Stratigraphy of the Bluff Formation (Miocene-Pliocene) and the newly defined Brac Formation (Oligocene), Cayman Brac, British West Indies. Carib J Sci 30:30–51

Jones B, Hunter IG, Ng KC (1990). Geological evolution of the Oligocene-Miocene Bluff Formation, Grand Cayman. In: Transactions of the 12th Caribbean Geological Conference. pp 125–132

Jones B, Lockhart EB, Squair C (1984) Phreatic and vadose cements in the Tertiary Bluff Formation of Grand Cayman Island, British West Indies. Bull Can Pet Geol 32:382–397

Jones B, Luth RW (2003) Temporal evolution of Tertiary dolostones on Grand Cayman as determined by $^{87}Sr/^{86}Sr$. J Sediment Res 73: 187–205

Klaus JS, Lutz BP, McNeill DF, Budd AF, Johnson KG, Ishman SE (2011) Rise and fall of Pliocene free-living corals in the Caribbean. Geology 39:375–378

Liang T, Jones B (2014) Deciphering the impact of sea-level changes and tectonic movement on erosional sequence boundaries in carbonate successions: a case study from Tertiary strata on Grand Cayman and Cayman Brac, British West Indies. Sed Geol 305: 17–34

Liang T, Jones B (2015) Ongoing, long-term evolution of an unconformity that originated as a karstic surface in the Late Miocene: a case study from the Cayman Islands, British West Indies. Sed Geol 322:1–18

Mather JD (1972) The geology of Grand Cayman and its control over the development of lenses of potable groundwater. In: VI

Conferencia Geologica Del Caribe-Margarita Venezuela Memorias. pp 154–157
Matley CA (1926) The geology of the Cayman Islands (British West Indies) and their relation to the Bartlett Trough. Q J Geol Soc Lond 82:352–387
McArthur JM, Howarth RJ, Shields GA (2012) Strontium isotope stratigraphy. In: Gradstein FM, Ogg JG, Schmitz M, Ogg G (eds) The geologic time scale. Elsevier, pp 127–143
McCormick CA, Jones B (2021) On the efficacy and limitations of isolated carbonate platforms as "oceanic dipsticks" to reconstruct subsidence histories, a case study from the Paleogene to Neogene strata on Grand Cayman and Cayman Brac, B.W.I. Mar Geol 436:106470
Miller KG, Browning JV, Schmelz WJ, Kopp RE, Mountain GS, Wright JD (2020a) Cenozoic sea-level and cryospheric evolution from deep-sea geochemical and continental margin records. Sci Adv 6:1–15
Miller KG, Kominz MA, Browning JV, Wright JD, Mountain GS, Katz ME, Sugarman PJ, Cramer BS, Christie-Blick N, Pekar SF (2005) The Phanerozoic record of global sea-level change. Science 310:1293–1298
Miller KG, Schmetz WJ, Browning JV, Kopp RE, Mountain GS, Wright JD (2020b) Ancient sea levels as key to the future. Oceanography 33:32–41
Pleydell SM (1987) Aspects of diagenesis and ichnology in the Oligocene-Miocene Bluff Formation of Grand Cayman Island, British West Indies. MSc thesis, University of Alberta, Edmonton
Pleydell SM, Jones B (1988) Boring of various faunal elements in the Oligocene-Miocene Bluff Formation of Grand Cayman, British West Indies. J Paleontol 62:348–367
Raymo ME, Mitrovica JX, O'Leary MJ, DeConto RM, Hearty PJ (2011) Departures from eustasy in Pliocene sea-level records. Nat Geosci 4:328–332
Reid RP, MacIntyre IG (1988) Foraminiferal-algal nodules from the Eastern Caribbean; growth history and implications on the value of nodules as paleo-environmental indicates. Palaios 3:424–435
Ren M (2017) Origin of island dolostone: case study of Cayman Formation (Miocene). British West Indies, Univesity of Alberta, Edmonton, Aberta, Grand Cayman
Ren M, Jones B (2016) Diagenesis in limestones-dolostone successions after 1 million years of rapid sea-level fluctuations: a case study from Grand Cayman, British West Indies. Sed Geol 342:15–30
Ren M, Jones B (2018) Genesis of island dolostones. Sedimentology 65:2002–2033
Rigby JK, Roberts HH (1976) Grand Cayman Island: geology, sediments and marine communities 4. Brigham Young University, Geology Studies. Special Publication no. 4, pp 97
Spencer T (1985) Marine erosion rates and coastal morphology of reef limestones on Grand Cayman Island, West Indies. Coral Reefs 4:59–70
Stoddart DR (1980) Geology and geomorphology of Little Cayman. Atoll Res Bull 241:11–16
Vahrenkamp VC, Swart PK (1988) Constraints and interpretation of $^{87}Sr/^{86}Sr$ ratios in Cenozoic dolomites. Geophys Res Lett 15:385–388
Vail PR, Mitchum RM, Todd RG, Widmier JM, Thompson S, Sangree JB, Bubb JN, Hatlelid WD (1977) Seismic stratigraphy and global changes of sea level. Seismic stratigraphy—applications to hydrocarbon exploration. American Association of Petroleum Geologists Memoir 26
Vaughan TW (1919) Fossil corals from Central America, Cuba, and Porto Rico, with an account of the America Tertiary, Pleistocene, and recent coral reefs. US Natl Musm Bull 103:189–524
Wang R, Jones B, Yu K (2019) Island dolostones: genesis by time-transgressive or event dolomitization. Sed Geol 390:15–30
Willson EA (1998) Depositional and diagenetic features of the Middle Miocene Cayman Formation, Roger's Wreck Point, Grand Cayman, British West Indies. MSc thesis, University of Alberta
Woodroffe CD (1981) Mangove swamp stratigraphy and Holocene transgression, Grand Cayman Island, West Indies. Mar Geol 41:271–294
Woodroffe CD, Stoddart DR, Giglioli MEC (1980) Pleistocene patch reefs and Holocene swamp morphology, Grand Cayman, West Indies. J Biogeogr 7:103–113
Woodroffe CD, Stoddart DR, Harmon RS, Spencer T (1983) Coastal morphology and late quaternary history, Cayman Islands, West Indies. Quatern Res 19:64–84
Wright CA, Murray JW (1972) Comparisons of modern and Palaeogene foraminiferal distributions and their environmental implications. Memoires Du Bureau De Recherches Geologiques Et Minieres, Colloque Sur La Geologie De La Manche, France 79:87–96
Zhao H, Jones B (2012a) Genesis of fabric-destructive dolostones: a case study of the Brac Formation (Oligocene), Cayman Brac, British West Indies. Sed Geol 267–268:36–54
Zhao H, Jones B (2012b) Origin of island dolostones: a case study from the Cayman Formation (Miocene), Cayman Brac, British West Indies. Sed Geol 243–244:191–206

5 Bluff Group—Pedro Castle Formation

Abstract

The Pliocene Pedro Castle Formation, the uppermost unit in the Bluff Group, is found in scattered outcrops on the western part of Grand Cayman, Cayman Brac, and Little Cayman. Although exposed successions are generally <5 m thick, subsurface sequences on the western peninsula of Grand Cayman are up to 21.5 m thick. The formation rests unconformably on the Cayman Formation and is unconformably overlain by the Ironshore Formation. In many areas, much of the Pedro Castle Formation has been lost to erosion. The Pedro Castle Formation is formed of limestones, dolostones, and limestones that have been dolomitized to varying degrees. Although the biota is similar to that in the Cayman Formation, the Pedro Castle Formation commonly contains more free-living corals and fewer large colonial corals. The original sediments were deposited on an open bank with water depths generally <25 m. Weathered phytokarst surfaces on the Pedro Castle Formation are generally more subdued than those on the Cayman Formation because the limestones–dolostones are not as hard as those in the Cayman Formation. The dolostones are typically formed of anhedral to euhedral dolomite crystals that are mostly <50 µm long. Porosity is commonly high with fossil-mouldic porosity being high due to leaching of the aragonitic skeletons of the corals, bivalves, and gastropods. Permeability and porosity are variable throughout the formation. Many of the original limestones underwent complex diagenetic changes that involved dissolution, dolomitization, precipitation of calcite and dolomite cements, and/or deposition of internal sediments. Caves are found in some parts of the formation.

5.1 Introduction

The Pedro Castle Formation is exposed around Pedro Castle on the south coast of Grand Cayman (Fig. 5.1a), on the west end of Cayman Brac (Fig. 5.1b), in the quarry on Little Cayman and possibly in scattered, isolated exposures over the rest of Little Cayman. These exposures, however, offer a limited view of the succession because the exposed vertical succession is typically <5 m high. Accordingly, most of the information on this formation comes from wells that have been drilled on the western part of Grand Cayman with some wells in the Safe Haven area (Fig. 5.2) providing a succession through the formation that is up to 19 m thick.

The Pedro Castle Formation is formed of limestones that have been dolomitized to varying degrees. It commonly contains an abundant, diverse biota that includes corals, bivalves, echinoids, gastropods, foraminifera, red algae, and green algae (Figs. 5.3 and 5.4). Distinction between rocks of the Pedro Castle Formation and Cayman Formation is commonly difficult, especially when dealing with isolated outcrops where the overall, larger-scale stratigraphic sequence is not apparent. In general, the limestones and dolostones of the Pedro Castle Formation are "softer" than the dolostones of the Cayman Formation and the phytokarst developed on their weathered surfaces tend to be more subdued than those on the dolostones of the Cayman Formation. Free-living corals are common in the Pedro Castle Formation (Fig. 5.3e, f), but rare in the Cayman Formation.

Following the initial designation and description of the Pedro Castle Formation (Jones et al. 1994a, b), detailed studies of the formation were undertaken on Grand Cayman (Wignall 1995) and Cayman Brac (MacNeil 2001; MacNeil

B. Jones, *Geology of the Cayman Islands*,
https://doi.org/10.1007/978-3-031-08230-6_5

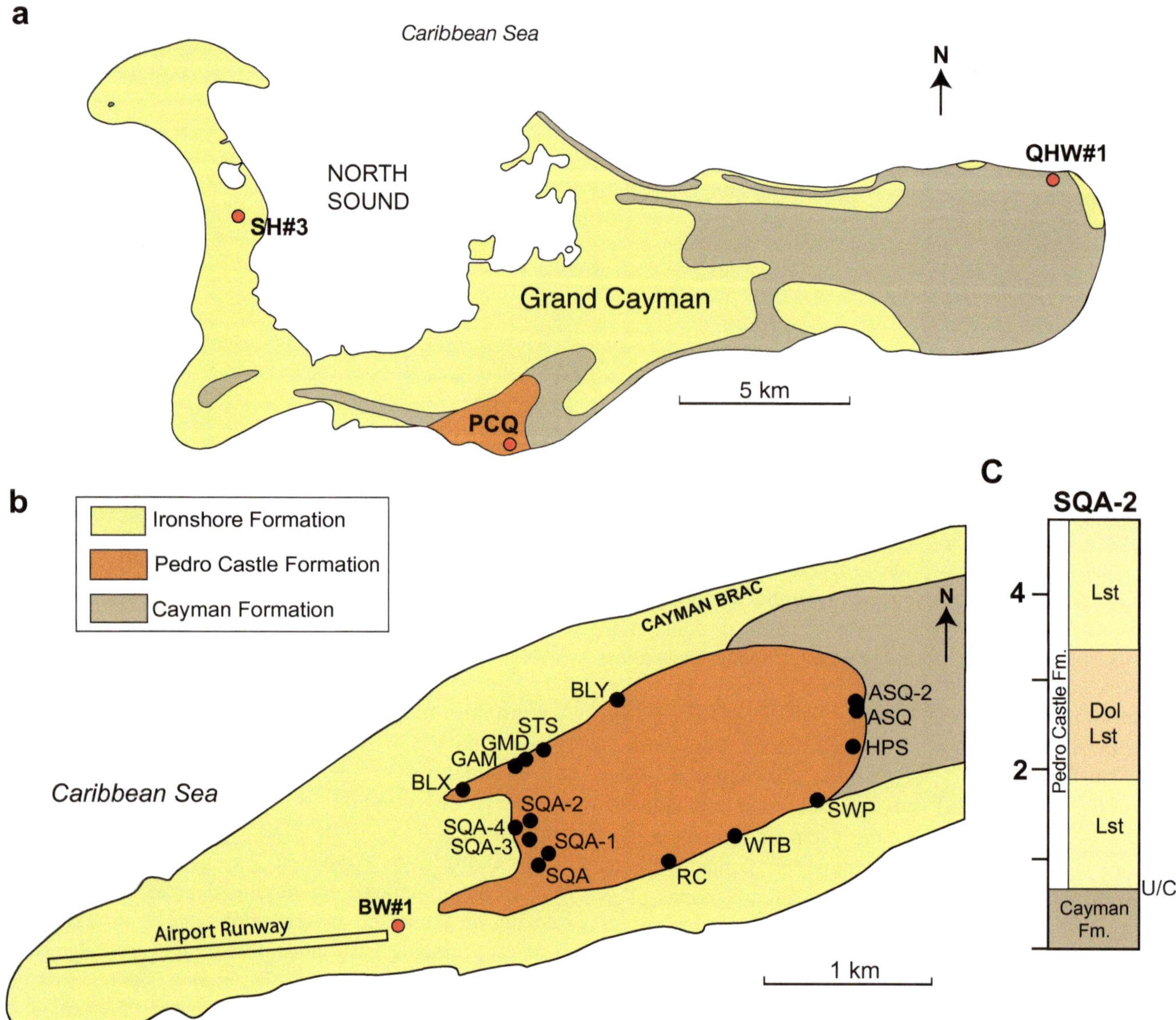

Fig. 5.1 **a** Map of Grand Cayman showing distribution of Pedro Castle Formation in area around Pedro Castle Quarry (PCQ). Reference section for formation came from well SH#3. **b** Map of west end of Cayman Brac showing location of measured sections through the Pedro Castle Formation and location of well BW#1 that was drilled at the Water Authority site. **c** Simplified lithological succession through the Pedro Castle Formation exposed in section SQA#2, located in Scott Quarry (modified from MacNeil 2001, his Fig. 2.6)

and Jones 2003). On Little Cayman, exposures of the Pedro Castle Formation are sparse with no more than 2 m of vertical section evident at any given locality (Jones 2019). Although these studies provided overviews of the sedimentology of the formation, the work on Grand Cayman focused largely on the sedimentological aspects, whereas the study on Cayman Brac focused largely on the diagenesis of the succession with stress on the dolomitization. On Little Cayman, the research focused largely on the unconformity that capped the strata that were tentatively assigned to the Pedro Castle Formation. Accordingly, there is presently an imbalance in the range of information that is available for this formation on each island.

5.2 Distribution

On Grand Cayman, the Pedro Castle Formation is largely restricted to the western part of the island (Fig. 5.1a). Exposures of the formation are rare, being restricted to Pedro Castle Quarry where it rests on the Cayman Unconformity and evident in the upper parts of the quarry walls (Fig. 5.5).

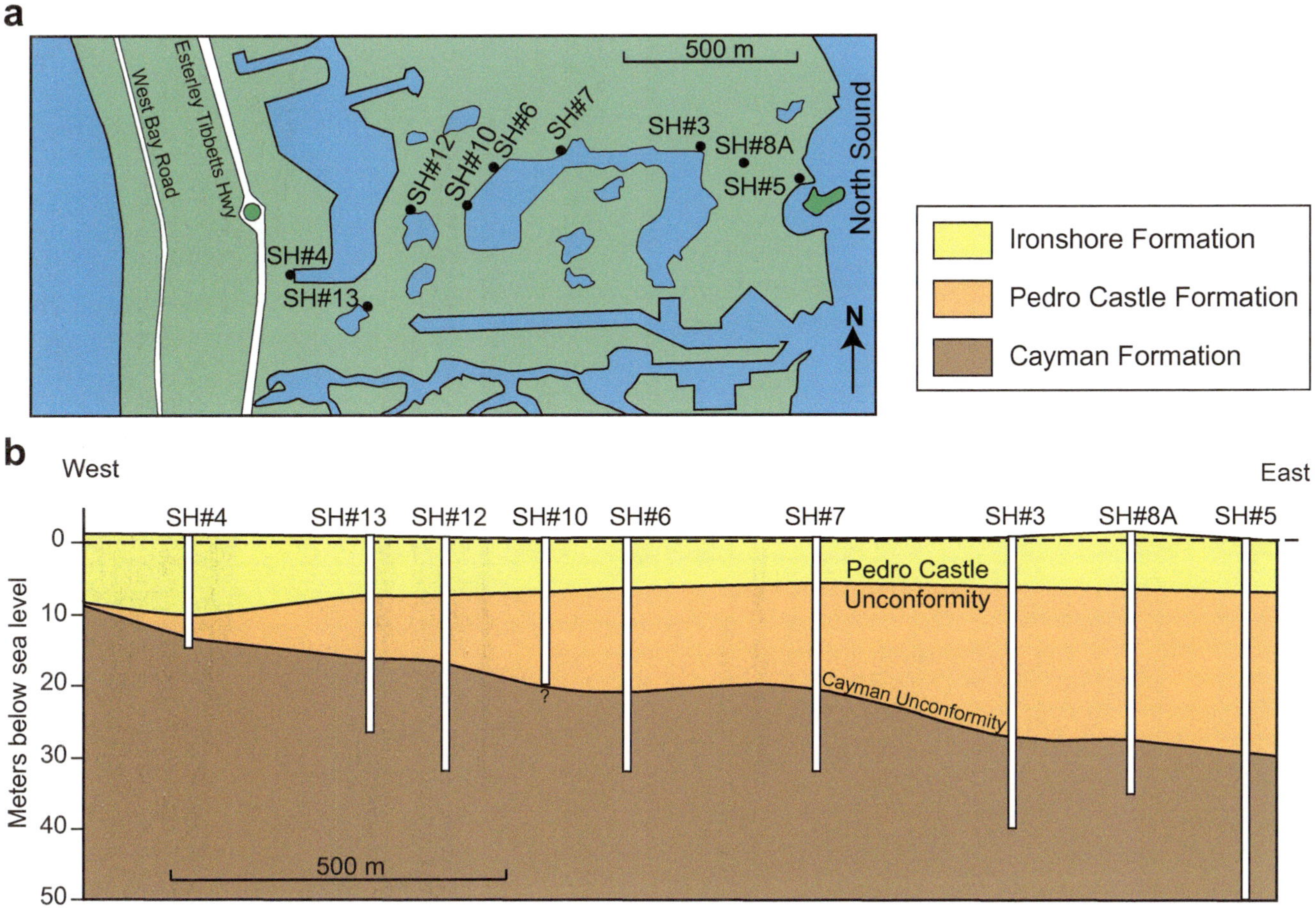

Fig. 5.2 **a** Map showing location of wells drilled a Safe Haven on the western peninsula of Grand Cayman. **b** East–West cross-section through Safe Haven area based on wells shown in panel A. Note that the thickness of the Pedro Castle Formation progressively increases to the east

The weathered surface of the formation is evident along the coastal area around Pedro Castle and can be traced westward as far as Spotts Bay before it disappears. On the western peninsula, the Pedro Castle Formation is buried beneath the Ironshore Formation, and information on its lithology and geochemical characteristics can only be gained from well samples. The thickest part of the Pedro Castle Formation is found in well SH#3, where it is 21.5 m thick.

The eastern part of Grand Cayman is devoid of surface exposures of the Pedro Castle Formation and has not been found in most of the wells that have been drilled in this area. In the northeast part of the island, colour air photographs around well QHW#1 show a series of circular to ovate depressions, up to 100 by 50 m in size, that are roughly aligned parallel to the nearly coastline. Each depression is surrounded by a rim, up to 5 m above the floor of the depression, that is formed of dolostones of the Cayman Formation, which are typical for the northeastern part of the island. Well QHW#1 (Fig. 5.6), located in one of these depressions, penetrated 40.6 m of the Pedro Castle Formation before reaching the Cayman Formation (Jones and Hunter 1994b). In that well, the Pedro Castle Formation is formed entirely of dolostones with only minor amounts of calcite being present in a few samples near the surface (Fig. 5.6). Jones and Hunter (1994b) suggested the depressions represent the locations of sinkholes that developed in the Cayman Formation prior to deposition of the sediments that now form the Pedro Castle Formation.

On Cayman Brac, the Pedro Castle Formation is restricted to the western end of the island (Fig. 5.1b). Vertical sequences through the formation are limited because the core of the island in that area is generally <10 m asl. Fortunately, there is a large active quarry in that area that provides clean vertical sections through the upper part of the Cayman Formation, the unconformity at the top of the Pedro Castle Formation, and the basal part of the Pedro Castle Formation (Fig. 5.7).

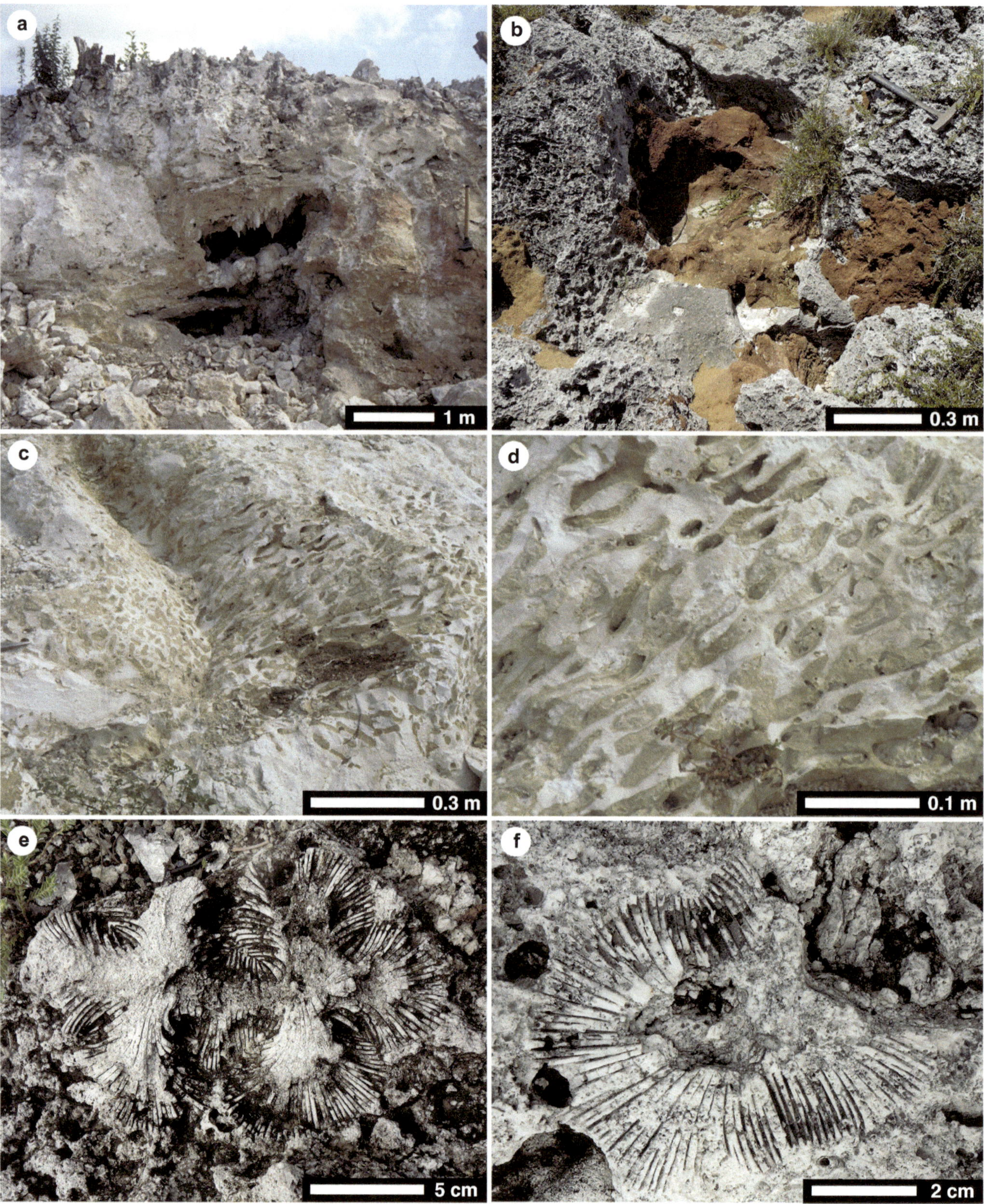

Fig. 5.3 Field photographs showing typical aspects of the Pedro Castle Formation. **a** Outcrop of basal part of Pedro Castle Formation in Scott Quarry on west end of Cayman Brac. Note cave developed in limestones, associated speleothems, and local terra rossa. **b** Surface exposure of Pedro Castle Formation near Pedro Castle on the south coast of Grand Cayman. Note subdued phytokarst and potholes filled with lithified terra rossa. **c, d** Leached branching coral (*Stylophora*?) in limestone, Scott Quarry, west end of Cayman Brac. **e, f** Examples of free-living corals in Pedro Castle Formation exposed in coastal area near Pedro Castle

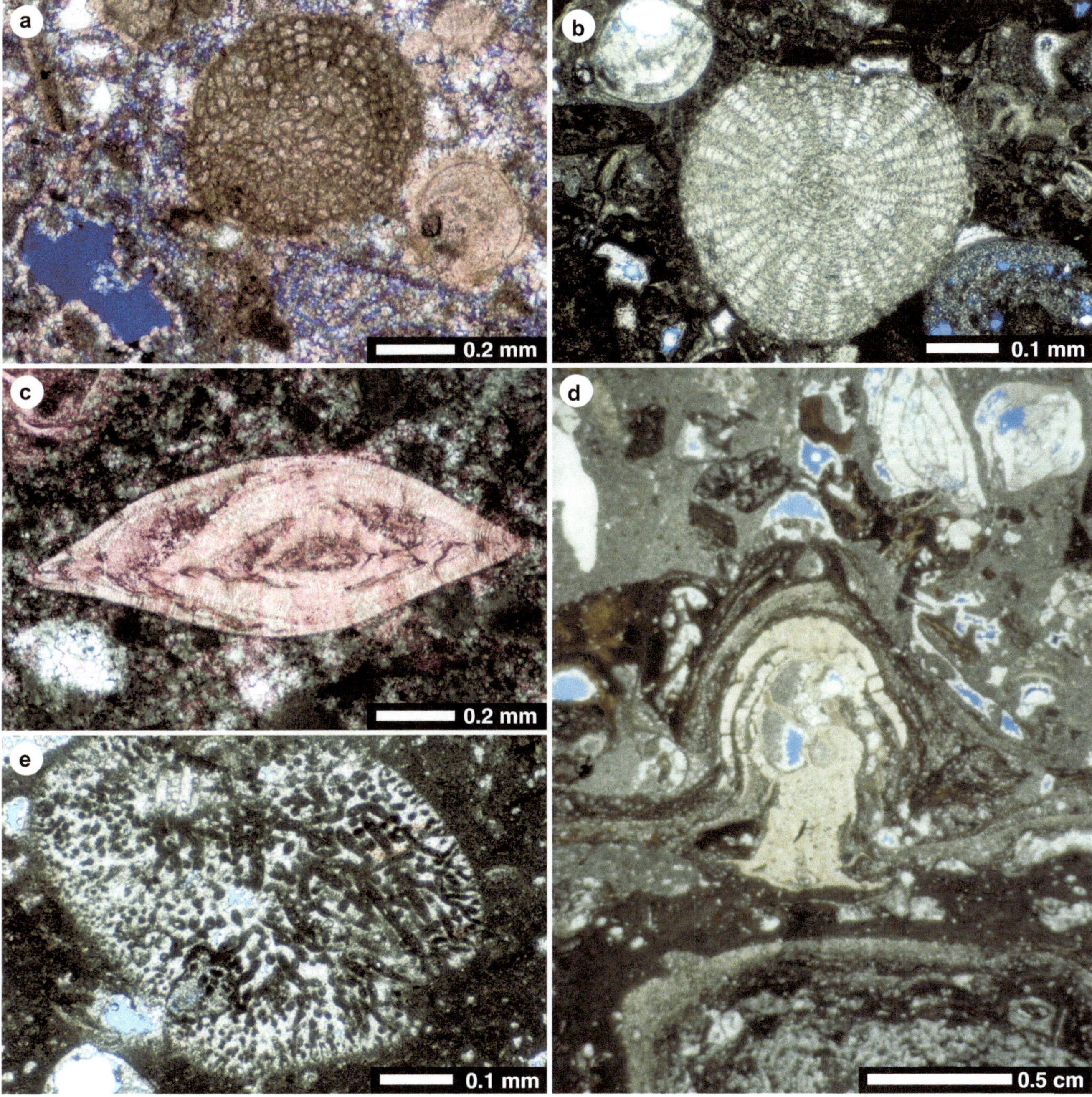

Fig. 5.4 Thin section photomicrographs showing textures common in the Pedro Castle Formation—all samples from west end of Cayman Brac. All thin sections impregnated with blue epoxy. Panels A and C stained with Alizarin Red S solution (red—calcite). **a, b** Foraminifera embedded in calcitic matrix with variable porosity. **c** Calcitic foraminifera embedded in dolomitized matrix. **d** Margin of rhodolite (lower part of image) with encrusting *Homotrema*(?)—embedded in foraminifera-rich limestone (upper part of image). **e** *Halimeda* plate in mudstone matrix

5.3 Lower Boundary

The lower boundary of the Pedro Castle Formation, first defined by Jones et al. (1994a), is the unconformity that is clearly evident in Pedro Castle Quarry (Fig. 5.5) and at other localities along the coast to the west until it disappears in the Spotts Bay area. In the subsurface of the western part of Grand Cayman, this boundary is usually marked by a sudden decrease in drilling rates as the drill passes from the "softer" dolostones/limestone of the Pedro Castle Formation into the dense, hard dolostones of the underlying Cayman Formation.

Fig. 5.5 Oblique view of south wall of Pedro Castle quarry showing Cayman Unconformity (white arrow) that separates the Pedro Castle Formation from the underlying Cayman Formation. Quarry wall is ~12 m high

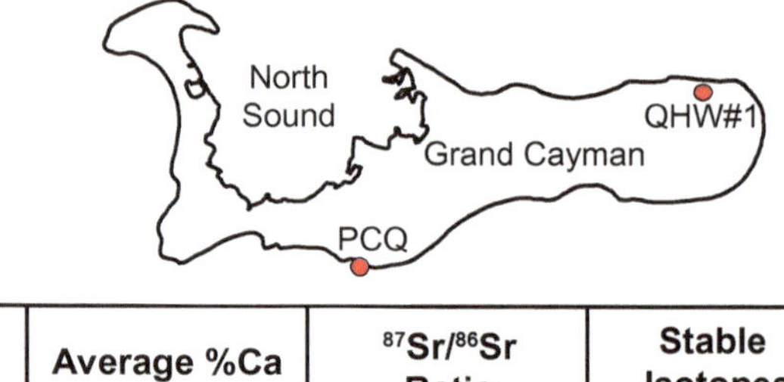

Fig. 5.6 Lithological succession in well QHW#1, northeast corner of Grand Cayman, showing percentage of low-calcian dolomite (LCD), high-calcian dolomite (HCD), and calcite, %Ca in the LCD and HCD, average %Ca, $^{87}Sr/^{86}Sr$ ratios, and $\delta^{18}O$ and $\delta^{13}C$ stable isotopes. At this locality the sequence is thought to be filling a sinkhole that existed in the Cayman Formation

On Cayman Brac, the basal unconformity of the Pedro Castle Formation is readily visible in the fresh cut walls of the quarry on the west end of the island (Fig. 5.7). In the weathered surfaces along the southern and northern margins of the uplifted core, however, the boundary is commonly masked by the grey to black patina of the weathered

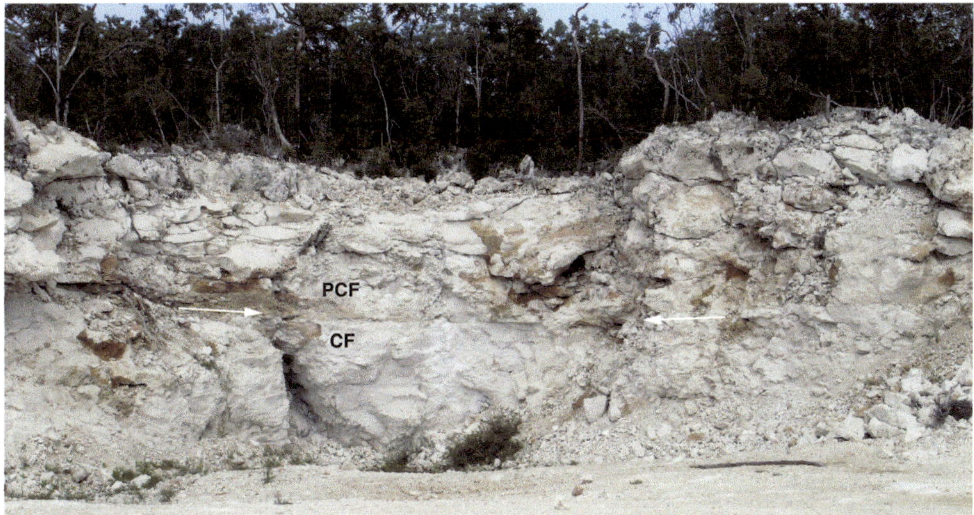

Fig. 5.7 View of wall, north side of Scott Quarry on Cayman Brac, showing unconformity that separates the Pedro Castle Formation (PCF) from the underlying Cayman Formation (CF). Note small caves developed in various parts of the outcrop

surfaces. In the quarry, only local vertical variations in the position of the boundary are evident. In a well drilled at the Water Authority site, ~1 km to the SSW (Fig. 5.1), however, that boundary is 10 m below sea level and serves to illustrate the topography that exists on that surface.

The basal boundary of the Pedro Castle Formation has not yet been found on Little Cayman.

5.4 Upper Boundary

The upper boundary is poorly exposed on Grand Cayman. In the subsurface on the western part of the island, the boundary is an unconformity that is marked by the change from the hard, well-lithified limestones–dolostones of the Pedro Castle Formation to the poorly lithified limestones of the Ironshore Formation.

On Cayman Brac, the upper boundary of the formation is not exposed anywhere on the island. In well BW#1, at the Water Authority site (Fig. 5.1), the boundary between the Pedro Castle Formation and the Ironshore Formation is at a depth of 10 m bsl.

On Little Cayman, vertical walls in the quarry exposed the boundary between the hard, dolostones of the Pedro Castle Formation and the friable limestones of the overlying Ironshore Formation (Jones 2019). There, the boundary was characterized by well-defined phytokarst with pinnacles as high as 2 m. The spaces between the phytokarst pinnacles were filled with fossiliferous limestones of the Ironshore Formation. Unfortunately, quarrying destroyed those exposures in 2019 and the unconformity, although still there, is difficult to see on the fresh cut surfaces that are masked with debris.

5.5 Thickness

In the type section in Pedro Castle Quarry on Grand Cayman, the Pedro Castle Formation is only 2.5 m thick and capped by the present-day weathering surface (Fig. 5.5). In well SH#3, which was the designated reference section for the formation (Jones et al. 1994a), it is 21.5 m thick (Fig. 5.8). In well QHW#1, the Pedro Castle Formation, which is thought to be formed of marine sediments that accumulated in a sinkhole, is 40.6 m thick. On the western part of Grand Cayman. Etherington (2004) suggested that the Pedro Castle Formation was up to 21 m thick, but highly variable from locality to locality because of the topography on the Cayman Unconformity that defines the base of the formation.

The exact thickness of the Pedro Castle Formation on Cayman Brac is poorly known because of the lack of exposures that include the basal and upper boundaries of the formation. In well BW#1, located at the Water Authority site, the Pedro Castle Formation is only 6 m thick. The thickness of the formation on Little Cayman is unknown.

5.6 Lithology

The lithologies in the Pedro Castle Formation were first described by Jones et al. (1994a, b) and Jones (1994). More detailed information on the facies in the Pedro Castle Formation were subsequently provided by Wignall (1995) and Etherington (2004) who focused on the western peninsula on Grand Cayman, and MacNeil (2001), who provided detailed information on the Pedro Castle Formation on Cayman Brac.

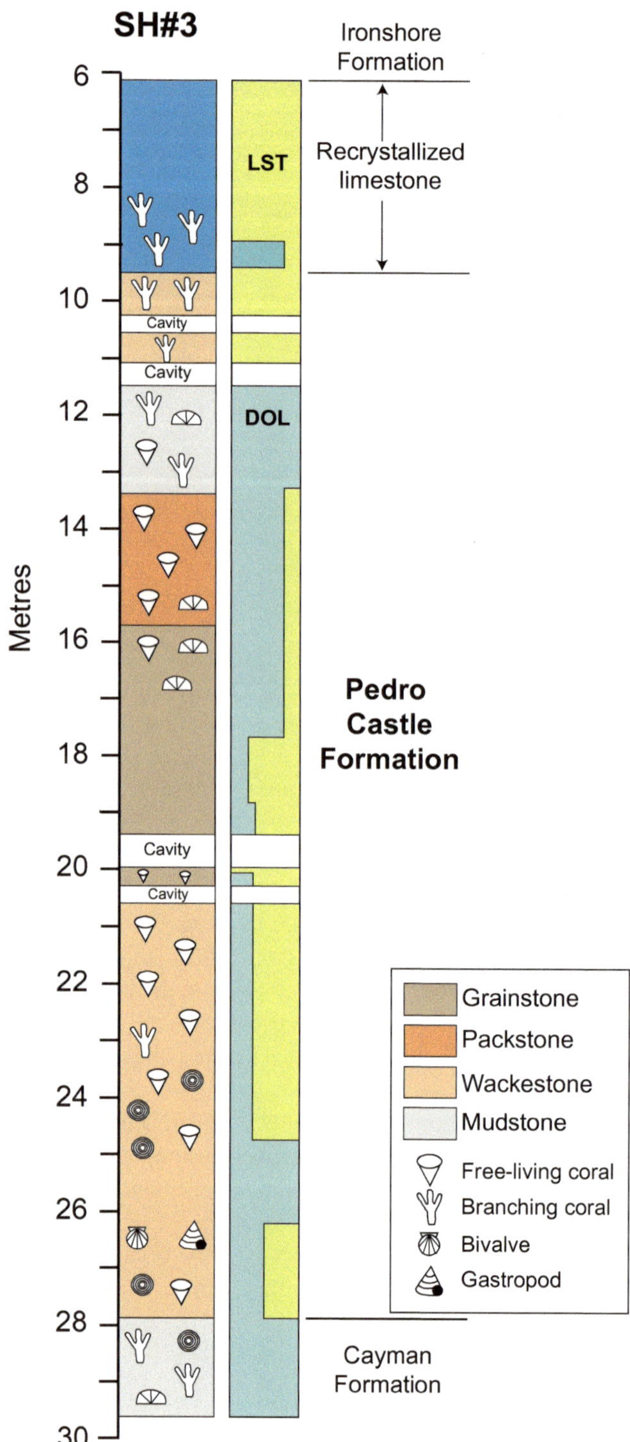

Fig. 5.8 Lithological succession through the Pedro Castle Formation in section SH#3 (Fig. 5.1) that Jones et al. (1994a, their Fig. 4) designated as the reference section for the formation. Note variable dolomite content through the formation

5.6.1 Grand Cayman

In Pedro Castle Quarry, the Pedro Castle Formation is formed of off-white to cream coloured dolostones that are relatively "soft" when compared to the "hard" dolostones in the underlying Cayman Formation. This difference is readily apparent on the weathered surface along the coastline near Pedro Castle where the phytokarst that developed on the Pedro Castle Formation is more subdued than that on the Cayman Formation.

In the reference section in well SH#3, the formation is formed of white to grey dolostone, dolomitic limestone, and limestone (Fig. 5.8). The dolostones, like those in the underlying Cayman Formation, are commonly fabric-retentive. There is no recognizable pattern to the distribution of the dolostone through the Pedro Castle Formation—some limestones contain no dolomite whereas nearby limestones have commonly been entirely replaced by dolomite. Diagenetic leaching of aragonitic skeletons means that many fossils are now represented by moulds (Fig. 5.3c–f) with the result that precise identification of those fossils is difficult.

The skeletal wackestones to grainstones evident in SH#3 contain free-living corals, branching corals, bivalves, and gastropods (Fig. 5.8). Rhodoliths are present throughout the succession. Skeletal grains in the matrix, which are commonly micritized, include foraminifera (e.g., *Amphistegina*, soritids), *Halimeda*, red algae, and various indeterminate forms that are held in a mudstone matrix.

The limestones and dolostones of the Pedro Castle Formation commonly contain an abundant, diverse biota that includes numerous large free-living corals (*Trachyphyllia*, *Thysanus*, *Teleiophilia*, *Antillocyathus*) (Fig. 5.3e, f), foraminifera (mainly *Amphistegina*), *Halimeda,* and red algae (Fig. 5.4) along with scattered colonial corals (Fig. 5.3c, d), echinoids, bivalves, and gastropods. Rhodoliths, up to 5 cm in diameter, are locally common—as in the lower part of the type section in Pedro Castle Quarry.

Based on the core that came from wells drilled through the Pedro Castle Formation in the Safe Haven area, Wignall (1995) delineated the Echinoid–red algae, the Algae-Mollusc, Red Algae, Rhodolith-foraminifera-*Halimeda*, Foraminifera-Mollusc, and Foraminifera-*Halimeda* facies (Fig. 5.9). Some facies, like the Rhodolith-Foraminifera-*Halimeda* were widespread, whereas others were more localized in their distribution (Fig. 5.10). By considering all of the available evidence, Wignall (1995) noted that bivalves, gastropods, benthic foraminifera

	FACIES	FABRIC	Allochems (major)	Allochems (minor)
1	Mollusc-*Halimeda* -Foram	Wackestone	Bivalves, Gastropods, *Halimeda*	Amphisteginids,echinoids, *Stylophora, Porites,* rhodoliths, red algae
	Low energy environment, protected from high enery conditions			
2	*Stylophora*-mollusc -foram	Wackestone to packstone	Stylophora, bivalves, gastropods, *Amphistegina*	*Halimeda*, *Porites*, red algae, rhodoliths
	Shallow (<10 m) marine, low to moderate energy			
3	Free-living coral, foram, *Halimeda*	Wackestone - packstone	*Trachyphyllia, Amphistegina, Halimeda*	*Porites, Stylophora*, *Archaias*, miliolinids, gastropods, echinoids, *Gorgonia,* rhodoliths
	Shallow (<10 m) marine, higher energy than for Facies 2			
4	*Halimeda* - mollusc	Wackestone - packstone	*Halimeda*, bivalves, *Amphistegina*,	Corals, rhodoliths
	High energy, shallower than for Facies 1, 2, 3, and 5			
5	Foram-Mollusc	Packstone	*Amphistegina*, bivalves, gastropods	Corals, rhodoliths, intraclasts
	Shallow (<10 m), low to moderate energy			
6	Rhodolith -foram - *Halimeda*	Packstone - Grainstone	Rhodoliths, *Amphistegina, Halimeda*	Cibicidids, Texularids, Nodosarids, Asteriginids, Miliolids, Glogigerinids, peloids, Rhodolites, *Porites*, *Stylophora*
	Shallow (<10 m), moderate energy			

Fig. 5.9 Summary descriptions of facies in the Pedro Castle Formation found in the Safe Haven area on the western peninsula of Grand Cayman. Data from Wignall (1995, his Table 2)

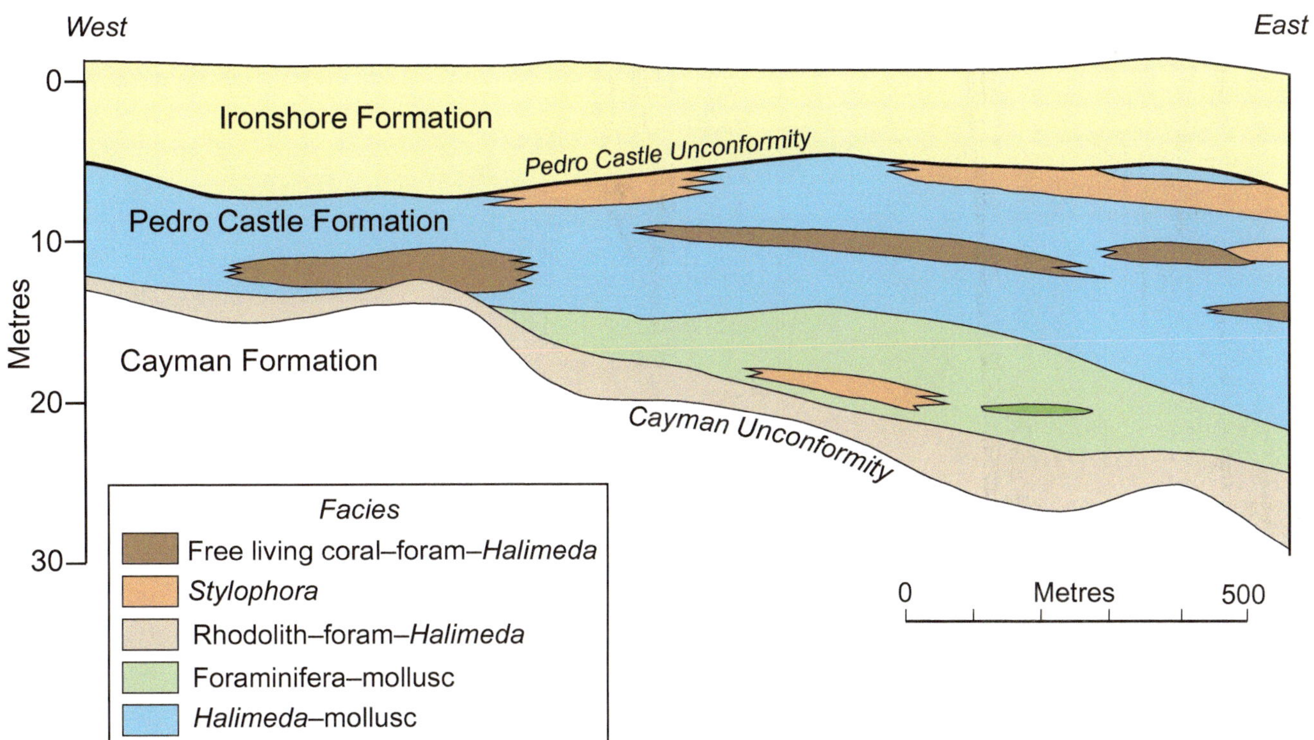

Fig. 5.10 East–west cross-section through Safe Haven area on western peninsula of Grand Cayman showing distribution of facies (as defined in Fig. 5.9). Modified from Wignall (1995, his Fig. 15)

(including *Amphistegina*, *Archasis*, and miliolid foraminifera) are most common in the lower part of the formation, bivalves and gastropods are most common in the lower and middle parts, and pelagic foraminifera are most common in the upper 10 m of the formation. Based on these observations, Wignall (1995) argued that the fossil record pointed to a deepening-upward succession.

Etherington (2004, her Fig. 3.1), based on her examination of the Pedro Castle Formation in the sequences that she examined from across Grand Cayman defined the Rhodo lith-foraminifera, *Trachyphyllia*-foraminifera-*Halimeda*, Foraminifera, Foraminifera-*Halimeda*-Bivalve, Foraminifera-Mollusc, *Halimeda*, Foraminifera-*Halimeda*, and *Stylophora*-*Amphistegina* facies. To a large extent these facies parallel those defined by Wignall (1995). She argued that the original sediments were deposited in water that was 20–30 m deep with the topography on the underlying Cayman Unconformity and a continually rising sea level being the main controls on facies development.

5.6.2 Cayman Brac

On Cayman Brac, the Pedro Castle Formation is formed of limestones, dolostones, and limestones that have been dolomitized to varying degrees. There is no obvious pattern to the distribution of the dolomite, although MacNeil (2001) showed that it tended to be most common in the middle part of the sequence that is exposed in the quarry (Fig. 5.1). Locally, the limestones and dolostones of the uppermost part of the Cayman Formation and the overlying Pedro Castle Formation have experienced extensive diagenetic change, so that little of the original rock is now evident (Fig. 5.11). Such diagenesis greatly complicates interpretation of the original strata. Dolostone samples taken at various levels above the quarry floor in section SQA-2 are characterized by variable mol %$CaCO_3$ but with modal values between 57 and 58 mol% $CaCO_3$ (Fig. 5.12).

MacNeil (2001) identified seven facies in the Pedro Castle Formation that were based on the rock type (as per Dunham's classification) and the biota that is dominated by foraminifera, red algae, green algae, and corals (Fig. 5.13). Numerous echinoid plates and rhodolites are present in some parts of the succession. According to MacNeil (2001), Amphisteginids (35%), Soritids (20%), and encrusting Homotrematids (25%) dominate among the foraminifera. Other benthic foraminifera include miliolid, textularids, cibicidids, nodosarids, asteriginids, and ataxoohragmids. Rare planktonic Globigerinidae and Globorotaliidae foraminifera are also present in some samples.

Sections BLX, GAM, GND, and STS on the north coast are dominated by Facies 1 to 6 with corals being rare. There is no vertical consistency to the facies as small-scale facies changes characterize this area (Fig. 5.14). Along the south coast, the array of facies is similar but scattered corals are present in sections RC, WTB, and ASQ (Jones and Hunter 1994a). As on the north coast, vertical and lateral facies changes are common (Fig. 5.14). The Pedro Castle Formation exposed in Scott's Quarry, which is located midway between the exposures along the north and south coasts, is characterized by the coral facies that is found in sections SQA-2, SQA-3, and SQA-4 (Fig. 5.14). The corals were

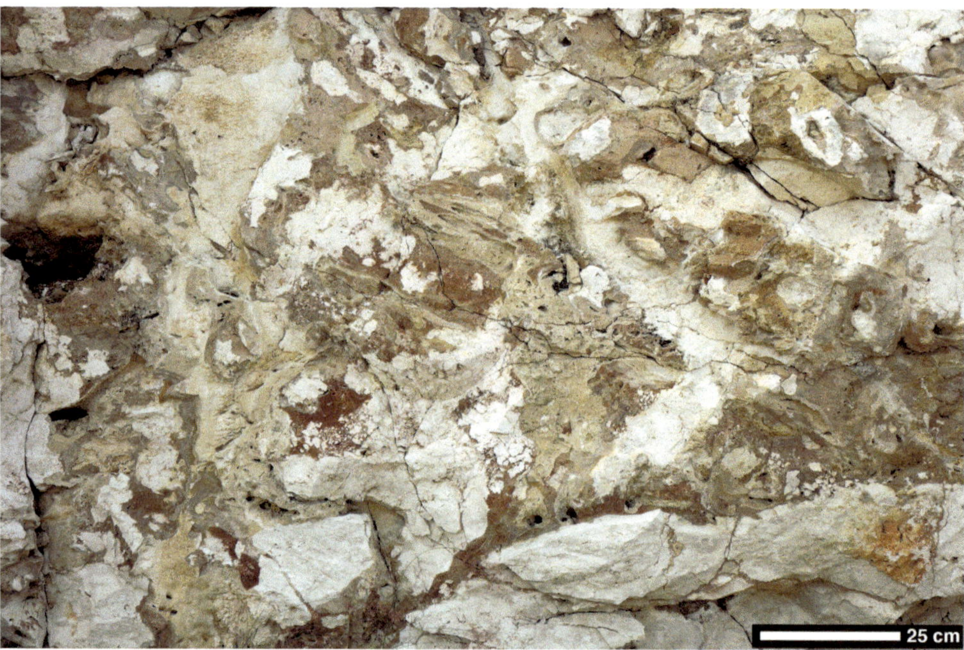

Fig. 5.11 View of quarry wall in Scott Quarry, west end of Cayman Brac, showing large-scale diagenetic alteration of the original limestones (white)

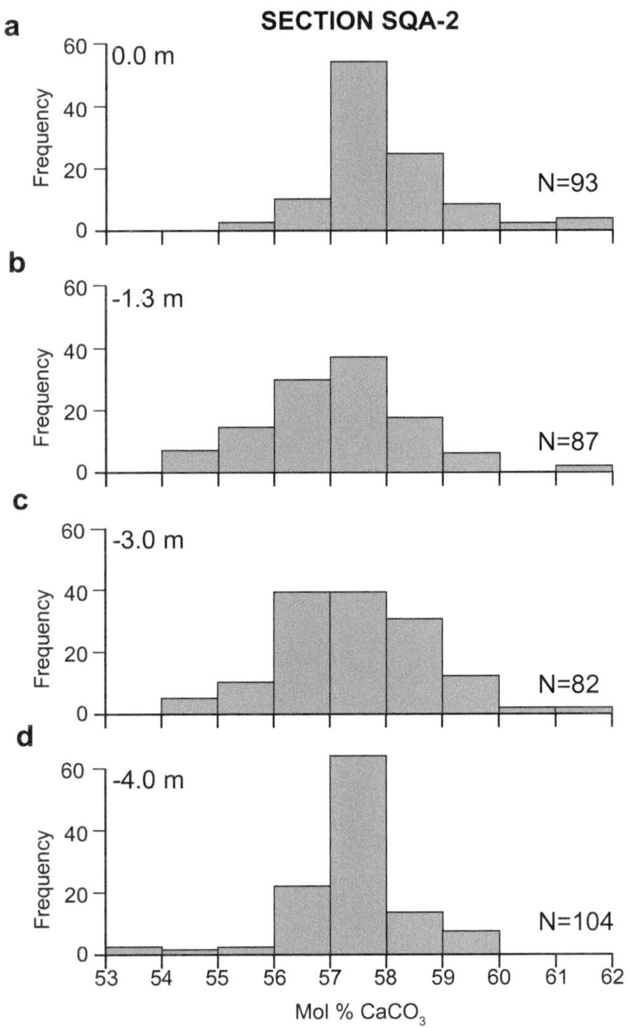

Fig. 5.12 mol% $CaCO_3$ in dolostones from the Pedro Castle Formation at various levels in section SQA-2, which is located in Scott Quarry, west end of Cayman Brac

buried in carbonate sands that are similar to the sands that dominate Facies 3, 5, and 6.

Hunter (1994), based on analogy with modern coral communities found in the Bahamas (Rose and Lidz 1977), suggested that the corals and the associated biota found in the Pedro Castle Formation in Scott's Quarry probably grew in water that was 10–20 m deep. Facies 1–6, which are characterized by various types of carbonate sands, probably reflect minor changes in water depths and/or subtle changes in the bank morphology. This suggestion is best illustrated by sections GAM and GMD that are characterized by different facies successions even though they are only ~10 m apart (Fig. 5.14).

In the dolostones of the Pedro Castle Formation, MacNeil and Jones (2003) recognized two types of replacive dolomite crystals, namely RI that is microcrystalline with subhedral rhombs that are <2 μm long with rounded corners and texture–preserving, and RII that comprises euhedral crystals, 15–35 μm long with sharp corners. The dolomite cement is divided into C1, which is a clear cement that overgrew some of the RII replacive crystals (producing crystals with "dirty cores" and clear rims), and C2 that is a limpid dolomite that lines moldic cavities and intergranular pores.

5.7 Stable Isotopes

For the Pedro Castle Formation exposed on the west end of Cayman Brac, MacNeil (2001) reported $\delta^{18}O$ values for the calcite of −6.18 to −2.41‰ with an average of −4.33‰ and $\delta^{13}C$ values for the calcite of −8.40 to −2.25‰ with an average of −7.45‰ (Fig. 5.15). For the dolomite, the $\delta^{18}O$ values are −0.08 to +2.16‰ with an average of +1.25‰ and $\delta^{13}C$ values are −1.81 to +1.42‰ with an average of +0.27‰ (MacNeil 2001; MacNeil and Jones 2003). The stable isotopes of samples from the Pedro Castle Formation on Little Cayman yielded similar values (Fig. 5.15; Jones 2019).

Overall, there is a linear trend in the stable isotopes for the calcite and dolomite from the limestones and dolostones of the Pedro Castle Formation (Fig. 5.15). The stable isotope values for the calcite points to a freshwater influence (MacNeil 2001).

5.8 Porosity and Permeability

For six samples of the Pedro Castle Formation from wells GTH#1, SHT#4, and GET#1 on the western part off Grand Cayman, the porosity ranged from 3.4 to 44.6% (average 24.9%), the maximum permeability varies from 0.7 to 7640 mD (average of 3893.3 mD), and the permeability at 90° to the maximum, ranges from 0.2 to 6870 mD (average 3247 mD), and the vertical permeability ranges from 0.1 to 5210 mD (average 1460.7 mD).

5.9 $^{87}Sr/^{86}Sr$ Geochronology

For 13 samples of dolostones from sections GAM, SQA-2, and SQA-4 on Cayman Brac, the $^{87}Sr/^{86}Sr$ ratios range from 0.709032 to 0.709108 with an average of 0.709067 (MacNeil and Jones 2003, their Table 1). For Little Cayman, 6 samples from the quarry and the north coast of the island that probably come from the Pedro Castle Formation yielded $^{87}Sr/^{86}Sr$ ratios ranging from 0.709070 to 0.709158 with an average of 0.709102. Based on the $^{87}Sr/^{86}Sr$ curve of Farrell et al. (1995), MacNeil and Jones (2003) suggested that the $^{87}Sr/^{86}Sr$ ratios from the Pedro Castle Formation on Cayman

Fig. 5.13 Summary descriptions of facies in the Pedro Castle Formation found on the western end of Cayman Brac. Data from (MacNeil 2001, his Table 2.1)

	FACIES	FABRIC	Allochems (major)	Allochems (minor)
1	Echinoid–red algae	Mudstone – wackestone	Red algae Echinoids	Molluscs Amphisteginids Rhodolites (< 1 cm)
	Low-energy environment – protected from high energy conditions			
2	Algae – Mollusc	Wackestone	Red algae Green algae Molluscs	Amphisteginids, soritids, Echinoids, peloids, coral fragments, sprong spicules
	Shallow (<10 m) marine, low to moderate energy conditions			
3	Red Algae	Wackestone - packstone	Red algae Molluscs	Amphisteginids, soritids, textularids, miliolids, Globigerinids, Green algae, peloids, coral fragments
	Shallow (<10 m) marine, higher energy than for Facies 2			
4	Rhodolite	Wackestone	Red algae, peloids, Rhodolites	Soritids, peloids,echinoid plates
	High energy, shallower than for facies 1, 2, 3 and 5			
5	Foram-Mollusc	Wackestone	Red algae, Molluscs, Amphisteginids, Soritids	Miliolids, Globigerinids, Green algae, echinoid plates
	Shallow (<10 m) , moderate energy			
6	Foraminifera	Wackestone - packstone	Red algae, Soritids, Amphisteginids, Homotrematids, Molluscs	Cibicidids, Texularids, Nodosarids, Asteriginids, Miliolids, Glogigerinids, peloids, Rhodolites, *Porites*, *Stylophora*
	Water to 10 to 20 m deep, moderate energy			
7	Coral	Wackestone - packstone	Red algae, Soritids, Amphisteginids, Homotrematids, Molluscs deposited between corals	Branching (*Porites*, *Stylophora*),*massive* (*Montastrea*),*free-living* (*Leptoseris*) *corals*
	Water to 10 to 20 m deep, moderate energy			

Brac indicated that dolomitization took place 4.4–1.2 Ma ago. If the McArthur et al. (2012) look-up tables are used, these $^{87}Sr/^{86}Sr$ values indicate ages of 5.13–1.41 Ma with an average age of 2.73 Ma.

Although the $^{87}Sr/^{86}Sr$ ratios from Little Cayman overlap with those from Cayman Brac, they tend to be higher and also have a higher average value. Nevertheless, they may indicate that dolomitization took place at the same time as on Cayman Brac. Much of this time span indicated by the Cayman $^{87}Sr/^{86}Sr$ ratios, however, overlaps with the plateau in the $^{87}Sr/^{86}Sr$ temporal curve that makes it impossible to give a more precise date.

5.10 Age

The biota in the Pedro Castle Formation is similar to that in the underlying Cayman Formation apart from the fact that free-living corals are more common. No definitive age-diagnostic fossils have been found in the formation. Jones and Hunter (1989) suggested that it was Miocene because it overlies the Cayman Formation, which at that time was considered to be of Late Oligocene age. Following a subsequent reassessment of the age of the Cayman Formation as Lower to Middle Miocene, Jones (1994) noted that the Pedro Castle Formation must also be younger than originally suggested. Jones et al. (1994b) argued that the Pedro Castle Member (as it was then known) must be Pliocene or older because it overlay the Cayman Member and it contained *Stylophora*—a coral that became extinct in the Caribbean at the end of the Pliocene (Frost 1977). It should be noted, however, that van Woesik et al. (2012) have suggested that *Stylophora* survived into the Pleistocene in the Caribbean region. The fact that the boundary between the two formations is an unconformity with significant topography implies that there was an extended period of subaerial exposure of the Cayman Formation before deposition of the sediments that now form the Pedro Castle

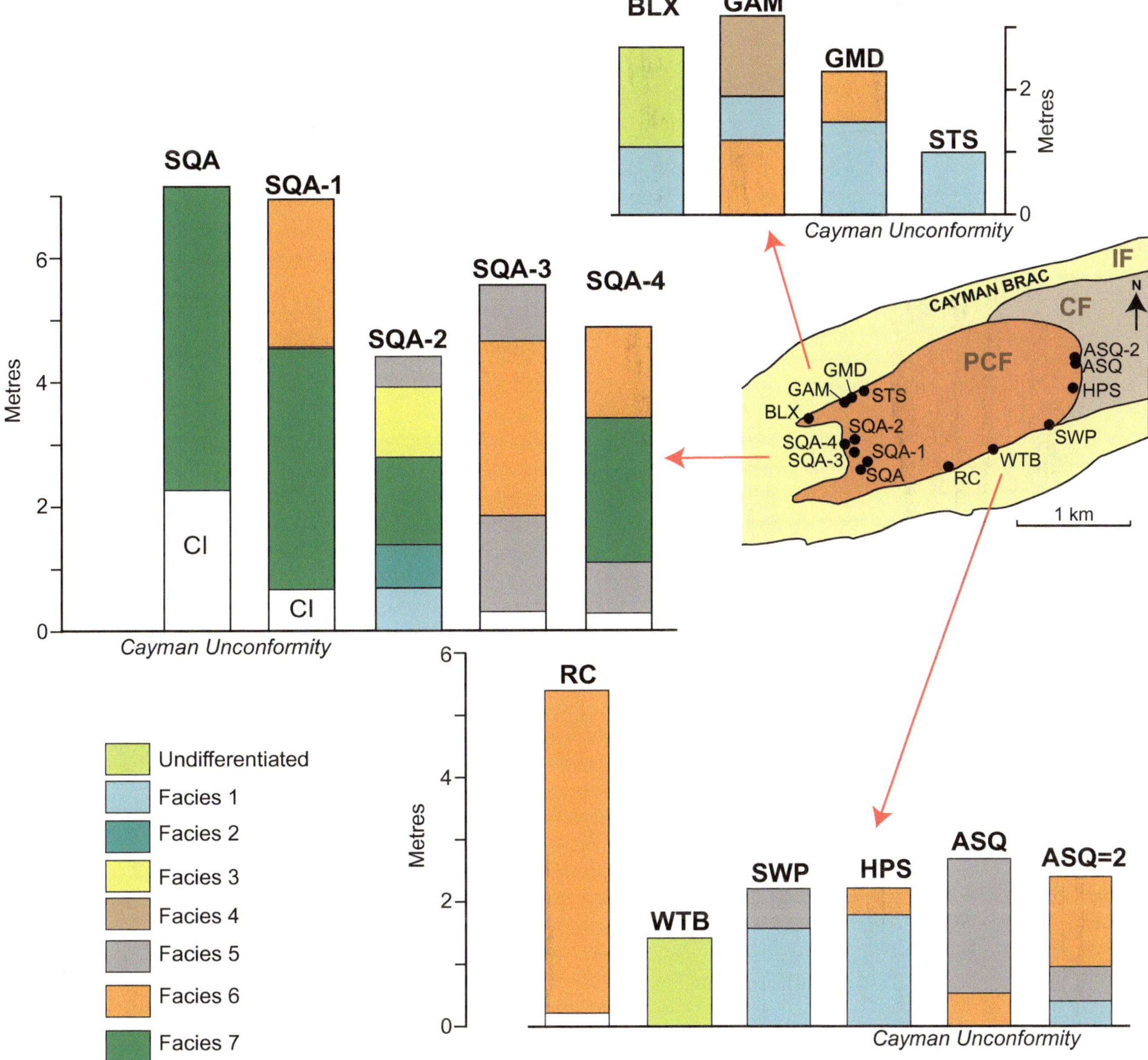

Fig. 5.14 Measured sections through Pedro Castle Formation on the west end of Cayman Brac showing localized nature of constituent facies with frequent vertical and lateral changes in the facies. Figure based on information in MacNeil (2001, his Figs. 2.4, 2.5 and 2.6)

Formation took place. The $^{87}Sr/^{86}Sr$ ratios from the Pedro Castle Formation indicate a minimum age of ~2 Ma.

Today, the highest known outcrop of the Pedro Castle Formation is found just to the east of Pedro Castle Quarry, where it is ~15 m asl. In Pedro Castle Quarry, the unconformity is ~10 m asl and dips at 1–2° to the west. Given that there is no evidence of tectonic uplift in the area, the transgression that flooded the Cayman Unconformity and led to deposition of the sediments that now form the Pedro Castle Formation must have risen to at least 15 m above present-day sea level. Such a rise in sea-level would undoubtedly have flooded the entire island. Dowsett and Cronin (1990) suggested that during the Middle Pliocene (3.5–3.0 Ma) sea-level was probably 35 ± 18 m higher than today, whereas Raymo et al. (2011) noted that estimates for sea level rise during the middle Pliocene range from 10 to 40 m, with a value of 25 m commonly being adopted. Although the specific sea level is open to debate, it is evident that such sea-levels would have led to flooding of Grand Cayman, renewal of the carbonate factory, and deposition of sediments that now form the Pedro Castle Formation.

The $^{87}Sr/^{86}Sr$ ratios obtained from 17 samples from the Pedro Castle Formation on Grand Cayman ranges from 0.709014 to 0.709125 with an average of 0.709065. These

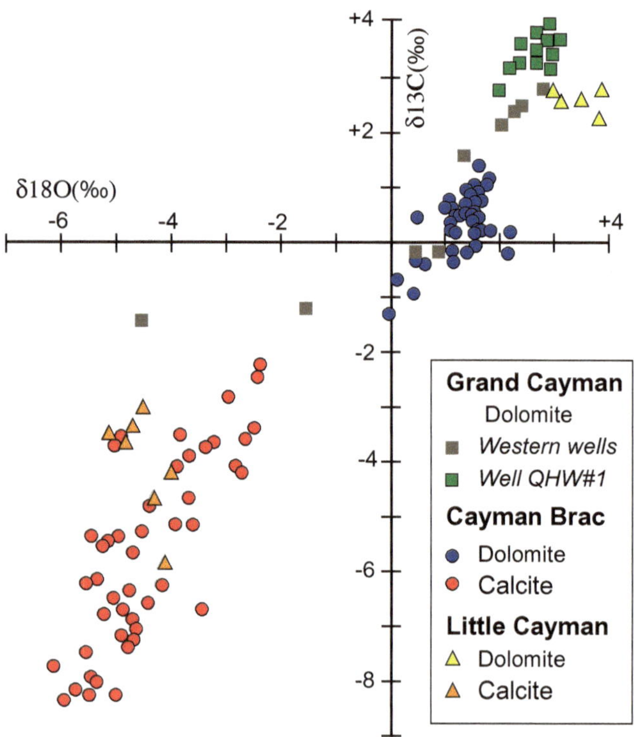

Fig. 5.15 $\delta^{18}O$ versus $\delta^{13}C$ for calcite and dolomite from Pedro Castle Formation on the western part of Cayman Brac (data from MacNeil 2001) and Little Cayman (data from Jones 2019)

numbers are comparable to the $^{87}Sr/^{86}Sr$ ratios that MacNeil and Jones (2003) reported from the Pedro Castle Formation on Cayman Brac, which were 0.709032 to 0.709108 with an average of 0.709067. The distribution and average values obtained from the Pedro Castle Formation on Grand Cayman are similar to those from the underlying Cayman Formation.

In well QHW#1, $^{87}Sr/^{86}Sr$ ratios of the dolostones (N = 11) of the Pedro Castle Formation ranged from 0.709034 to 0.709085 with an average of 0.709056, whereas those from the dolostones in the underlying Cayman Formation (N = 9) ranged from 0.708997 to 0.709066 with an average of 0.709022. In this well, the six lowermost $^{87}Sr/^{86}Sr$ ratios from the Cayman Formation are significantly lower than those in the overlying Pedro Castle Formation.

5.11 Depositional Regimes

Even though corals are a significant part of the biota, no reefs have yet been found in the Pedro Castle Formation. Only scattered thickets of *Stylophora* and *Porites* have been found in some areas. Relative to the Cayman Formation, the Pedro Castle Formation contains more mudstones and wackestones, less grainstone and packstone, fewer rhodolites, and fewer corals (Jones and Hunter 1994a). Based on these comparisons, Jones and Hunter (1994a) suggested that the depositional conditions associated with the development of the sediments that now form the Pedro Castle Formation were quieter.

For the Pedro Castle Formation exposed on the west end of Cayman Brac, MacNeil and Jones (2003) suggested that the original sediments were deposited in normal marine conditions when the water was <20 m deep and characterized by low to moderate energy conditions where wackestones were the dominant sediment. A diverse biota that included algae, foraminifera, molluscs, scattered echinoids, and corals grew in those conditions.

5.12 Conclusions

Analysis of the Pedro Castle Formation, found on the western parts of Grand Cayman and Cayman Brac and scattered outcrops on Little Cayman, have yielded the following conclusions.

- The maximum known thickness of 21.5 m for the Pedro Castle Formation is found in the subsurface of the central part of the western peninsula of Grand Cayman.
- Karstic unconformities define the lower and upper boundaries of the Pedro Castle Formation.
- The formation is formed of limestones, dolostones, and dolomitic limestones. There is no obvious pattern to the dolomitization.
- Although the biota is similar to that in the underlying Cayman Formation, free-living corals are more common.
- The original sediments of the Pedro Castle Formation were deposited on an open bank with water depths less than 25 m deep.
- Porosity and permeability are highly variable throughout the formation. Much of the porosity was created by leacheing of the aragonitic skeletons of the corals, bivalves, and gastropods.
- The original limestones of the Pedro Castle Formation underwent complex diagenetic alteration that involved dolomitization, leaching of aragonitic components, precipitation of calcite and/or dolomite cements, and deposition of internal sediments. Karst development led to the formation of caves and solution-widened joints/fractures.
- Dolomitization of the limestones developed under near-surface conditions from normal seawater or slightly evaporated seawater.

References

Dowsett HJ, Cronin TM (1990) High eustatic sea level during the middle Pliocene: evidence from the southeastern U.S. Atlantic Coast Plain. Geol 18:435–438

Etherington SS (2004). Sedimentology and depositional architecture of the Cayman (Miocene) and Pedro Castle (Pliocene) Formations on western Grand Cayman, British West Indies. MSc thesis, University of Alberta, Edmonton

Farrell JW, Clemens SC, Gromet LP (1995) Improved chronostratigraphic reference curve of late Neogene seawater $^{87}Sr/^{86}Sr$. Geology 23:403–406

Frost SH (1977) Cenozoic reef systems of Caribbean—prospects for paleoecologic synthesis. Am Assoc Pet Geol Stud Geol 4:93–109

Hunter IG (1994) Modern and ancient coral associations of the Cayman Islands. PhD thesis, University of Alberta, Edmonton

Jones B (1994) Geology of the Cayman Islands. In: Brunt MA, Davies JE (eds) The Cayman Islands: natural history and biogeography. Kluwer, Dordrecht, The Netherlands, pp 13–49

Jones B (2019) Diagenetic processes associated with unconformities in carbonate successions on isolated oceanic islands: case study of the Pliocene to Pleistocene sequence, Little Cayman, British West Indies. Sediment Geol 386:9–30

Jones B, Hunter IG (1989) The Oligocene–Miocene Bluff Formation on Grand Cayman. Caribb J Sci 25:71–85

Jones B, Hunter IG (1994a) Evolution of an isolated carbonate bank during Oligocene, Miocene, and Pliocene times, Cayman Brac, British West Indies. Facies 30:25–50

Jones B, Hunter IG (1994b) Messinian (Late Miocene) karst on Grand Cayman, British West Indies: an example of an erosional sequence boundary. J Sediment Res 64:531–541

Jones B, Hunter IG, Kyser TK (1994a) Revised stratigraphic nomenclature for Tertiary strata of the Cayman Islands, British West Indies. Caribb J Sci 30:53–68

Jones B, Hunter IG, Kyser TK (1994b) Stratigraphy of the Bluff Formation (Miocene–Pliocene) and the newly defined Brac Formation (Oligocene), Cayman Brac, British West Indies. Caribb J Sci 30:30–51

MacNeil AJ (2001) Sedimentology, diagenesis, and dolomitization of the Pedro Castle Formation on Cayman Brac, British West Indies. MSc thesis, Alberta, Edmonton

MacNeil AJ, Jones B (2003) Dolomitization of the Pedro Castle Formation (Pliocene), Cayman Brac, British West Indies. Sediment Geol 162:219–238

McArthur JM, Howarth RJ, Shields GA (2012) Strontium isotope stratigraphy. In: Gradstein FM, Ogg JG, Schmitz M, Ogg G (eds) The Geologic Time Scale. Elsevier, Amsterdam, pp 127–143

Raymo ME, Mitrovica JX, O'Leary MJ, DeConto RM, Hearty PJ (2011) Departures from eustasy in Pliocene sea-level records. Nat Geosci 4:328–332

Rose PR, Lidz B (1977) Diagnostic foraminiferal assemblages of shallow-water modern environments: South Florida and the Bahamas. Sedimenta VI. University of Miami, p 55

van Woesik R, Franklin EC, O'Leary J, McLanahan TR, Klaus JS, Budd AF (2012) Hosts of the Plio-Pleistocene past reflect modern-day coral vulnerability. Proc R Soc B 279:2448–2456

Wignall BD (1995) Sedimentology and diagenesis of the Cayman (Miocene) and Pedro Castle (Pliocene) Formations at Safe Haven, Grand Cayman, British West Indies. MSc thesis, University of Alberta, Edmonton

6 Diagenesis–Bluff Group

Abstract

The carbonate successions in the Brac Formation, Cayman Formation, and Pedro Castle Formation are characterized by complex diagenetic fabrics that have involved pervasive leaching of the original limestones, dolomitization, repeated episodes of limpid dolomite and calcite cement precipitation, and deposition of internal sediments. Karst development that occurred during the times between active carbonate deposition led to the development of cavities and caves at all scales, and large-scale surface weathering that sculpted the complex, high relief weathering profiles. The dolostones are formed of anhedral to euhedral dolomite crystals that are up to 1.5 mm long in the Brac Formation but typically less than 50 μm long in the Cayman Formation and Pedro Castle Formation. Many of these crystals have a "dirty" core formed of high-calcian dolomite that is encased by a clean outer layer that is formed of low-calcian dolomite. Hollow dolomite crystals developed locally, with some of them being subsequently filled with calcite or dolomite cements. The relative percentages of low-calcian and high-calcian dolomites vary stratigraphically and geographically. The Cayman Formation on Cayman Brac and the western part of Grand Cayman are formed entirely of dolostone. In contrast, on the eastern part of Grand Cayman, the formation is formed entirely of dolostone in the coastal areas but of limestones and limestones capped by dolostones in the central part of the island. Although available evidence from the petrography and geochemistry of these "island dolostones" indicates that dolomitization was probably mediated by normal or slightly evaporate seawater at low temperatures (~ 30 °C), the exact conditions that triggered the pervasive replacement of the original limestones is still open to debate.

6.1 Introduction

Limestones and dolostones in the Bluff Group (Brac Formation, Cayman Formation, Pedro Castle Formation) are characterized by complex diagenetic histories that reflect multiple cycles of diagenesis that collectively reflect the ever-changing surface and subsurface environmental conditions that have affected the succession ever since the sediments in each formation were first deposited. The fact that diagenetic processes have been variable through time and space is amply demonstrated by the geographic and stratigraphic contrasts that are evident throughout the carbonate successions in the Bluff Group on each of the Cayman Islands. Pervasive dolomitization, for example, is common throughout much of the Cayman Formation but variable throughout the Brac Formation and Pedro Castle Formation. Similarly, it is difficult to find consistent geographic and/or stratigraphic patterns in other diagenetic aspects, including for example, the development of calcite cements, leaching of aragonite skeletons, and development of cavity-filling precipitates and cements.

The "dolomite problem", which refers to the processes by which limestones are transformed to dolostone, has existed ever since the term "dolomite" was first coined by de Saussure in 1792 (de Saussure 1792), following the description of the mineral/rock found in the Dolomite Mountains of northern Italy by Déodat de Dolomieu in 1791 (de Dolomieu 1791). The fact that dolomite has not been produced in the laboratory under the low temperature and pressure conditions that exist in modern sedimentary environments also poses problems for any model proposed for its formation. "Island dolostones"—a term applied to Cenozoic dolostone successions found on isolated islands (Budd 1997; Ren and Jones 2018) throughout the oceans of the world

B. Jones, *Geology of the Cayman Islands*,
https://doi.org/10.1007/978-3-031-08230-6_6

have attracted attention in this respect because, as noted by Budd (1997), (1) the hydrologic systems and fluid chemistry can be reasonably inferred, (2) the temperature range is constrained to modern surface or near-surface values, (3) the dolostones are young and can be dated with reasonable accuracy, and (4) the successions have not experienced deep burial. As such, this avoids many of the problems associated with the interpretation of older dolostones that have experienced prolonged diagenetic histories.

The dolostone successions in the Bluff Group of the Cayman Islands are similar to those found elsewhere in the Caribbean region (e.g., Barbados, Bonaire, Curacao, Puerto Rico), the Pacific Ocean (e.g., Mururoa Atoll, Niue Island, Funafuti Atoll) (Budd 1997, his Table 2), the Xisha Islands in the South China Sea (Wang et al. 2018, 2019, 2021), and various islands throughout southeast Asia (Epting 1980; Carnell and Wilson 2004). The Cayman dolostones can be characterized using parameters derived from individual dolomite crystals and whole rock parameters that are, in effect, averages of all the crystals in the sample. For ease of reference, herein, the term "dolomite" refers to the mineral whereas the term "dolostone" refers to the rock that is formed of dolomite.

6.2 Dolostones of the Bluff Group

Dolostones in the Brac Formation, Cayman Formation, and Pedro Castle Formation are characterized by complex fabrics that offer mute testimony to the many episodes of diagenesis that they have experienced (Fig. 6.1). Dolostones in the Brac Formation on Cayman Brac are formed largely of sucrosic dolomite with euhedral crystals, up to 1.5 mm long (Fig. 6.2), that typically have a "dirty" core encased by a clear, outer rim (Fig. 6.2a, c). Ghost structures of red algae and foraminifera are commonly evident in these dolomite crystals (Fig. 6.2c, d).

Dolostones in the Cayman Formation and Pedro Castle Formation are typically finely crystalline with crystals rarely being more than 0.5 mm long (Figs. 6.3 and 6.4a). Fossil mouldic porosity created by the dissolution of the aragonitic skeletons of corals, bivalves, and gastropods and ghost structures of foraminifera, algae, and other organisms offer testimony to the fossiliferous nature of the original marine limestones (Fig. 6.3a, b). Cavities of various sizes and shapes are commonly lined by limpid dolomite cement (Figs. 6.3c–e, g and 6.4c) with many of the larger cavities being later filled with internal sediments (Fig. 6.3c) and/or calcite cement (Fig. 6.3f, g).

6.2.1 Dolomite Crystals

The individual dolomite crystals, which are the building blocks of the Cayman dolostones, can be characterized by their size, internal crystallographic structures, and compositional zoning.

Crystal Size

Dolostones from the Cayman Formation are so finely crystalline that individual crystals are impossible to accurately measure from hand samples, thin sections, or fractured surfaces viewed on a scanning electron microscope (SEM). Such crystals can, however, be measured from SEM images of cut and polished surfaces that have been etched in concentrated HCl because the acid preferentially attacks the boundary areas between the crystals and thereby highlights the crystal outlines (Fig. 6.4a). Some caution has to be used with this method because there is no guarantee that the flat surface cut through randomly oriented crystals will show the maximum length of each crystal. Nevertheless, accurate digital measurement of 50 crystals from each image, like that shown in Fig. 6.4a, does provide a good estimate of the minimum, maximum, and average crystal lengths. This is amply demonstrated by data from 15 samples from a depth range of 0 to 48 m in the Cayman Formation in well HRQ#2 (Figs. 6.4a, and 6.5a), which is located near the southwest corner of High Rock Quarry in the central eastern part of Grand Cayman. Measurement of 200 crystals from four (50 from each sample) of those samples (Fig. 6.5b–e) shows that (1) all of the crystals are < 50 μm long, and (2) the crystals from the upper part of the succession are larger than those from the lower part of the succession. Similar crystal sizes have been obtained from other samples from the Cayman Formation in other wells on Grand Cayman and Cayman Brac.

Internal Structures and Compositional Zoning in Dolomite of the Cayman Islands

Dolomite crystals are commonly characterized by complex internal structures (Fig. 6.6) that can be described using the terminology of Jones (2013, his Fig. 3). Etching of cut and polished surfaces, like those used to measure the crystal sizes, clearly highlight the complex internal structures of these dolomite crystals (Jones 2005, his Figs. 6 to 9). The etch patterns, which reflect solubility differences on a micron scale, highlight many different features, including (1) the core and cortex (Fig. 6.7a–c), (2) twin planes (Fig. 6.7b) (3) growth zones in the core and cortex that are defined by compositional differences (Fig. 6.7d, e), (4) dissolution slots

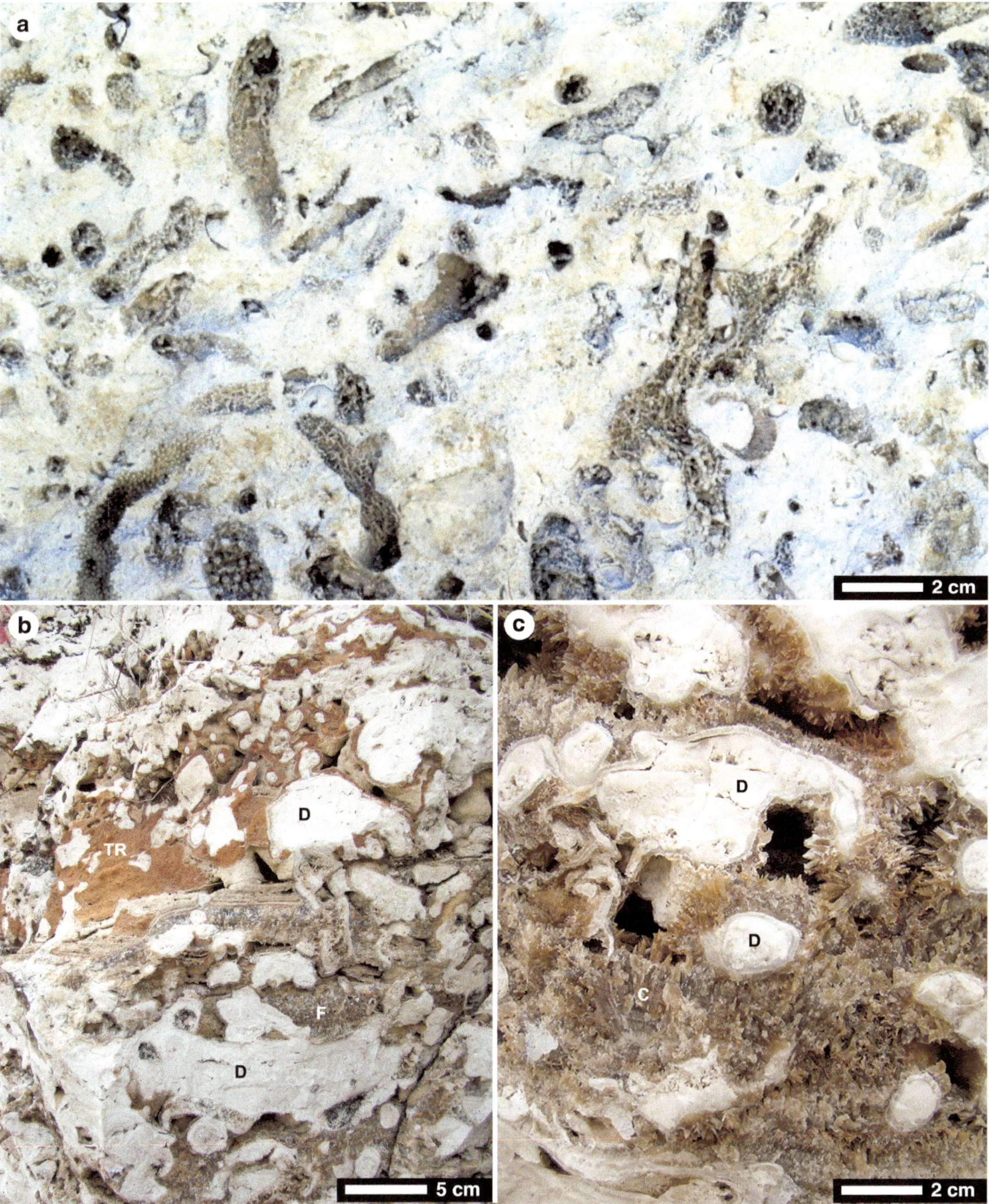

Fig. 6.1 Outcrop photographs of dolostones of the Cayman Formation. **a** White, finely crystalline dolostones with high porosity due to leaching of the branching coral *Stylophora.* High Rock Quarry, east end of Grand Cayman. **b** Cavities in white dolostones (D) of Cayman Formation filled with terra rossa (TR) and brown, crystalline calcite flowstone (F). **c** Close up view of crystalline calcite (C) flowstone filling cavities in dolostone (D), Cayman Formation C and D – Anne Tatum Drive, Cayman Brac

(Fig. 6.7d), (5) growth sector boundaries (Fig. 6.7E), and/or (6) the cortical boundary (Fig. 6.7f).

The fact that the internal features of the dolomite crystals are largely a reflection of variations in the mol % $CaCO_3$ content (hereafter referred to as %Ca) is difficult to verify through the use of Energy Dispersive X-ray analysis (EDX) on the SEM because of analytical issues associated with the micro-scaled analyses that are required. Electron microprobe (EMP) analyses of polished but not etched surfaces, however, allows mapping of the %Ca content through

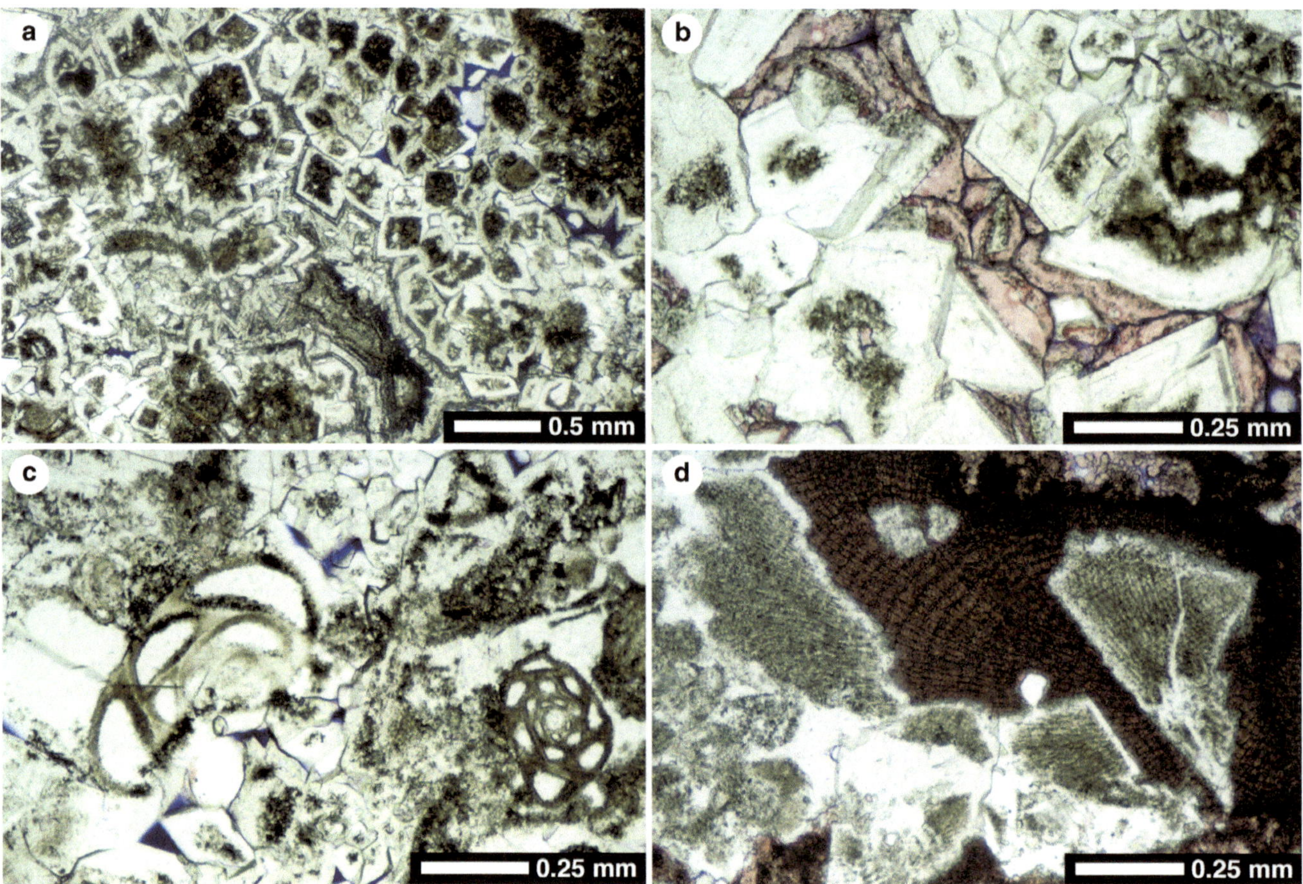

Fig. 6.2 Thin Section photomicrographs of dolostones from upper part of Brac Formation, west end of Cayman Brac. All with plane polarized light; red = calcite stained with Alizarin Red S solution. **a** General view of sucrosic dolostone showing euhedral dolomite crystals with each crystal having a dirty core encased by a clear rim. **b** Enlarged view of dolomite crystals with dirty cores and clear rims. Note calcite cement (red) filling pore. **c** Sucrosic dolomite with ghost structures of foraminifera. **d** Red algae partly replaced by dolomite. Note ghost structures of red algae preserved in the replacive dolomite

the use of backscatter imaging (which reflects differences in the average atomic weights) and spot analyses that provide accurate measurements of the %Ca on a microscale. Such analyses also show that the dolomite is characterized by low concentrations (<250 ppm total) of other elements (e.g., Fe, Na, Mn) that could potentially affect the backscatter images.

Backscatter images of dolostones from the Cayman Formation in well DHB#1 (located on western peninsula of Grand Cayman, just north of George Town) clearly show dolomite crystals that are zoned with respect to their %Ca content (Figs. 6.8 and 6.9). Such images show euhedral (Fig. 6.8a) or irregular (Fig. 6.8c, d) cores formed of HCD that are encased by a cortex that is formed of a succession of zones of LCD that are of variable widths (Fig. 6.8). Calcite inclusions are common in some of the HCD cores (Fig. 6.8a, b). In some crystals, the cortical zones are characterized by very fine scale zones within the thicker zones (Fig. 6.8b). Compositional transects across these zoned crystals (Fig. 6.9) clearly show the variations in the %Ca that match the zones highlighted by backscatter mapping.

In the dolostones of the Cayman Formation, many of the dolomite crystals are also characterized by complex syntaxial overgrowths (Jones, 1989). Such overgrowths involve various combinations of aggregate trigonal crystals, needle, wall and sheet overgrowths, and needle crystals. The aggregate trigonal crystals commonly include four orders of subcrystals. Jones (1989) suggested that most of these overgrowths developed from rapid evaporation of mixed freshwater and seawater while the crystals were exposed in near-surface cavities.

Hollow Dolomite Crystals

Hollow dolomite crystals, formed by the preferentially leaching of their HCD cores, are common in some dolostones from the Cayman Formation (Fig. 6.10). Analyses of these hollow crystals shows that they are formed of LCD and comparison with complete crystals, like those shown in Fig. 6.8, suggests that they formed through the preferential leaching of the HCD that formed their cores (Fig. 6.11). Further diagenetic modifications may result in the hollow

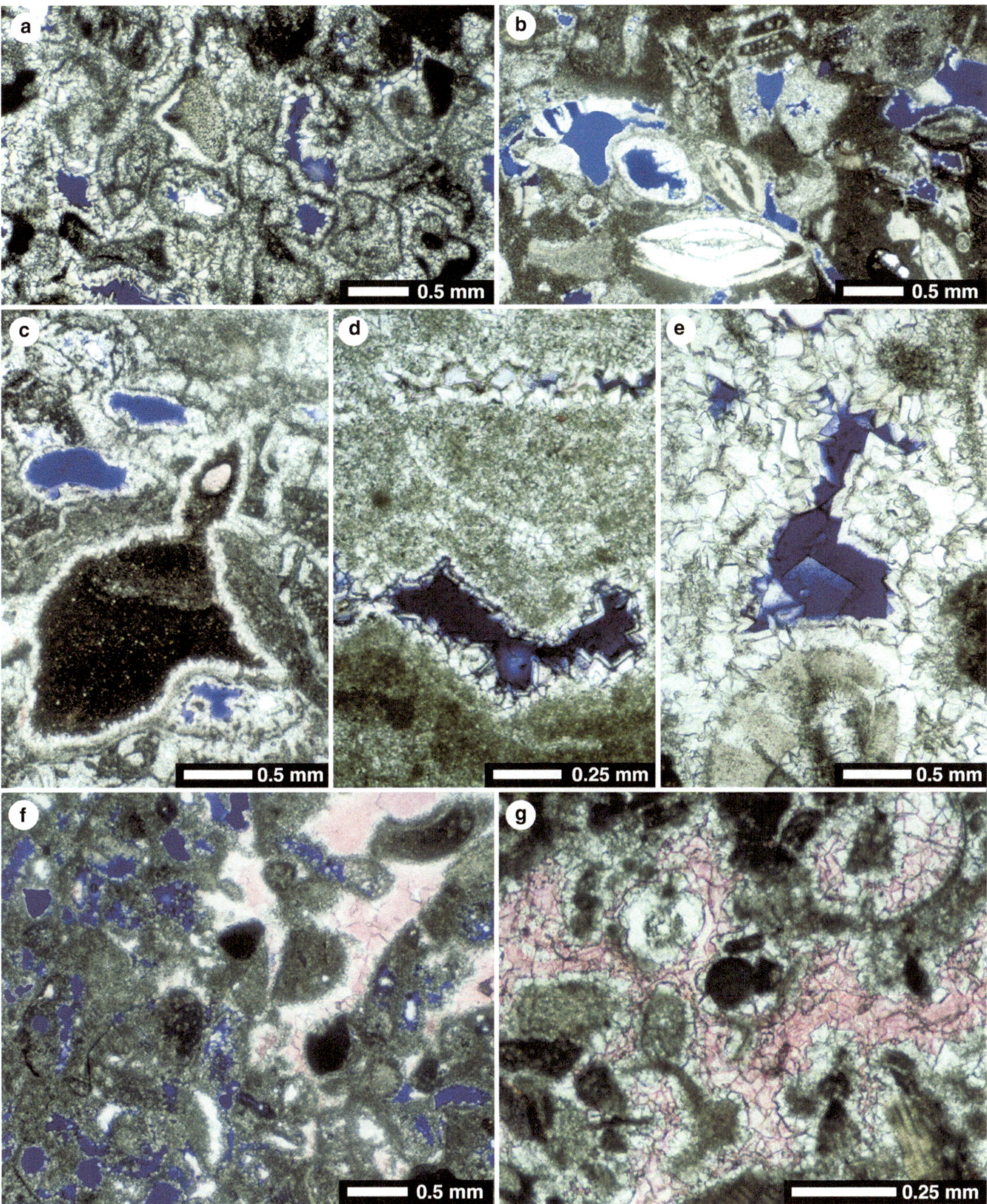

Fig. 6.3 Thin section photomicrographs showing variety of diagenetic fabrics in dolostones from upper part of Cayman Formation, west end of Cayman Brac. All with plane polarized light; red = calcite stained with Alizarin Red S solution. **a** Porous dolomitized grainstone with individual grains highlighted by outer micrite envelopes and cavities lined with limpid dolomite cement. **b** Porous foraminifera-rich dolostone. **c** Dolostone with cavities lined by limpid dolomite cement; larger cavity filled with dolomitized mudstone and calcite cement. **d, e** Porous dolostone with cavities lined with limpid dolomite cement. Note zoning in some of the larger limpid dolomite crystals shown in panel E. **f** Porous dolomitized grainstone with some pores occlude by calcite cement. **g** Dolostone with pores occluded by limpid dolomite and calcite cements

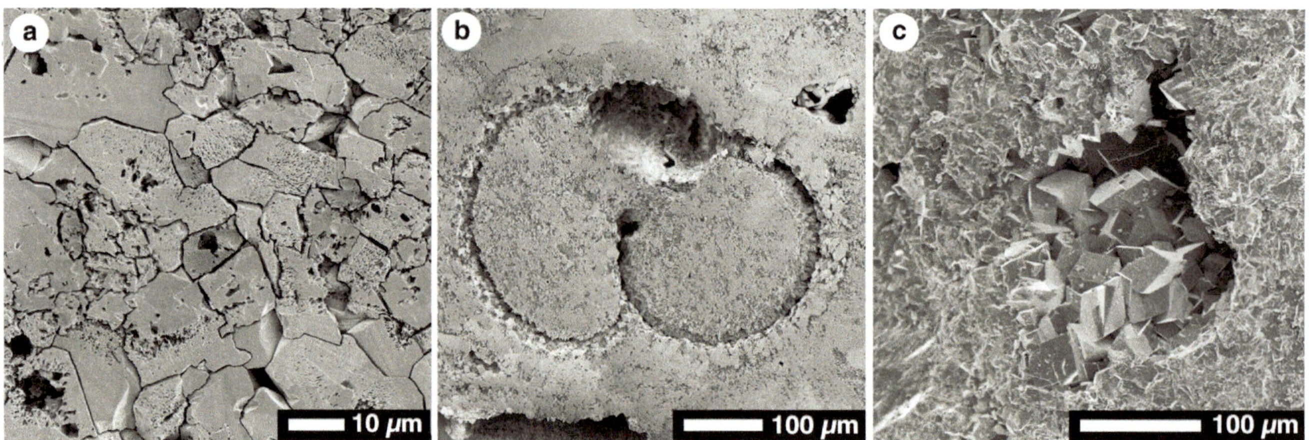

Fig. 6.4 SEM photomicrographs of dolostones from Cayman Formation, Grand Cayman. All depths are below ground level. Samples shown in panels a and b were cut, polished, and etched with HCl for 10 secs, and coated with carbon. Well BAC#1, 9.1 m. **a** General view of finely crystalline dolostone, showing crystals of various sizes and shapes. Well BAC#1, 9.1 m. **b** Example of poorly preserved foraminifera in dolostone. **c** General view of dolostone showing finely crystalline groundmass with small cavity lined with euhedral limpid dolomite crystals. Well CRQ#2, 19.5 m

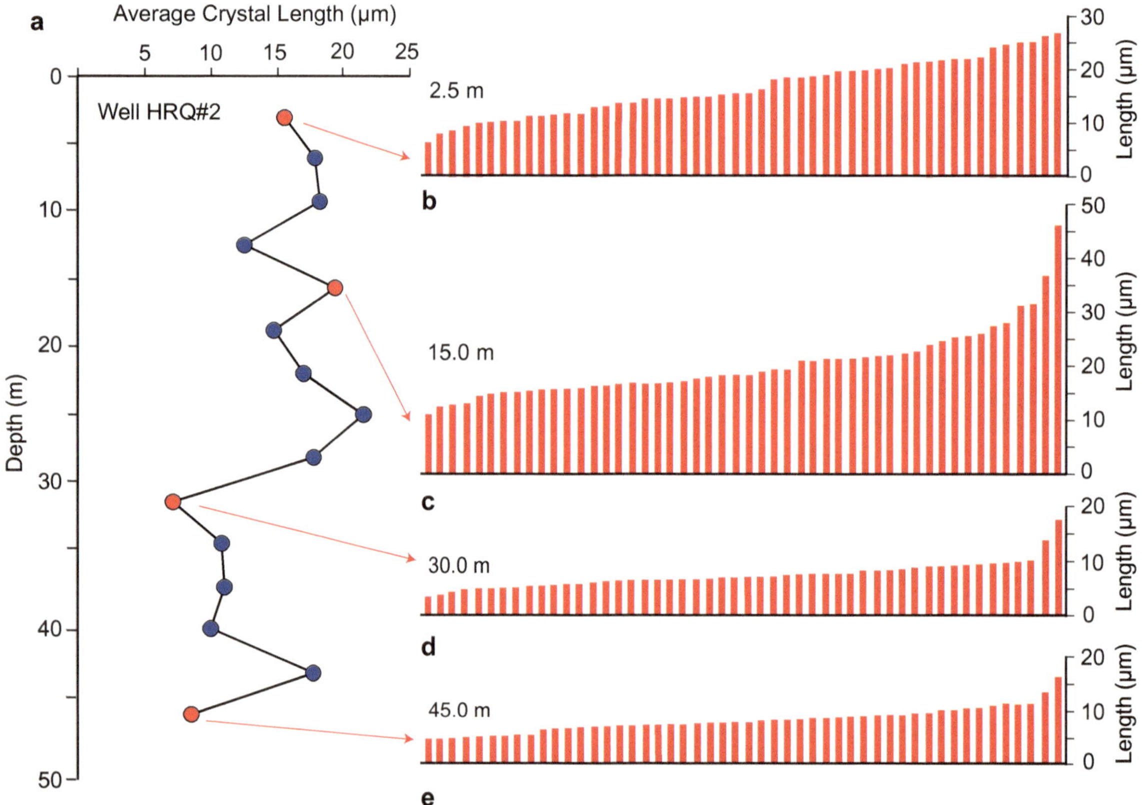

Fig. 6.5 Dolomite crystal sizes in Cayman Formation, well HRQ#2, east end of Grand Cayman. For each sample, maximum lengths of 50 different crystals were measured from SEM images of polished dolostone sample that had been etched with HCl for 10 s. **a** Comparison of average maximum crystal lengths from various depths in well HRQ#2. **b** to **e** Comparison of maximum crystal lengths (50 crystals per sample), for samples from 2.5 m, 15.0 m, 30.0 m, and 40.0 m below ground level. For each sample, maximum lengths are ordered from shortest (left) to longest (right)

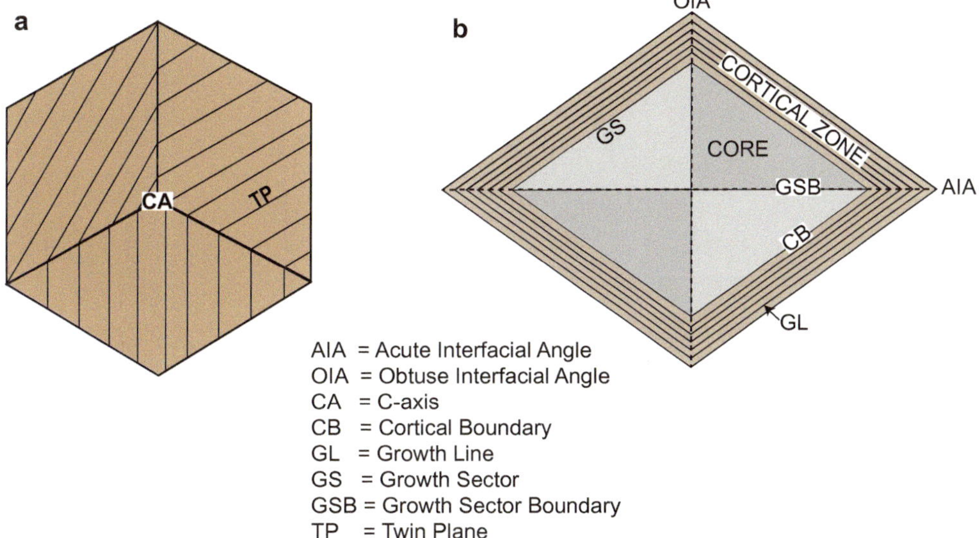

Fig. 6.6 Schematic diagrams showing main structural elements of a dolomite crystal. **a** View down c-axis of dolomite crystal showing traces of twin planes on crystal faces. Modified from Rogers (1929, his Fig. 5). **b** Terminology for internal structures of a dolomite crystal. Modified from Jones (2013, his Fig. 3)

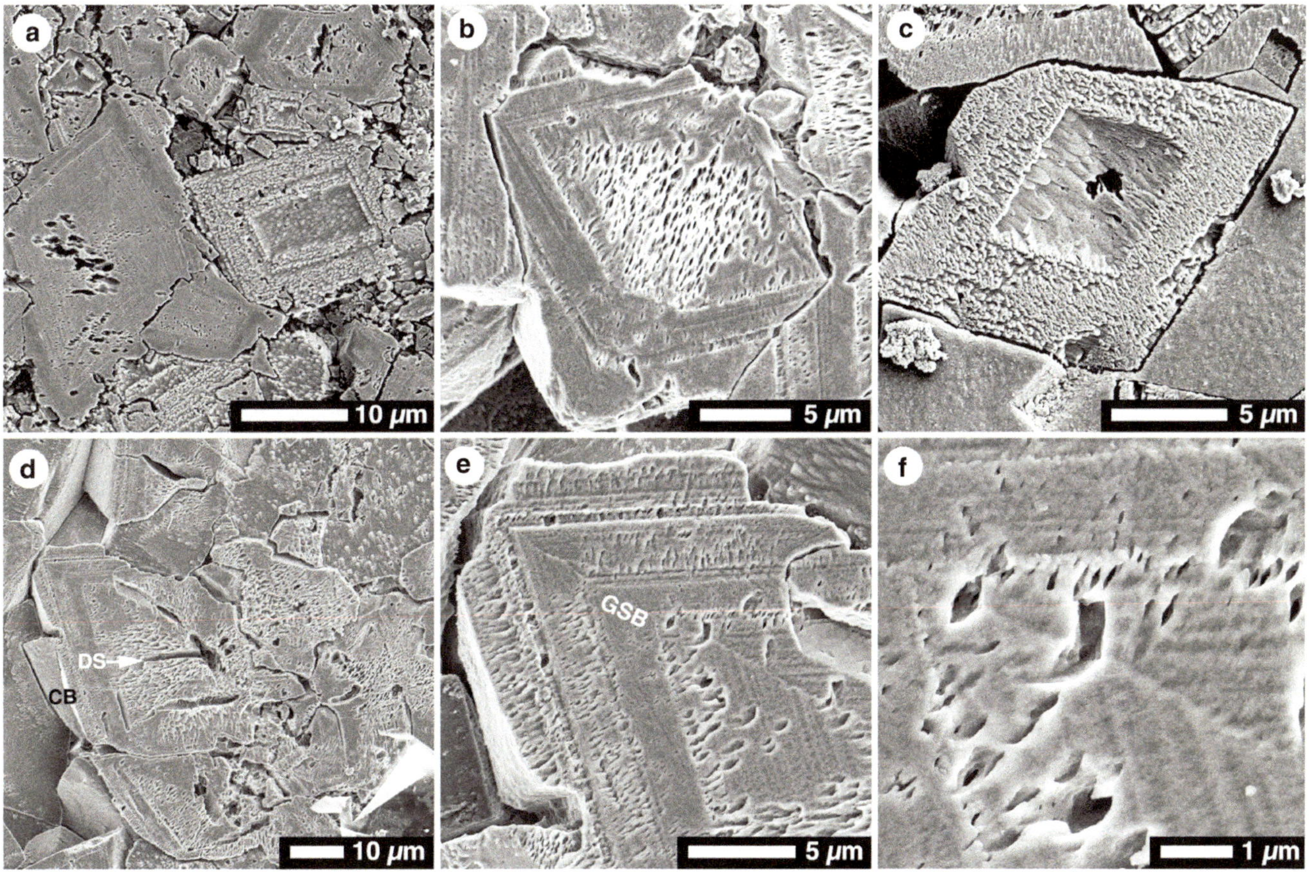

Fig. 6.7 SEM images of etched dolomite crystals from Cayman Formation. **a** Dolostone formed of variously zoned dolomite crystals. Well HRQ#2, 26.2 m. **b** Zoned dolomite crystal with porous core (pores possibly caused by dissolution of calcite inclusions) encased by zoned rim of inclusion-free dolomite. Well HRQ#2, 26.2 m. **c** Dolomite crystal with core encased by less-soluble dolomite. Well HRQ#2, 26.2 m. **d** Zoned dolomite crystal with porous core separated from non-porous outer cortical zone by well-defined cortical boundary (CB). Note large, elongate pores (DS) in core that are probably dislocation slots. Well HRQ#2, 3.6 m. **e** Enlarged view of corner of dolomite crystal shown in panel G. Note numerous zones of various thicknesses and alternation of porous and non-porous dolomite zones. Pores probably formed by leaching of calcium-rich inclusions. Well HRQ#2, 3.6 m. **f** Enlarged view of apex of dolomite crystal shown in panel H

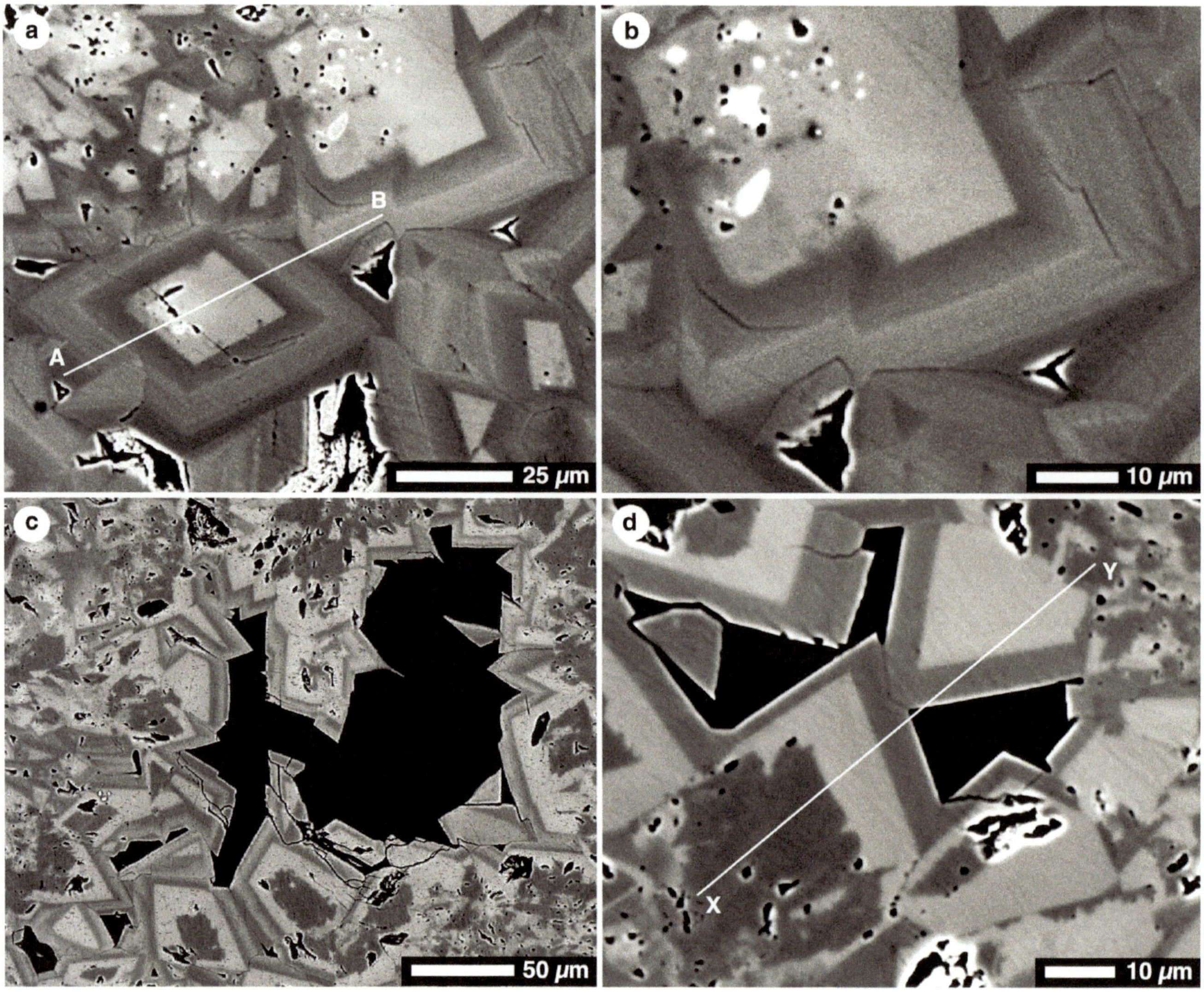

Fig. 6.8 Backscatter electron (BSE) images of dolomite crystals from dolostones of the Cayman Formation. All images generated from polished thin sections on a JEOL 8900R electron microprobe. Light grey = low-calcian dolomite; dark grey = high calcian dolomite. All samples from well DHB#1, southern part of western peninsula of Grand Cayman. **a** Zoned dolomite crystals with zones defined by variable Ca content: core has lower Ca content than outer layers. See Fig. 6.9a for %Ca content along transect A to B. Sample from depth of 24.4 m. Modified from Jones (2007, his Fig. 3a). **b** Enlarged view of crystal from upper right corner of panel A, showing high-Ca dolomite around low-Ca dolomite core that contains calcite inclusions (white). **c** Pore in dolostone (from depth of 36.6 m) lined with zoned dolomite crystals. **d** Zoned dolomite crystals lining pore in dolostone from depth of 36.6 m. See Fig. 6.9b for %Ca content along transect X to Y. Modified from Jones (2007, his Fig. 3F)

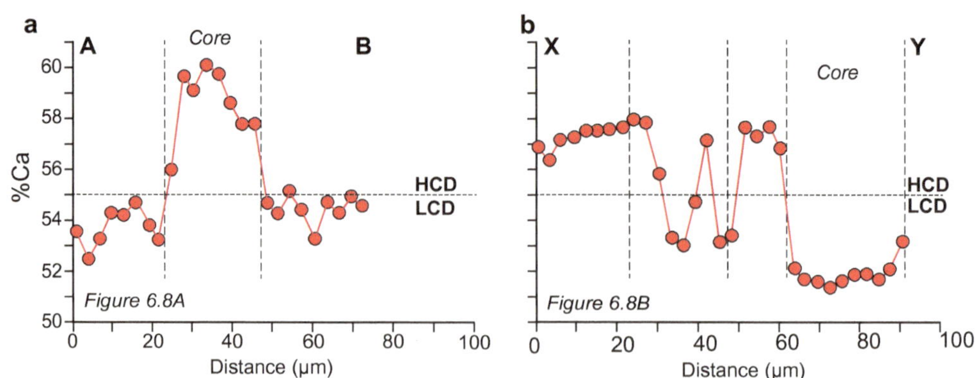

Fig. 6.9 Percent Ca (%Ca) along transect **a** A–B (Fig. 6.8a), and **b** transect X–Y (Fig. 6.8c). Modified from Jones (2007), his Figs. 4b and f, respectively

Fig. 6.10 SEM photomicrographs of dolomite crystals from Cayman Formation, Grand Cayman. Fracture sample. Depths in well are below ground level. **a** Euhedral limpid dolomite crystals lining cavity wall. Well ST#5, 46.0 m. **b** Hollow dolomite crystal. Well ST#5, 46.0 m. **c** Hollow dolomite crystal. Well BAC#1, 8.5 m

crystals being filled with calcite, LCD, or HCD cement (Fig. 6.11). Given that the resultant crystal will have a core that is younger than its outer rim, Jones (2007) used the term "inside-out crystals" to reflect the fact that their zonation is the reverse of normal crystal growth. Hollow dolomite crystals that are later filled with calcite cement are important because they resemble "dedolomite" that has generally been attributed to the dissolution of dolomite and concomitant precipitation of calcite (e.g., Budai et al. 1984; Jones et al. 1989). According to that model, there is no cavity phase involved in the process. Jones (2007) also pointed out that hollow dolomite crystals could also be filled with either LCD or HCD and thereby produce dolomite crystals with cores that are younger than their rims (Fig. 6.11).

6.2.2 Low and High Calcian Dolomite

Electron microprobe analyses and scanning electron microscopy clearly show that the dolostones in the Bluff Group are formed of mixtures of LCD and HCD (Fig. 6.8). Jones et al. (2001) demonstrated that X-Ray diffraction analyses also highlighted the two types of dolomite and that such analyses allowed calculation of the %Ca in each dolomite phase. These analyses involved (1) spiking the sample with quartz so that there is standard against which the positions of the dolomite peaks can be measured, (2) a scan over a small range of 2θ that includes the quartz d_{101} peak and the dolomite d_{104} peak, and (3) application of peak-fitting techniques to the X-ray diffractogram that are capable of

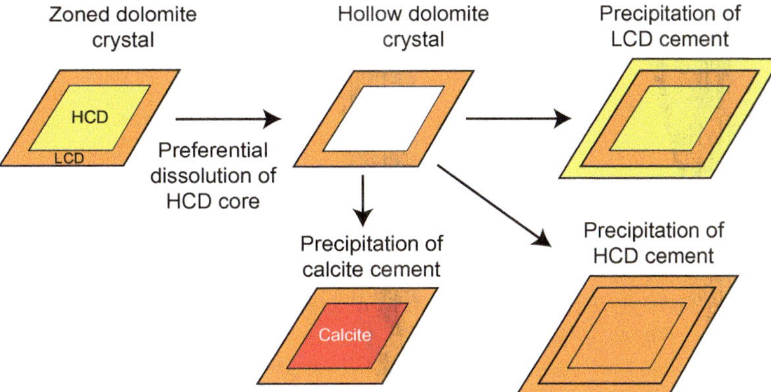

Fig. 6.11 Schematic diagram showing potential evolutionary pathways from a zoned dolomite crystal with a core of HCD encased by a cortex formed of LCD to a hollow crystal that may then be later filled with calcite, LCD, or HCD. Modified from Jones (2007, his Fig. 9)

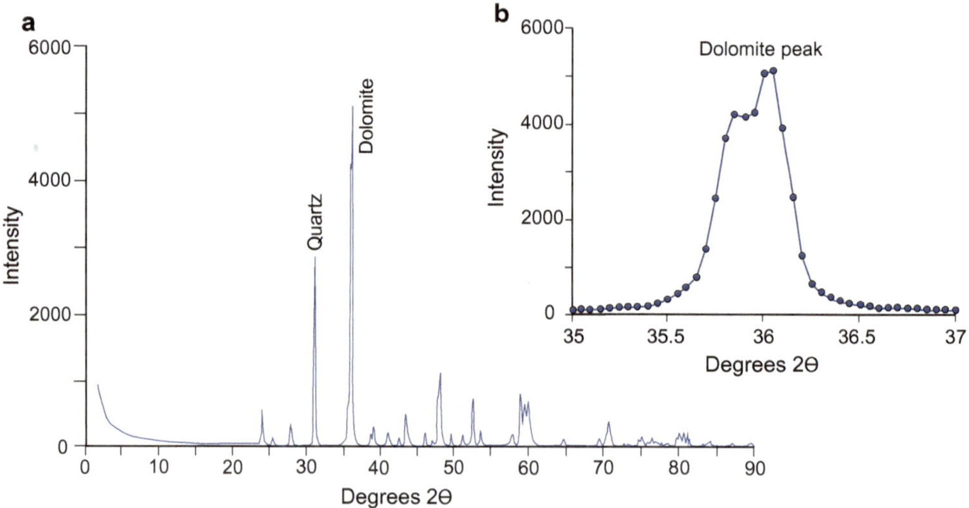

Fig. 6.12 X-ray diffraction pattern for dolostone from the Cayman Formation. Quartz added to sample for reference purposes. **a** General X-ray diffraction pattern for dolostone sample, showing location of quartz peak and main dolomite peak. **b** Enlarged view of dolomite peak showing presence of two peaks that indicate dolomite with different percentages of calcium. Modified from Jones et al. (2001, their Fig. 4)

resolving composite peaks (Jones et al. 2001). Given that the application of peak-fitting techniques requires closely-spaced points on the diffractogram, an XRD sampling interval of 0.004°2θ/step over the range of 35 to 37°2θ was used for these analyses. Jones et al. (2001) showed that a single summed Gaussian–Lorentzian peak provides a good fit for resolving the composite peaks.

Peak-fitting analysis of the XRD diffractograms produces a synthetic diffractogram that allows calculation of the centroid position of each peak relative to the quartz peak (Fig. 6.12b). Those values are then used in the Lumsden Equation (Lumsden, 1979) to calculate the %Ca of each dolomite population. The relative abundance of each dolomite population is calculated from the areas beneath each peak (Jones et al. 2001).

XRD analyses of numerous dolostone samples from the Cayman Formation shows that most are formed of two dolomite populations that are defined by differences in the %Ca (Fig. 6.12b). Samples that yield a single XRD peak are rare. A histogram based on the %Ca derived from 525 samples from 8 different wells on Grand Cayman is characterized by two distinct modes, one between 52 and 53 % Ca and the other between 57 and 58 %Ca (Fig. 6.13). In comparison, values between 54 and 56 %Ca are rare. Given this distribution of values, Jones et al. (2001) and Jones and Luth (2002) divided the dolostones into low-Ca dolomite (LCD < 55 mol % Ca) and high-Ca dolomite (HCD > 55 mol %Ca).

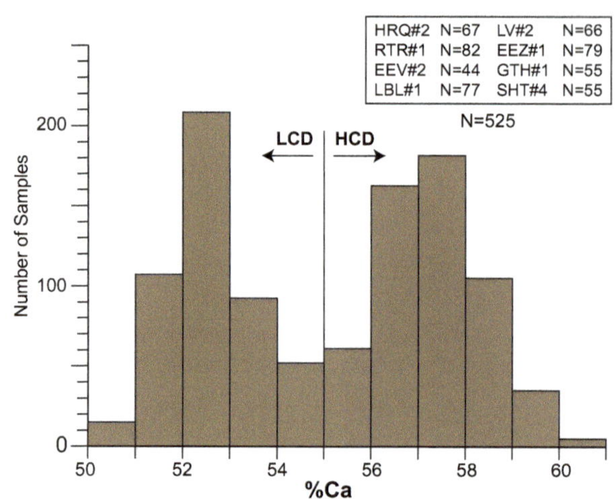

Fig. 6.13 Histogram of %$CaCO_3$ values for dolostones from the Cayman Formation, Grand Cayman, as calculated from X-ray diffraction analyses of samples using method outlined in Jones et al. (2001). Based on 515 samples, with most samples containing both low-Ca dolomite (LCD) and high-Ca dolomite (HCD). Division between LCD and HCD placed at 55 %Ca—as proposed by Jones et al. (2001) and Jones and Luth (2002)

6.2.3 Distribution of Low and High Calcian Dolomite

Cayman Brac

Dolostones in the Brac Formation are formed of non-stoichiometric dolomite that contains 50 to 57.5 %Ca with an average of 56.7 %Ca (N = 68) (Zhao and Jones 2012; Zhao 2013). Backscatter images of the sucrosic dolostones show that these crystals and cements are formed of dolomite with relatively uniform %Ca (Zhao and Jones 2012, their Fig. 9).

On Cayman Brac, section CRQ#1 includes the lower boundary of the Cayman Formation, whereas section SQW#1, located 7.9 km to the southwest, includes the upper boundary of the formation (Fig. 6.14). The lack of any marker beds in the dolostones of the Cayman Formation makes it impossible to precisely correlate the upper part of section CRQ#1 with the lower part of section SQW#1. It is estimated, however, that the stratigraphic gap between the two sections is probably <15 m. The Cayman Formation throughout well CRQ#1 contains more LCD than HCD (Fig. 6.14). Minor amounts of calcite are found only in the uppermost part of the succession. At that locality, the %Ca content of the LCD ranges from 50.5 to 53.4 %Ca with an average of 51.9 %Ca (N = 35), whereas the %Ca content of the HCD ranges from 55.1 to 58.5 %Ca with an average of 56.8 %Ca (Fig. 6.14b). Overall, the succession is dominated by LCD with HCD being absent from many parts (Fig. 6.14b). Although the average %Ca of the dolomite remains relatively constant with depth, it is higher in the lowermost part of the Cayman Formation (Fig. 6.14b).

For the sequence in well SWQ#1 and exposures in the nearby quarry wall, which includes the boundary with the overlying Pedro Castle Formation (~3 m asl), the ratio between LCD and HCD remains relatively constant throughout (Fig. 6.14a). In that well, the %Ca content of the LCD ranges from 51.1 to 54.2 %Ca with an average of 52.6 %Ca, whereas the %Ca content in the HCD ranges from 55.5 to 59.7 with an average of 57.6 %Ca (Fig. 6.14a). In well BW#1, drilled 765 m to the SW of SWQ#1 where the top of the Cayman Formation is ~ 23 m below the boundary position in SWQ#1, the dolostones in the uppermost 30 m of the Cayman Formation are also dominated by LCD. In that well, the %Ca content of the LCD ranges from 51.2 to 54.8 %Ca with an average of 52.0 %Ca (N = 35), whereas the %Ca content of the HCD ranges from 55.0 to 60.1 %Ca with an average of 57.0 %Ca.

Based on 615 electron probe measurements, the dolomite in the Pedro Castle Formation on Cayman Brac is formed of HCD with an average 57.4 ± 1.2%Ca (MacNeil and Jones 2003). The %Ca values are relatively constant from sample to sample with no indication of vertical or lateral changes.

Grand Cayman

The lack of core and the general paucity of good well cutting samples from the Brac Formation on Grand Cayman means that it has been difficult to accurately assess its main lithological traits. None of the cutting samples available from Grand Cayman are characterized by coarse sucrosic dolomite crystals and no *Lepidocylina* have been identified in any of them. From a lithological perspective, these well cuttings are very similar to those obtained from the overlying Cayman Formation.

Dolostones in the Cayman Formation on Grand Cayman, like those on Cayman Brac, are formed of various combinations of LCD and HCD with calcite being present in some parts of the sequences (Fig. 6.14c). In well LV#2 on Grand Cayman, which is used as an informal type section for the Cayman Formation, HCD dominates in the lower part of the formation whereas LCD dominates in the upper part of the formation (Fig. 6.14c). The %Ca in the LCD and HCD and the average %Ca is relatively constant through the Cayman Formation in this well (Fig. 6.14c).

Comparison of successions from various wells across Grand Cayman shows that there are some systematic variations in the distributions of the LCD and HCD and the presence/absence of calcite (Fig. 6.15):

- Sequences on the north and north-east coasts of Grand Cayman are dominated by LCD whereas those on the south and west coasts typically contain more HCD.
- Wells EER#1 and HRQ#1 are distinctive because of the calcite in the lower parts of their successions.
- Wells GFN#1, FFM#1, and NSC#1/3, located in the east-central part of the island are distinctive because their successions include significant thicknesses of limestone, virtually no LCD, and variable amounts of HCD.

The variability in the lithologies in the Cayman Formation are further emphasized by comparison between a selection of 19 wells that have been drilled and sampled on the eastern part of Grand Cayman (Fig. 6.16). Collectively, those wells show that LCD dominates in the successions that are close to the coastline and especially those on the northeast corner of the island (Fig. 6.16). The inland wells, such as those in High Rock Quarry, generally have a high content of calcite (limestone) in their lower parts that are overlain by LCD (Fig. 6.16). Limestone is evident in the

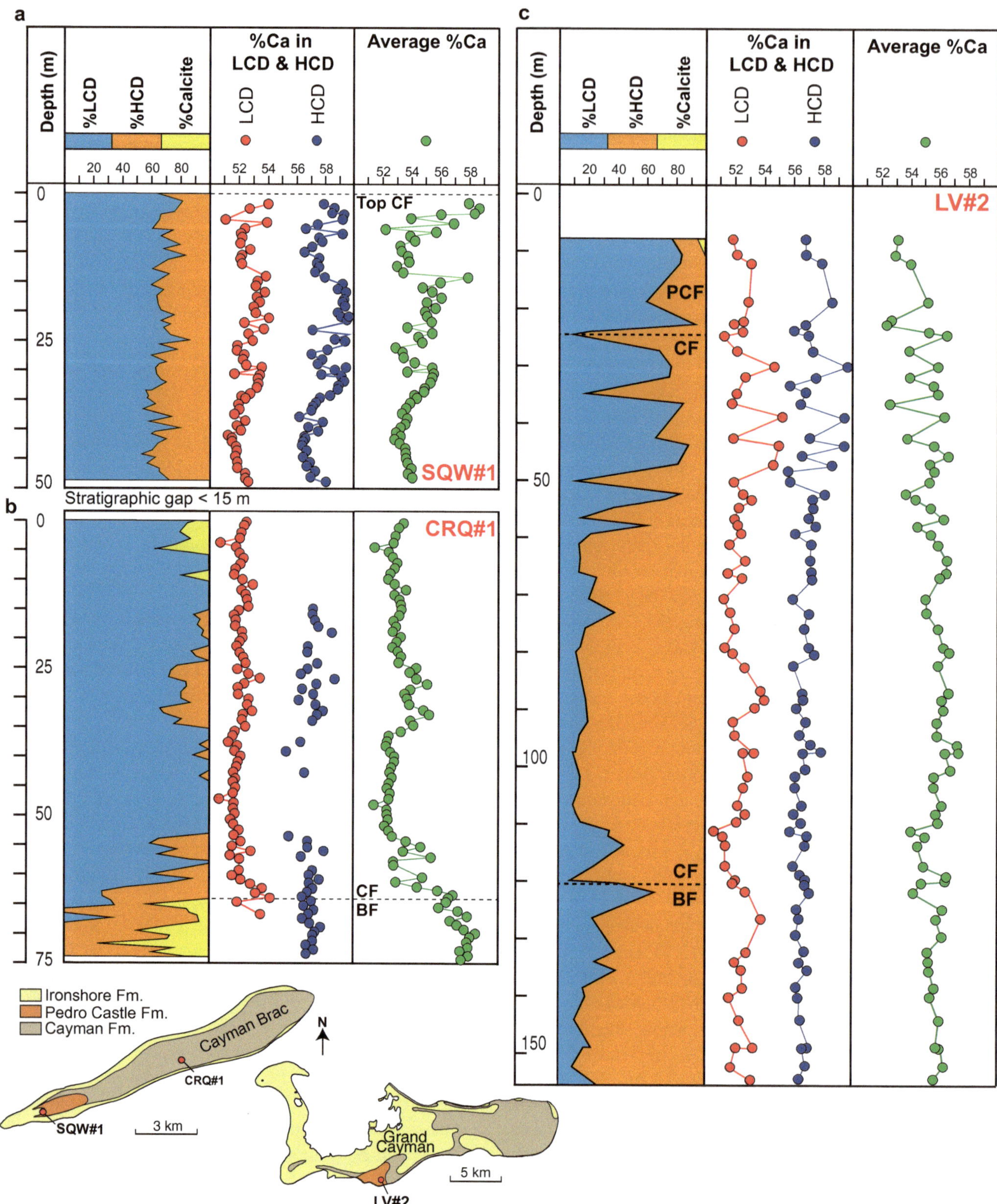

Fig. 6.14 Succession through the Cayman Formation in wells **a** SQW#1 and **b** CRQ#1 on Cayman Brac and **c** LV#2 on Grand Cayman showing percentages of LCD and HCD, the %Ca in the LCD and HCD, and the average %Ca. Although the stratigraphic gap between the top of section CRQ#1 and the base of section SWQ#1 is unknown, it is probably < 15 m. BF = Brac Formation, CF = Cayman Formation. Data derived from XRD analyses

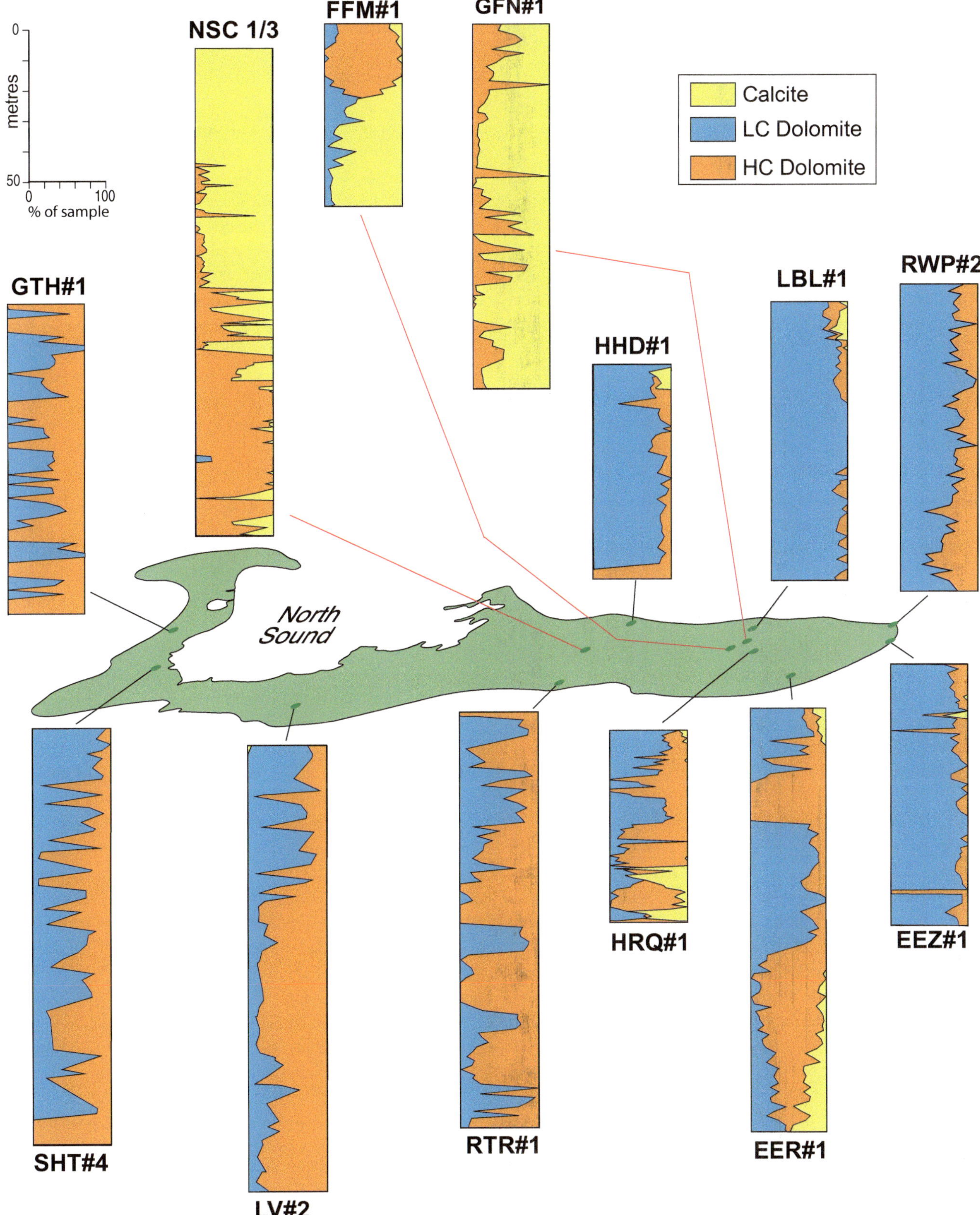

Fig. 6.15 Stratigraphic variations in percentage of calcite, LCD, and HCD (based on XRD analyses) in the Cayman Formation from various wells on Grand Cayman. Note dominance of calcite in central wells on eastern part of Grand Cayman and higher percentages of LCD in wells on northeastern part of the island

lower part of well FFM#1 and forms most of the succession in GFN#2 (Fig. 6.16). EEV#2, located in the southeast quadrant of this area, is dominated by HCD and has limestone in its uppermost part (Fig. 6.16).

6.2.4 Stable Isotopes

Brac Formation

The $\delta^{18}O$ values of 138 limestone samples from the Brac Formation on Cayman Brac range from −6.11% to +1.24‰ (average −1.82‰) and the $\delta^{13}C$ values range from −10.68‰ to + 1.16‰ (average −2.43‰) (Fig. 6.17a). Although there is an overall positive covariant trend between the $\delta^{18}O$ and $\delta^{13}C$, the range of $\delta^{18}O$ values for a given $\delta^{13}C$ is high (Fig. 6.17a). In comparison, the $\delta^{18}O$ values of 156 samples of dolostone range from −4.64‰ to +4.73‰ (average +1.39‰), whereas the $\delta^{13}C$ values range from −4.53‰ to +3.80‰ (average +1.36‰) (Fig. 6.17b). For the dolostones from wells EOR#1 and APL#1 there is positive covariant correlation between the $\delta^{18}O$ and $\delta^{13}C$, whereas the $\delta^{18}O$ and $\delta^{13}C$ values from sections WOJ-2 and WOJ-7 (equivalent to SCD) are clustered between +1 and + 3‰ $\delta^{18}O$ and 0 to +3‰ $\delta^{13}C$. In wells EOR#1 and APL#1, the $\delta^{18}O$ and $\delta^{13}C$ are variable whereas in sections CRQ#1, KEL#1, WOJ-2 and WOJ-7, the $\delta^{18}O$ and $\delta^{13}C$ values are relatively constant throughout the depth of the well.

Zhao and Jones (2012) and Zhao (2013) pointed out that the $\delta^{13}C$ values of the dolostones are different from those of the limestones with the negative values from the limestones indicating that freshwater was probably involved in their diagenesis.

Cayman Formation

The $\delta^{18}O$ values of 435 samples of dolostones from the Cayman Formation in 13 wells on Grand Cayman range from −4.33 to +7.93‰, whereas the $\delta^{13}C$ values range from −1.70 to +4.07‰ (Fig. 6.17b). It is important to note, however, that the negative $\delta^{18}O$ and negative $\delta^{13}C$ values come from only 14 samples that came from wells GTH#1, NSC#1/3, and SHT#4. If those outlier values are removed, the remaining 421 dolomite samples yielded $\delta^{18}O$ values from +0.5 to + 5.0‰ and $\delta^{13}C$ values from + 0.5 and +3.5‰. For the calcite, the $\delta^{18}O$ values range from −7.93 to + 3.97‰ and the $\delta^{13}C$ values range from −6.36 to +2.12‰ (Fig. 6.17a). All of the positive $\delta^{13}C$ values for calcite come from well NSC#1/3 with most from depths >110 m. It is important to note, however, that the range of $\delta^{18}O$ and $\delta^{13}C$ calcite values from that well is the largest that has been found from any well on the Grand Cayman. In most wells, the $\delta^{18}O$ and $\delta^{13}C$ values tend to be relatively constant. Major changes in these values are only apparent in those wells (e.g., NSC#1/3) where there is a significant vertical change in the proportions of calcite and dolomite.

On Cayman Brac, the $\delta^{18}O$ from the dolostones in section BW#1, based on 20 samples, ranges from +1.09 to +2.96‰ with an average of +2 .21‰, whereas the $\delta^{13}C$ ranges from +0.47 to +3.06‰ with an average of +1.82‰. In comparison, the $\delta^{13}C$ from the dolostones in section CRQ#1, based on 33 samples, ranges from +1.91 to +3.19‰ with an average of +2.54‰, whereas the $\delta^{13}C$ ranges from +1.15 to +3.33‰ with an average of +2.45‰. Overall, the values from both sections are similar. Although the $\delta^{18}O$ and $\delta^{13}C$ values fluctuate between their extremes, they are relatively consistent throughout the successions in sections CRQ#1 and BW#1. These values overlap with the range of values for the Cayman Formation on Grand Cayman.

Pedro Castle Formation

Variations apparent in the $\delta^{18}O$ and $\delta^{13}C$ values from the limestones and dolostones from the Pedro Castle Formation (Fig. 6.17) are largely related to the calcite content of the samples, which ranges from 0 to 100%. For 13 samples of dolostones from the Pedro Castle Formation in well QHW#1 on Grand Cayman, the $\delta^{18}O$ ranges from +2.70 to +3.85‰ with an average of +3.32‰ and the $\delta^{13}C$ ranges from +1.96 to +3.02 with an average of +2.62‰. Forty-eight samples of limestone from the Pedro Castle Formation on Cayman Brac yielded $\delta^{18}O$ from -8.29 to -2.28‰ (average −4.58‰) and $\delta^{13}C$ from −8.20 to −0.05‰ (average −4.31‰) (MacNeil 2001, his Table 4.1). In contrast, the 36 samples of dolomite yielded $\delta^{18}O$ values from −0.22 to + 2.02‰ (average + 1.09‰) and $\delta^{13}C$ values from −1.81 to + 1.42‰ (average +0.29‰) and (MacNeil 2001, his Table 5.1). For Little Cayman, five samples of dolostones collected from the north coast of the island yielded $\delta^{18}O$ values from +2.05 to +4.12‰ (average + 3.01‰) and $\delta^{13}C$ values from +2.25 to + 3.15‰ (average + 2.65‰). In contrast, limestones from the quarry yielded $\delta^{18}O$ values of −3.53 to −4.86‰ (average −4.36‰) and $\delta^{13}C$ values from −3.05‰ to −5.96‰ (average −4.21‰).

There is no consistency in the $\delta^{13}C$ and $\delta^{18}O$ values relative to their positions in the measured sections. All of the isotope values from the limestones and dolostones of the Pedro Castle Formation from Grand Cayman, Cayman Brac, and Little Cayman plot along the same trend line (Fig. 6.17). Comparison of the $\delta^{13}C$ and $\delta^{18}O$ isotopes of the dolomite with the %Ca content of the dolomite shows that there are no obvious relationships between these variables (Fig. 6.18).

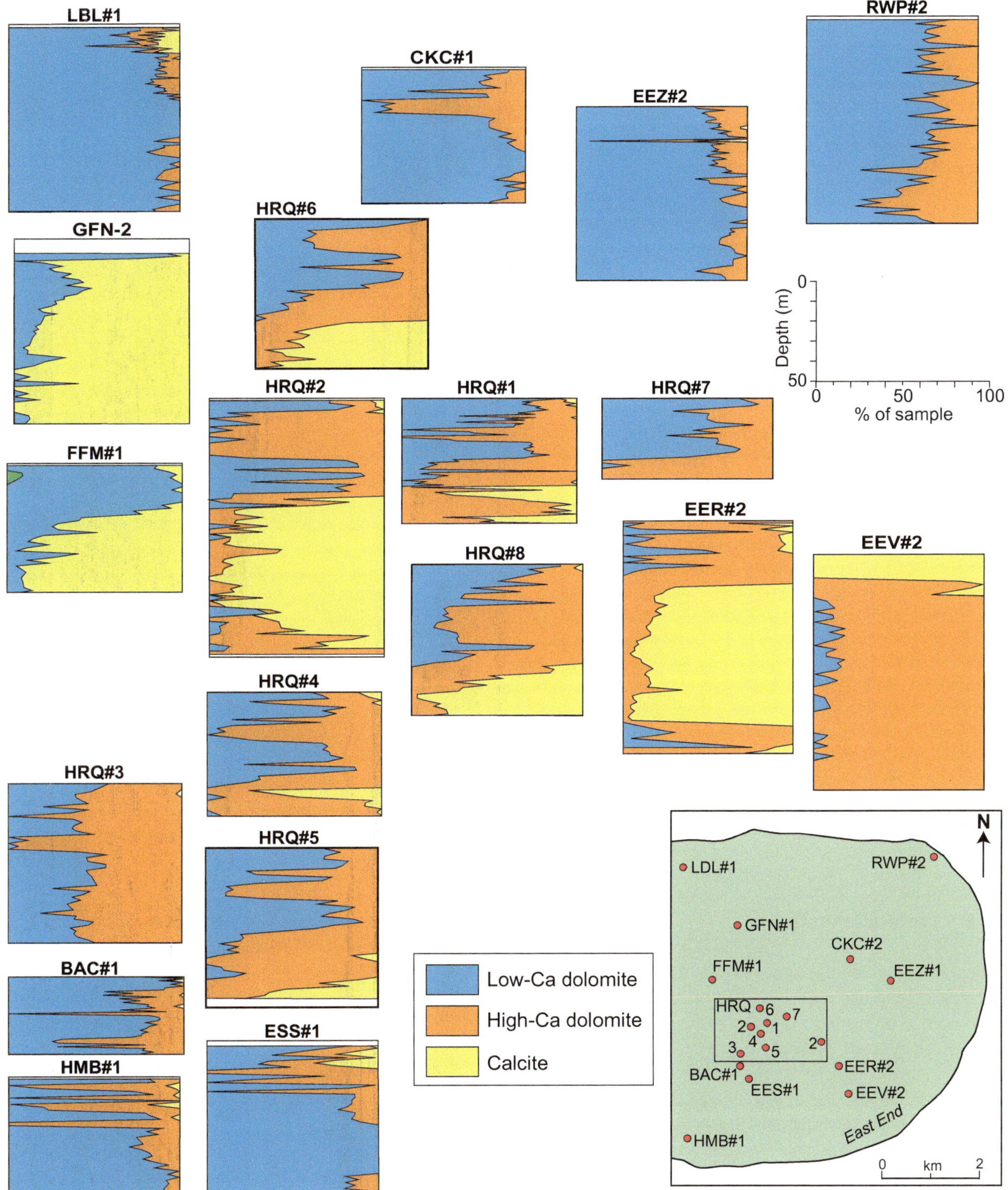

Fig. 6.16 Stratigraphic variations in percentage of calcite, LCD, and HCD in the Cayman Formation from wells on the eastern part of Grand Cayman. Note calcite (limestone) in central part of island and dominance of LCD in coastal areas

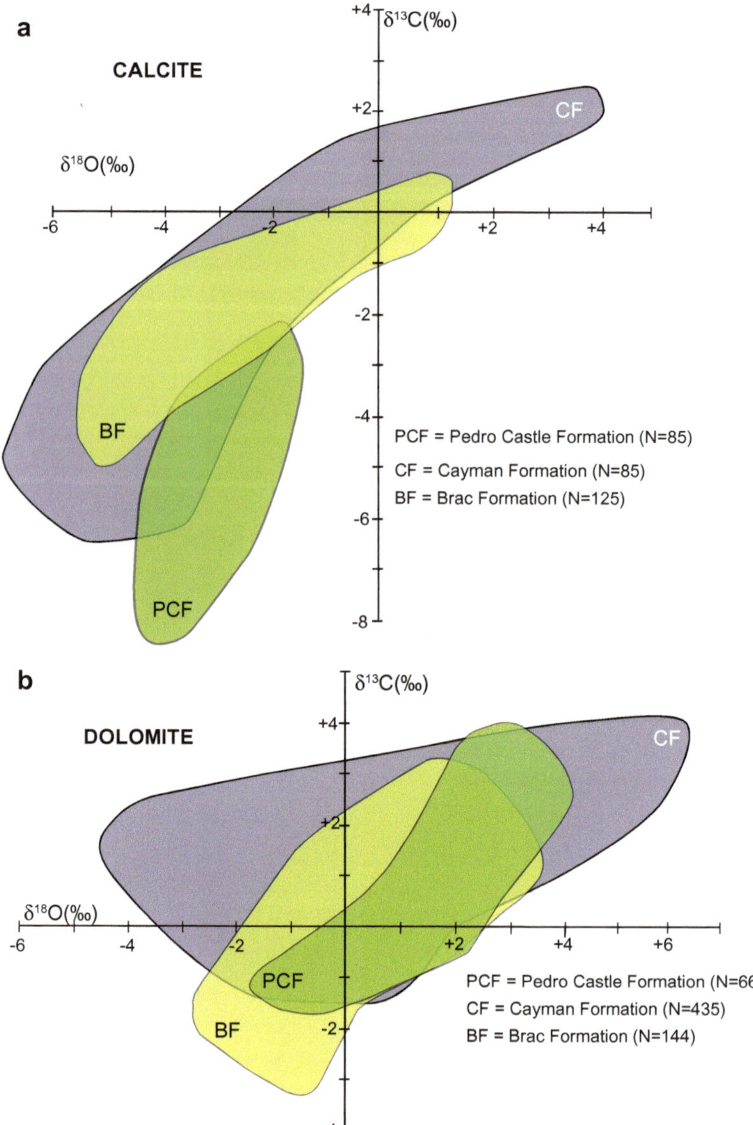

Fig. 6.17 Comparison of $\delta^{18}O$ and $\delta^{13}C$ for dolomite and calcite in the Brac Formation, Cayman Formation, and Pedro Castle Formation. Isotope data for the Brac Formation is from Uzelman (2009, her Fig. 4.1); data for the Cayman Formation on Grand Cayman came from wells SHT#4, HMB#1, RTR#1. NSC#1/3, LV#2, HRQ#1, EEV#1, GTH#1, RG#1, CKC#1, EER#1, DTE#1, and FSR#11; and data for the Pedro Castle Formation on Grand Cayman is from well QHW#1 and Pedro Castle Quarry and on Cayman Brac, data are from MacNeil (2001, his Fig. 4.1), MacNeil and Jones (2003, their Fig. 9), and Jones (2019)

6.2.5 Trace Elements

Critical to any dolomitization model is the composition and nature of the fluids that mediated the transformation of limestones to dolostones. Little is known about the significance of many trace elements (e.g., F, Cu, Zn, Ni, Cd, Ba) because their low concentrations (<10 ppm) greatly hinders their interpretations (Budd 1997). Trace elements found in dolomite have the potential, at least in theory, of providing insights into the chemistry of the formative fluid include Na, Fe, Mn, and Sr (Budd 1997). Although theoretically viable, the interpretation of the trace element contents in dolostones is severely hampered by the following issues.

- The fact that dolomite has not yet been formed at low temperatures in controlled laboratory settings means that the Distribution Coefficient (D_{Me}) of each element remains uncertain (Budd 1997). This, coupled with the wide range in concentrations of each element means that a large range of parent fluid compositions can be inferred from any given data set.
- Inclusions in dolomite crystals can inflate the concentration of various elements. In the Cayman dolostones, for examples, Ca-rich inclusions like those commonly evident on backscattered images, can inflate the Ca content of the whole dolomite crystal and also account for high values of some trace elements, such as Sr.

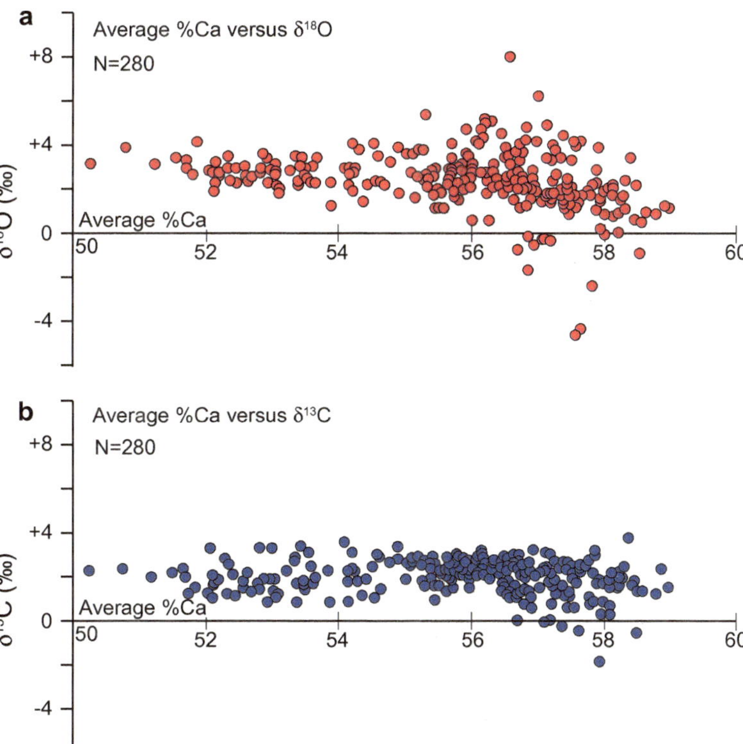

Fig. 6.18 Comparison of $\delta^{18}O$ and $\delta^{13}C$ dolomite isotopes with %Ca in dolomite for 280 samples from the Cayman Formation from wells EER#1, EEV#2, AIR#1, CKC#1, FFM#1, FSR#11, HMB#1. HRQ#1, NSC#1/3, RRTR#1, and LV#2 from the central and western part of Grand Cayman

- In theory, trace elements in dolostones can reside within the crystal lattice of individual dolomite crystals, inclusions in the dolomite crystals, along the boundaries between neighbouring crystals, and/or within individual microcrystals in the dolostone matrix. Although some of these possibilities can be detected through careful microprobe analyses, it is generally impossible to fully resolve this issue.
- Most analyses of the very finely crystalline island dolostones assume that the dolostones are compositionally uniform. Despite their small size (mostly < 100 μm), each of the formative crystals in the Cayman dolostones are typically zoned with a core encased by a laminated cortex. These zones are defined by variations in their %Ca content that are accompanied by variations in their trace element contents. Most analyses of such dolostones provide results that are, in reality, averages of all of these different components.
- Limpid dolomite, which typically forms as a cement in the Cayman dolostones, either encases the replacive dolomite crystals or lines/fills voids that exist in the dolostones (e.g., Jones et al. 1984, their Figs. 3, 4 and 5). Although obviously different from the dolomite that replaced the original limestone groundmass, it is commonly difficult to completely separate these two components.

Despite these potential problems, trace elements have been used to infer the types of fluids that were responsible for the genesis of island dolostones, like those found on the Cayman Islands.

Na in Dolomite/Limestones

The Na content of the dolostones and limestones of the Cayman Islands is variable but relatively consistent from formation to formation (Figs. 6.19 and 6.20).

- Brac Formation: In sections SCD, WOJ#3/7, KEL#1, and CRQ#1, 44 samples of dolostones from the Brac Formation yielded Na concentrations of 111 to 5705 ppm (average 872.7 ppm) (Fig. 6.19).
- Cayman Formation: On Grand Cayman, the Cayman Formation in wells EER#2, LV#2, and RTR#1 is formed of dolostones, whereas in wells NSC#1 and GFN#2, limestones and limestones with minor amounts of dolomite dominate. On Cayman Brac, the Cayman Formation in wells CRQ#1, SQW#1, BW#1, and KEL#1 is formed entirely of dolostone. The Na concentrations in these wells (Fig. 6.20) are highly variable with many samples containing >2000 ppm Na (Fig. 6.20). There is no readily apparent difference between the Na content in the limestones and dolostones.
- In some wells, there is a suggestion that the Na content increases with depth (Fig. 6.20).

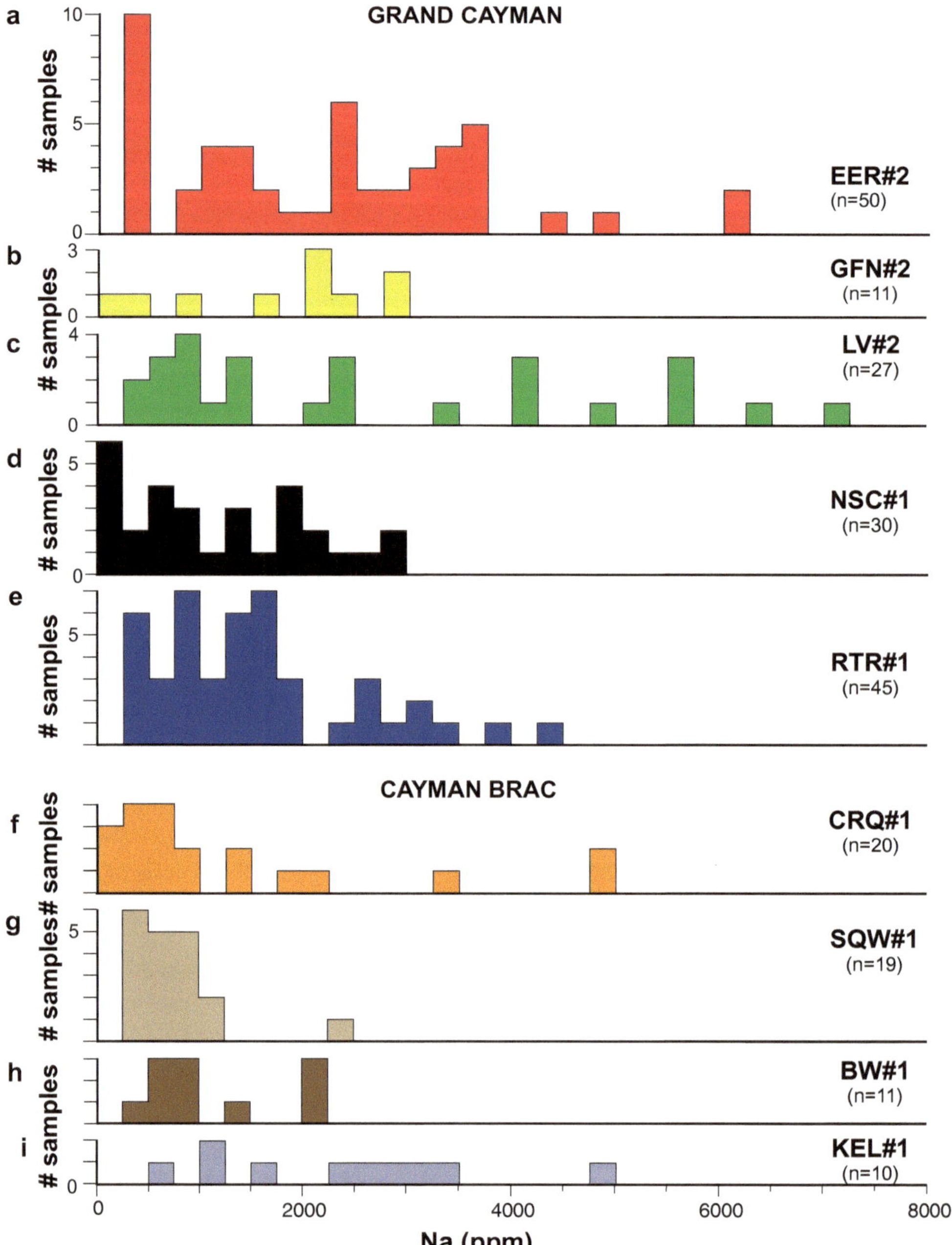

Fig. 6.19 Variation in Na content in dolostones and limestones of the Cayman Formation and Pedro Castle Formation, in wells EER#2, LV#2, RTR#1, GFN#2, NSC#1 on Grand Cayman and wells CRQ#1, SQW#1, BW#1, and KEL#1 on Cayman Brac

- Pedro Castle Formation: The few analyses available from the limestones/dolostones from this formation have Na contents that are as high as 1848 ppm with 8 samples giving an average of 940 ppm (Fig. 6.19).

There is no obvious consistent pattern to the distribution of the Na in any of the formations. In the Cayman Formation, for example, the Na of the limestones in wells NSC#1 and GFN#2 are no different from the Na content of the dolostones in wells LV#2, RTR#1, and EER#1 (Fig. 6.20). If all of the Na contents in these five wells are compared, then collectively there appears to be a general increase in the Na content with depth. Nevertheless, it is important to note that this apparent relationship is inconsistent because (1) the Na content in RTR#1 is generally lower than that in LV#2, and (2) the Na concentrations in the lowermost part of well LV#2 are low and similar to those found in shallow depths of other wells (Fig. 6.20). There is no obvious correlation between the Na content of the limestones and dolostones and the modern-day hydrological zones. The Na content in the dolostones of RTR#1, which is located near the coastline and has no freshwater zone in it is the same as that in wells NSC#1 and EER#1 where the upper part of the succession resides in freshwater lenses.

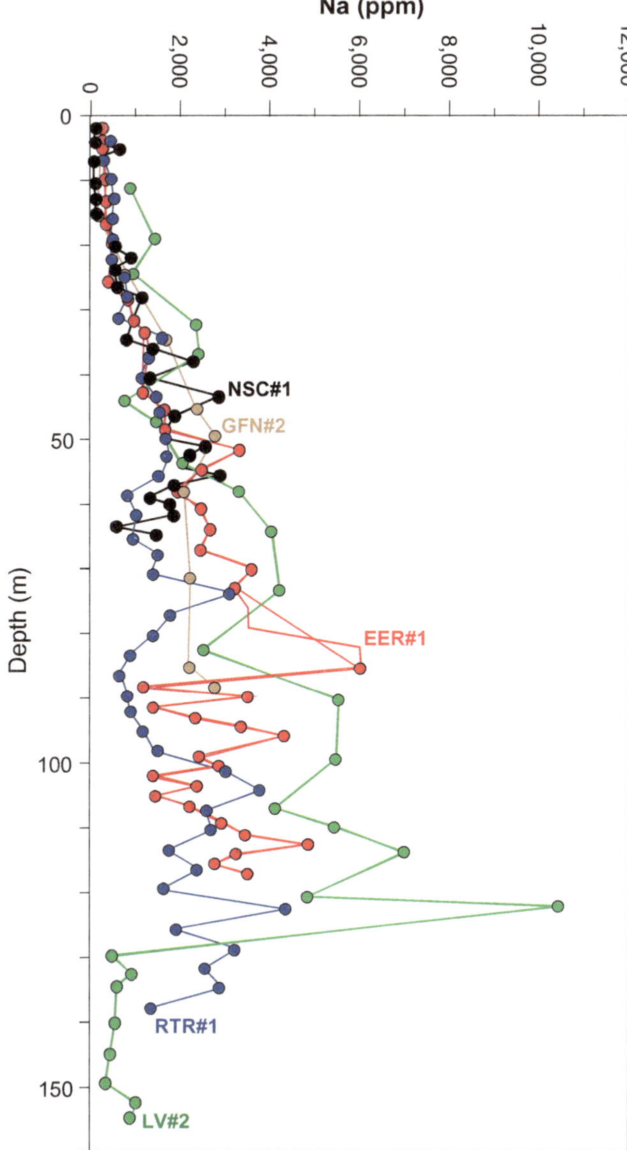

Fig. 6.20 Variation in Na concentrations with depth in Cayman Formation in wells NSC#1 and GFH#2 (limestones) and wells LV#2, RTR#1, and EER#1 (dolostones)

According to Budd (1997, his Fig. 15 and Table 4), the reported Na concentrations in island dolostones range from 48 to 1500 ppm with mean values between 256 and 711 ppm and values > 900 ppm being rare. Relative to these values, the Na concentrations in the dolostones from the Cayman Islands are very high. Apart from well SQW#1, all of the wells yielded average Na concentrations that exceed the 256 to 711 ppm range noted by Budd (1997). Although some of the highest Na concentrations in Cayman values are outliers, many are above the 2000 ppm that Budd (1997) considered to the highest values associated with island dolostones.

As noted by Budd (1997), the variable Na contents of dolostones has produced conflicting conclusions as dolostones attributed to mixing-zone dolomitization and dolomitization mediated by normal or hypersaline seawater have essentially the same ranges of Na. Budd (1997) pointed out that interpretations of the Na content can be debatable because the Na in dolostones may be located in fluid or solid inclusions, in crystal defects, and/or as absorbed ions. Staudt et al. (1993) suggested that these potential problems could be overcome by the use of Na, Cl, and SO_4 content of dolostones, which would avoid the issue of Na contamination (Budd 1997). Although a promising technique, most data sets for island dolostones, including those from the Cayman Islands, lack the Cl and SO_4 concentrations.

Mn in Dolostones/Limestones

On Cayman Brac, the average Mn content of the dolostones and limestones in the Cayman Formation are generally higher than those in the Brac Formation (Fig. 6.21). The Mn contents in the Cayman Formation on Grand Cayman are generally similar to those on Cayman Brac (Fig. 6.21). The only exceptions are from wells GFN#2 and LV#2 where the Mn has a greater range and higher mean values than for equivalent rocks in other wells on Grand Cayman and Cayman Brac (Fig. 6.21).

Budd (1997, his Table 4 and Fig. 17B) showed that the Mn content in island dolostones are generally low with mean values typically between 8 and 35 ppm apart from the dolostones in the Seroe Domi Formation on Curacao that have a mean value of 442 ppm. For the formations on the Cayman Islands, the Mn content of the dolostones from any given location generally have mean values of < 50 ppm (Fig. 6.21a). The only exceptions are from the limestones in well GFN#2 with an average of 88 ppm and the limestones/dolostones from the Pedro Castle Formation that yielded an average of 52 ppm with values up to 263 ppm (Fig. 6.21c, d).

The source of the Mn that is incorporated into the dolomite is open to debate. Given that seawater has a low Mn content, the high Mn content found in some limestones and dolostones (e.g., those in wells LV#2, GFN#2, Pedro Castle Formation) on the Cayman Islands must come from an external source such the precursor limestones, soils, detrital siliciclastic, and/or volcanic rocks (cf., Budd 1997). Given that there are no siliciclastic or volcanic rocks on the Cayman Island, it seems that most of the Mn probably came

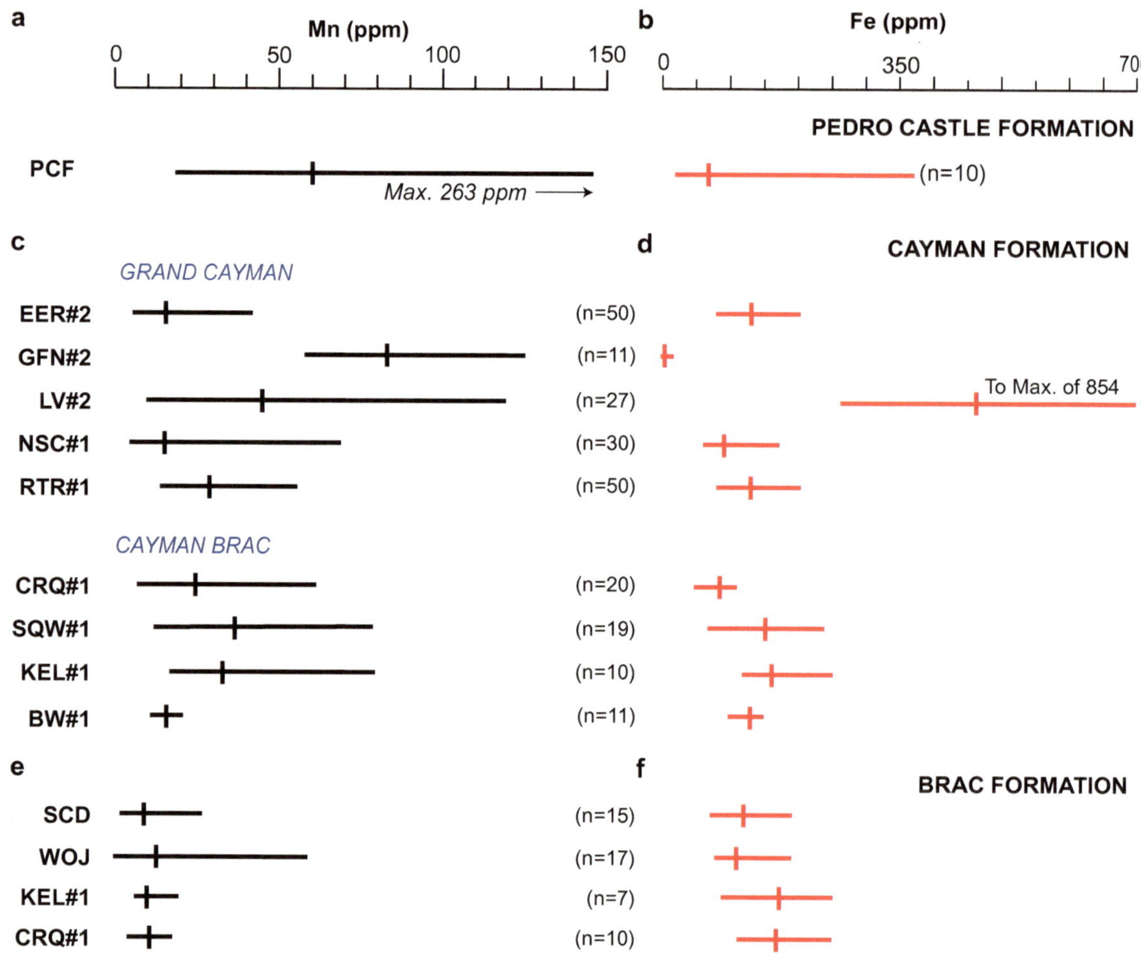

Fig. 6.21 Mn and Fe contents in limestones from the Ironshore Formation (**a**, **b**) and limestones and dolostones from the Pedro Castle Formation (**c**, **d**), Cayman Formation (**e**, **f**), and Brac Formation on Grand Cayman and Cayman Brac (**g**, **h**)

from the terra rossa soil, which commonly has a high Mn content (Jones 2021).

Fe in Dolostones/Limestones

The Fe content of limestones and dolostones from the Brac Formation, Cayman Formation, Pedro Castle Formation and Ironshore Formation tend to vary from location to location (Fig. 6.21). The Fe content is similar on Grand Cayman and Cayman Brac, apart from well LV#2 that has a high Fe content. The Fe content of the limestones/dolostones in the Pedro Castle Formation are similar to those in the Cayman Formation (Fig. 6.21).

With the exception of samples from well LV#2, all of the limestones/dolostones from the Cayman Islands yielded Fe mean values of <300 ppm (Fig. 6.22). For well LV#2, the mean value is 495 ppm with maximum values up to 854 ppm. The Fe content of island dolostones, as determined by Budd (1997, his Table 4 and Fig. 17a), varies from 10 to 5740 ppm with most values from 23 to 1627 ppm and modes of < 300 ppm and >1627 ppm.

The Fe found in the dolomite, like the Mn, probably originated from the terra rossa soil that is common on each of the Cayman Islands.

Sr in Dolostones/Limestones

The average Sr content of the limestones and dolostones from the Brac Formation (115–224 ppm), Cayman Formation (186–429 ppm), and Pedro Castle Formation (115–224 ppm) are relatively consistent (Table 6.1, Figs. 6.22 and 6.23) with most being within the range of mean values that Budd (1997, his Table 4, Fig. 13) documented for other island dolostones. None of the Cayman values approach the high Sr concentrations known from the Falmouth Formation of Jamaica or the Golden Grove and subsurface dolomite on Barbados.

The Sr content of limestones from the Cayman Formation on Grand Cayman ranges from 176 to 729 ppm (Table 6.1). There is no obvious relationship between the Sr content and depth (Fig. 6.24a). For limestones in the upper part of well NSC#1, the Sr content decreases with depth, whereas in the

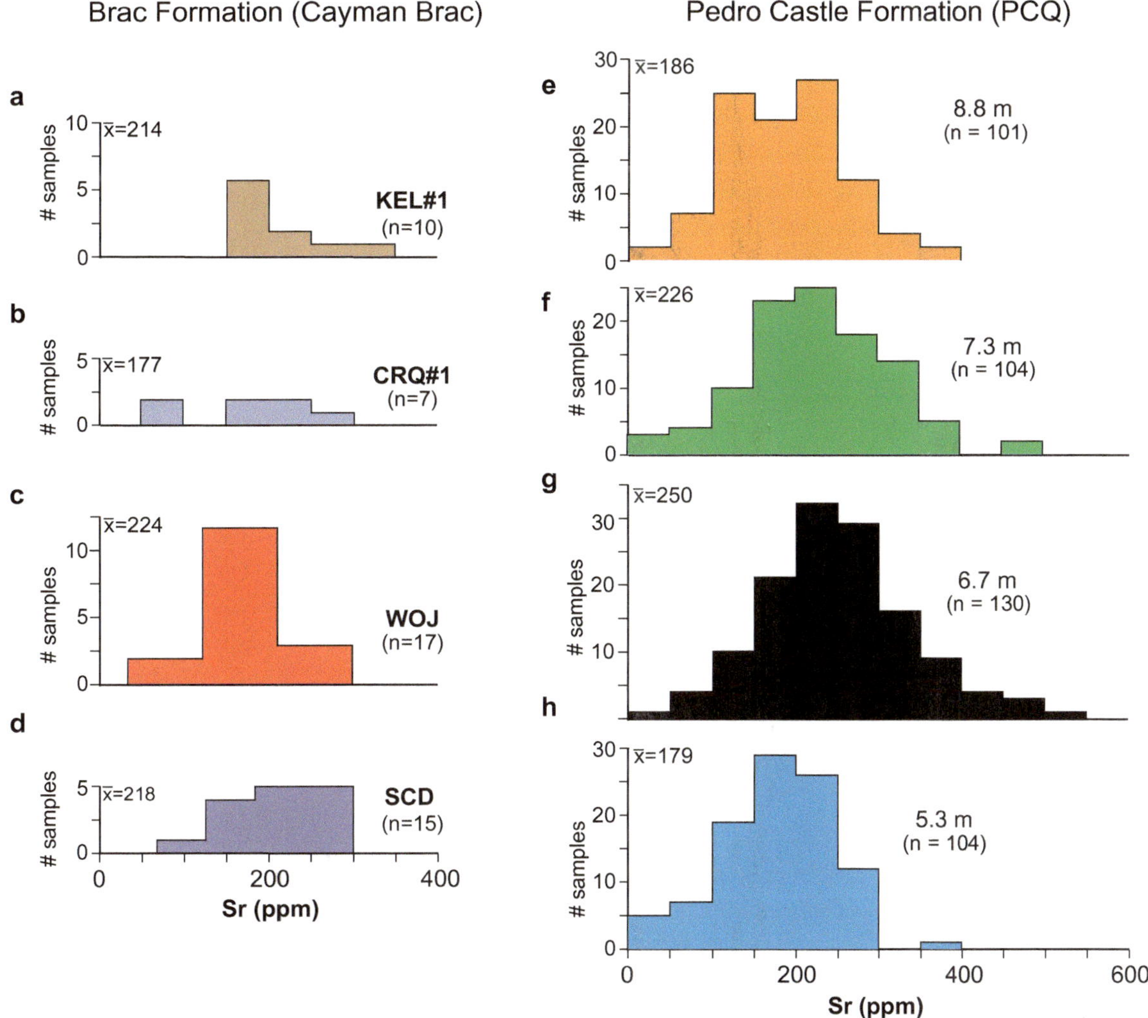

Fig. 6.22 Sr concentrations in limestones and dolostones of the (A-D) Brac Formation in sections SCD, WOJ, CRQ#1, and KEL#1, Cayman Brac, determined by ICPMS analyses, and (E–H) Pedro Castle Formation from Pedro Castle Quarry (heights are above quarry floor which is approximately at sea level), Sr concentrations determined by electron microprobe analyses, with n being the number of points analyzed for each sample. x = average valve for each section

upper part of well GFN#2, the Sr increases with depth (Fig. 6.23b). There is no obvious reason for this difference, although it should be noted that the uppermost part of the succession in NSC#1 (to a depth of ~16 m) lies within the freshwater lens. The Sr content of the limestones in wells NSC#1 and GFN#2 are generally higher than the Sr content in the dolostones from wells RTR#1 and LV#2 (Fig. 6.23).

The Sr content of the dolostones of the Cayman Formation varies from location to location and, in some cases, with depth (Figs. 6.22 and 6.23). The Sr content of the dolostones from well RTR#1, for example, are relatively constant (108 to 226 ppm) with depth, whereas the Sr content in samples from well EER#1 range from 145 to 583 ppm with high Sr values between 38 to 92 m that contrast with the lower Sr values from strata above and below that interval (Fig. 6.23 a). The interval with the high Sr content corresponds to a zone that is dominated by LCD (Fig. 6.15). The Sr content of dolostones in well LV#2, which are generally higher than those in RTR#1, fluctuate without any readily apparent cause (Fig. 6.23a).

Dawans and Swart (1988) and Vahrenkamp and Swart (1990) suggested that the Sr content covaries with the %Ca of dolomite, possibly by as much as 20 ppm per mole of excess Ca. As noted by Budd (1997) this could account for variations of 40 to 160 ppm Sr given the range of Ca content in island dolomite. For 116 dolostone (all formed of 100% dolomite) samples from five different wells on Grand Cayman and Cayman Brac, the Sr content (determined by ICPMS analyses) increases as the average %Ca (determined by XRD analysis of same samples using method outlined by Jones et al. (2001)) increases (Fig. 6.24). Calculations based on the equation that relates these two variables indicates that the Sr

Table 6.1 Summary information for Sr content of limestones and dolostones in the Brac Formation, Cayman Formation, Pedro Castle Formation, and Ironshore Formation of the Cayman Islands. All values given to nearest whole number. Numbers derived from ICPMS analyses (number indicates number of samples analyzed) with exception of those from the Pedro Castle Formation that were determined by Electron Microprobe analysis (number of points analyzed in single sample—indicated by asterisk)

			Sr content (ppm)			
Island	Section	Formation	Minimum	Mean	Maximum	Number
GC		Ironshore-F	1206	5257	7856	23
GC		Ironshore-E	2301	5553	6997	9
GC		Ironshore-D	602	4204	8353	84
GC		Ironshore-C	856	4540	7407	14
GC		Ironshore-B	1200	2743	7325	13
GC		Ironshore-A	703	2994	7922	35
GC	PCQ-1 (5.3 m)	Pedro Castle	34	179	373	99*
GC	PCQ-1 (6.7 m)	Pedro Castle	42	250	507	130*
GC	PCQ-1 (7.3 m)	Pedro Castle	34	226	465	104*
GC	PCQ-1 (8.8 m)	Pedro Castle	42	287	389	100*
GC	RTR#1 (Dol)	Cayman	108	186	226	45
GC	LV#2 (Dol)	Cayman	160	242	306	27
GC	EER#1 (Dol)	Cayman	145	346	583	50
GC	GFN#2 (Lst)	Cayman	206	429	589	10
GC	NSC#1 (Lst)	Cayman	176	348	729	27
CB	SCD	Brac	149	218	290	15
CB	WOJ	Brac	195	224	275	17
CB	CRQ#1	Brac	80	115	181	7
CB	KEL#1	Brac	109	136	165	10

content increases by 22.5 ppm for each increase of 1 %Ca in the dolomite. This value, similar to that proposed by Vahrenkamp and Swart (1990), explains much of the variation seen in the Sr content of the dolostones in the Bluff Group.

With the exception of the Falmouth Formation (Jamaica) and the Golden Grove Dolomite and other subsurface dolostones on Barbados, the mean Sr values for island dolostones are 152 to 306 ppm (Budd 1997). It should be noted, however, that the maximum Sr value for many island dolostones are between 400 and 600 ppm (Fig. 6.25). In the Bluff Group of the Cayman Islands the Sr content for the dolostones in the Brac Formation (80 to 290 ppm, mean 216 ppm), Cayman Formation on Grand Cayman (108 to 583, mean 122 ppm), Cayman Formation on Cayman Brac (80 to 290 ppm, mean 143 ppm), and Pedro Castle Formation (34 to 507, mean 213 ppm) are consistent with the Sr contents of most other island dolostones (Fig. 6.25). For the limestones in the Cayman Formation in well NSC#1, the Sr content ranges from 176 to 729 ppm (mean 349 ppm).

Budd (1997) noted that the Sr content of island dolomite might be due to inheritance from the fluid that mediated dolomitization, kinetic effects, and/or contamination from Sr-rich calcite/aragonite inclusions. For the Cayman dolostones, the variance in the Sr content of the dolostones is due largely to kinetic effects as is evident from the relationship between the Sr content and the average %Ca of the dolomite (Fig. 6.24). Budd (1997) argued that dolostones with a Sr content of <300 ppm probably indicate that seawater mediated dolomitization.

Rare Earth Elements in Dolostones/Limestones

The rare earth elements (REE) and yttrium (Y) content of the limestones and dolostones form the Brac Formation, Cayman Formation, and Pedro Castle Formation on Grand Cayman and Cayman Brac are all similar with a maximum $\sum$REE value of ~7 ppm and a maximum $\sum$REE+Y value of ~ 10 ppm (Fig. 6.26). For most samples, the maximum $\sum$REE values are commonly <5 ppm and the $\sum$REE+Y values <7 ppm (Fig. 6.26). In general, the range of values and mean values of the $\sum$REE and $\sum$REE+Y are similar. Notable exceptions are samples from LV#2 that have the largest ranges of values and the highest mean values of $\sum$REE and $\sum$REE+Y (Fig. 6.26).

Comparison of the REE content of samples from various wells on Grand Cayman and Cayman Brac shows that Y and light REE (HREE—La, Ce, Pr, Nd) dominate with many of the medium REE (MREE—Sm, Eu, Gd, Tb, Dy) and heavy REE (LREE—Ho, Er, TR, Yb, Lu) being absent or present in low quantities in only a few samples (Fig. 6.27). Overall, all the samples are dominated by Y, La, Ce, and Nd (Fig. 6.27).

PAAS (Post-Archean Average Shale) normalization of the REE data (cf., McLennan 1989, 2001) produces similar

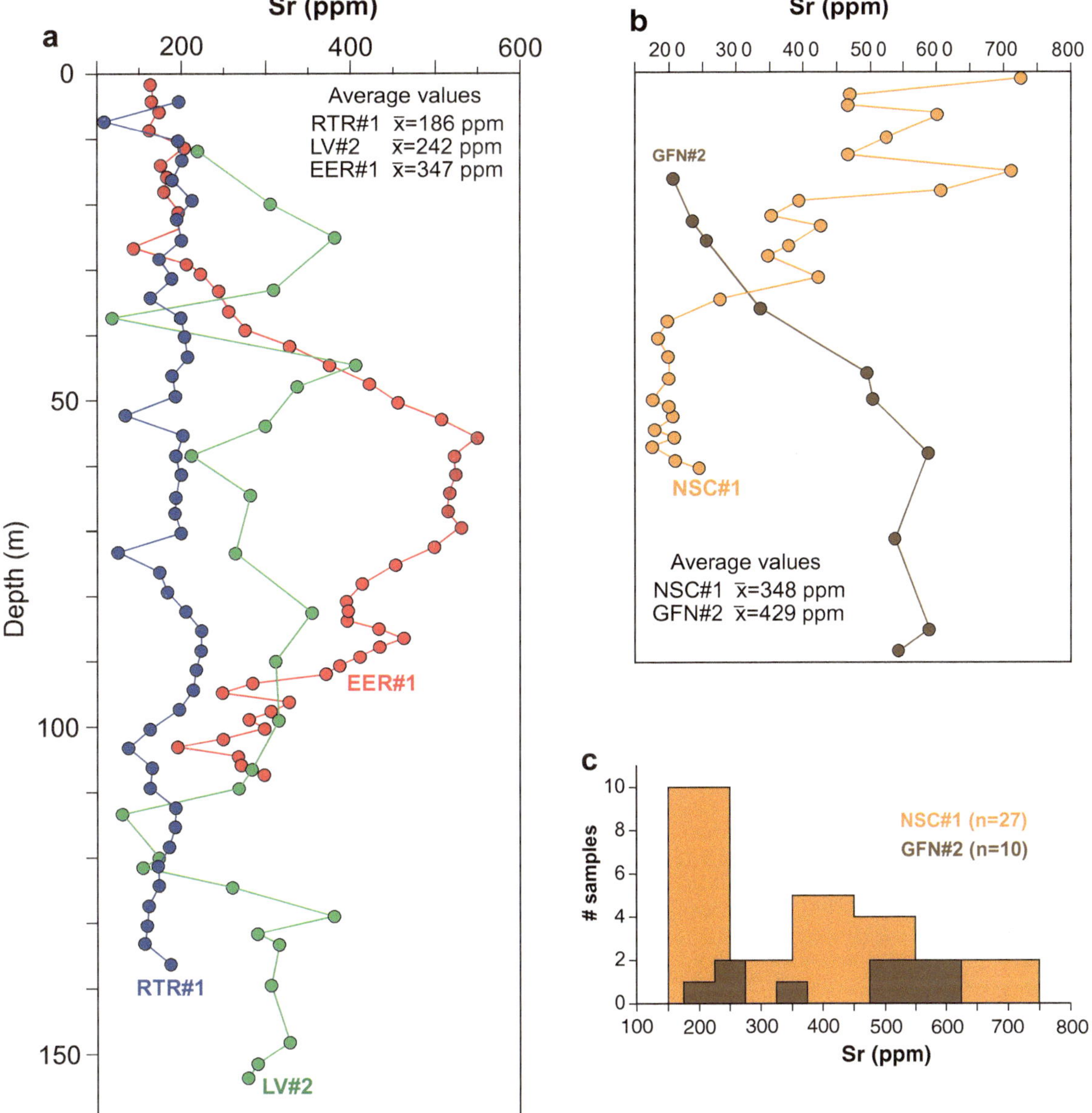

Fig. 6.23 Sr concentrations relative to depth in Cayman Formation. **a** Dolostones in wells LV#2, RTR#1, and EER#1, average value (x) for each well given in upper right corner. Average values given in upper right corner. **b** Limestones in wells NSC#1 and GFN#2, average value (x) for each well given in lower left corner. **c** Histogram showing comparison between Sr concentrations in wells NSC#1 and GFN#2

profiles for all of the limestones and dolostones of the Pedro Castle Formation, Cayman Formation, and Brac Formation on Grand Cayman and Cayman Brac (Fig. 6.28). All of the profiles include a negative Ce anomaly and a positive Y anomaly. Comparison between these profiles and seawater profiles shows their similarity (Fig. 6.31). The profiles for seawater from Misteriosa Bank (10 m depth), which is located ~400 km west of Grand Cayman (data from Osborne et al. 2015) and from BATS-15 m (The Bermuda Atlantic Time-series Study), located ~80 km southeast of Bermuda (van der Flierdt et al. 2012) are both characterized by a negative Ce anomaly (Fig. 6.29) like the samples from the limestones and dolostones of the Cayman Islands. Although their seawater analyses did not include Y, profiles generated from data for seawater at various depths from elsewhere in the world show the negative Ce anomaly and a positive Y anomaly (Fig. 6.29b).

6.2.6 $^{87}Sr/^{86}Sr$ Ratios

Importance has been attached to the $^{87}Sr/^{86}Sr$ ratios of limestones and dolostones because they can, in theory, provide the youngest possible age of the limestones or the

400
300
200
100
0
Sr (ppm)
52
54
56
58
60
Average %Ca
y=22.46x -1047.83
KEL#1 (n=10)
LV#2 (n=27)
EER#2 (n=20)
RTR#1 (n=45)
BW#1 (n=14)

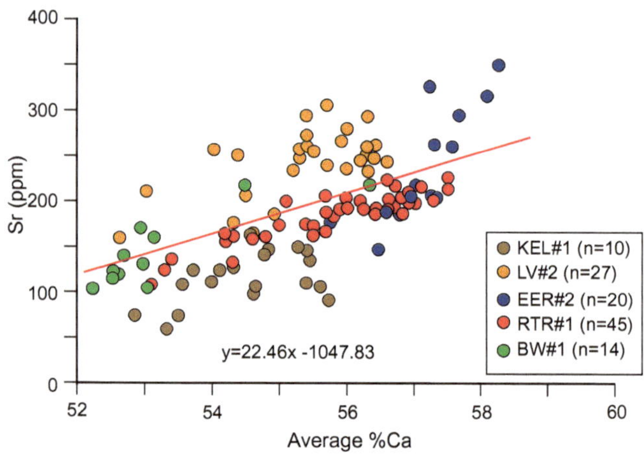

Fig. 6.24 Graph showing relationship between the Sr (ppm) in dolostone (as determined by ICPMS analyses) and the %Ca of the dolostone (as determined by the method developed by Jones et al. 2001)

oldest possible age of dolomitization. This is predicated on the known temporal change in the $^{87}Sr/^{86}Sr$ ratio in seawater over the last 40 million years (Burke et al. 1982; DePaolo and Ingram 1985; Koepnick et al. 1985; Hodell et al. 1991; McArthur et al. 2012). For the Cayman Islands, 532 $^{87}Sr/^{86}Sr$ ratios have been determined from the dolostones and limestones of the Bluff Group with the view of determining the age of the limestones and the time(s) when dolomitization took place (Fig. 6.30).

Brac Formation

The $^{87}Sr/^{86}Sr$ ratios of 81 samples of dolostones, partly dolomitized limestones, and limestones from the Brac Formation on Grand Cayman ranged from 0.708037 to 0.709155. Uzelman (2009) suggested that the wide range of $^{87}Sr/^{86}Sr$ ratios was due to dolomitization being a time-transgressive process. Zhao and Jones (2012), however,

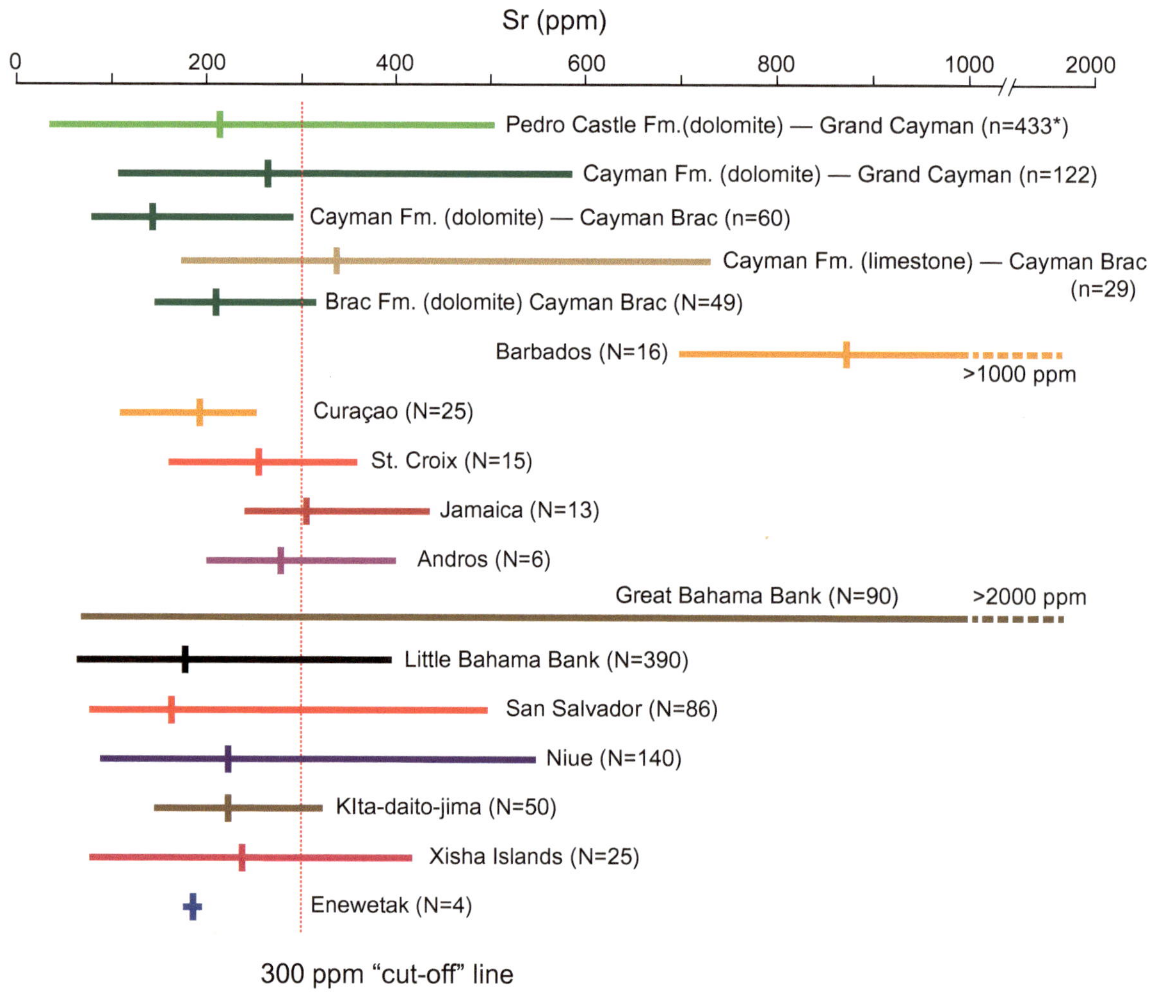

Sr (ppm)
0
200
400
600
800
1000
2000
Pedro Castle Fm.(dolomite) — Grand Cayman (n=433*)
Cayman Fm. (dolomite) — Grand Cayman (n=122)
Cayman Fm. (dolomite) — Cayman Brac (n=60)
Cayman Fm. (limestone) — Cayman Brac (n=29)
Brac Fm. (dolomite) Cayman Brac (N=49)
Barbados (N=16)
>1000 ppm
Curaçao (N=25)
St. Croix (N=15)
Jamaica (N=13)
Andros (N=6)
Great Bahama Bank (N=90)
>2000 ppm
Little Bahama Bank (N=390)
San Salvador (N=86)
Niue (N=140)
Klta-daito-jima (N=50)
Xisha Islands (N=25)
Enewetak (N=4)
300 ppm “cut-off” line

Fig. 6.25 Comparison of Sr content of dolostones of the Pedro Castle Formation, Cayman Formation, and Brac Formation and limestones from the Cayman Formation with other island dolostones. Sr content for the Pedro Castle Formation based on electron microprobe analyses whereas Sr content for the Brac Formation and Cayman Formation determined by ICMPS analyses. The Sr content of limestone from the Cayman Formation is included for comparison. Modified from Wang et al. (2019, their Fig. 4)

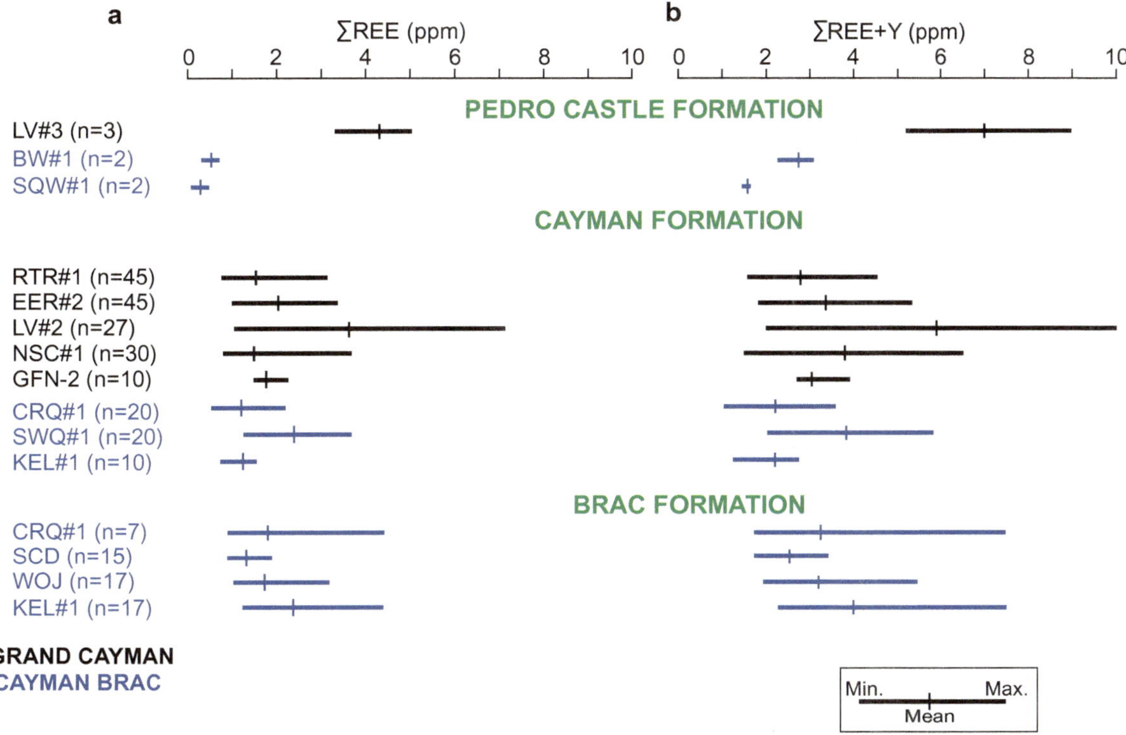

Fig. 6.26 Comparison of range and mean values of **a** ∑REE and **b** ∑REE+Y content of Ironshore Formation (units A to F) from Grand Cayman, Pedro Castle Formation on Grand Cayman and Cayman Brac, Cayman Formation from Grand Cayman and Cayman Brac, and Brac Formation from Cayman Brac. Black = dolostone; Blue = limestone

Fig. 6.27 Comparison of average REE content in different sections of the Cayman Formation on Grand Cayman and Cayman Brac. Note dominance of LREE in all samples

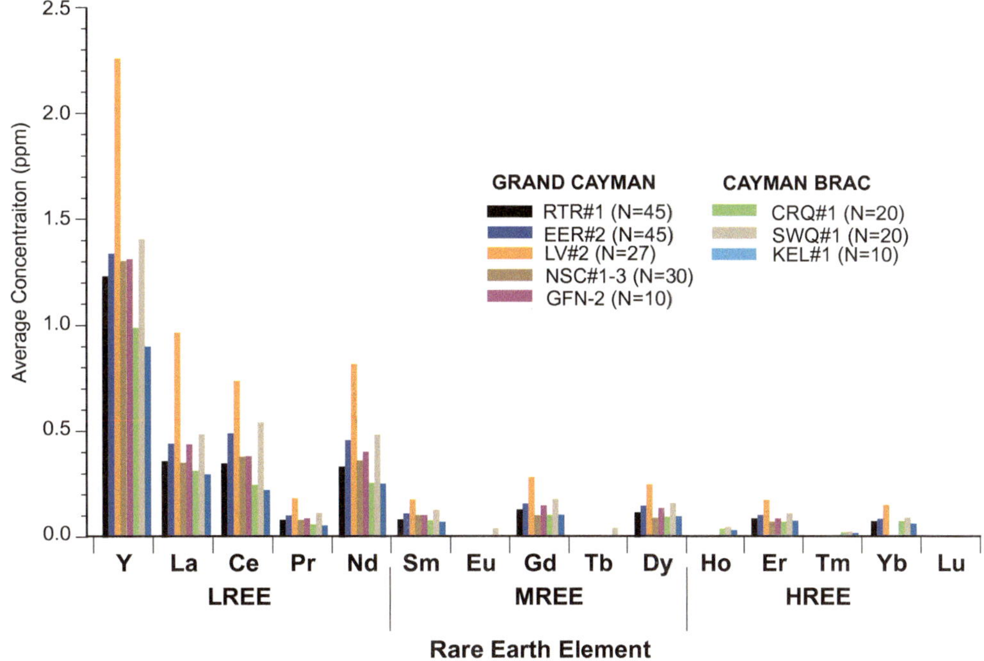

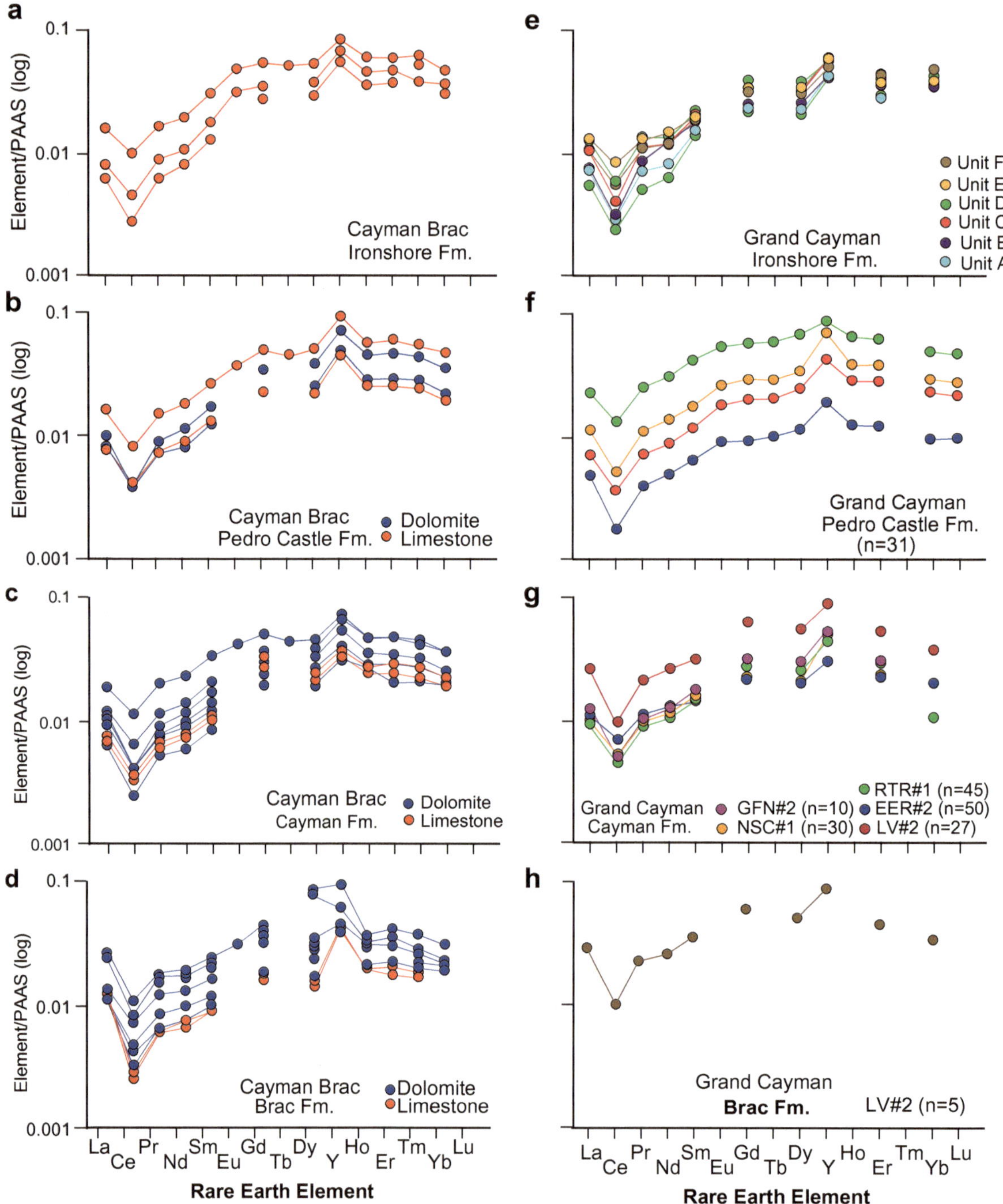

Fig. 6.28 Comparison of REE+Y in the Ironshore Formation, Pedro Castle Formation, Cayman Formation, and Brac Formation on Grand Cayman (**a–d**) and Cayman Brac (**e–h**). Based on representative samples from each succession that have been PAAS normalized with Y inserted between Dy and Ho. A to D modified from Zhao and Jones (2013, their Fig. 5); E modified from Li and Jones (2014, their Fig. 10a, b, c). Note that for panel F, analyses excluded Tm because of technical issues

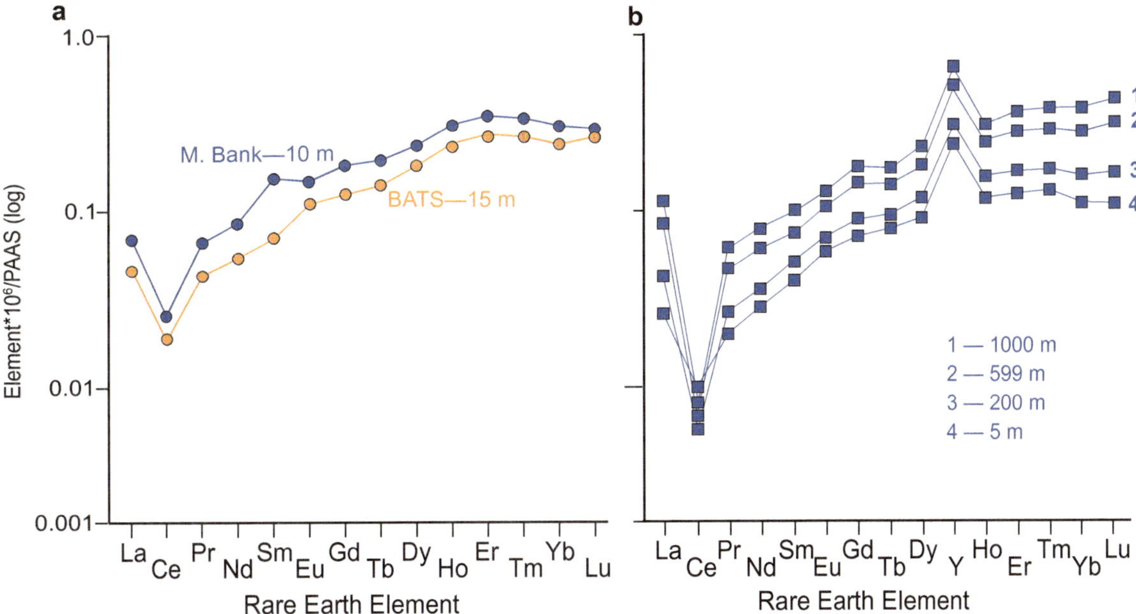

Fig. 6.29 PAAS normalized REE profiles of ocean water from **a** Misteriosa Bank (10 m depth), which is located ~400 km west of Grand Cayman (data from Osborne et al. 2015) and from BATS-15 m (Bermuda Atlantic Time-series Study), located ~80 km southeast of Bermuda (van der Flierdt et al. 2012). Note that Y was not included in these analyses. **b** Ocean water at various depths from different parts of the world (including Y), based on analyses given in Alibo and Nozaki (1999), modified from Zhao and Jones (2013, their Fig. 5E). Given the low REE concentrations in seawater, the original REE values have been multiplied by 10^6 in order to compare their overall shapes with the REE profiles shown in Fig. 6.28

cautioned against that suggestion given that the $^{87}Sr/^{86}Sr$ ratios had a high correlation with the dolomite content of the mixed calcite-dolomite rocks in that data set (Fig. 6.31a).

Comparison of limestones and dolostones, exposed in cliffs on the southeast coast of Cayman Brac, shows the difference between the $^{87}Sr/^{86}Sr$ ratios of the limestones and dolostones (Fig. 6.31b) and clearly indicates the time gap between deposition of the limestones and their subsequent dolomitization. The $^{87}Sr/^{86}Sr$ ratios of the dolostones are similar to those obtained from the dolostones in the basal part of the overlying Cayman Formation (Fig. 6.31b).

Cayman Formation

Seventy-one samples of limestones from the Cayman Formation in wells NSC-1, DTE-1, FSR-10, and GFN-2 on Grand Cayman yielded $^{87}Sr/^{86}Sr$ ratios ranging from 0.708988 to 0.709170 with an average of 0.709066 (Fig. 6.30). In contrast, 52 samples of dolostones from the Cayman Formation on Cayman Brac yielded $^{87}Sr/^{86}Sr$ ratios ranging from 0.708610 to 0.70180 with an average of 0.709062, whereas 269 samples of dolostone from the Cayman Formation on Grand Cayman yielded $^{87}Sr/^{86}Sr$ ratios ranging from 0.708930 to 0.709127 with an average of 0.709026 (Fig. 6.30). Collectively, the dolostones have the same range of $^{87}Sr/^{86}Sr$ ratios as the limestones (Fig. 6.30).

Critical aspects of these data include (1) the $^{87}Sr/^{86}Sr$ ratios of the limestones and dolostones of the Cayman Formation being higher than those derived from the Brac Formation, (2) some of the higher $^{87}Sr/^{86}Sr$ ratios of the limestones are akin to the $^{87}Sr/^{86}Sr$ ratio of modern seawater, (3) the average value of the $^{87}Sr/^{86}Sr$ ratios of the dolostones from Grand Cayman is lower than the average $^{87}Sr/^{86}Sr$ ratio of the limestones, and (4) many of the $^{87}Sr/^{86}Sr$ ratios of the dolostones and limestones indicate a post-Pliocene age (Fig. 6.30).

Pedro Castle Formation

Fifty-one samples of dolostone from the Pedro Castle Formation on Grand Cayman and Cayman Brac yielded $^{87}Sr/^{86}Sr$ ratios from 0.708980 to 0.709125 with an average of 0.709059 (Fig. 6.30). This range overlaps the range of $^{87}Sr/^{86}Sr$ ratios derived from the limestones and dolostones of the Cayman Formation (Fig. 6.30). The average $^{87}Sr/^{86}Sr$ ratio derived from the Pedro Castle Formation is the same as that of the limestones in the Cayman Formation on Grand Cayman, the same as that of the dolostones on Cayman Brac, but higher than the average $^{87}Sr/^{86}Sr$ ratio of the dolostones on Grand Cayman (Fig. 6.30).

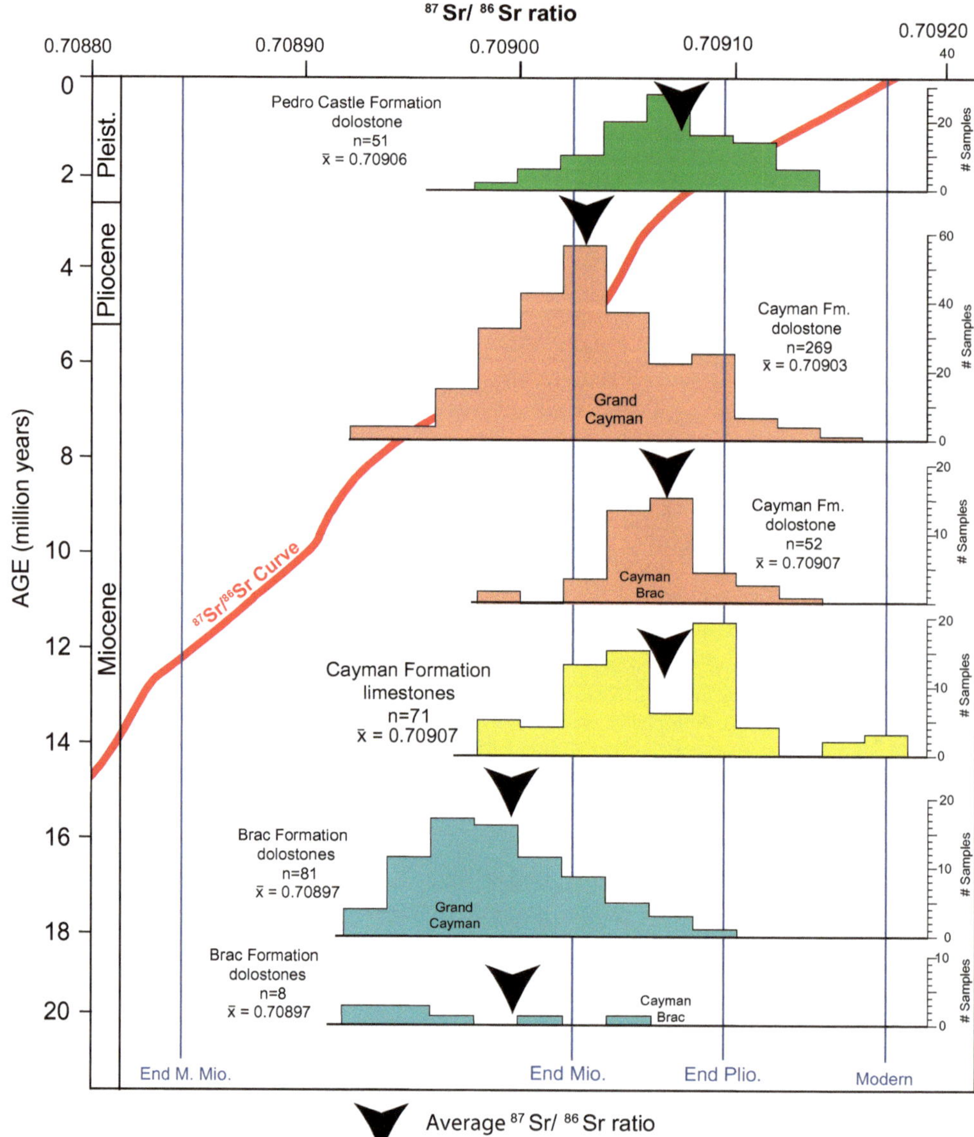

Fig. 6.30 Histograms for $^{87}Sr/^{86}Sr$ ratios for the Brac Formation (limestone and dolostone), Cayman Formation (limestone and dolostone), and Pedro Castle Formation from Grand Cayman and Cayman Brac. Red line shows the $^{87}Sr/^{86}Sr$ curve for seawater for the Miocene to Pleistocene (from McArthur et al. 2012). The vertical blue lines indicate the $^{87}Sr/^{86}Sr$ values for seawater at the end of the Middle Miocene, end of Miocene, end of Pliocene (from McArthur et al. 2012) and modern seawater (from Pleydell et al. 1990)

Modern Seawater

The $^{87}Sr/^{86}Sr$ ratios determined from four samples of seawater from East End, Pease Bay, North Sound, and Rum Point on Grand Cayman ranged from 0.709150 to 0.709180 with an average of 0.709170 (Pleydell et al. 1990, their Table 1). The average value is similar to the value of 0.709160 that Land (1989) determined for seawater from Jamaica.

Potential Problems with $^{87}Sr/^{86}Sr$ Ratios

The use of $^{87}Sr/^{86}Sr$ ratios to determine when dolomitization took place assumes that the $^{87}Sr/^{86}Sr$ ratios were inherited directly from the seawater that mediated the limestone to dolostone transformation. It also assumes that the $^{87}Sr/^{86}Sr$ of that water had not been modified by (1) inheritance of $^{87}Sr/^{86}Sr$ from the aragonite and/or high Mg-calcite of the original limestone (Saller 1984; Vahrenkamp and Swart 1988; Budd 1997), (2) interaction with other carbonates (Land 1991; Budd 1997), and/or (3) interaction with siliciclastic sediments, volcanic deposits, and/or volcanoclastic sediments (Fouke 1994; Machel and Burton 1994; Gill et al. 1995; Budd 1997). Any of these processes would produce $^{87}Sr/^{86}Sr$ ages that are too old.

Other issues that may affect the $^{87}Sr/^{86}Sr$ ratios derived from dolostones include:

- Recrystallization of the dolostones that may produce a false age (Wang et al. 2019).
- Zoned dolomite crystals (Fig. 6.2a), like many in the Cayman Formation that have a "dirty" cores that formed through replacement and clear rims that may have formed later as cements ("overdolomitization" of Choquette and

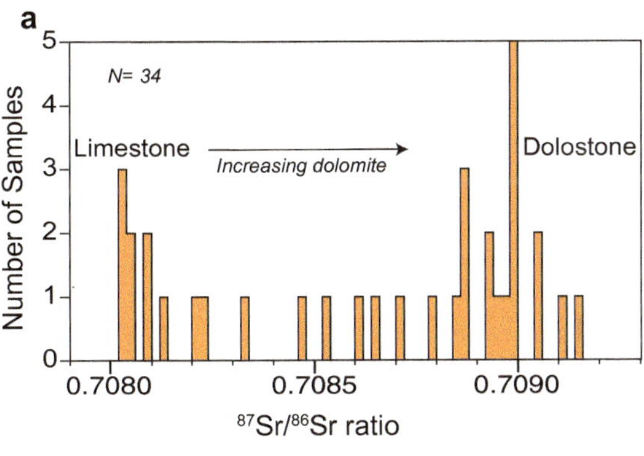

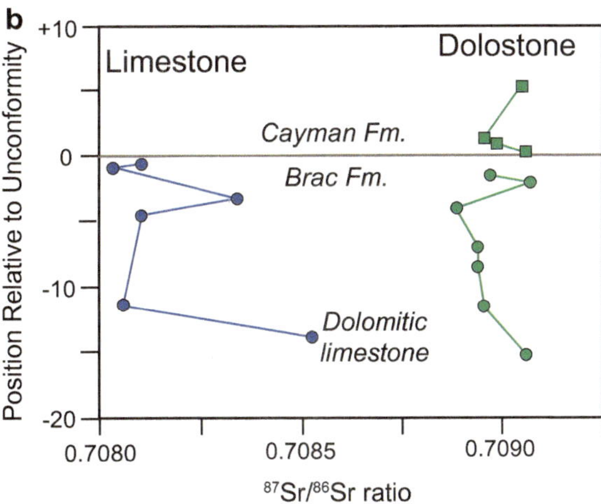

Fig. 6.31 $^{87}Sr/^{86}Sr$ ratio from Brac Formation on Cayman Brac. **a** Bar graph showing progressive increase in $^{87}Sr/^{86}Sr$ values as amount of dolomite in limestones increases. **b** Graph showing comparison of $^{87}Sr/^{86}Sr$ values from limestones and dolostones in upper part of the Brac Formation exposed on southeast coast of Cayman Brac. Modified from Jones and Luth (2003, their Fig. 9)

Hiatt 2008). $^{87}Sr/^{86}Sr$ ratios determined from samples with these two components may represent values that are an average of the two components.

- Limpid dolomite cements, found in many of the Cayman dolostones (Fig. 6.3c, d, e, g), represent a later phase of dolomite formation that may have occurred at any time after dolomitization of the original limestones. Analysis of samples with high percentages of dolomite cements may produce $^{87}Sr/^{86}Sr$ values that are "too young" and hence not truly representative of the time when the matrix was pervasively dolomitized.
- The $^{87}Sr/^{86}Sr$ ratios of the limestones and dolostones are typically determined from small powdered samples and therefore represent an average of the numerous crystals that form that powder, irrespective of whether or not they all formed at the same time.

Age determination from the $^{87}Sr/^{86}Sr$ ratios of the dolostones is based on comparison with the $^{87}Sr/^{86}Sr$–age curves for seawater over the last 30–40 million years. Given that at least seven different $^{87}Sr/^{86}Sr$–age curves have been developed for this period (Burke et al. 1982; DePaolo 1986; Hess et al. 1986; Hodell et al. 1991; Ohde and Elderfield 1992; Farrell et al. 1995; McArthur et al. 2012), different ages can be derived from the same $^{87}Sr/^{86}Sr$ ratio. Although predicated on a continued overall increase in the $^{87}Sr/^{86}Sr$ ratio over the last 40 million years, there have been periods when $^{87}Sr/^{86}Sr$ ratios remained relatively constant for extended periods of time. These points need to be considered when using the $^{87}Sr/^{86}Sr$ ratios to establish the time(s) when dolomitization of the limestones in the Bluff Group took place.

Brac Formation

The distinct offset of the $^{87}Sr/^{86}Sr$ ratios of the dolostones relative to the limestones in the Brac Formation on Cayman Brac (Fig. 6.31b) is consistent with the time lapse between deposition of the limestones and later dolomitization. On Cayman Brac, the paleontological evidence and $^{87}Sr/^{86}Sr$ ratios point to an upper Oligocene age for the limestones in the upper part of the Brac Formation. The mean $^{87}Sr/^{86}Sr$ values of the dolostones from the Brac Formation on Cayman Brac and Grand Cayman suggests that dolomitization took place during the Late Miocene ($\sim$ 8 Ma) if the $^{87}Sr/^{86}Sr$ trend of McArthur et al. (2012) is used.

Cayman Formation and Pedro Castle Formation

The $^{87}Sr/^{86}Sr$ ratios from limestones and dolostones of the Cayman Formation on Cayman Brac and Grand Cayman provide problematic results, despite the large number of values (321 dolostones, 71 limestones) that are available (Fig. 6.30). The $^{87}Sr/^{86}Sr$ values include some that are equal to or higher than that of modern seawater (Fig. 6.30). The mean $^{87}Sr/^{86}Sr$ for the dolostones from Cayman Brac is the same as the mean value for the limestones on Grand Cayman. In contrast, the mean $^{87}Sr/^{86}Sr$ of the dolostones on Grand Cayman is lower than either the mean value of the dolostones on Cayman Brac or the limestones on Grand Cayman (Fig. 6.30). Similarly, the range of the $^{87}Sr/^{86}Sr$ values derived from the dolostones on Grand Cayman include numerous samples with lower values than those from Cayman Brac. The range of $^{87}Sr/^{86}Sr$ ratios from limestones and dolostones of the Pedro Castle Formation is similar to that of the underlying Cayman Formation (Fig. 6.30). The average $^{87}Sr/^{86}Sr$ ratio of 0.70905, however, is slightly higher than that from the dolostones of the Cayman Formation but slightly lower than the average $^{87}Sr/^{86}Sr$ ratio of the limestones in the Cayman Formation (Fig. 6.30).

Relative to the $^{87}Sr/^{86}Sr$ curve of McArthur et al. (2012), the full $^{87}Sr/^{86}Sr$ range from the limestones and dolostones of the Cayman Formation indicate an upper Miocene to late Pleistocene ($\sim$7 to 1 Ma) age (Fig. 6.30). The histogram, which includes all of the dolostone $^{87}Sr/^{86}Sr$ values for the Cayman Formation on Grand Cayman, has a main mode that corresponds to the Miocene–Pliocene boundary and a minor mode that corresponds to an early Pleistocene age (Fig. 6.30). The mode for the dolostones on Cayman Brac lies between the two modes evident in the Grand Cayman data. The range of $^{87}Sr/^{86}Sr$ from the dolostones of the Pedro Castle Formation is similar to that for the Cayman Formation with the mode and mean "values" being similar to those of the dolostones on Cayman Brac.

Assessment of the $^{87}Sr/^{86}Sr$ ratios based on solely on the histograms compiled from all of the samples, irrespective of their geographic and stratigraphic locations, tacitly assumes that dolomitization of the strata was uniform through time and space. The wide spread in the $^{87}Sr/^{86}Sr$ ratios from the dolostones and their similarity to those from the limestones suggests that these ratios may not be solely a function of time.

There is little evidence that the $^{87}Sr/^{86}Sr$ ratios are linked to stratigraphic level or the type of dolostone and/or limestone from which they were derived, as is evident from comparison of the $^{87}Sr/^{86}Sr$ ratios from wells in the central (Fig. 6.32) and eastern (Fig, 6.33) parts of Grand Cayman. For the central cross-section, which comprises five wells (RTR#1, AIR#1, DTE#1, NSC#1, BOG#1) over a distance of 3 km inland from the south coastline, the following points are evident.

- In each well, the $^{87}Sr/^{86}Sr$ ratios vary stratigraphically with no evidence of a trend that could be allied with a systematic temporal change in the $^{87}Sr/^{86}Sr$ ratios.
- In wells DTE#1, NSC#1, and BOG#1, located in the central part of the island where the Cayman Formation is formed largely of limestone, the $^{87}Sr/^{86}Sr$ ratios fluctuate stratigraphically and from well to well (Fig. 6.32). In well NSC#1, the $^{87}Sr/^{86}Sr$ ratios fluctuate about the datum value of 0.70903 (value for seawater at end of the Miocene—from McArthur et al. 2012), whereas in wells BOG#1 and DTE#1, the $^{87}Sr/^{86}Sr$ ratios are all higher than 0.70903.
- In wells RTR#1 and AIR#1, located near the coast, the $^{87}Sr/^{86}Sr$ ratios fluctuate stratigraphically with values similar to those in wells DTE#1, NSC#1, and BOG#1. There is no evidence that the $^{87}Sr/^{86}Sr$ ratios are related to fluctuations in the LCD:HCD ratios that vary throughout the sequence (Fig. 6.32).
- In RTR#1, the Cayman Formation is formed largely of LCD and HCD with only minor amounts of calcite in the upper part of the well. The $^{87}Sr/^{86}Sr$ ratios in RTR#1 fluctuate about the datum of 0.70903 whereas those in AIR#1, which is only 700 m from RTR#1, are all above the datum value (Fig. 6.32). The $^{87}Sr/^{86}Sr$ ratios in RTR#1 are similar to those in NSC#1, even though the sequences are lithologically different (Fig. 6.32).

For the eastern cross-section (Fig. 6.33), which includes wells EEV#2, EER#2, HRQ#1, and FFM#1 that are spread over a distance of $\sim$ 4.5 km, the following points are evident.

- There is considerable variation in the $^{87}Sr/^{86}Sr$ ratios – in wells EEV#2 and EER#2, all of the values except two in EER#2, are higher than the datum value of 0.70903; whereas in well HRQ#1 all of the values except one, are below 0.70903, and in FFM#1, the values oscillate around the datum (Fig. 6.33).
- In some wells, there is a hint that the higher $^{87}Sr/^{86}Sr$ ratios may be related to an increased content of LCD, but this is not universally the case.
- Comparison of the four wells show that there is a geographic change inland with the $^{87}Sr/^{86}Sr$ ratios in the near-coast wells being higher than those of the inland wells (Fig. 6.33).

Evident from the wells shown in Figs. 6.32 and 6.33 and other wells not included in those figures is the fact that the stratigraphic fluctuations in the $^{87}Sr/^{86}Sr$ ratios make it difficult to identify systematic trends for each well. This issue, however, can be partly mitigated by fitting a polynomial equation to the $^{87}Sr/^{86}Sr$ ratio data from each well (Fig. 6.34). Some of those $^{87}Sr/^{86}Sr$ versus depth profiles are characterized by a general increase in the $^{87}Sr/^{86}Sr$ values towards the surface (Fig. 6.34a), whereas other profiles are relatively consistent with depth or fluctuate with little consistency from one well to another (Fig. 6.34b). Relative to the $^{87}Sr/^{86}Sr$ values that McArthur et al. (2012) gave for seawater at the end of the Miocene, end of Pliocene, and end of Pleistocene, some wells have $^{87}Sr/^{86}Sr$ trends that range from the Upper Miocene to Upper Pliocene, whereas others have $^{87}Sr/^{86}Sr$ trends that are largely in the Pliocene (Fig. 6.34). For example, the $^{87}Sr/^{86}Sr$ trend for wells EEV#2, is largely within the Pleistocene whereas the trend for EER#2 oscillates between the Pliocene and the Pleistocene (Fig. 6.34).

Jones and Luth (2003) linked dolomitization of the Cayman Formation to transgressive seas that followed lithification and karst development of the original limestones. Wang et al. (2019) reviewed the use of $^{87}Sr/^{86}Sr$

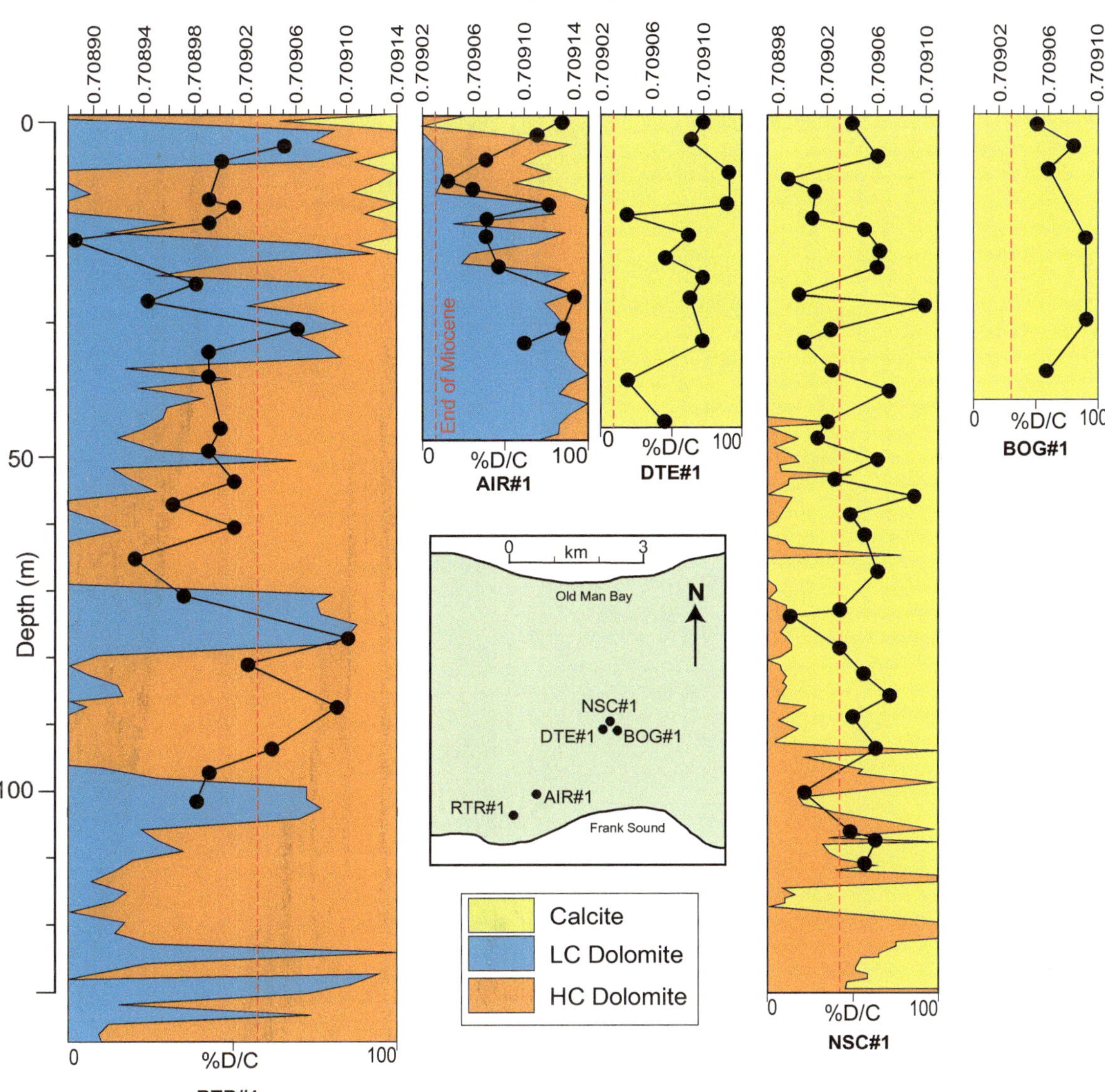

Fig. 6.32 Comparison of $^{87}Sr/^{86}Sr$ values and host lithologies (calcite, LCD, HCD) in Cayman Formation in wells RTR#1, AIR#1, DTE#1, and NSC#1 (see inset map for locations), central part of Grand Cayman. Width of each column reflects the range of $^{87}Sr/^{86}Sr$ values in each well

ratios for dating dolomitization on the Cayman Islands and concluded that numerous problems arise because of the many different $^{87}Sr/^{86}Sr$ – age curves that have been produced for seawater. Although all of these curves show the same general trends, subtle differences between them can translate into significant time differences for the same Sr ratio. They argued, however, that the "time-transgressive" dolomitization that occurred between the late Miocene and late Pleistocene produced the dolomitized sequences on the Cayman Islands.

Comparison with Other Island Dolostones

Vahrenkamp et al. (1991) examined the developmental history of the dolostones on the Bahama platform where the Pleistocene Lucayan Formation rests unconformably on top of an older succession that comprises a lower Miocene unit formed of coarsely crystalline dolostone that is uncomfortably overlain by a Pliocene unit that is formed of cream to white, poorly to well lithified dolostone. $Sr^{87}/^{86}Sr$ ratios from the two dolostone units (Vahrenkamp et al. 1991, their Table 2), to a depth of 180 m (equivalent to depths of

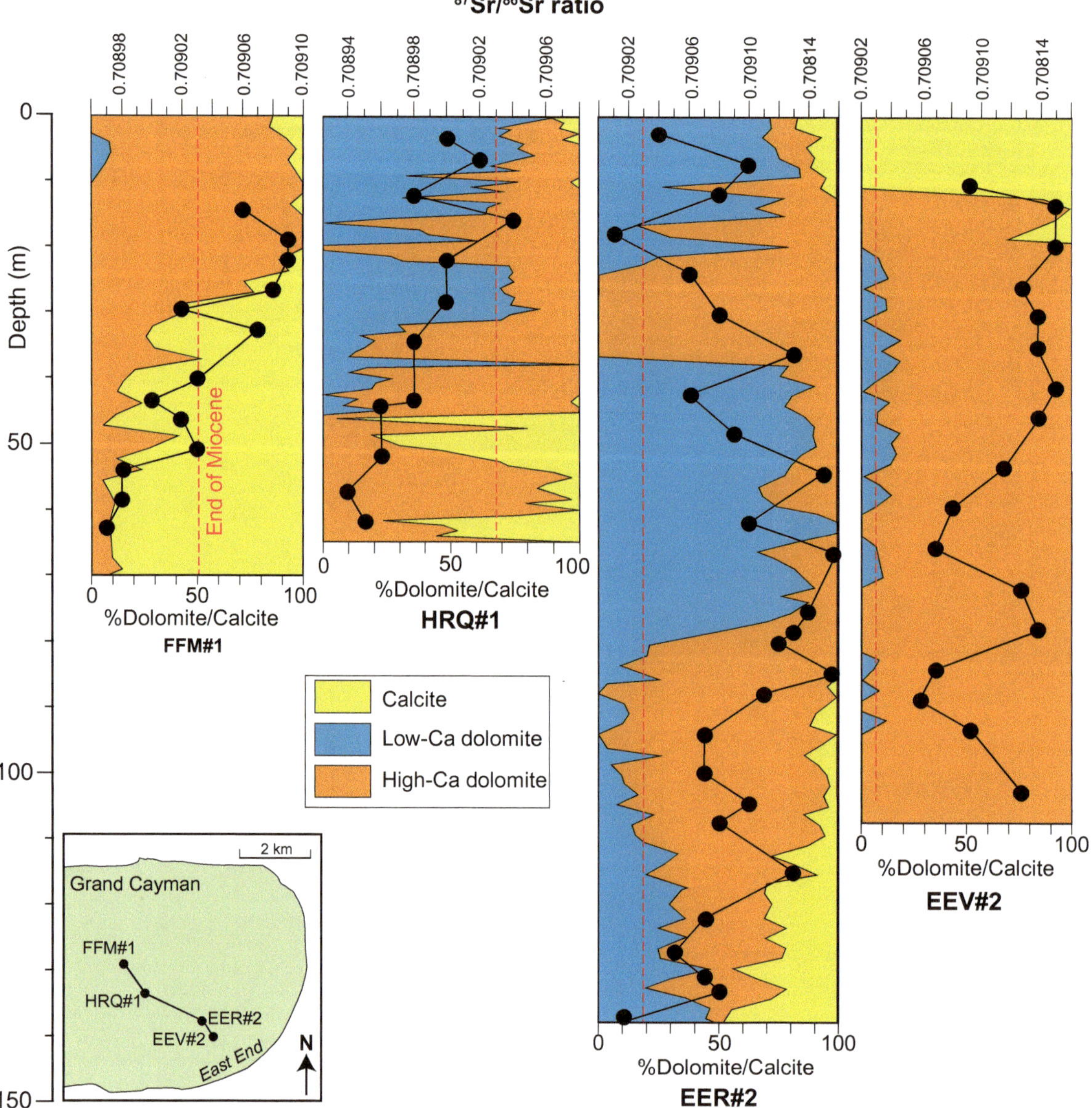

Fig. 6.33 Comparison of $^{87}Sr/^{86}Sr$ values and host lithologies (calcite, LCD, HCD) in Cayman Formation in wells EEV#2, EER#2, HRQ#1, and FFM#1 (see inset map for locations), east end of Grand Cayman. Width of each column reflects the range of $^{87}Sr/^{86}Sr$ values in each well

samples from the Cayman Islands), yields a range of values and an average value that are similar to those for the unconformity bounded Pedro Castle Formation and Cayman Formation on the Cayman Islands (Fig. 6.35). Vahrenkamp et al. (1991) argued that the $Sr^{87}/^{86}Sr$ from the Bahamian dolostones, which they considered relative to the DePaolo (1986) $Sr^{87}/^{86}Sr$ seawater-age curve, indicated multiple episodes of dolomitization, with phases I, II, and III that occurred during the Middle Miocene, end of Pliocene, and the late Pliocene to Pleistocene. Minor phases included Phase A (late Early Miocene) and Phase IV (Late Pleistocene). They argued that the spread of $^{87}Sr/^{86}Sr$ ratios reflected the fact that different parts of the succession were dolomitized during different time periods.

Budd (1997, his Fig. 19), based on ~140 $Sr^{87}/^{86}Sr$ ratios from various islands throughout the world (excluding those deemed to have extraneous Sr), suggested that there have been six phases of dolomitization that occurred during the (A) late Early Miocene, (B) late Middle Miocene to Early Late Miocene, (C) late Late Miocene, (D) Pliocene to early Pleistocene, (E) Pliocene/Early Pleistocene, (F) early Pleistocene, and (G) Middle to Late Pleistocene. He noted that

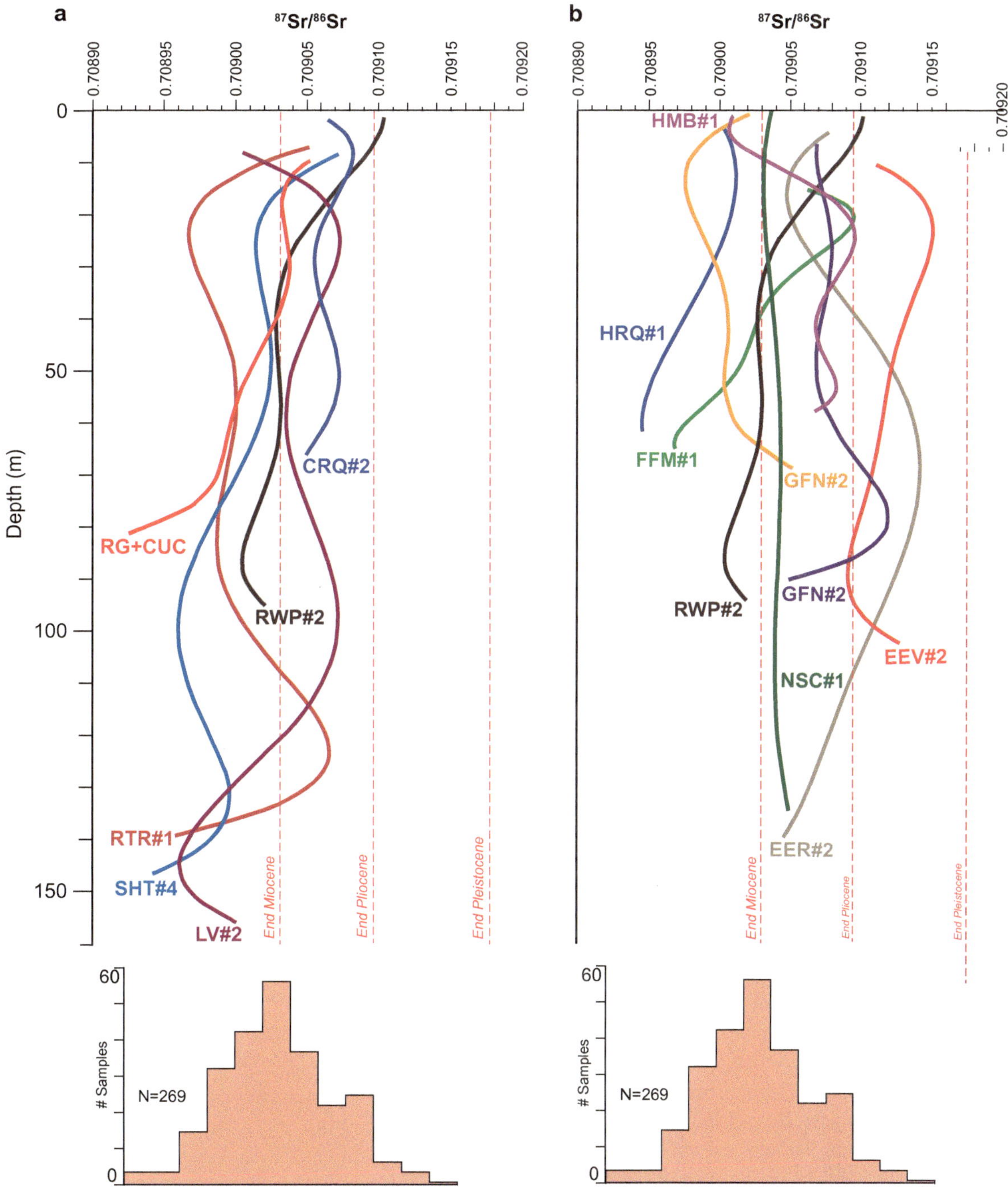

Fig. 6.34 Comparison of $^{87}Sr/^{86}Sr$ values versus depth for **a** various wells on the western part of Grand Cayman and well CRQ#2 on Cayman Brac, and **b** wells in central and eastern part of Grand Cayman. The trend line shown for each well is a polynomial fit to the stratigraphic trend in the $^{87}Sr/^{86}Sr$ values (like those Figs. 8.37 and 8.38). The histograms at the bottom of each graph show the overall distribution of $^{87}Sr/^{86}Sr$ values for the Cayman Formation on Grand Cayman. The $^{87}Sr/^{86}Sr$ values for seawater at the end of the Miocene, Pliocene, and Pleistocene (red dashed lines), from McArthur et al. (2012) are shown for comparative purposes

those events matched the five intervals of dolomitization that had been postulated for the Bahamas by Vahrenkamp et al. (1991), Ohde and Elderfield (1992), and the Philippine Sea and northeast Australia by McKenzie et al. (1993). Budd (1997) suggested that the events E, F, and G were local events whereas the older events A, B, and C produced massive and more widespread dolomite bodies. He also noted that in some cases, as on the Cayman Islands, the $Sr^{87/86}Sr$ ratios became intermingled as parts of the successions were dolomitized during different events.

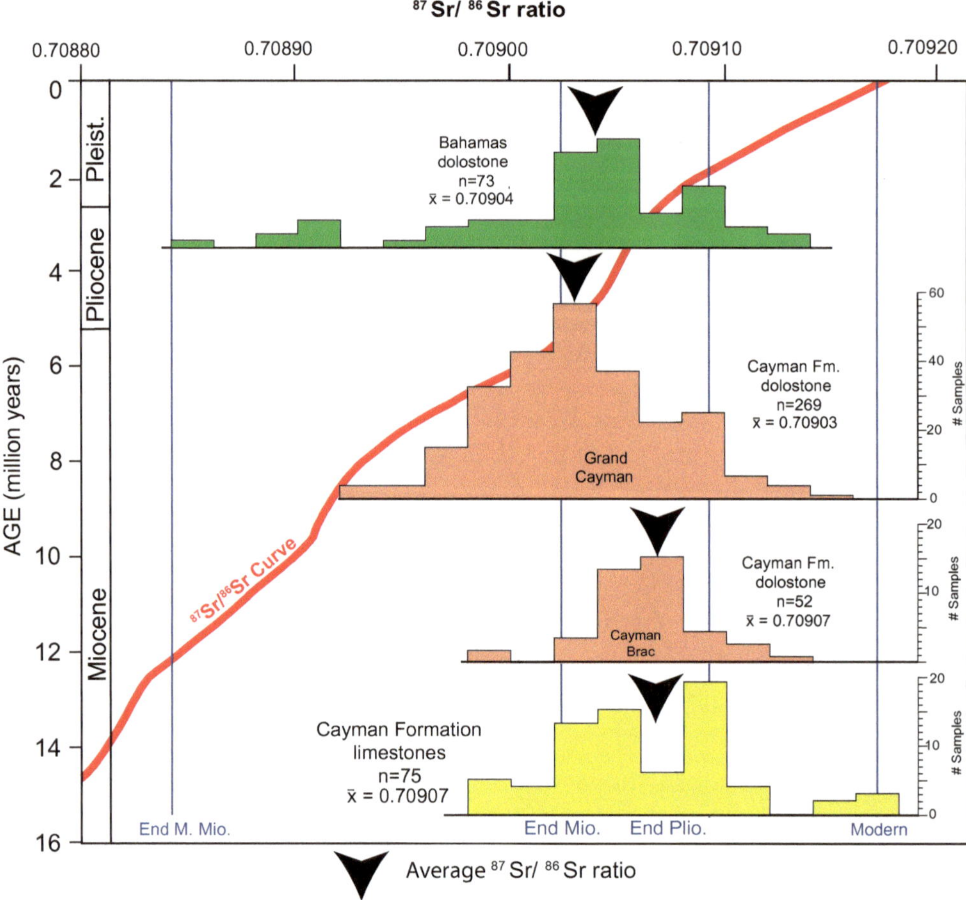

Fig. 6.35 Comparison of ^{87}Sr/^{86}Sr values from limestones and dolostones of the Cayman Formation on Grand Cayman and Cayman Brac relative to the ^{87}Sr/^{86}Sr values of dolostones on the Bahamas (data from Vahrenkamp et al. 1991, their Table 2). Red line shows the ^{87}Sr/^{86}Sr curve for seawater for the Miocene to Pleistocene (from McArthur et al. 2012). The vertical blue lines indicate the ^{87}Sr/^{86}Sr values for seawater at the end of the Middle Miocene, end of Miocene, end of Pliocene (from McArthur et al. 2012) and modern seawater (from Pleydell et al. 1990)

6.3 Discussion

Many of the theoretical dolomitization models proposed to explain the large-scale replacement of limestone by dolomite are not viable for small islands that are surrounded by deep oceanic waters. For example, the notion that the dolostones of the Cayman Islands may have originated on expansive intertidal and/or supratidal flats is improbable and there is no evidence to indicate that hydrothermal fluids were involved. Even with the rejection of various dolomitization models like these, "island dolostones" have been variously attributed to many different processes including tidal pumping, seepage reflux, brine reflux, mixing zone, ocean current pumping, and/or Kohout convection models (see Tucker and Wright 1990, their Fig. 8.31; Budd 1997, his Fig. 1).

Irrespective of the "dolomitization model" invoked, the transformation of thick limestone successions to dolostones on the Cayman Islands required a large supply of Mg and a mechanism for freely circulating that Mg through the bedrock. Seawater around the islands is the most obvious source for the Mg. Today, on Grand Cayman, circulation of seawater through the bedrock is readily apparent from the short lag time between high tide at the coastline and its effect on the water table in the central part of the east end of Grand Cayman, which is as little as 2 hours (Ng 1990). The notion that seawater mediated dolomitization of the limestones in each formation in the Bluff Group is supported by the following considerations.

- The δ^{13}C values (most between +0.5 and + 3.2‰) of the Cayman dolostones are consistent with the range that Budd (1997) considered typical of island dolostones with inheritance from the marine carbon of the precursor limestone.
- The average δ^{18}O values between +1.5 and + 3.5‰ are consistent with the range that Budd (1997) considered "normal" for island dolostones and consistent with late Cenozoic seawater (0 to + 2‰ SMOW) or slightly evaporated seawater (+2.8‰ SMOW).
- The Mn, Fe, and Sr concentrations in the dolostones are similar to those in seawater.
- The PAAS normalized REE+Y patterns (Fig. 6.28) of the dolostones are similar to those from seawater on the Misteriosa Bank, ~400 km of Grand Cayman and the BATS seawater from the North Atlantic Ocean (Fig. 6.34).

- Although the high Na content in the Cayman dolostones may indicate that evaporated seawater was involved with the dolomitization, the Na may have been trapped between the dolomite crystals and not bound into the lattices of the dolomite crystals (cf., Budd 1997).

Given the above characteristics and the fact that that each of the Cayman Islands have, at least for the last 30 million years, been surrounded by deep oceanic waters, it is logical to argue that the Mg needed for dolomitization came from seawater. Thus, dolomitization of the strata would have been controlled by circulation of that seawater through the limestones via the operative hydrological regime, which would have been linked to sea level and its oscillations through time.

Establishing a dolomitization model for the Bluff Group is inexorably linked to the issue of time given that the geological setting of the Cayman Islands has constantly evolved and changed over the last 25 million years. Although it is clear that dolomitization of the limestones can be no older than the original sediments in each formation, the precise ages of dolomitization are difficult to establish. Such ages have typically been assessed by the $^{87}Sr/^{86}Sr$ ratios of the dolostones with the premise that their $^{87}Sr/^{86}Sr$ ratios were inherited directly from seawater with the notion that those $^{87}Sr/^{86}Sr$ ratios of have progressively increased over the last 40 million years (Vahrenkamp et al. 1991; Budd 1997; Wang et al. 2019).

The ages of the limestones in the Brac Formation, Cayman Formation, and Pedro Castle Formation on the Cayman Islands are known from their fossil content that can, in some cases be supported by the $^{87}Sr/^{86}Sr$ ratios. Such dates can also be supported, to some extent, by consideration of eustatic sea-level curves that show the times when lowstand and highstand conditions existed. Although fluctuations in sea level over the last 30 million years have been significant, different schemes offer vastly different ideas regarding the absolute ranges of those changes (Fig. 6.36). The sea-level curve of Vail et al. (1977a) shows lowstands that probably correspond to the unconformities in the Cayman succession and Haq et al. (1987) suggested that some of those unconformities formed when sea level was 100 to 200 m below present day sea level (Fig. 6.36). Conversely, Miller et al. (2005) suggested that the magnitudes of the lowstands were much lower than those suggested by Haq et al. (1987) (Fig. 6.36).

The smoothed sea-level curves of Miller et al. (2020a, their Fig. 1) and Miller et al. (2020b, their Fig. 2) show highstand positions during the late Late Oligocene, the late Early and early Middle Miocene, and the middle Pliocene with lowstands in the intervening periods (Figs. 6.36 and 6.37). Based on this sea level curve, the sediments of the Brac Formation, Cayman Formation, and Pedro Castle Formation were probably deposited during these three highstand periods (Figs. 6.36 and 6.37). Such dates are consistent with the

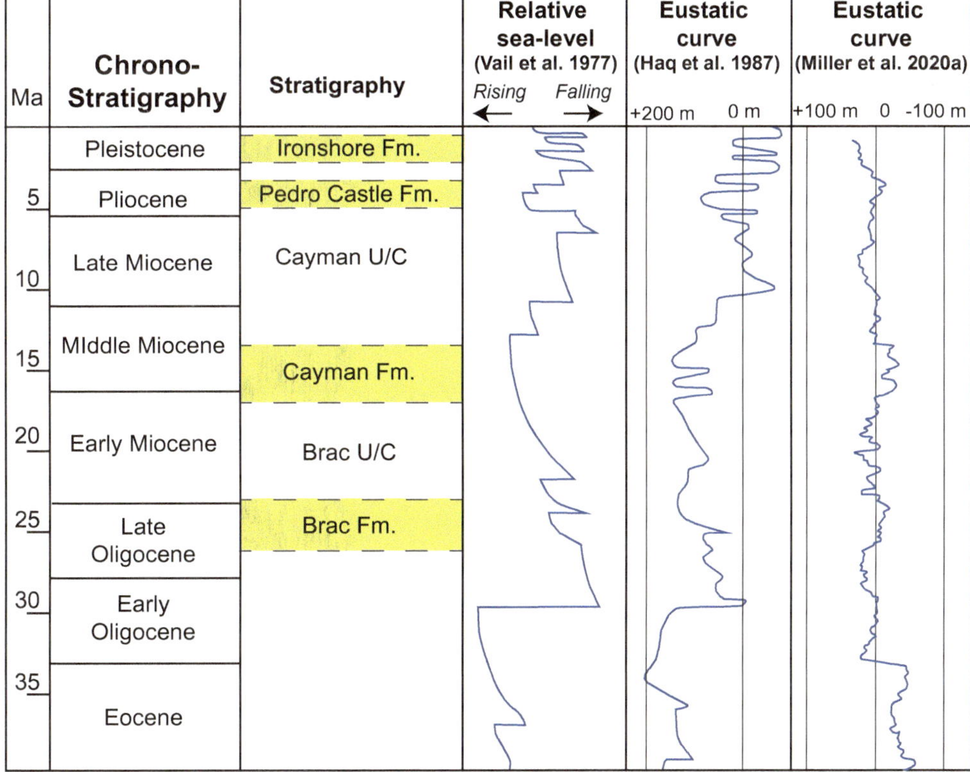

Fig. 6.36 Relationship of unconformity bounded Brac Formation, Cayman Formation, and Pedro Castle Formation relative to sea levels suggested by Vail et al. (1977a), Haq et al. (1987) and Miller et al. (2020a; b). Modified from McCormick and Jones (2021, their Fig. 5)

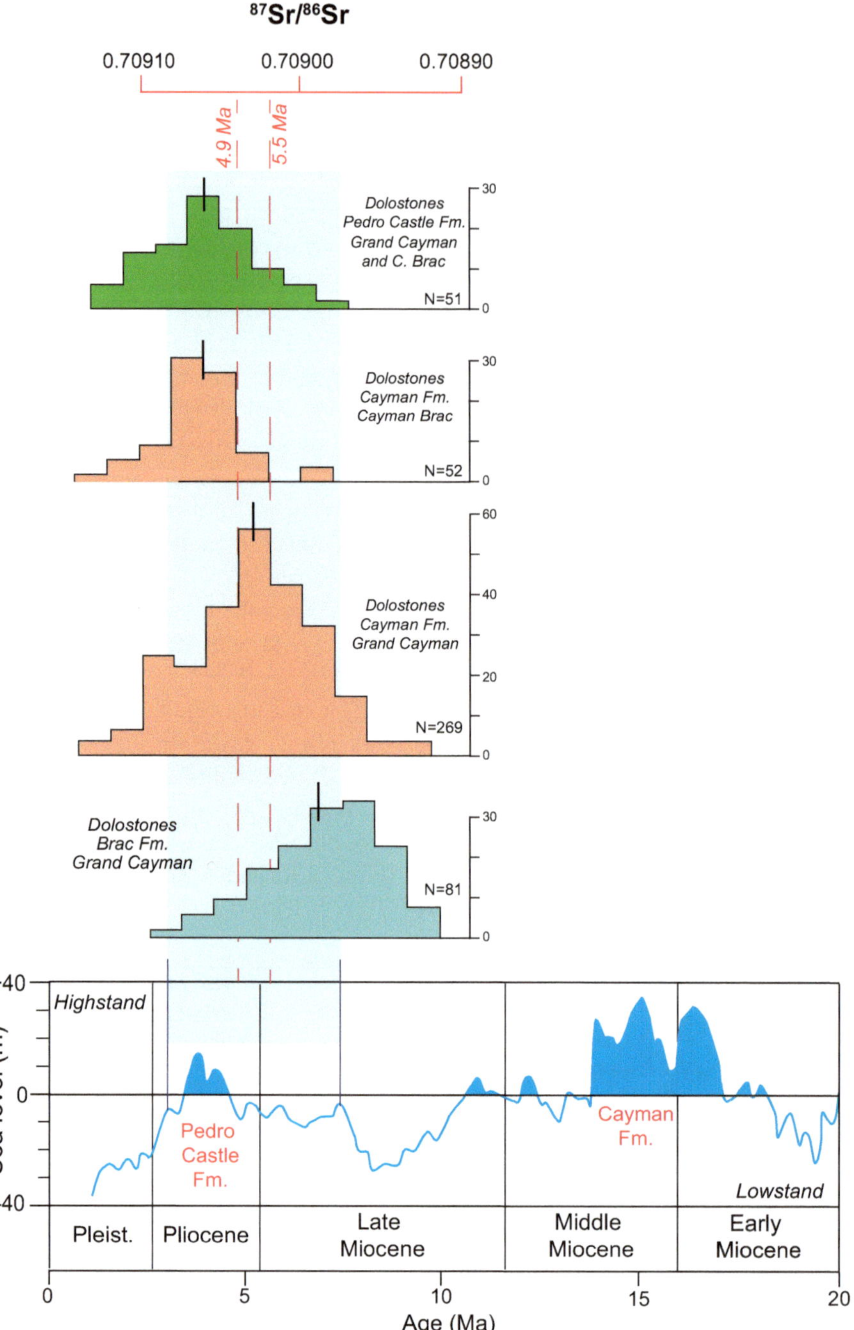

Fig. 6.37 Histograms for $^{87}Sr/^{86}Sr$ ratios from the Pedro Castle Formation, Cayman Formation, and Brac Formation on Grand Cayman and Cayman Brac plotted relative to the eustatic sea level curve of Miller et al. (2020b, their Fig. 2). The histograms (black vertical line indicates mean value) are correlated to the sea level curve on the basis of the ages for $^{87}Sr/^{86}Sr$ ratios of 0.709010 and 0.709020 as determined from the "look-up" table of McArthur et al. (2012). Ages of the Cayman Formation and Pedro Castle Formation are superimposed on the eustatic sea level curve

biostratigraphic evidence and the $^{87}Sr/^{86}Sr$ ratios of the limestones in each of the formations. Although the absolute sea levels associated with each lowstand and highstand phase must be treated with some caution, they generally seem to have oscillated between ~ 30 m below and ~ 30 m above present-day sea level. The unconformities that form the boundaries between the formations developed during the lowstand periods when exposure to the atmosphere resulted in lithification, weathering, and karst development of sediments that had been deposited in the previous depositional cycle. Available evidence indicates that lithification of the original sediments with leaching of the aragonite skeletons (e.g., corals, bivalves, gastropods) took place before or during dolomitization.

For other islands, including the Bahamas, Bonaire, and Curacao, pervasive dolomitization has been attributed to numerous, time-specific events that have been developed on the basis of the $^{87}Sr/^{86}Sr$ ratios derived from the dolostone (e.g., Vahrenkamp et al. 1991; Budd 1997). This model is difficult to apply to the Cayman Formation on the Cayman Islands because of the stratigraphic variability that is evident in the $^{87}Sr/^{86}Sr$ ratios of individual wells and the geographic variability from well to well (Figs. 6.32, 6.33 and 6.34). Budd (1997), based on data in Pleydell et al. (1990) suggested that the $^{87}Sr/^{86}Sr$ ratios in the dolostones of the Grand Cayman were "intermingled" because the strata had experienced more than one phase of dolomitization. For many islands, assessment of the age(s) of dolomitization have been based on a limited number of $^{87}Sr/^{86}Sr$ ratios from a single well or outcrop succession and therefore offers no assessment of possible geographic variations in the $^{87}Sr/^{86}Sr$ ratios. It is possible that these different time-specific events are a product of small sample sizes rather than different "dolomitization events".

The $^{87}Sr/^{86}Sr$ ratios of the dolostones from the Cayman Formation collectively indicate that dolomitization occurred during the Late Miocene, Pliocene, and possibly Pleistocene (Fig. 6.37). According to the look-up tables in McArthur et al. (2012), $^{87}Sr/^{86}Sr$ values of 0.70902 and 0.70904 correspond to ages of 5.46 and 4.90 million years, respectively. If these ages are accepted, then the histograms for the $^{87}Sr/^{86}Sr$ ratios of the dolostones of the Brac Formation on Grand Cayman, the dolostones of the Cayman Formation on Cayman Brac and Grand Cayman, and the dolostones of the Pedro Castle Formation on Cayman Brac and Grand Cayman can be linked to the sea level curve of Miller et al. (2020a, their Fig. 1) and Miller et al. (2020b, their Fig. 2) (Fig. 6.37). If it is accepted that the $^{87}Sr/^{86}Sr$ ratios of these dolostones collectively reflect the time periods when dolomitization took place, then the following important points are apparent.

- For the Brac Formation, most of the dolomitization occurred during the Late Miocene.
- For the Cayman Formation and Pedro Castle Formation, dolomitization took place during the late Late Miocene and Pliocene.
- The range of $^{87}Sr/^{86}Sr$ values for the dolostones from each formation suggests that dolomitization took place over extended periods of time.
- For each formation there is no clear evidence of multiple, time-separated phases of dolomitization.

The smoothed sea level curve of Miller (2020a, b) shows that sea level was up to 30 m below modern sea level between the late Middle Miocene and upper Late Miocene (Fig. 6.37). During that lowstand period, the carbonates of the Cayman Formation would have been subaerially exposed and subject to weathering and karst development. During the late Late Miocene (Fig. 6.37), sea level rose to ~10 m below modern sea level before stabilizing until the middle Pliocene when sea level rose to ~ 15 m above modern sea level (Fig. 6.37). Under that scenario, the lithified strata of the Cayman Formation would have been consistently bathed in marine waters in a manner akin to the present data situation. The highstand during the middle to late Pliocene allowed deposition of carbonate sediments that now form the Pedro Castle Formation. The $^{87}Sr/^{86}Sr$ ratios from the dolostones of the Pedro Castle Formation and the underlying Cayman Formation, suggest that dolomitization took place during and soon after deposition of the sediments that now form the Pedro Castle Formation.

Dolomitization of the Bluff Group on the Cayman Islands has been framed in the context of the smoothed eustatic sea level developed by Miller et al. (2020a; b). It should be noted, however, that other sea level curves (Fig. 6.36), like those provided by Vail et al. (1977b) and Haq et al. (1987), for example, show some lowstands during the Miocene and Pliocene that may have been 100 to 200 m below present day sea level (Fig. 6.36). Similarly, Dowsett and Cronin (1990) suggested that the middle Pliocene highstand, based on information from the U.S. Atlantic coastal plain, was 35 ± 18 m above present day sea level. Raymo et al. (2011), however, pointed out that estimates for the Middle Pliocene sea level range from 10 to 40 m above present day sea level with a figure of 25 m commonly being adopted. They attributed much of the variance in these estimates to the fact that post-depositional subsidence and/or uplift had not been considered. Although the periods of lowstands and highstands during the Miocene and Pliocene are relatively consistent from study to study, the magnitudes of sea level rise or fall are open to debate.

Available data indicate that seawater mediated dolomitization as it moved through the bedrock of the Cayman Islands. These hydrological regimes would have been linked to the position of sea level relative to the elevation of each island (Jones and Luth 2003; Ren and Jones 2018; Wang et al. 2019). For the Cayman Islands, assessment of those sea level positions is complicated by the fact that each island is located on different fault blocks and evolved with different tectonic histories. McCormick and Jones (2021) suggested that (1) during the Miocene, Grand Cayman and Cayman Brac both subsided by 100 to 175 m but with the rate of sediment accumulation keeping pace with that subsidence, and (2) during the Pliocene, the tectonic histories diverged

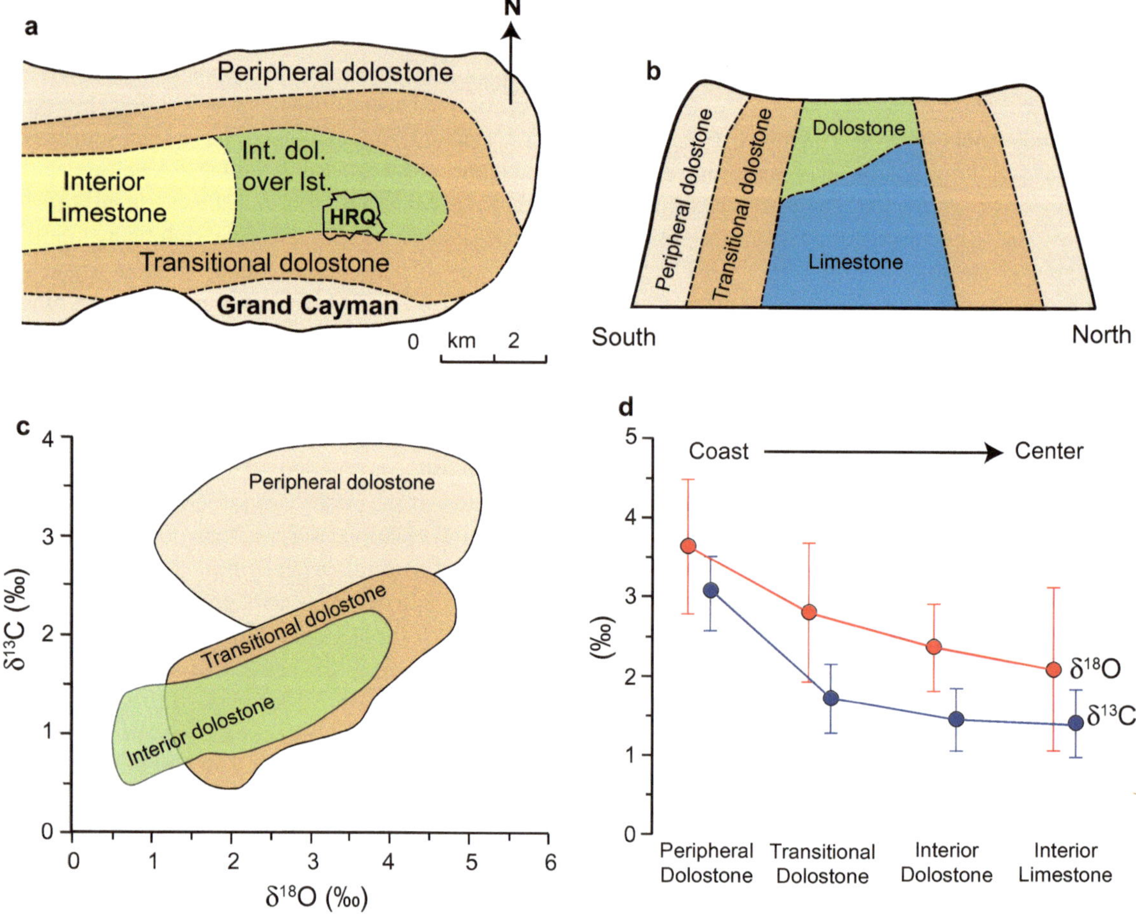

Fig. 6.38 Schematic map showing distribution of dolostone zones on eastern part of Grand Cayman. **a** Distribution of peripheral and transitional dolostones zones around central core that is formed of limestone that is locally capped by dolostone. HRQ – outline of High Rock Quarry. **b** Schematic north–south cross-section showing arrangement of dolostone and dolostone capping limestone in central part of island. **c** Comparison of $\delta^{13}C$ and $\delta^{18}O$ isotopes for dolostones in the peripheral, transitional, and interior dolostones/limestones on eastern part of Grand Cayman. **d** Trends in range and average $\delta^{13}C$ and $\delta^{18}O$ isotopes for peripheral, transitional, and interior dolostones/limestones on eastern part of Grand Cayman. Modified from Ren and Jones (2017, their Figs. 6 and 18)

with Cayman Brac experiencing uplift and tilting, whereas uplift of Grand Cayman occurred with no rotational component. Liang and Jones (2014) argued that the rotational axis on Cayman Brac was located near the southwest end of the island, such that the northeast end of the island was uplifted by as much as 165 m whereas the southwest end was uplifted by ~ 10 m. The fact the Cayman Formation on the east end of the island is formed entirely of dolostone indicates that dolomitization must have taken place before uplift of Cayman Brac.

For the dolostones in the Cayman Formation on Grand Cayman, Ren and Jones (2018) stressed their geographic and stratigraphic variabilities. From a geographical perspective, Ren and Jones (2017) divided the Cayman Formation on eastern Grand Cayman, from the coastline inland, into the (1) peripheral dolomite zone, formed largely of LCD, (2) transitional zone, formed of LCD and HCD, and (3) interior zone formed of limestones and HCD dolostones (Fig. 6.38a, b). In some parts of the interior zone, the Cayman Formation is formed largely of limestones (e.g., wells NSC#1, DTE#1) whereas in other parts (e.g., High Rock Quarry), surface dolostones overlie limestones and calcareous dolostones (Fig. 6.38B). The lack of a geographic zonation on Cayman Brac was attributed to the fact that Cayman Brac (average ~2 km wide) is significantly narrower than Grand Cayman, which is up to 13 km wide at its eastern end. Ren and Jones (2017) demonstrated that the stable isotopes of the dolostones varied with the geographic zonation Grand Cayman (Fig. 6.38c, d). Although the range of $\delta^{13}C$ and $\delta^{18}O$ values between the different zones overlap, there is a progressive inland decrease in values (Fig. 6.38d). These geographic changes were attributed to the fact that the saline groundwater, as it migrated inland, underwent chemical changes due to water–rock interactions and/or

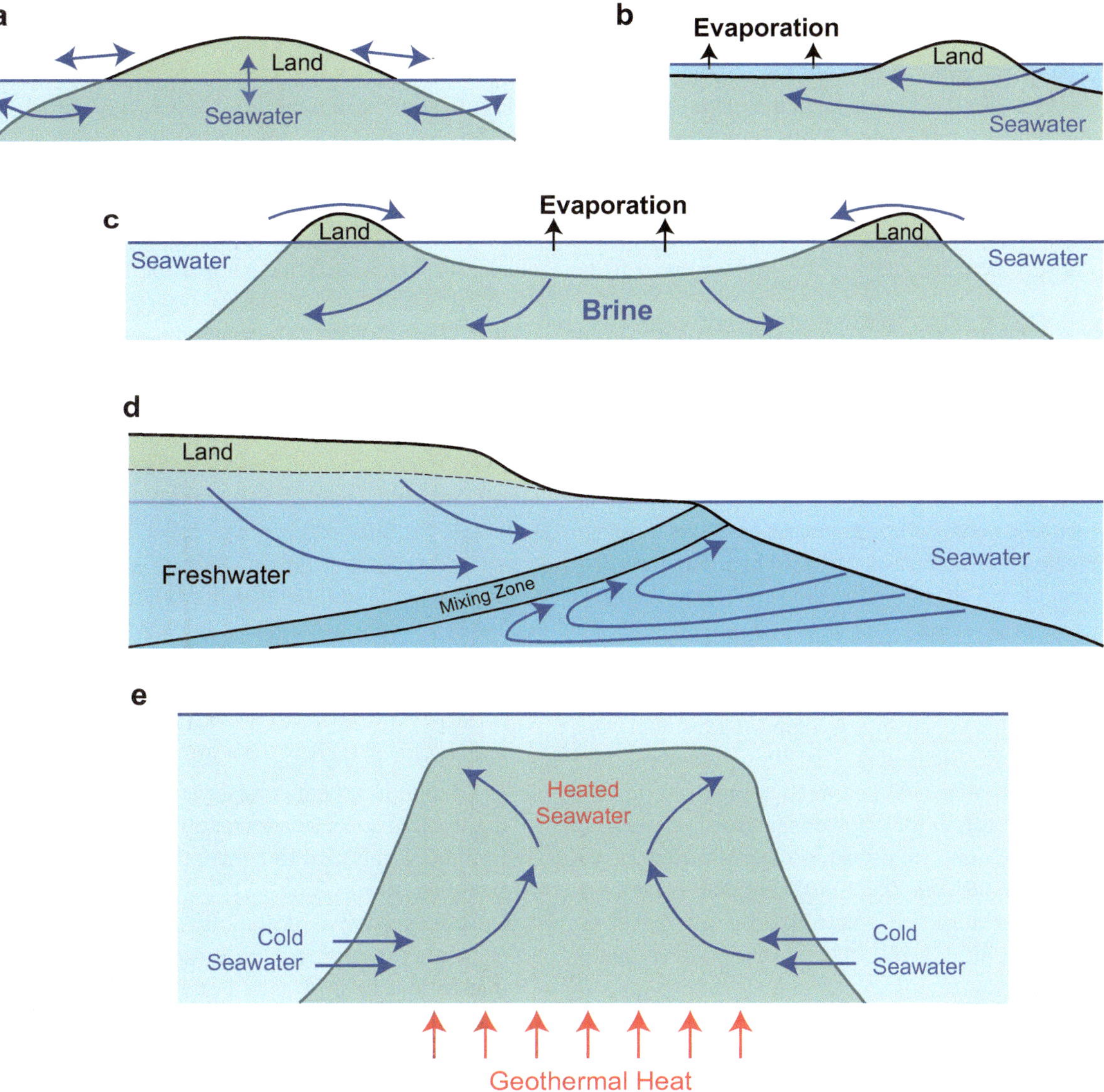

Fig. 6.39 Schematic diagrams showing possible mechanisms for transporting Mg from ocean waters to sites of dolomitization on isolated carbonate islands. **a** Tidal pumping, **b** seepage influx, **c** brine reflux, **d** coastal mixing, and **e** thermal convection. Modified from Budd (1997, his Fig. 1)

mixing with freshwater with the extent of dolomitization being limited by reaction kinetics and seawater supply.

Budd (1997, his Fig. 1) showed that seawater deemed responsible for dolomitization can be circulated through islands via many different pathways (Fig. 6.39). The exact flow regime that governed dolomitization on the Cayman Islands can only be inferred from available data because the ancient flow regimes left no obvious physical markers in the bedrock. For example, if marine waters flowed inland from the coastal areas, as it does today, then the progressive loss of Mg to dolomitization in the coastal areas may have meant that there was insufficient Mg available for dolomitization by the time the waters reached the interior of the island.

Following the pioneering work of Vahremkamp et al. (1991) and Budd (1997), island dolostones have commonly been attributed to multiple "events" with dolomitization mediated by seawater. In those schemes, however, it is notable that in many cases, there is little or no time gap between the events and collectively, they could be considered indicative of a single, long–lived event. This is certainly the conclusion that could be drawn from the $^{87}Sr/^{86}Sr$ data from the dolostones in the Bluff Group of the Cayman Islands (Fig. 6.37). The critical difference between the "event" model and the "Cayman model" is that of time. With the Cayman model, the strata would have been bathed in seawater for ~4 .5 million years if the ages derived from the $^{87}Sr/^{86}Sr$ are accepted. During that time, the lithified limestones would have been slowly dolomitized with the spatial and temporal patterns of dolomitization being controlled by the hydrological systems and the patterns of water migration through the limestones. With such a model, different parts of the limestones may have been dolomitized at different times—a process that would produce the geographic and stratigraphic variations in the distribution of the $^{87}Sr/^{86}Sr$ values seen throughout the dolostones of the islands.

The process that triggers dolomitization of thick successions on isolated oceanic islands remains debatable. Interpretation of the dolostone fabrics indicates that the original carbonate sediments were lithified prior to dolomitization and that dissolution of aragonitic skeletons (e.g., corals, bivalves) took place before or during dolomitization. As on other islands, the dolostone sequences on the Cayman Islands are capped by unconformities, which suggests that dolomitization took place while the limestones were close to sea level. Available evidence from the dolostones on the Cayman Islands and other islands consistently point to dolomitization being mediated by normal or slightly evaporated seawater. This conclusion is supported by many different geochemical signatures (stable isotopes, trace elements, rare earth elements) and stratigraphic setting. The time when dolomitization took place is a critical element of the puzzle because time may enable determination of the other factors such as sea level and climate conditions. Although the general age of dolomitization can be determined by stratigraphic criteria, $^{87}Sr/^{86}Sr$ ratios have generally been used for this purpose given that the $^{87}Sr/^{86}Sr$ ratios of seawater has progressively increased over the last 40 million years. Such use, however, assumes that the $^{87}Sr/^{86}Sr$ ratios truly represent the original seawater and have not been modified by various processes that occurred before and/or after dolomitization. Analysis of $^{87}Sr/^{86}Sr$ ratios from various islands has led to the notion that dolomitization of various limestone successions, such as those on the Bahama Islands, involved a series of distinct dolomite events that collectively dolomitized the succession. An alternative explanation, as developed for the Cayman Islands is that dolomitization was a single event that took place over an extended period of time.

The fact that dolomite has not been produced in the laboratory under low temperature conditions means that the precise factors that control its development remains unknown. Likewise, it is difficult to interpret many of the geochemical attributes of the island dolostones, like those on the Cayman Islands. As yet, there is no clear evidence regarding the factor(s) that may trigger dolomitization in natural settings, like those on the Cayman Islands. From available evidence, it seems that the transformation of thick successions of limestones over large geographical areas must have taken place over extended periods of time. If it is assumed that the natural transformation of calcite by dolomite is a slow process, then perhaps time is the key to the dolomitization problem. Evidence developed from the Cayman Islands, although debatable in many respects, suggests that dolomitization of a thick succession of limestone that extended over the entire island took place while the strata were bathed in seawater for nearly 4.5 million years during the late Late Miocene and the Pliocene.

6.4 Conclusions

The limestones and dolostones in the Brac Formation, Cayman Formation, and Pedro Castle Formation of the Bluff Group are characterized by complex diagenetic fabrics that reflect, to various degrees, the collective effects of dolomitization, dissolution, precipitation of cements, and/or deposition of internal sediments. Karst development that followed each depositional cycle had a major effect on the carbonate successions. Integration of currently available information has led to the following important conclusions.

- Dolostones in the Brac Formation are formed of large (up to 1.5 mm long) euhedral crystals whereas the dolostones in the Cayman Formation and Pedro Castle Formation are formed of anhedral to euhedral crystals that are typically less than 50 μm long.
- Despite pervasive dolomitization, most of the original depositional fabrics were preserved in the Cayman Formation and Pedro Castle Formation.
- The dolomite crystals, irrespective of their size, are commonly zoned with a core of "dirty" high-calcian dolomite encased by a rim of clear, low-calcian dolomite.
- Some dolomite crystals were transformed into hollow crystals due to the preferential dissolution of their high-calcian dolomite cores. In some cases, these crystals were subsequently filled with calcite or dolomite cements.
- The trace element concentrations (Fe, Mn, Sr, REE) are consistent with dolomitization being mediated by marine waters.
- The Na concentrations in the Cayman dolostones are high, possibly due to some of the Na being trapped between the crystals rather than being incorporated into the lattices of the dolomite crystals.
- The stable isotopes of the limestones from the Brac Formation, Cayman Formation, and Pedro Castle Formation are generally negative whereas those from the dolostones in these formations are generally positive.
- The stable isotopes of the dolostones are consistent with the notion that dolomitization was mediated by normal seawater or seawater that had undergone some evaporation.

- The precise factors that controlled dolomitization of the carbonate sediments remain open to debate.
- Weathering and dissolution of the bedrock that took place during periods of subaerial exposure between deposition of the sediments that now form the Brac Formation, Cayman Formation, and Pedro Castle Formation led to the development of rugged, complex phytokarst surface terrains and the development of subsurface caves and tunnels of various dimensions. During these periods, significant but unknown thicknesses of strata were lost from each formation.
- Although open to some debate, the $^{87}Sr/^{86}Sr$ ratios indicate that dolomitization of the Brac Formation occurred during the Late Oligocene whereas dolomitization of the Cayman Formation and Pedro Castle Formation probably took place during the Pliocene.

References

Alibo DS, Nozaki Y (1999) Rare earth elements in seawater: particle association, shale-normalization and Ce oxidation. Geochim Cosmochim Acta 63:363–372

Budai JM, Longman KC, Owen RM (1984) Burial dedolomitization in the Mississippian Madison Limestone, Wyoming and Utah Thrust Belt. J Sediment Petrol 54:276–288

Budd DA (1997) Cenozoic dolomites of carbonate islands: their attributes and origin. Earth Sci Rev 42:1–47

Burke WH, Denison RE, Hetherington EA, Koepnick RB, Nelson HF, Otto JB (1982) Variation of seawater $^{87}Sr/^{86}Sr$ throughout Phanerozoic time. Geology 10:516–519

Carnell AJH, Wilson MEJ (2004) Dolomites in SE Asia—varied origins and implications for hydrocarbon exploration. In: Braithwaite CJR, Rizzi G, Drake G (eds) The geometry and petrogenesis of dolomite hydrocarbon reservoirs. Geological Society, London, England, pp 255–300

Choquette PW, Hiatt EE (2008) Shallow-burial dolomite cement: a major component of many ancient sucrosic dolomites. Sedimentology 55:423–460

Dawans JM, Swart PK (1988) Textural and geochemical alterations in Late Cenozoic Bahamian dolomites. Sedimentology 35:385–403

DePaolo DJ (1986) Detailed record of the Neogene Sr isotopic evolution of seawater from DSDP site 590B. Geology 14:103–106

DePaolo DJ, Ingram BL (1985) High-resolution stratigraphy with strontium isotopes. Science 227:938–941

de Dolomieu D (1791) Sue un genre de pierres calcarires trespeu effervescentes avec acides et phosphorescentes pare la collision. J Phys 39:3–10

de Saussure NT (1792) Analyse de la dolomie. J Phys 40:161–172

Dowsett HJ, Cronin TM (1990) High eustatic sea level during the middle Pliocene: Evidence from the southeastern U.S. Atlantic Coastal Plain. Geology 18:435–438

Epting M (1980) Sedimentology of Miocene carbonate buildups, central Luconia, offshore Sarawak. Geol Soc Malaysia Bull 12:17–30

Farrell JW, Clemens SC, Gromet LP (1995) Improved chronostratigraphic reference curve of late Neogene seawater $^{87}Sr/^{86}Sr$. Geology 23:403–406

Fouke BW (1994) Deposition, diagenesis and dolomitization of Neogene Seroe Domi Formation coral reef limestones on Curaçao. Publication Foundation for Scientific Research in the Caribbean Region, Netherlands Antilles, p 182

Gill I, Moore CH, Aharon P (1995) Evaporitic mixed-water dolomitization on St. Croix, U.S.V.I. J Sediment Petrol A65:591–604

Haq BV, Hardenbol J, Vail PR (1987) Chronology of fluctuating sea levels since the Triassic. Science 235:1156–1167

Hess J, Bender ML, Schilling JG (1986) $^{87}Sr/^{86}Sr$ evolution from Cretaceous to present - applications to paleoceanography. Science 231:979–984

Hodell DA, Muller PA, Garrido JR (1991) Variations in the strontium isotopic composition of seawater during the neogene. Geology 19:24–27

Jones B (1989) Syntaxial overgrowths on dolomite crystals in the Bluff Formation, Grand Cayman, British West Indies. J Sediment Petrol 59:839–847

Jones B (2005) Dolomite crystal architecture: genetic implications for the origin of the Tertiary dolostones of the Cayman Islands. J Sediment Res 75:177–189

Jones B (2007) Inside-out dolomite. J Sediment Res 77:539–551

Jones B (2013) Microarchitecture of dolomite crystals as revealed by subtle variations in solubility: implications for dolomitization. Sed Geol 288:66–80

Jones B (2019) Diagenetic processes associated with unconformities in carbonate successions on isolated oceanic islands: case study of the Pliocene to Pleistocene sequence, Little Cayman, British West Indies. Sed Geol 386:9–30

Jones B (2021) Formation, dispersal and accumulation of terra rossa on the Cayman Islands. Sedimentology 68:1–45

Jones B, Luth RW (2002) Dolostones from Grand Cayman, British West Indies. J Sediment Res 72:560–570

Jones B, Luth RW (2003) Temporal evolution of Tertiary dolostones on Grand Cayman as determined by $^{87}Sr/^{86}Sr$. J Sediment Res 73:187–205

Jones B, Lockhart EB, Squair C (1984) Phreatic and vadose cements in the Tertiary Bluff Formation of Grand Cayman Island, British West Indies. Bull Can Pet Geol 32:382–397

Jones B, Pleydell SA, Ng KC, Longstaffe FJ (1989) Formation of poikilotopic calcite-dolomite fabrics in the Oligocene-Miocene Bluff Formation of Grand Cayman, British West Indies. Bull Can Pet Geol 37:255–265

Jones B, Luth RW, MacNeil AJ (2001) Powder X-ray analysis of homogeneous and heterogeneous dolostones. J Sediment Res 71:791–800

Koepnick RB, Burke WH, Denison RE, Hetherington EA, Nelson HF, Otto JB, Waite LE (1985) Construction of the seawater $^{87}Sr/^{86}Sr$ curve for the Cenozoic and Cretaceous: supporting data. Chem Geol (isotope Geoscience Section) 58:55–81

Land LS (1989) Dolomitization and dolomite recrystallization, Hope Gate Formation, North Jamaica – reassessment of mixing zone dolomite. Geological Society of America, Abstracts with Programs, pp 220–221

Land LS (1991) Dolomitization of the Hope Gate Formation (north Jamaica) by seawater: reassessment of mixing-zone dolomite. In: Taylor HP, ONeil JR, Kaplan IR (eds) Stable Isotope Geochemistry: A tribute to Samuel Epstein, pp 121–133

Li R, Jones B (2014) Evaluation of carbonate diagenesis: a comparative study of minor elements, trace elements, and rare-earth elements (REE+Y) between Pleistocene corals and matrices from Grand Cayman, British West Indies. Sed Geol 314:31–46

Liang T, Jones B (2014) Deciphering the impact of sea-level changes and tectonic movement on erosional sequence boundaries in carbonate successions: a case study from Tertiary strata on Grand

Cayman and Cayman Brac, British West Indies. Sed Geol 305:17–34

Lumsden DN (1979) Discrepancy between thin-section and X-ray estimates of dolomite in limestone. J Sediment Petrol 49:429–436

Machel HG, Burton EA (1994) Golden Grove dolomite, Barbados: origin from modified seawater. J Sediment Res A64:741–751

MacNeil AJ, Jones B (2003) Dolomitization of the Pedro Castle Formation (Pliocene), Cayman Brac, British West Indies. Sed Geol 162:219–238

MacNeil AJ (2001) Sedimentology, diagenesis, and dolomitization of the Pedro Castle Formation on Cayman Brac, British West Indies. M.Sc. thesis, Alberta, Edmonton

McArthur JM, Howarth RJ, Shields GA (2012) Strontium isotope stratigraphy. In: Gradstein FM, Ogg JG, Schmitz M, Ogg G (eds), The Geologic Time Scale. Elsevier, pp 127–143.

McCormick CA, Jones B (2021) On the efficacy and limitations of isolated carbonate platforms as "oceanic dipsticks" to reconstruct subsidence histories, a case study from the Paleogene to Neogene strata on Grand Cayman and Cayman Brac. B.W.I. Marine Geol 436, 106470

McKenzie JA, Isern A, Elderfield H, Williams A, Swart PK (1993) Strontium isotope dating of paleoceanographic, lithographic and dolomitization events on the northeastern Australia margin, Leg 133. In: McKenzie JA, Davies PJ, Palmer-Judson A (eds), Proceedings of Ocean Drilling Program, Scientific Results, pp 489–498

McLennan SM (1989) Rare earth elements in sedimentary rocks: influence of provenance and sedimentary processes. Rev Mineral Geochem 21:169–200

McLennan SM (2001) Relationships between the trace element composition of sedimentary rocks and upper continental crust. Geochem Geophys Geosyst 2:1–24

Miller KG, Kominz MA, Browning JV, Wright JD, Mountain GS, Katz ME, Sugarman PJ, Cramer BS, Christie-Blick N, Pekar SF (2005) The Phanerozoic record of global sea-level change. Science 310:1293–1298

Miller KG, Browning JV, Schmelz WJ, Kopp RE, Mountain GS, Wright JD (2020a) Cenozoic sea-level and cryospheric evolution from deep-sea geochemical and continental margin records. Sci Adv 6:1–15

Miller KG, Schmetz WJ, Browning JV, Kopp RE, Mountain GS, Wright JD (2020b) Ancient sea levels as key to the future. Oceanography 33:32–41

Ng KC (1990) Diagenesis of the Oligocene-Miocene Bluff Formation of the Cayman Islands—a petrographic and hydrogeochemical approach. (Ph.D. thesis), University of Alberta, Edmonton

Ohde S, Elderfield H (1992) Strontium isotope stratigraphy of Kita-daito-jima Atoll, North Phillipine Sea, implications for Neogene sea-level change and tectonic history. Earth Planet Sci Lett 113:473–486

Osborne AH, Haley BA, Hathorne EC, Plancherel Y, Frank M (2015) Rare earth element distribution in Caribbean seawater: continental input versus lateral transport of distinct REE compositions in subsurface water masses. Mar Chem 177:172–183

Pleydell SM, Jones B, Longstaffe FJ, Baadsgaard H (1990) Dolomitization of the Oligocene-Miocene Bluff Formation on Grand Cayman, British West Indies. Can J Earth Sci 27:1098–1110

Raymo ME, Mitrovica JX, O'Leary MJ, DeConto RM, Hearty PJ (2011) Departures from eustasy in Pliocene sea-level records. Nat Geosci 4:328–332

Ren M, Jones B (2017) Spatial variations in the stoichiometry and geochemistry of Miocene dolomite from Grand Cayman: implications for the origin of island dolomite. Sed Geol 348:69–93

Ren M, Jones B (2018) Genesis of island dolostones. Sedimentology 65:2002–2033

Rogers AF (1929) Polysynthetic twinning in dolomite. Am Miner 14:245–250

Saller AH (1984) Petrologic and geochemical constraints on the origin of subsurface dolomite, Enewetok Atoll: an example of dolomitization by normal seawater. Geology 12:221–225

Staudt WJ, Oswald EJ, Schoonen MAA (1993) Determination of sodium, chloride and sulfate in dolomites: a new technique to constain the composition of dolomitizing fluids. Chem Geol 107:97–109

Tucker ME, Wright VP (1990) Carbonate Sedimentology. Blackwell Scientific Publications, Oxford, p 282

Uzelman BC (2009) Sedimentology, diagenesis, and dolomitization of the Brac Formation (Lower Oligocene), Cayman Brac, British West Indies. (M.Sc. thesis), University of Alberta.

Vahrenkamp VC, Swart PK (1988) Constraints and interpretation of $^{87}Sr/^{86}Sr$ ratios in Cenozoic dolomites. Geophys Res Lett 15:385–388

Vahrenkamp VC, Swart PK (1990) New distribution coefficient for the incorporation of strontium into dolomite and its implications for the formation of ancient dolomites. Geology 18:387–391

Vahrenkamp VC, Swart PK, Ruiz J (1991) Episodic dolomitization of Late Cenozoic carbonates in the Bahamas: evidence from strontium isotopes. J Sediment Petrol 61:1002–1014

Vail PR, Mitchum RM, Thompson S (1977a) Seismic stratigraphy - applications to hydrocarbon exploration. Am Assoc Pet Geol Mem 26:99–116

Vail PR, Mitchum RM, Todd RG, Widmier JM, Thompson S, Sangree JB, Bubb JN, Hatlelid WD (1977b) Seismic stratigraphy and global changes of sea level. Seismic stratigraphy – applications to hydrocarbon exploration. American Association of Petroleum Geologists Memoir 26.

van der Flierdt T, Pahnke K, Amakawa H, Andersson P, Basak C, Coles B, Colin C, Crocket K, Frank M, Frank N, Goldstein SL, Goswami V, Haley BA, Hathorne EC, Hemming SR, Henderson GM, Jeandel C, Jones K, Kreissig K, Lacan F, Lambelet M, Martin EE, Newkirk DR, Obata H, Pena L, Piotrowski AM, Pradoux C, Scher HD, Schoberg H, Singh SK, Stichel T, Tazoe H, Vance D, Yang JJ, Partici GI (2012) GEOTRACES intercalibration of neodymium isotopes and rare earth element concentrations in seawater and suspended particles. Part 1: reproducibility of results for the international intercomparison. Limnol Oceanogr Methods 10:234–251

Wang R, Yu K, Jones B, Wang Y, Zhao J-X, Feng LB, Bian L, Xu S, Fan T, Jiang W, Zhang Y (2018) Evolution and development of Miocene "island dolostones" on Xisha Islands, China South Sea. Mar Geol 406:142–158

Wang R, Jones B, Yu K (2019) Island dolostones: genesis by time-transgressive or event dolomitization. Sed Geol 390:15–30

Wang R, Yu K, Jones B, Jiang W, Xu S, Fan D, Zhang Y (2021) Dolomitization micro-conditions constraint on dolomite stoichiometry: a case study from the Miocene Huangliu Formation, Xisha Islands, South China Sea. Mar Pet Geol 133(1–12):105286

Zhao H (2013) Origin of island dolostones: case study based on Tertiary dolostones from Cayman Brac. Univerity of Alberta, Edmonton, British West Indies

Zhao H, Jones B (2012) Genesis of fabric-destructive dolostones: a case study of the Brac Formation (Oligocene), Cayman Brac, British West Indies. Sed Geol 267–268:36–54

Zhao H, Jones B (2013) Distribution and interpretation of rare earth elements and yttrium in Cenozoic dolostones and limestones on Cayman Brac, British West Indies. Sed Geol 284–285:26–38

7 Ironshore Formation

Abstract

On the Cayman Islands, the Ironshore Formation unconformably overlies the Pedro Castle Formation or the Cayman Formation. Exposed surfaces of the limestones in this formation are typically dark grey to black in colour due to microbially mediated weathering. On Grand Cayman, large areas of the formation are covered by mangrove peats. The formation comprises Units A, B, C, D, E, and F that formed during highstand conditions that existed ~400 ka, 345 ka, ~229 ka, ~125 ka, 95–110 ka, and 73–87 ka, respectively. Unit D is widely exposed across the islands, whereas units A, B, and C are known only from cores from the northeast area of Grand Cayman; Unit E has only been found in cores from offshore the west coast of Grand Cayman, and Unit F is known from the offshore cores and scattered localities on the western peninsula of Grand Cayman. U/Th dating of fossils from Units E and F yielded ages of ~104 ka and ~84 ka, respectively. The Ironshore Formation is formed of limestones, with the percentage of aragonite generally decreasing as the age of the limestones increases. The limestones of the Ironshore Formation, (especially Unit D) contains an abundant, diverse biota of well-preserved corals, bivalves, and gastropods, and various other organisms. Fossil preservation is generally very good, with many organisms preserved in life position. Original shell colours are evident in some bivalves and gastropods and the original aragonite skeletons are still intact in many of the corals, bivalves, and gastropods. Rhizoliths are commonly evident in some lithified skeletal sands and oolitic sands in the upper units of the Ironshore Formation on Little Cayman and Cayman Brac. Diagenesis of the limestones in the Ironshore Formation included precipitation of calcite and aragonite cements and, in some cases, neomorphic replacement of the aragonite skeletons by calcite. Only minor amounts (<5%) of dolomite are present in the oldest limestones of Unit A. The stable isotopes are generally negative, with only a few samples yielding positive $\delta^{18}O$ and $\delta^{13}C$ values. The limestones of the Ironshore Formation are characterized by low concentrations of Mn and Fe, variable concentrations of Na, and Sr contents that generally decrease as the percentage of aragonite in the limestones decreases. Exposed surfaces of the Ironshore Formation are commonly capped by well-developed calcrete crusts.

7.1 Introduction

The Ironshore Formation, present on each of the Cayman Islands, is typically formed of poorly lithified limestones that commonly contain numerous well-preserved fossils. The name came from the local Caymanian term “Ironshore” that was used to describe the low rocky shore area of the coastal platform (Matley 1926; Warthin 1959). No type section has ever been designated for this formation. The information provided on the Pleistocene sequence in this chapter has been derived from the examination and collection of samples from over 150 surface exposures, with most coming from Grand Cayman (Fig. 7.1). Many sites are no longer accessible because of the extensive development that has taken place on Grand Cayman over the last 20 years. Obtaining information on the vertical succession of facies in this formation is a challenge because exposures of the formation are typically low-lying with minimal thicknesses exposed. This situation is exacerbated by the fact that attempts to drill and core this succession are generally an exercise in futility because the friable rocks typically disintegrate during drilling. Fortunately, cores obtained from the Ironshore Formation at (1) Roger’s Wreck Point on the northwest corner Grand Cayman and (2) offshore from the west coast of Grand Cayman just north of George Town (Fig. 7.1a) have

B. Jones, *Geology of the Cayman Islands*,
https://doi.org/10.1007/978-3-031-08230-6_7

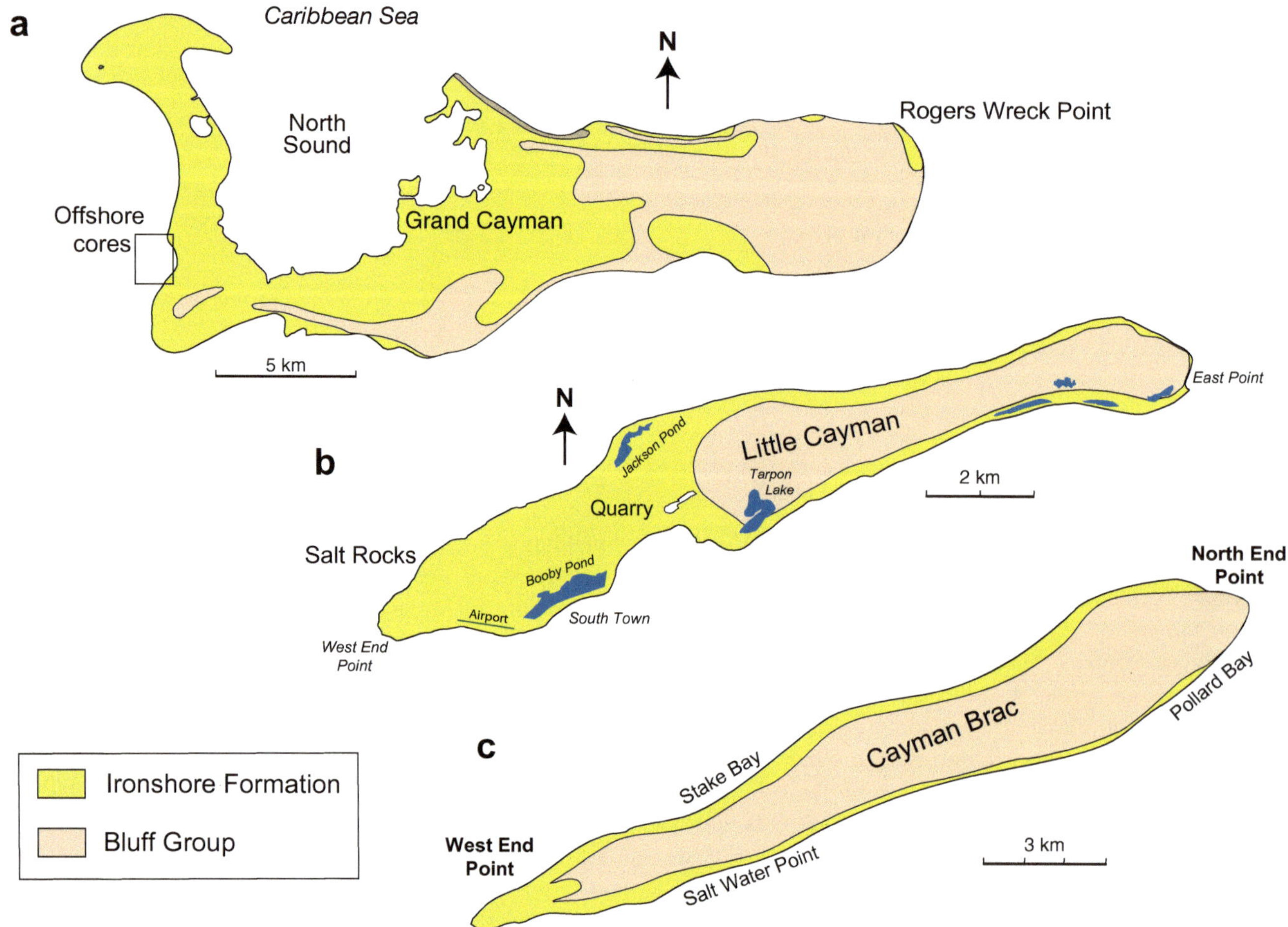

Fig. 7.1 Maps showing distribution of the Ironshore Formation on **a** Grand Cayman, **b** Little Cayman, and **c** Cayman Brac

allowed determination of the internal stratigraphy of the formation (Vézina 1997; Vézina et al. 1999; Coyne 2003; Coyne et al. 2007).

7.2 Distribution

On Grand Cayman, the geographically extensive Ironshore Formation covers large tracts of land on the eastern part of the island in addition to the typical coastal exposures (Fig. 7.1a). Natural exposures of the Ironshore Formation in the inland areas are rare because they are typically covered by mangrove swamps (e.g., Woodroffe 1982). Its presence can usually be established where canals have been cut through the swamps with the excavated material being used to construct "roads" that provide access to the interior parts of the island. Various quarries in the central part of the island (mostly around Bodden Town) also provide excavated material that can be used to establish the presence of the formation as well as samples of the formation. On Little Cayman, the Ironshore Formation is evident in coastal exposures and in the quarry (Fig. 7.1b). On Cayman Brac, the Ironshore Formation is restricted to coastal exposures (Fig. 7.1c) where the limestones may be up to 6 m above sea level (asl). Although the limestone is "soft", weathering has produced a hard, black crust over its surface. Exposures around the coastlines of these islands is patchy with many areas covered by modern storm deposits, vegetation, and/or coastal buildings.

7.3 Lower Boundary

The lower boundary of the Ironshore Formation, which is only exposed a few places, is easy to recognize because the soft friable limestones of the formation contrasts sharply with the hard, well-lithified limestones and dolostones of the underlying Pedro Castle Formation or Cayman Formation. The nature of the basal unconformity is highly variable in terms of its profile and position relative to modern-day sea level. In Paul Bodden Quarry, the boundary is nearly horizontal over most of the area but in the northeast corner it rises sharply for ~1 m and then gradually for another 3–4 m (Fig. 7.2a). Throughout that area, the unconformity is characterized by minor low and high areas. To the west of the jetty at Spotts Bay, the unconformity is nearly vertical because limestones of the Ironshore Formation accumulated against an ancient cliff face (Fig. 7.2b). In the quarry on Little Cayman (Fig. 7.1b), the Ironshore Formation lies on top of the Pedro Castle Formation which is characterized by a rugged phytokarst surface with up to 2 m of relief (Fig. 7.2c).

7.4 Upper Boundary

The modern-day weathering surface is generally the upper boundary of the formation. In many areas, including the western part of Grand Cayman, limestones of the Ironshore Formation are covered with peat that formed in the mangrove swamps or sands that are formed by modern sedimentation processes.

7.5 Thickness

It is difficult to establish the true maximum thickness of the Ironshore Formation because (1) there are no exposures where the complete formation is exposed, (2) the upper surface of the Ironshore Formation is commonly the modern surface where unknown thicknesses of the formation have been lost to erosion, and (3) obtaining core from the formation is difficult because the rocks usually disintegrate as they are being drilled. The only exceptions found so far are (1) near Roger's Wreck Point on the northeast corner of the island where cores have been successfully obtained to a depth of 19 m before the top of the Cayman Formation was reached (Figs. 7.1a and 7.3), and (2) offshore from the west coast of Grand Cayman just north of George Town where the formation is at least 20 m thick (Fig. 7.3b). In many areas, the thickness of the formation is apparent during drilling because the rapid rates of drilling decrease when the top of the underlying Pedro Castle Formation or Cayman Formation is reached. On the southwestern part of Grand Cayman, Brunt et al. (1973) reported thicknesses of up to 17 m for the Ironshore Formation.

Information derived from wells drilled on the western peninsula of Grand Cayman clearly illustrates the variable thicknesses of the Ironshore Formation (Fig. 7.4) that are also a reflection of the paleotopography on the underlying Cayman Formation and Pedro Castle Formation. In some areas to the south of George Town, for example, there are reported thicknesses of 17 m (Fig. 7.4). There are no identifiable patterns to these thicknesses (Fig. 7.4).

On Cayman Brac, a combination of information obtained by drilling and the thickness of exposed strata at the Water Authority site on the west end of the island, shows that the formation is at least 12 m thick. On Little Cayman, the situation is more complex because of the scattered nature of outcrops of the Ironshore Formation and the lack of well information. At Salt Point on the west coast of the island, ~6 m of the formation is exposed. To the north of that site, limestones that are thought to belong to the Ironshore Formation are found at an altitude of 16 m asl (Fig. 7.1c). Unfortunately, there are no exposures between that isolated exposure and the nearby shoreline and it is therefore impossible to determine the relationship between this exposure and the coastal exposures.

7.6 Internal Stratigraphy

At Roger's Wreck Point, near-vertical outcrops of the Cayman Formation along the south side of the Queen's Highway are up to 4 m high with a well-developed wave-cut notch at 6 m above sea level (asl). Data from a series of wells drilled northwards from the highway towards East Sound shows that the erosional surface that defines the top of the Cayman Formation becomes progressively deeper as it changes from 3 m near the road to a maximum depth of 17 m in well RWP#13 (Fig. 7.3b). Sediments that now form the Ironshore Formation were deposited on that erosional bench. By combining identified unconformities that are highlighted by thin layers of caliche or terra rossa with U/Th dating of well-preserved corals in the 15 wells drilled at Rogers Wreck Point, Vézina (1997) and Vézina et al. (1999) showed that the succession comprised Units A, B, C, and D (Fig. 6.3b) that were dated at ~400 ka, 345 ka, ~229 ka, and ~131 ka, respectively. They linked the highstands associated with the deposition of these units to marine isotope stages 5, 7, 9, and 11, respectively. The wave-cut notch evident in the cliff face by the road probably formed when the sediments of Unit D were deposited.

At various localities on the western peninsula of Grand Cayman, including those on the west coast of North Sound north of Salt Creek, it is readily apparent that younger units of poorly fossiliferous, cross-bedded limestones overlie the fossiliferous limestones of Unit D. Dating of scattered corals and conch shells found in these beds at some localities yield a variety of dates that are younger than the 125,000 year date associated with unit D (Fig. 7.5a). In 1996, four wells were drilled offshore from the coastline just north of George Town where the water is about 15 m deep (Fig. 7.5b). U/Th dating of corals from those units coupled with recognition of unconformities in the succession led to the identification of units E (~104 ka) and and F (~84 ka) (Coyne 2003; Coyne et al. 2007). The dates of these units correspond to

Fig. 7.2 Field photographs showing lower boundary of the Ironshore Formation. **a** Ironshore Formation lying unconformably on top of the Cayman Formation. Boundary indicated by white arrows. Note topography on upper surface of the Cayman Formation. Paul Bodden Quarry #1. **b** Near vertical boundary (white arrows) between the Ironshore Formation (IF) and Cayman Formation (CF), on east side of jetty at Spotts Bay. **c** Highly irregular phytokarst boundary between the Pedro Castle Formation (PCF) and Ironshore Formation (IF) in quarry on Little Cayman. Black line superimposed to show boundary

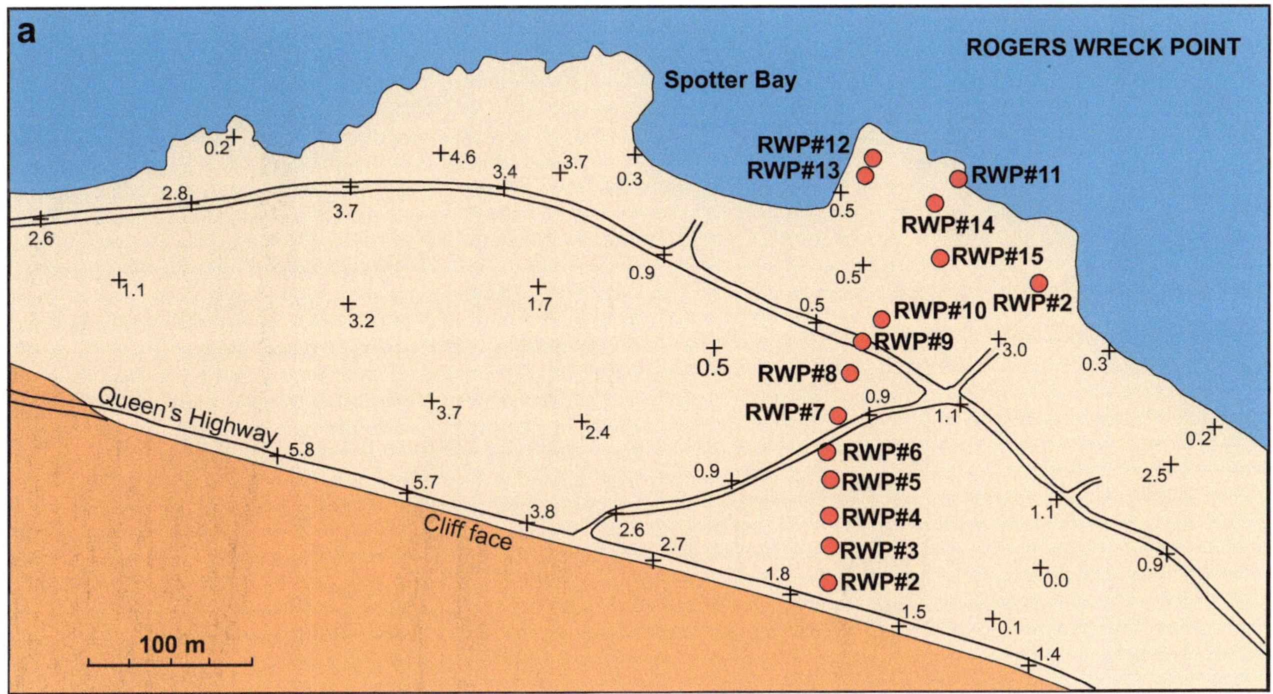

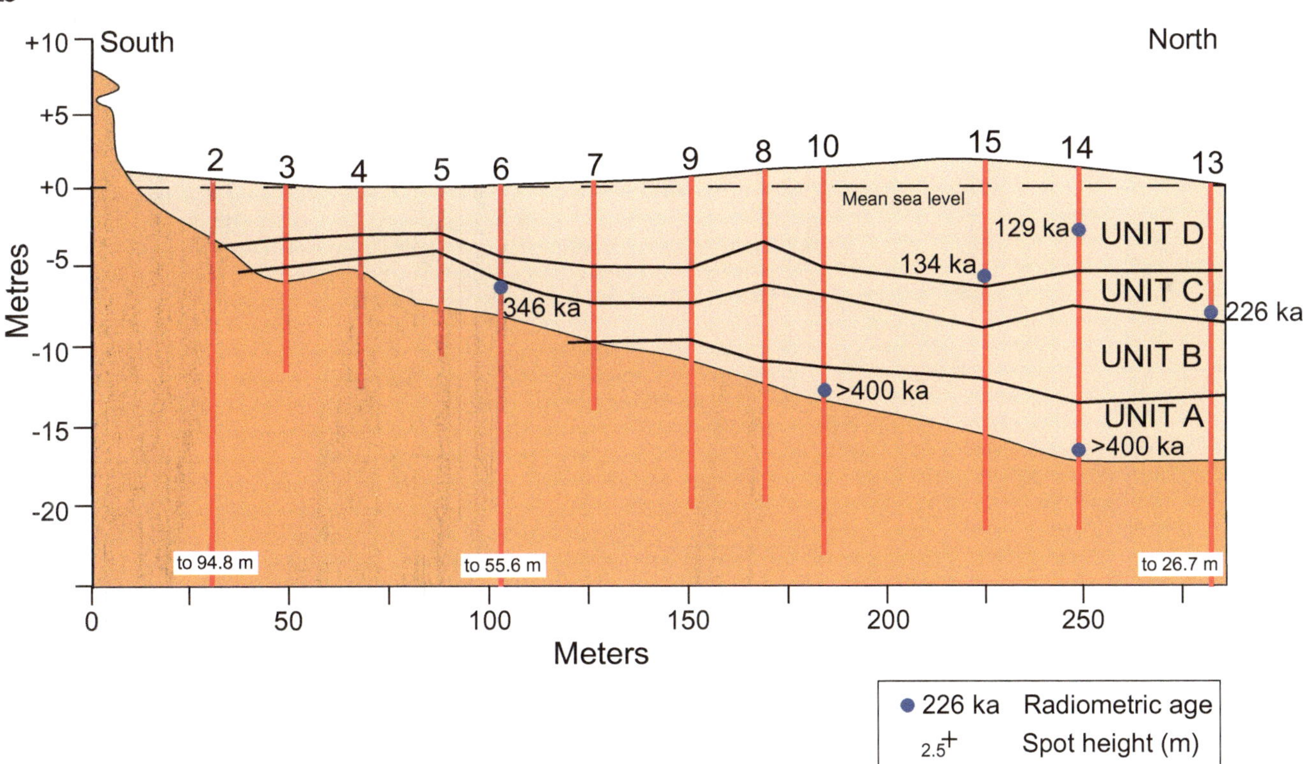

Fig. 7.3 Geology of Rogers Wreck Area, Grand Cayman. **a** Map of area around Rogers Wreck Point (Fig. 1a) showing relationship between the Ironshore Formation and Cayman Formation and locations of wells RWP#1 to RWP#15. **b** Cross-section showing relationship between the Cayman Formation and Ironshore Formation with internal division into units A–D and location of corals with their radiometric ages. Modified from Vézina et al. (1999, their Fig. 3)

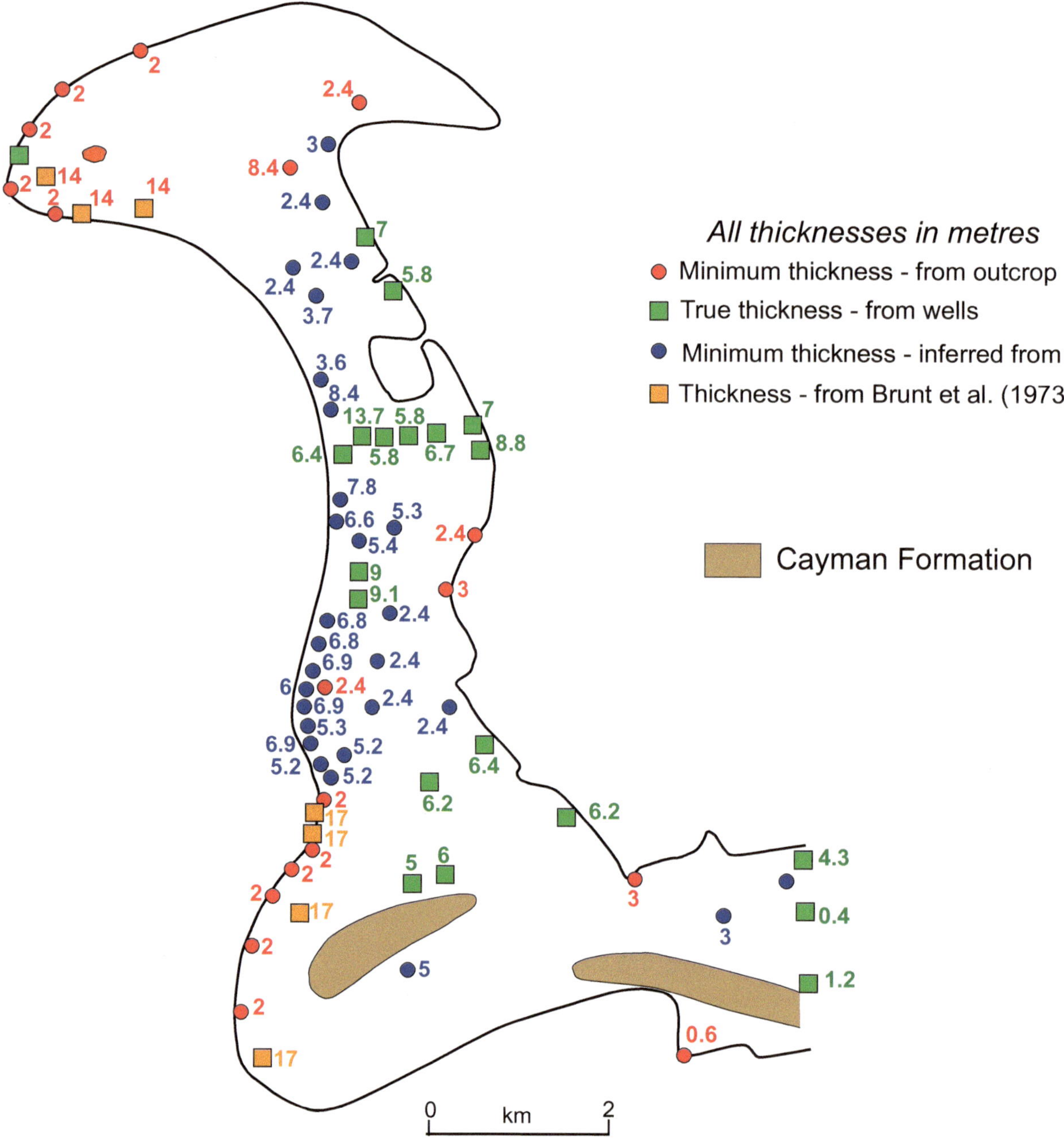

Fig. 7.4 Map of western peninsula of Grand Cayman showing thicknesses of Ironshore Formation as determined from information in Brunt et al. (1973), outcrops, wells, and dredging. Modified from Shourie (1993, his Fig. 4.1)

the younger dates found at various on-land localities elsewhere on the western peninsula (Fig. 7.5b).

Collectively, data from the Rogers Wreck Point area and the area offshore from George Town shows that the Ironshore Formation comprises six units (A–F) that formed in association with recognized sea-level highstands (Fig. 7.6). This stratigraphic framework is used as a basis for evaluation of the various outcrops of the Ironshore Formation found on each of the Cayman Islands.

7.6.1 Lithological and Biological Characteristics of Units A to F

Vézina (1997) and Vézina et al. (1999) outlined the lithological and biological characteristics of Units A to D in the Ironshore Formation, whereas Coyne (2003) and Coyne et al. (2007) provided detailed information on Units E and F.

Unit A: Although absent in the south part of the Rogers Wreck Point transect, this unit is 4.1 m thick in the northern

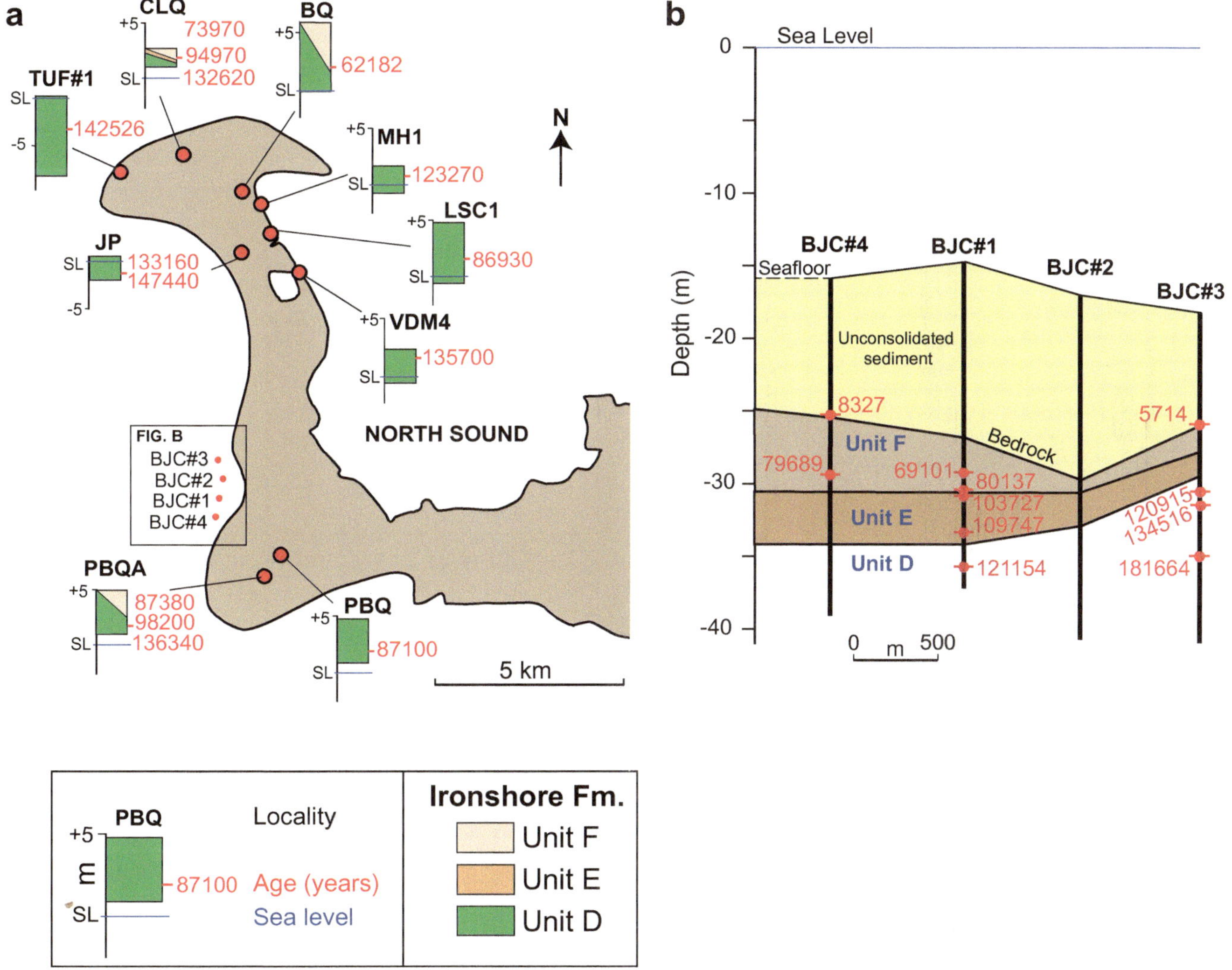

Fig. 7.5 Ages of Ironshore Formation as determined by radiometric dating. **a** Dates determined from samples collected from outcrop on western peninsula of Grand Cayman (Coyne 2003; Coyne et al. 2007, their Fig. 3). **b** Dates for Ironshore Formation in wells drilled offshore from Grand Cayman. Modified from Coyne et al. (2007, their Fig. 5)

part. The dense, well-indurated limestones are formed of grainstones and framestones with subordinate amounts of packstones and wackestones. Grains in the skeletal sands, which are typically micritized, were derived from foraminifera, *Halimeda*, red algae, bivalves, and gastropods. Massive (*Diploria*, *Montastraea*, *Siderastrea*) and branching (*Acropora palmata*) corals are commonly covered by a 2–3 cm thick crust formed of red algae and foraminifera, and bored by *Lithophaga*.

Unit B: Up to 6.7 m (average 3.7 m) thick, this unit is formed of packstones and grainstones with lesser amounts of wackestones with patchy recrystallization. The skeletal sands are formed of grains derived from foraminifera, *Halimeda*, *Homotrema*, red algae, echinoid spines, bivalves, and gastropods with many grains being partly micritized. *M. annularis*, *Porites astreoides*, *Dichocenia stokesii*, *Siderastrea*, *Diploria strigosa*, *D. clivosa*, *Goniopora*, *Porites porites*, and *Acropora cervicornis*(?) dominate the coral fauna. Many corals are encrusted by a 3–4 cm thick layer that was formed by red algae and foraminifera.

Unit C: Up to 3.7 m (average 2.3 m) thick, this unit is formed of rudstones, packstones, grainstones, and bafflestones. The skeletal sands are formed of grains (commonly micritized) derived from benthic foraminifera, *Halimeda*, red algae, *Homotrema*, echinoid spines, bivalves, and gastropods. Scattered head corals and numerous branching corals (*A. cervicornis*, *P. porites*) are present in this unit.

Unit D: Up to 5.8 m (average 4 m) thick, this unit is formed of grainstones, with packstones, wackestones, and mudstones common in some areas. Framestones and bafflestones are locally common. The skeletal sands are formed

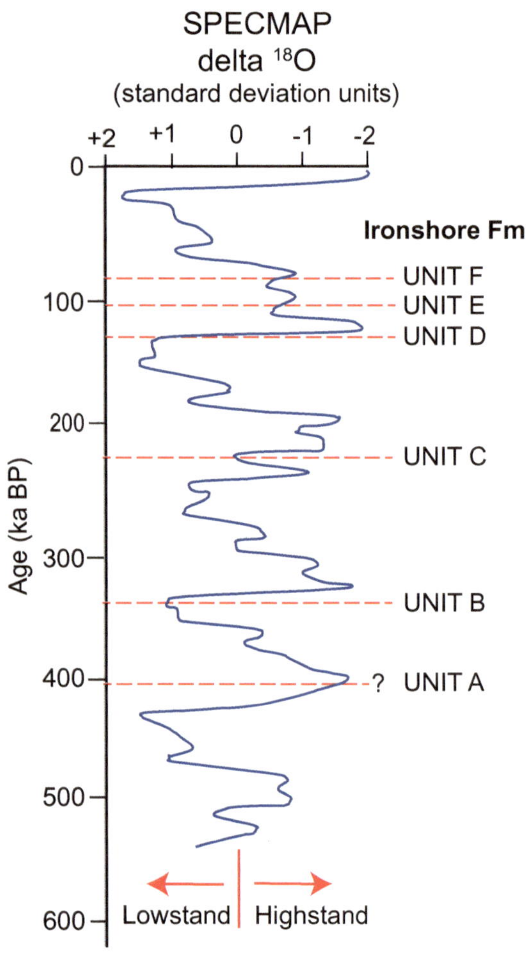

Fig. 7.6 Correlation of Units A–F of Ironshore Formation, based on radiometric dates, with SPECMAP time scale (modified from Imbrie et al. 1984). The δ^{18}O scale represents the fractionation of $\delta^{18}O/\delta^{16}O$ against standard ocean value of 0. Positive valves indicate glacial periods whereas negative values indicate interglacial periods. Modified from Coyne (2003, her Fig. 2.1)

of grains (commonly micritized) derived from gastropods, red algae, benthic foraminifera, *Halimeda*, echinoid spines, corals, serpulid worm tubes, and barnacles. Head corals include *Dendrogyra cylindricus*(?), *Dichocoenia stokesii*, *Diploria* sp. *D. labrinthiformis*, *D. strigosa*, *Favia fragum*, *Goniopora* sp., *Montastraea annularis*, *Porites astreoides*, *P. porites*, *Siderastrea* sp., and *S. siderea*.

Unit E: This unit is poorly known because it has only been found in the cores from the wells that were drilled offshore from the west coast of Grand Cayman (Fig. 7.5a), where it is <5 m thick. It is formed of head coral floatstones (dominated by *M. annularis* with fewer *Diploria* sp. and *M. cavernosa*) and mixed coral floatstones (*A. cervicornis*, *A. palmata*, *M. annularis*, *Porites* sp., *D. strigosa*, *D. labyrinthiformis*, *D. clivosa*, *Agaricia fragilis*(?), *Siderastrea* sp.) with fragmented corals being variously bored and encrusted. The matrices in these facies are skeletal grainstones, packstones, or wackestones that are formed of poorly sorted fine to coarse grains derived from red algae, benthic foraminifera, echinoderm spines, mollusc, *Halimeda,* peloids, and micritized allochems.

Unit F: Up to 6 m think in well BJC#4, this unit is formed of cross-bedded and planar-bedded poorly fossiliferous grainstones. On the western part of Grand Cayman, this unit rests directly on top of Unit D, with the boundary between the two units being topographically irregular (Fig. 7.7). The unconformity at the top of unit D commonly has large tabular lithoclasts resting on it. The basal unconformity does not appear to have formed subaerially because there are no rhizoliths, calcrete, or terra rossa associated with it. At Salt Creek, *Skolithus* and *Conichnus* have their apertures at the unconformity (Jones and Pemberton 1989).

7.7 The 6 m Sea Level Notch

One of the most conspicuous features in the sea cliffs on Cayman Brac is the wave-cut notch that is 6 m above modern day sea level (Fig. 7.8a–c). Although common on the western part of the island, it is rare on the eastern part of the island because the collapse of the high cliffs above the notch means that it is no longer evident. At one locality on the northeast corner of the island, two wave-cut notches are apparent—the lower one being associated with modern-day sea level whereas the other is 6 m above modern sea level (Fig. 7.8a). On the western part of Cayman Brac, the wave-cut notch can easily be traced along the cliff facies on the north and south sides of the uplifted core of the island (Fig. 7.8b). Woodroffe et al. (1983) showed that the mid-point of the wave-cut notch was 5.5–7.0 m with an average of 6.4 m above sea level. Woodroffe et al. (1983) and Woodroffe (1988), however, implied that the wave-cut notch had formed during a highstand that predated deposition of the limestones 125,000 years ago. In some areas the notch is up to 3.0 m deep and locally decorated with stalactites, stalagmites, and columns (Fig. 7.8c) (Zheng 2017; Jones et al. 2018).

The 6 m wave cut notch is not found on Little Cayman because there are no vertical coastal sea cliffs in which it could have been preserved. On Grand Cayman the 6 m wave-cut notch is evident at only two localities on the eastern part of the island (Fig. 7.8d). On the south side of the Queen's Highway at Rogers Wreck Point, the wave-cut notch is still evident in the low cliff face that is cut in the dolostones of the Cayman Formation. There, erosion has destroyed parts of the wave-cut notch and other parts are commonly covered by dense vegetation. The wave-cut notch is also evident at the north end of Frank Sound Road where it is cut into the dolostones of the Cayman Formation that is located on private property on the east side of the road.

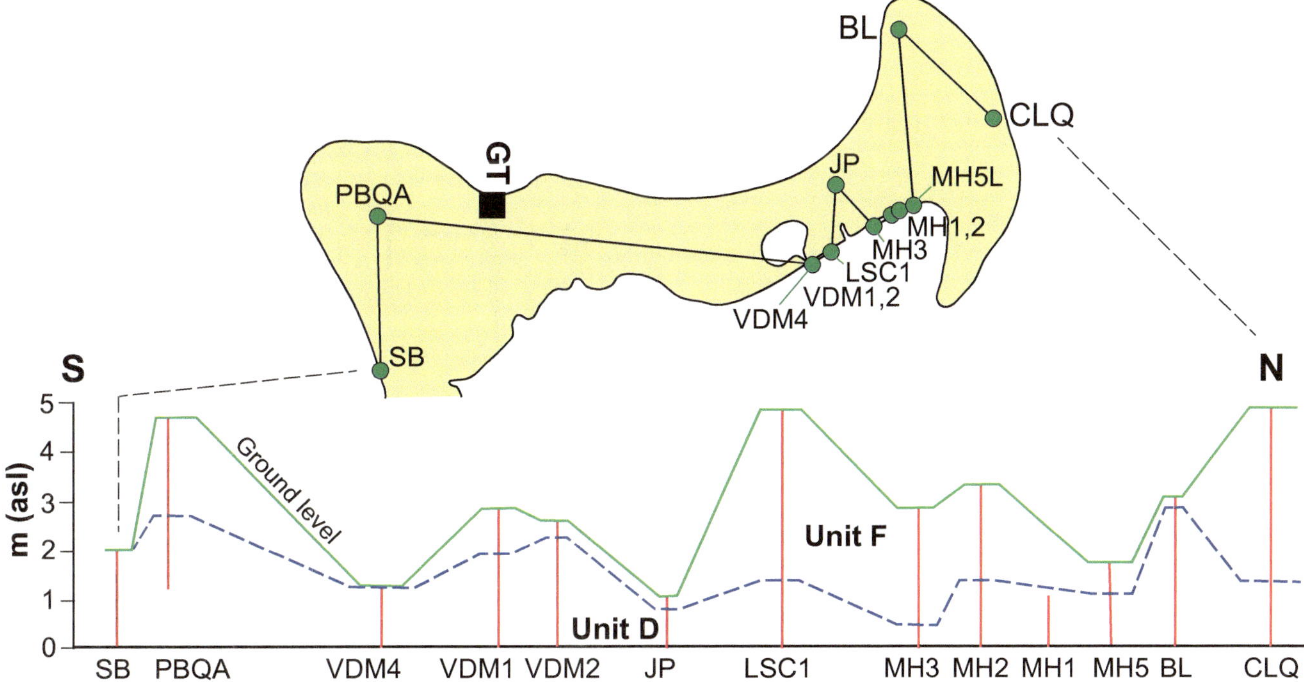

Fig. 7.7 Stratigraphic relationship between units D and F on the western peninsula of Grand Cayman. Note absence of unit E and topographically irregular boundary between units D and F. Modified from Li and Jones (2014b, their Fig. 2c)

7.8 Key Exposures of the Ironshore Formation

Although rare, there are some key exposures of the Ironshore Formation on Grand Cayman, Little Cayman, and Cayman Brac that provide important information about the formation.

7.8.1 Grand Cayman

Salt Creek

An outcrop of the Ironshore Formation to the south of Salt Creek on the west coast of North Sound shows the complex array of facies that commonly characterize the upper part of the Ironshore Formation (Fig. 7.9). There, the lowermost unit of low-angle to parallel laminated ooid grainstones is overlain by cross-bedded and parallel laminated ooid grainstones. Channels cut into these units, at least 2 m deep, were subsequently filled with an array of cross-bedded ooid grainstones that, in the South Channel, contained corals and bivalves that are commonly broken and no longer in growth positions (Fig. 7.9). Large tabular lithoclasts (up to 0.75 m long), formed of oolitic grainstones, rest on the channel floors (Fig. 7.10a). Spectacular arrays of ichnofossils, including the *Ophiomorpha–Polykladichnus*, *Ophiomorpha–Polykladichnus–Skolithos*, *Ophiomorpha–Skolithos*, and *Polykladichnus–Skolithos* assemblages are evident in this outcrop (Jones and Pemberton 1989). *Ophiomorpha*, which is the most common ichnofossil, includes some with a vertical burrow that is up to 1.75 m long and stretches from the surface of the exposure to their main chamber. In many areas, burrowing was so intense that it homogenized the sediment and destroyed the original laminations (Fig. 7.10b).

The ooids in the Ironshore Formation in the Salt Creek area, up to 0.55 mm long, are typically ovoid with large nuclei, thin cortices, and matt surfaces (Jones and Goodbody 1984, their Figs. 6, 7 and 8). As such they contrast with the classical spherical ooids, like those from the Bahamas, that have small nuclei and white, shiny surfaces. The ooids from the Ironshore Formation have nuclei that originated as fecal pellets, possibly produced by the burrowing shrimp *Callianassa*, that underwent early lithification. Precipitation of the thin calcite cortical layers around those pellets may have been bacterially induced. These ooids probably developed in quiet-water environments rather than the high-energy environments where ooids typically form.

No reliable age could be obtained from the few fossils found in the outcrop at Salt Creek. The basal unit, which is largely below sea level, appears to belong to Unit D. Around Morgan's Harbour, located north of Salt Creek, there are patch reefs with numerous corals that are typical of Unit D. These are overlain by an upper unit that is formed of

Fig. 7.8 Wave-cut notch, 6 m above modern-day sea level, on Cayman Brac (**a–c**) and Grand Cayman (**d**). **a** Large block of Bluff Formation near cliff face on northeast corner of Cayman Brac with two wave-cut notches – the lower one associated with modern sea-level and the higher one (6 m asl) that formed ~125,000 years ago. **b** Deeply incised wave-cut notch cut into Cayman Formation on south coast, west end of Cayman Brac. **c** Speleothems in notch shown in panel C. **d** Wave-cut notch at Rogers Wreck Point, cut into Cayman Formation, south side of Queen's Highway, ~6 m asl

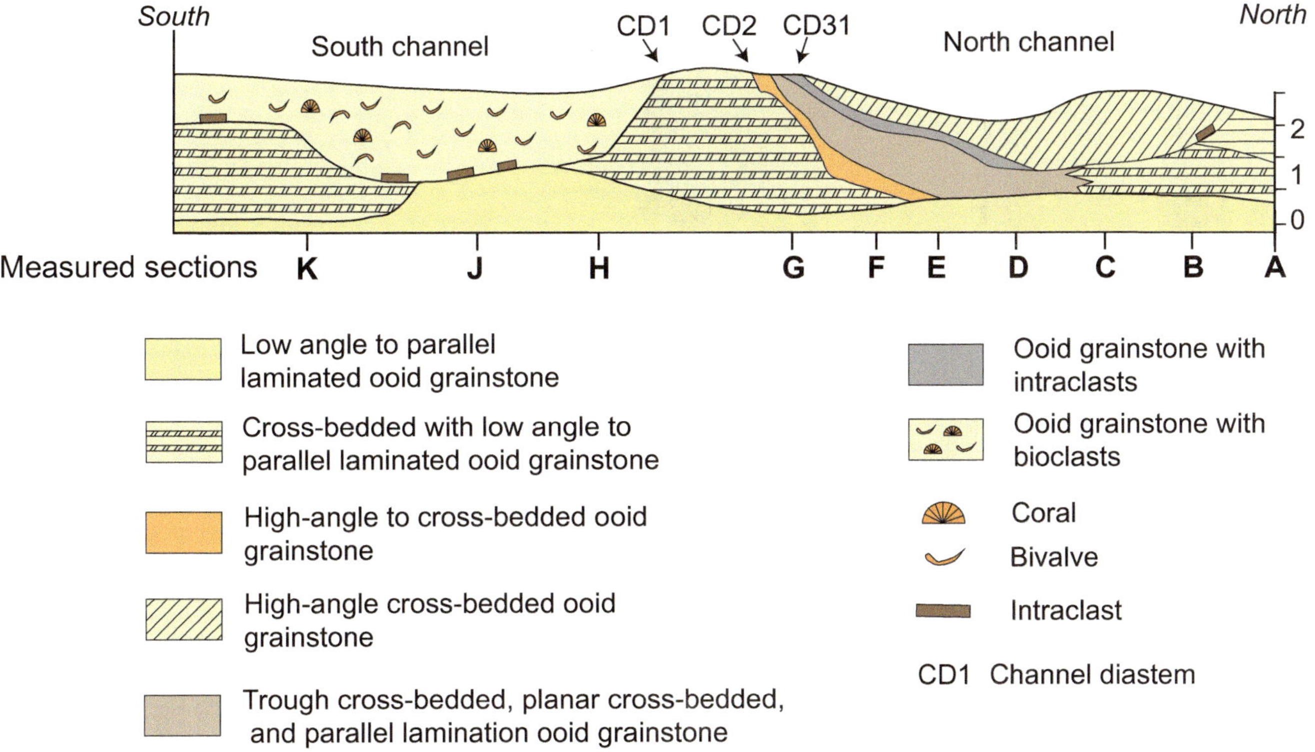

Fig. 7.9 Schematic section showing facies and structures in the Ironshore Formation exposed on west coast of North Sound just south of Salt Creek. Note complex facies architecture with the North Channel and South Channel, each filled with different facies. Based on measured sections A–K. Modified from Jones and Pemberton (1989, their Fig. 4)

cross-bedded and planar laminated oolitic limestones that contain scattered *Ophiomorpha* and *Conichnus*. Two corals from the patch reefs yielded ages of 123 ± 3 and 136 ± 5 ka, whereas a conch shell from the upper unit gave an age of 87 ± 3 ka (Coyne et al. 2007). These ages are consistent with the succession being formed of Unit D and overlain by Unit F (Fig. 7.5a).

Canary Lane Quarry

Prior to the construction of houses that now occupy this site, it was evident that the floor of the Canary Lane Quarry was formed of limestones, belonging to Unit D of the Ironshore Formation, that lay on top of dolostones that belong to the Cayman Formation (Fig. 7.11a, b). The limestones, which contain numerous head corals and conch shells held in an oolitic grainstone matrix, probably belong to Unit D. From the walls of the quarry, it was evident that the fossil-rich unit was overlain by low-angle, cross-bedded oolitic limestones (Fig. 7.11a). These limestones, which are characterized by scattered *Strombus gigas* along with *Ophiomorpha nodosa* and *Conichnus* (Fig. 7.11c) up to ~3 m above sea level, probably belong to Unit F of the Ironshore Formation. No fossils were found above that level. One coral from the basal limestone provided an age of 133 ± 3.5 ka whereas three conch shells from the overlying unit yielded ages of 74 ± 1, 79 ± 2, and 95 ± 3 ka.

Boat Launch, Northwest Coast

Coastal exposures along the coastline to the north of the public boat launch facility on the northwest coast of Grand Cayman provide a spectacular view of the coral reefs that existed as the sediments of Unit D were accumulating (Fig. 7.12). There, a diverse array of almost perfectly preserved corals that are mostly in life positions, provide a clear view of what the seafloor must have looked like when growth occurred 125,000 years ago. The diverse coral assemblage includes numerous specimens of *Acropora palmata, Porites porites, and Montastraea annularis,* common *Acropora cervicornis, Diploria strigosa,* and *Montastraea cavernosa*, scarce *Agaricia agaricities, Porites astreoides, Favia fragum, Diploria labyrinthiformis, Manicina areolata,* and rare *Meandrina meandrites, Isophyllastrea rigida, and Myceptophyllia* spp. Scattered conch and bivalve shells are present between the corals. This unit is overlain by poorly fossiliferous, laminated to cross-bedded limestones that probably belong to Unit F.

Fig. 7.10 Ironshore Formation near Salt Creek, as shown in Fig. 7.9. **a** Large lithoclast resting on erosional surface that separates the lowermost burrowed oolitic limestones from the overlying cross-bedded limestones. Midway between sections J and K (Fig. 7.9). **b** Lowermost part of sequence in northern part of area showing numerous horizontal and vertical *Ophiomorpha* in oolitic grainstone. Most of the original sedimentary structures were destroyed by burrowing

7.8.2 Cayman Brac

On Cayman Brac, outcrops of the Ironshore Formation are rare, being restricted to a coastal platform on the south coast at the east end of the island and some small quarries around the airport on the west end of the island (Jones and Ng 1988). Only scattered outcrops are found along the south and north coasts.

One of the old quarries close to the east end of the airport runway is now the site of the Water Authority operations. Exposures on the north side of the entrance drive reveal a sequence of friable limestones (Fig. 7.13a) that overlie a basal bed that is formed of rounded coral cobbles and scattered gastropod shells, including *Cittarium pica* that still display their original shell colours (Fig. 7.13b). The friable limestones are poorly laminated apart from a few layers that are better cemented than others. Although no fossils were found in these limestones, rhizoliths are common (Fig. 7.13c, d).

Three corals from the basal conglomerate (Fig. 6.13b) yielded U/Th ages of 124,096 ± 0.009, 124,474 ± 0.008, and 120,873 ± 0.015 ka, which are consistent with the bed belonging to Member D, whereas the overlying friable limestones probably belong to Member F.

7.8.3 Little Cayman

The field relationships between outcrops on Little Cayman are difficult to establish because they are widely scattered across the low-lying island that is largely covered with dense vegetation. Rare outcrops of the Ironshore Formation include those on the west end of the island around Salt Rocks, and in the small quarry (Fig. 7.1). In addition, limestones akin to those in the Ironshore Formation are present on an unnamed hill on the west end of the island.

Fig. 7.11 Ironshore Formation exposed in Canary Lane Quarry, northwest Grand Cayman. **a** General view of quarry showing irregular floor where exposures of the Cayman Formation and Unit D of the Ironshore Formation are found. Quarry wall in background, which is ~2.5 m high (on average) is formed of units D and F of the Ironshore Formation. **b** View of quarry floor showing Unit D of the Ironshore Formation with scattered corals. Field of view is ~2 m wide. **c** Unit D of Ironshore Formation exposed in lower part of quarry wall showing scattered *Ophiomorpha* in laminated ooid grainstone

Little Cayman Quarry

The only quarry on Little Cayman, with walls no higher than 4 m, is important because it provided a clear view of the stratigraphic relationship between the Ironshore Formation and the older Pedro Castle Formation (Jones 2019). There, the boundary between the two formations is characterized by a variable but rugged phytokarst topography with a relief of up to 1.5 m. Locally, isolated corals grew on the unconformity that defines the upper boundary of the underlying Pedro Castle Formation (Fig. 7.14a). The low-lying areas between the karst pinnacles are filled with the bivalve-rich limestones of the Ironshore Formation (Fig. 7.14b). In addition to the bivalves, the lower beds (Unit D) of the Ironshore Formation also contained scattered gastropods and varicoloured micrite limestone lithoclasts (Fig. 7.14c).

When initially examined in 2018, the Ironshore Formation was characterized by a biota that was dominated by bivalves, scattered gastropods, and rare, scattered corals. Unfortunately, further quarrying in early 2019 destroyed those quarry walls with those showing the phytokarst and the overlying Ironshore Formation being lost. The renewed quarrying did, however, reveal the presence of numerous corals (mostly *Montastraea annularis*) in the Ironshore Formation that had not been previously evident. Even though the corals appeared to be well-preserved, attempts to obtain U/Th dates from them failed, presumably because they had undergone

Fig. 7.12 Ironshore Formation exposed near boat launch on northwest coast of Grand Cayman. **a** General view of outcrop showing densely packed corals in grainstone matrix. **b** *Porites porites* in growth position. **c** Large branch of *Acropora palmata*—near present-day coastline. **d** Large colony of *Diploria* in growth position. **e** Matrix between large coral heads formed of broken branches of *Acropora cervicornis*

Fig. 7.13 Ironshore Formation at Water Authority site, west end of Cayman Brac. **a** General view of Ironshore Formation with rhizoliths penetrating sequence formed of alternating poorly cemented and lithified grainstones. **b** Conglomerate at base of succession shown in (**a**), formed of rounded pieces of corals. Two corals from this bed yielded U/Th ages of 124,096 and 124,474 years. **c** General view of rhizoliths in lower part of succession shown in (**a**). **d** Cross-section through rhizolith showing central root tube and concentric zoning of carbonate sands with degree of cementation decreasing to outer rim

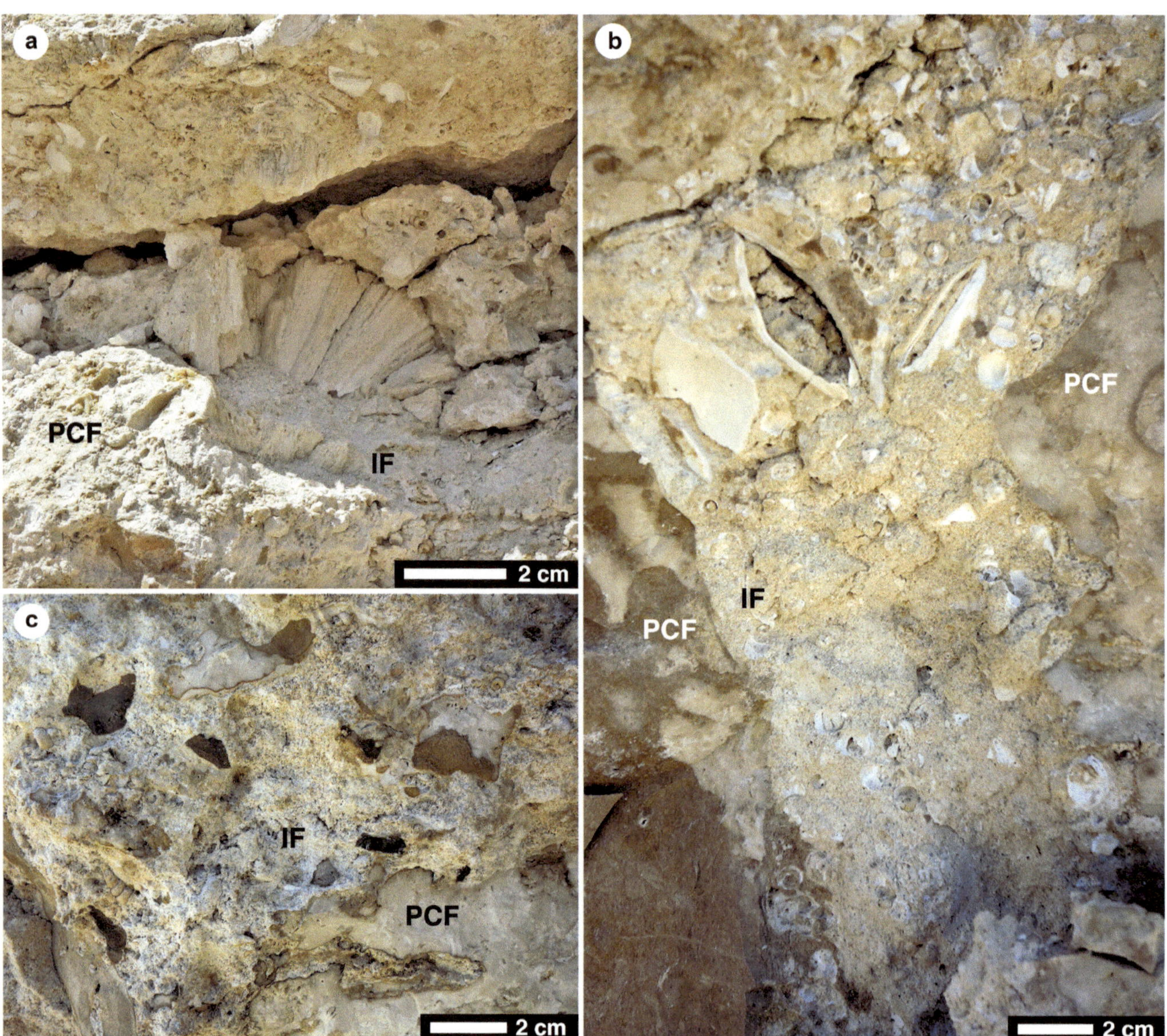

Fig. 7.14 Ironshore Formation in quarry on Little Cayman. The quarry faces shown in these images were destroyed by quarry operations in early 2019. PCF = Pedro Castle Formation; IF = Ironshore Formation. **a** Large colonial coral (*Meandrina meandrites*) that grew on the unconformity at the top of the Pedro Castle Formation was subsequently covered by bivalve-rich limestones. East end of quarry. **b** Phytokarst developed on top of the Pedro Castle Formation, filled-in by limestones of the Ironshore Formation that contain numerous bivalves and scattered gastropods. **c** Basal limestones of the Ironshore Formation with varicoloured carbonate lithoclasts overlying unconformity that caps the Pedro Castle Formation

diagenetic alteration that affected their geochemical properties. Further quarrying between 2020 and May,2022, in the northeast corner of the quarry where the land is the highest in the are, showed that the Ironshore Formation included two units that are separated by layer of terra rossa (Fig. 7.15). In this new quarry face, only scattered fossils are evident in Unit D that lies on top of the phytokarst that had developed on top of the Pedro Castle Formation (PCF). Unit F, which lies on top of the terra rossa, is a grainstone that is devoid of fossils. It is tentatively correlated with unit F that is found on Grand Cayman. The paucity of exposures in the area around the quarry means that it is impossible to determine the lateral extent of this unit. This exposure also highlights the fact that some periods of non-deposition led to the development of phytokarst,whereas other periods of exposure led to the accumulation of terra rossa.

Fig. 7.15 View of northeast corner of quarry on Little Cayman showing Ironshore Formation overlying phytokarst surface that defines the upper boundary of the Pedro Castle Formation (PCF). The Ironshore Formation is divided into two units by a layer of terra rossa (TR). The lower unit belongs to Unit D, whereas the upper unit may belong to Unit F of the Ironshore Formation. The quarry wall is 2.0 to 2.5 m high. The vertical lines on the face of the quarry wall are excavation marks created by quarrying.

Salt Rocks

Salt Rocks, located on the west coast of Little Cayman (Fig. 7.1b), offers an extensive view of the Ironshore Formation with outcrops rising up to ~5 m asl (Fig. 7.16a). Stoddart (1980) recognized the basal reef unit that included *Acropora palmata*, *Diploria labyrinthiformis*, *D. strigosa*, *Montasteria annularis*, *M. cavernosa*, *Porities porites*, and *Meandrina* sp., and numerous *Strombus gigas*. The corals are typically in growth position and the *Strombus* shells are commonly well preserved. This unit is overlain by bedded calcarenite that rises to ~5 m asl over a distance of 25–35 m inland (Fig. 7.16a), Stoddart (1980) noted that it comprised a basal coarse grit with blocks of oolite and *Strombus* shells, a bedded calcarenite with a seaward dip, and an upper unit formed of gently dipping calcarenite.

The basal, coral-rich unit is characterized by numerous well-preserved corals with conch shells lying between them (Fig. 7.16b). In the overlying unit, tabular lithoclasts, up to 1 × 0.75 × 0.25 m in size are found at the base and in an upper layer ~0.25 m above the base (Fig. 7.16c, d). The upper calcarenite units are various cross-bedded and devoid of fossils. Inland, the top of the unit is buried beneath vegetation and the coastal road. Scattered rhizoliths are present in some of the uppermost beds (Fig. 7.16e).

Metz (2011) reported the presence of *Conichnus conicus, Ophiomorpha nodosa, Planolites beverleyensis* from the Ironshore Formation found at Salt Rocks.

Unnamed Hill

Brunt et al. (1973, their Fig. 3) showed the location of calcarenite ridges, up to 7 m asl, on Little Cayman and Grand Cayman. Without giving specific information regarding their height, Stoddart (1980) drew attention to "peripheral ridges" around Little Cayman that are formed of fine-grained, massive, structureless, cream to buff limestones, which he assigned to the Ironshore Formation.

At the summit of an unnamed hill, ~2 km northeast of Salt Rocks, which is ~16 m asl (according to digital topographic maps of the area) there are roadside cuts, up to 3 m high, that expose dense, fine-grained limestone that is generally featureless or characterized by low-angle cross-bedding (Fig. 7.17a, b). The black weathered surfaces, however, hide many of the finer details of these deposits. Despite extensive searching, no fossils were found in these limestones. Scattered rhizoliths are present throughout the outcrop (Fig. 7.17c). The lack of outcrop along the road between the summit of the hill and the coastal road means that it is difficult to determine the relationship of those outcrops with the strata exposed at Salt Rocks. It is important to note, however, that these limestones lie at an altitude that is above any of the known Pleistocene sea levels.

Fig. 7.16 Coastal exposure of the Ironshore Formation near Salt Rocks, west end of Little Cayman. **a** General view of Ironshore Formation at Salt Rocks. Exposure in foreground and right side of image is Unit D. White arrow shows location of upper lithoclast unit with exposures above that being Unit F. **b** Uppermost part of Unit D with colonial corals, conch shells, and lithoclasts. **c** Exposure near ramp (to left of area shown in panel A) showing Unit D, formed of fossiliferous limestones, at base with lower lithoclast unit on top (L1). Note upper lithoclast unit (L2) that equates to unit denoted by white arrow in panel A. **d** Upper surface of Unit D with tabular lithoclasts of various sizes and shapes. **e** Rhizoliths in upper part of section (above upper lithoclast unit) that is devoid of fossils

Fig. 7.17 Outcrop of limestones at east end of Little Island Drive, west coast of Little Cayman, ~2 km northeast of Salt Rocks. Elevation ~16 m above sea level (derived from digital map of island). **a** General view of well-lithified non-fossiliferous limestones at highest point on road. Central part of outcrop is ~2 high. **b** Close-up view of outcrop showing thinly laminated white limestones. **c** Vertical rhizolith embedded in thinly laminated limestones. Opposite side of road from outcrop shown in panel A

7.9 Biota of the Ironshore Formation

Knowledge of the biota that existed in the waters around the Cayman Islands during the time when sediments of the Ironshore Formation were deposited is highly biased towards the fossils from Unit D simply because that unit is the most widely exposed, especially on Grand Cayman. In contrast, the biota in units A, B, and C is limited to that found in the cores from the Roger's Wreck Point area. Cores from offshore of the west coast of Grand Cayman provide some insights into the biota associated with units E and F. Fossils are very rare in unit E, which is not exposed on Grand Cayman and Cayman Brac.

For Grand Cayman, Jones and Hunter (1990, their Fig. 2) and Hunter and Jones (1996, their Fig. 2a) suggested that

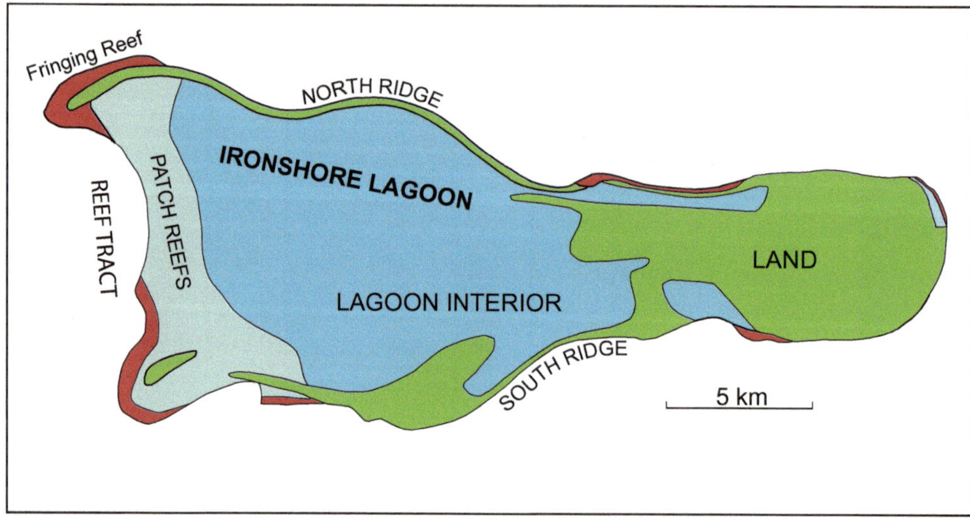

Fig. 7.18 Map showing paleogeography of Grand Cayman when sediments of Unit D of the Ironshore Formation were deposited. Modified from Hunter and Jones (1988, their Fig. 2a)

deposition of the sediments that now form Unit D took place in the "Ironshore Lagoon" that was rimmed to north, east, and south by high-standing outcrops of the Cayman Formation (Fig. 7.18). The north margin extended along the north coast so that North Sound, as we know it today did not exist. The Ironshore Lagoon opened to the west, with a zone of patch reefs developed in the area that now forms the western peninsula of Grand Cayman. Fringing reefs developed along the northwest and southeast coasts of Grand Cayman. During that time the most extensive depositional tract was large lagoon that covered much of what is now Grand Cayman (Fig. 7.18).

Limestones in the Ironshore Formation are commonly characterized by numerous well-preserved corals, bivalves, and gastropods that in Unit D generally show little evidence of diagenetic alteration (Figs. 7.12 and 7.14). Rhodolites formed by various encrusting communities are common in some parts of the formation. Although rare, echinoid spines have been identified in some of the limestones, and fragmentary specimens of *Leodia sexiesperforata* have been found in the Member F that is exposed just south of Morgans Harbour on the west coast of North Sound (Donovan et al. 2016).

7.9.1 Corals

Hunter (1994) and Hunter and Jones (1988, 1996) documented numerous species of corals that are common throughout Unit D of the Ironshore Formation (Figs. 7.12 and 7.19). Hunter and Jones (1996) assigned these corals to 10 associations that were based on the dominant species, overall composition, and faunal diversity (Fig. 7.19). Many of these associations are similar to those found in the modern environments around the Cayman Islands. These associations, defined by their dominant species and species diversity are:

Porites divaricata (4 species)
Montastrea annularis–Porites porites (6 to 7 species at any one locality)
Porites porites –Manicina areolata (5 to 8 species at any one locality)
Montastrea annularis–Mycetophyllia (8 to 15 species at any one locality)
Massive Head Coral (15 to 21 species at any one locality)
Montastrea annularis–Isophyllastrea rigida (27 species)
Pocillopora–Dendrogyra (10 to 12 species at any one locality)
Porites porites–Isophyllastrea rigida (22 species with 16 to 18 at any one locality)
Diploria clivosa (10 species with 7 to 8 at any one locality)
Acropora (19 species with 8 to 16 at any one locality)

Hunter and Jones (1996) suggested that the distribution of these coral associations in the Ironshore Lagoon was controlled by water clarity, sedimentation rates, light availability, and wave energy (Fig. 7.18). The *Porites divaricata* and *Montastraea annularis–Porites porites* associations that dominated the interior of the Ironshore Lagoon were both low-diversity associations with corals that were able to thrive in the restricted, muddy environments found there. Seagrass banks probably covered the seafloor between the corals. The patch reefs, 10–200 m in diameter, that

VC = Very common
C = Common
S = Scarce
R = Rare
VR = Very rare

SPECIES \ ASSOCIATION	P. divaricata	M.annularis – P. porites	P. porites – M. areolata	M. annularis – Mycetophyllia	Massive head coral	M. annularis – I. rigida	Pocillopora-Dendrogyra	P. porites – I. I. rigida	Acropora	D. clivosa
Porites divaricata	R	R								
Siderastrea radians	R	R	R		R	VR		VR	VR	
Manicini areolata	R	R	S	S	VC	R	S	S	VR	
Montastrea annularis	VR	S	S	VC	C	VC	C	C	C	S
Porites porites		S	VC	C	S	C	C	VC	S	
Favia fragum		R	R	S	C	S	S	S	R	R
Diploria strigosa		R	R	C	C	C	S	C	S	C
Siderastrea siderea		R	R	C	C	S		R	R	
Agaricia agaricites			R	C	S	S		S	R	
Porites astreoides			S	S	R	S				
Agaricia fragilis			R	R		VR				
Stephanocoenia michelini			VR	VR	VR	VR		VR		
Mycetophyllia spp.				S	R	R	R	VR	R	
Diploria labyrinthiformis				S	S	S	S	S	S	S
Monastrea cavernosa				VR	S	S	S	S	S	S
Colpophyllia natans				VR	R	S	R	C	R	VC
Diploria clivosa				VR	R	R		VR		
Eusmilia fastigiata				VR	R	VR				
Scolymia cubensis				VR	R				VR	C
Dichocoenia stokesi					R	R	R	R	R	
Isophyllastrea rigida					R	R		R		
Isohyllia sinuosa					R	VR			VC	
Acropora palmata					VR	C				
Dendrogyra cyclindrus					VR	R	VC	S	R	
Leptoseris cucullata					VR					
Oculina diffusa					VR					
Acropora cervicornis						C	S	S	VC	S
Meandrina meandrites						R	VR	R	VR	
Pocillopora cf. P. palmata						VR	VC			
Mussa angulosa						VR				
Madracis decactis						VR				
DIVERSITY	**4**	**9**	**11**	**17**	**24**	**27**	**15**	**22**	**19**	**10**

Fig. 7.19 Coral species and coral associations found in the Ironshore Formation on the Cayman Islands. Modified from Hunter and Jones (1988, their Table 1, their Fig. 4). The abundance of each species in the different Coral Associations is given on a qualitative scale based on the number of specimens of each species seen at each locality

developed in the western part of the Ironshore Lagoon (Fig. 7.18) in water that may have been up to 10 m deep, include the *Porites porites–Manicina areolata*, *Montastraea annularis–Mycetophyllia*, and Massive Head Coral associations (Fig. 7.19). In general, the patch reefs were characterized by a greater diversity of corals than in the lagoon interior because of more favourable growth conditions. Nevertheless, some patch reefs were monospecific whereas others have a high diversity of corals. Zoning is evident in some of the larger patch reefs with *M. annularis* and *A. agaricites* in the centers and *P. porites* in the outer parts. The reef tract, which encompasses the *Montastraea annularis–Isophyllastrea*, *Pocillopora–Dendrogyra*, *Porites porites–Isophyllastrea rigida*, *Acropora*, and *Diploria clivosa* associations flourished in water that was <6 m deep. The coral diversity ranged from nearly monospecific in some areas to other areas with high diversities (up to 22 species).

Many of the corals found in the patch reefs have been extensively bored by *Lithophaga* (Jones and Pemberton 1988b, 1988a). The borings (Fig. 7.19), which belong to the ichnospecies *Gastrochaenolites torpedo*, have a circular cross-section with the distal part tapering to a rounded terminus. The wide range in the size of the borings, which are 4–14 cm long and 0.3–4 cm in diameter, reflects the fact that all sizes from juveniles to adults were present. Many borings, which still contain the *Lithophaga* that produced them, are lined with laminated, fluorescent cryptocrystalline calcite. After death of the *Lithophaga*, many borings were filled with various combinations of porous to honeycombed cryptocrystalline calcite, pelsparite, and pelmicrite that contains small ostracod(?) shells. In the Ironshore Formation, *Lithophaga* are common in the large domal *Montastraea annularis* (Fig. 7.20) and *Diploria labyrinthiformis*, rare in *Porites asteroids* and seemingly absent from other corals. In extreme cases, the outer 14 cm of the coral heads have been so extensively altered by the borings that the original fabrics of the coral skeleton are hard to detect. The fact that some borings cross-cut other borings provides evidence of multiple generations of boring. These borings are important because they can weaken the corals and make them more prone to erosion (especially during storms), liberate substantial quantities of calcium carbonate into the surrounding habitat, increase the surface areas that are available for boring by other organisms, provide sites for sediment accumulation (once the formative shell is lost), and provide pathways by which diagenetic fluids can penetrate the interiors of corals.

7.9.2 Rhodolites

Rhodolites are rare in the limestones of Unit D of the Ironshore Formation, having been found only in the Crystal Harbour area on the western Peninsula, Grand Cayman (Hills 1997; Hills and Jones 2000). These spherical, discoidal, prolate and bladed rhodoliths (Fig. 7.21), up to 14 cm long, have a nucleus formed of coral (mostly *Siderastrea radians* and *Porites porites*) and bivalve shells (Fig. 7.21b). They are typically found around the patch reefs in Unit D that developed on the western part of Grand Cayman.

The nuclei are encased by the Coralline Algal Community, which forms a 2 to 10 mm thick layer around the

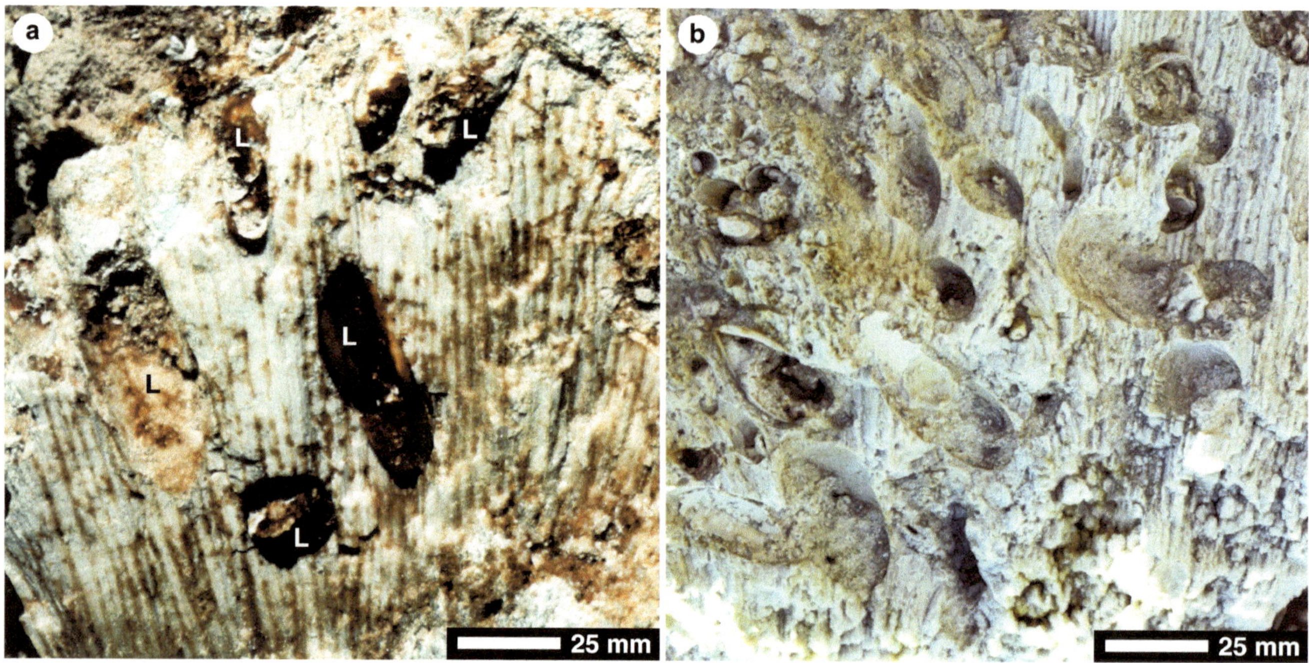

Fig. 7.20 *Lithophaga* (L) borings in *Montastraea annularis* from Member D of the Ironshore Formation, central part of western peninsula, Grand Cayman

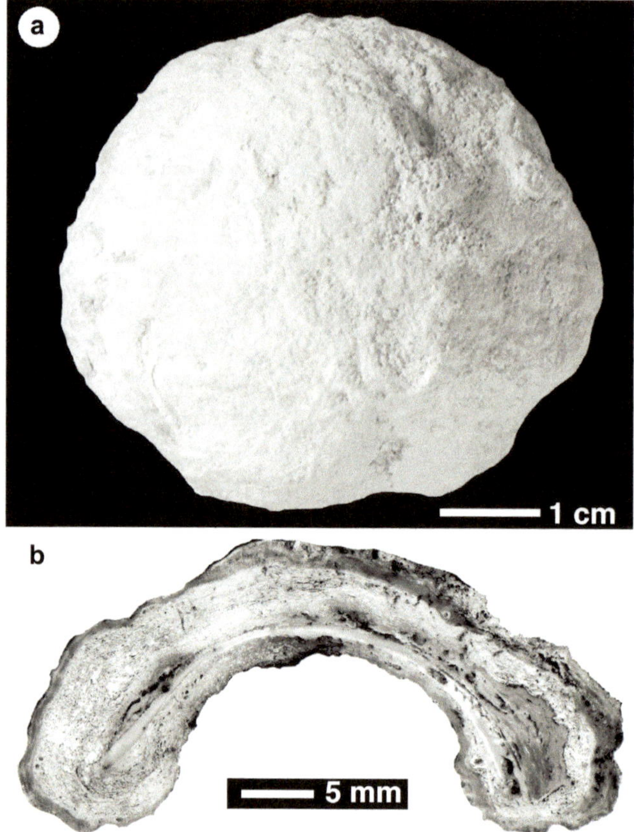

Fig. 7.21. Examples of rhodolite from Member D of the Ironshore Formation, western peninsula of Grand Cayman. **a** Large spherical rhodolite. **b** Cross-section through rhodolite that formed around a bivalve shell. See Hills and Jones (2000, their Fig. 3) for more examples of rhodolites from the Ironshore Formation

nucleus that includes *Lithoporella* sp., *Lithophyllum* sp., *Neogoniolithon* sp. with lesser numbers of *Peysonnelia rubra*, *Hydrolithon reinboldii*, *Lithothamnium* sp. and *Porolithon* sp. (Hills and Jones 2000, their Fig. 5). The outer part of the cortices, up 60 mm thick, are formed of the "Peyssonnelid Community" that is dominated by *Peysonnelia rubra* and scattered coralline algae, foraminifera, serpulid worms, and bivalves (Hills and Jones 2000, their Fig. 7). The change from the Peyssonnelid Community to the Coralline Community was probably a natural ecological succession that was controlled by local depositional conditions around the patch reefs.

7.9.3 Stromatolites

At some localities on Grand Cayman (e.g., DPQ, EOT, SBE), it is evident that corals became the nuclei of rhodolites that were subsequently encased by microbialites (Jones and Hunter 1991). The size and shape of these structures, which are commonly difficult to identify in the black weathered outcrops, are a function of the corals that became the nucleus for these complex structures. Some examples from DPQ are up to 20 cm high and up to 5 cm in diameter. Many of the coral branches, which were partly altered to calcite, were bored by worms, algae, and/or fungi before being encased by various genera of red algae (e.g., *Porolithon*, *Neogoniolithon*, *Mesophyllum*), foraminifera (e.g., *Carpenteria*, *Homotrema*), and serpulid worms. In some specimens, this zone is of even thickness around the coral, whereas in other specimens the coating is only developed on one side.

The coated coral branches were subsequently encased by microbialites that are formed of alternating light brown (0.5–4 mm thick) and black micrite laminae (0.1 mm thick). Peloids, small biofragments, and numerous tunicate spicules are found in some laminae (Jones 1990). Although the quantity of allochems varies from layer to layer, they are most common in the light-coloured laminae. The outermost parts of these coatings are typically pustulose or digitate. Some microbialite coatings were penetrated by borings (<1 mm long, 0.5 mm diameter) that were probably formed by worms and/or sponges.

Jones and Hunter (1991) suggested that these structures developed under regressive conditions that accompanied the falling sea level that led to termination of sediment deposition for Unit D. The red-algae-foraminifera coating probably developed while the coral branches lay on the seafloor and were periodically overturned during high-energy currents (storms?). In contrast, the microbialites developed under calm water conditions that accompanied the terminal phase of the highstand under which the sediments of Unit D had been deposited.

7.9.4 Bivalves and Gastropods

Bivalves and gastropods are common throughout the Ironshore Formation (Fig. 7.22). Exposures of the Ironshore Formation in the central part of Grand Cayman are scarce because the area is low-lying and covered with mangrove swamps. Material excavated during the construction of dyke roads through the mangrove swamps and from various quarries in the area, however, provide samples of the Ironshore Formation that commonly includes a diverse array of bivalves and gastropods. The superb preservation of most of the bivalves is evident from the fact that most still have intact aragonitic shells and many still show traces of their original colouration. Rehder (1962) compiled a list of 50 different bivalves and gastropods from the Ironshore Formation based on those previously noted by Matley (1926) and Richards (1955) and a collection that came from a marl pit located on the eastern part of the island (possibly along Frank Sound Road).

The bivalve and gastropod fauna in the Ironshore Formation is very diverse. A collection of 32,840 shells from 44 localities on the western part of Grand Cayman (Fig. 7.23; Table 7.1) included 83 species of marine bivalves and 90 species of marine gastropods (Cerridwen 1989; Cerridwen and Jones 1991). With the exception of localities, EOT, DPQ, TFB, and BTD, all locality collections contained more bivalves than gastropods (Fig. 7.24a). Samples with more gastropods than bivalves came from the fringing reef localities. Most of the bivalve and gastropod taxa in the Ironshore Formation are the same as those found in modern systems.

Cerridwen (1989) and Cerridwen and Jones (1991) used cluster analyses to define seven Faunal Associations that belong to Faunal Groups FG-I and FG-II and seven Locality Groups (LG1-7). Faunal Group I, dominated by *Arca*, *Barbatia*, and *Chama,* is typically associated with the fringing reefs and patch reefs on the western part of Grand Cayman (Table 7.2). In contrast, FG-II, dominated by *Cerithium*, *Chione*, and *Codakia*, is found in the lagoon interior and between the patch reefs (Table 7.2). These analyses showed that the distribution of the molluscs and gastropods was controlled primarily by the occurrence of hard and soft substrates throughout the Ironshore Lagoon, the patch reefs, and the fringing reefs (Cerridwen and Jones 1991).

The distribution of the bivalves and gastropods was controlled primarily by their preferred mode of life (Fig. 7.24b) and their food source (Fig. 7.24c), both of

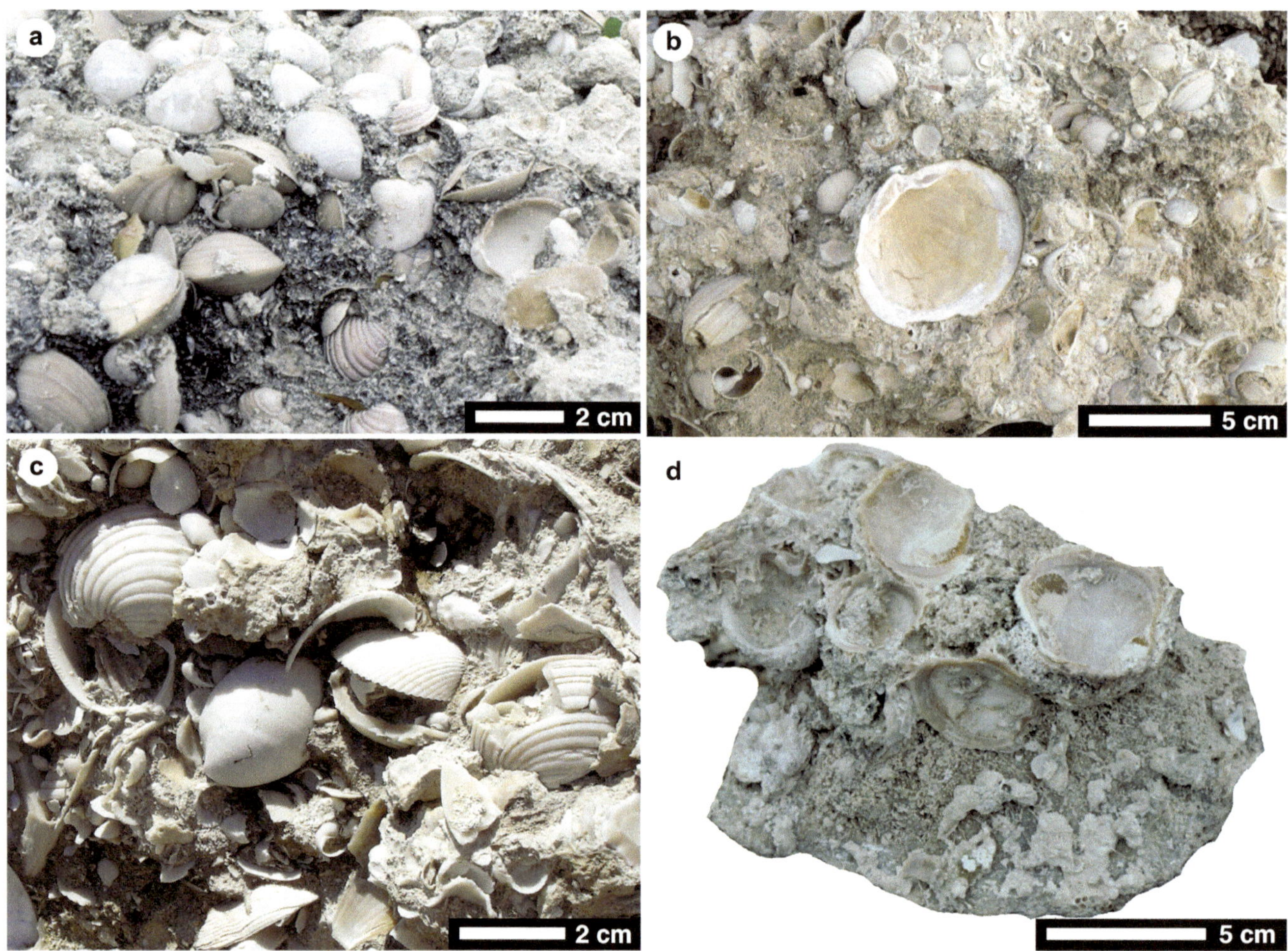

Fig. 7.22 Field photographs of bivalves and gastropods found at various locations in Member D of the Ironshore Formation on Grand Cayman. Localities shown in Fig. 7.23a. **a** Numerous articulated and disarticulated bivalve shells in limestone matrix. Locality TG. **b** Well-preserved articulated and disarticulated bivalves. Locality TG. **c** Well-preserved disarticulated bivalves from locality MA. **d** Bivalves attached to *Strombus gigas* shell (conch), lower part of image. Locality CYA

which are related to the environmental conditions where they lived. This is reflected in the Locality Groups and the areas where they developed within the Ironshore Lagoon.

LG-1: Mainly in the reef tract on the western margin of the Ironshore Lagoon

LG-2: This group included one locality in the fringing reef and four localities in the north part of the patch reef area.

LG-3: All localities in this group are transitional between the patch reefs and the surrounding lagoonal waters.

LG-4: All from lagoonal area with bivalves characteristic of FG-I and FA-1.

LG-5: With the exception of locality M, all the localities in this group are found in the southern part of the Ironshore Lagoon.

LG-6: Found only at localities E and F, this group lived in the northern part of the patch reef area.

LG-7: This group is poorly defined because most constituent localities only yielded small numbers of shells. These localities were located in the transitional areas between the lagoon and the patch reefs in the western part of the Ironshore Lagoon.

Table 7.1 List of bivalve and gastropod species from Unit D in the Ironshore Formation on Grand Cayman. Data from Cerridwen and Jones (1991, their Table 2). ASE = Articulated shell equivalents

Class Bivalvia	ASE		ASE		ASE
Family Nuculidae	1	**Family Spondylidae**		**Family Tellinidae**	
Nucula aegeensis	1	*Spondylus americanus*	6	*Leporimetis intastriata*	12
Family Mytilidae		**Family Anomiidae**		*Stringilla mirabilis*	20
Botula fusca	13	*Anomia simplex*	1	*Tellina aequistriata*	15
Brachidontes modiolus	37	**Family Ostreidae**		*T. candeana*	241
Gregariella coralliophaga	15	*Dendostrea frons*	37	*T. fausta*	18
Lithophaga antillarum	12	*C. orbicularis*	534	*T. gouldii*	11
L. nigra	16	*C. orbiculata*	291	*T. listeri*	101
Lithophaga sp.	7	*C. pectinella*	453	*T. mera*	33
Musculus lateralis	38	*Divaricella dentata*	15	*T. radiata*	18
Family Arcidae		*D. quadrisulcata*	95	*T. similis*	99
Anadara floridana	193	*Divaricella sp.*	7	*T. sp.*	10
A. notabilis	71	*Linga pensylvanica*	1186	*T. sybaritica*	421
Anadara sp.	15	**Family Ungulinidae**		**Family Semedidae**	
Arca imbricata	410	*Diplodonta punctata*	188	*Cumingia coarctata*	67
Arca asp.	8	*D. semiaspera*	7	Semele bellastriata	3
A. zebra	63	*Diplodonta sp.*	1	S. proficua	45
Arcopsis adamsi	113	**Family Chamidae**		**Family Solecurtidae**	
Barbatia cancellaria	485	*Chama congregata*	251	*Tagelus divisus*	11
B. candida	12	*C. macerophylla*	408	**Family Trapeziidae**	
B. domingensis	248	*C. sinuosa*	35	Coralliophaga coralliophaga	8
Family Glycymerididae		*Chama sp.*	21	**Family Veneridae**	
Glycymeris pectinata	5	*Pseudochama radians*	16	*?Agriopoma texasianum*	1
Family Pteridae		*P. radians variegata*	91	*Chione cancellata*	1095
Pinctada imbricata	5	Unidentified chamidae	65	*C. paphia*	823
Pteriid fragments	47	**Family Sportellidae**		*Gouldia cerina*	483
Family Limidae		Basterotia quadrata	8	*Periglypta listeri*	36
Lima lima	8	**Family Crassitellidae**		*Pitar fulminatus*	712
L. scabra	11	*Crassinella martininicensis*	1948	*Transennella gerrardi*	141
Family Pectinidae		**Family Cardiidae**		**Family Petricoliidae**	
Argopecten nucleus	80	*Americardia guppyi*	1119	*Petricola lapicida*	13
Bractechlamys antillarum	9	*A. media*	86	**Family Gastrochaenidae**	
Pectinid fragments	2	Laevicardium laevigatum	365	*Gastrochaena hians*	3
Family Plicatulidae		**Family Mesodesmatidae**		**Family Poromyidae**	
Plicatula gibbosa	11	*Ervilia concentrica*	316	*?Poromya sp.*	2
Family Fissurellidae		**Family Epitoniidae**		**Family Buccinidae**	
Diodora dysoni	9	*Epitoneum sp. A*	2	*Bailya parva*	3
D. jaumei	3	*Epitoneum sp. B*	3	*Pisania auritula*	2
D. listeri	91	**Family Epitoniidae**		**Family Epitoniidae**	
D. minuta	55	*Melanella sp.*	1	*Nassarius albus*	147
Diodora sp.	4	**Family Eulimidae**		**Family Fasciolariidae**	
Emarginula pumila	12	*Melanella sp.*	1	*Fasciolaria tulipa*	16

(continued)

Table 7.1 (continued)

Class Bivalvia	ASE		ASE		ASE
Fissurella barbadensis	6	**Family Strombidae**		*Latirua carinifer*	20
Hemitoma emarginata	3	Strombus gigas	19	*Leucozonia nassa leucozonalis*	43
Lucapina suffusa	24	*S. gigas samba*	2	**Family Olividae**	
Puncturella pauper	16	*Strombus sp.*	3	*Oliva reticularis*	2
Family Acmacidae		**Family Hipponicidae**		*Olivella sp. A*	156
Acmacea sp.	1	*Hipponix antiquatus*	92	**Family Mitridae**	
Patelloida pustulata	17	**Family Calyptraeidae**		*Mitra barbadensis*	2
Family Cyclostremadidae		*Cheilea equestris*	1	**Family Marginellidae**	
Arene cruentala	1	*Crepidula aculeata*	3	*Marginella apicina*	58
Family Turbinidae		**Family Triviidae**		*M. guttata*	4
Astralium phoebium	14	*Trivia pediculus*	1	*M. pruniosum*	8
Lithopoma tectum	64	*T. quadripunctata*	11	*Marginella sp.*	2
Lithopoma tuber	3	**Family Cypraeidae**		*Volvarina avena*	7
Family Phasianellidae		*Cypracea sp. A*	1	**Family Conidae**	
Tricolia thalassicola	255	*Cypracea sp. B*	11	*Conus jaspideus*	14
Family Neritidae		**Family Ovulidae**		*C. jaspideus verrucocus*	4
Smaragdia viridis	13	*Cyphoma gibbosum*	16	*C. mus*	3
Family Rissoidae		**Family Naticidae**		**Family Turridae**	
Rissoina cancellata	16	*Natica canrena*	55	*Brachycythara biconica*	9
Schwartziella bryerea	26	*N. livida*	2	*Crassispira fuscescens*	6
Zebina browniana	149	*Natica sp.*	11	*Ithycythara sp.*	2
Family Modulidae		*Polinices lacteus*	81	*Mangelia lastica*	1
Modulus modulus	200	*Sinum perspectivum*	1	*Pyrgocythara candidissima*	11
Family Cerithiidae		**Family Ranellidae**		**Family Pyramellidae**	
Bittium varium	11	*Cymatium pileare*		*Odostomia laevigata*	1
Cerithium eburneum	856	**Family Ranellidae**		*Pyramidella dolabrata*	28
C. eburneum algicola	489	*Cymatium pileare*	12	*Triptychus niveus*	8
C. litteratum	1162	**Family Muricidae**		*Turbonilla sp.*	1
C. lutosum	406	*Chicoreus florifer*	1	**Family Scaphandridae**	
Cerithium sp.	3	*?Dermomurex pauperculus*	1	*Acteocina sp.*	40
Family Cerithiopsidae		*Phyllonotus pomum*	1	**Family Bullidae**	
Cerithiopsis emersonii		*Thais deltoidea*	4	*Bulla striata*	511
Seila adamsi		**Family Coralliophilidae**		**Family Atyidae**	
Family Triphoridae		*Coralliophila abbreviata*	1	*Atys sp.*	114
Triphora melanura	1	**Family Columbellidae**			
Triphora turristhomae	1	*Anachis hoyessieriana?*	198		
		Columbella mercatoria	40		
		Mitrella sp.	2		

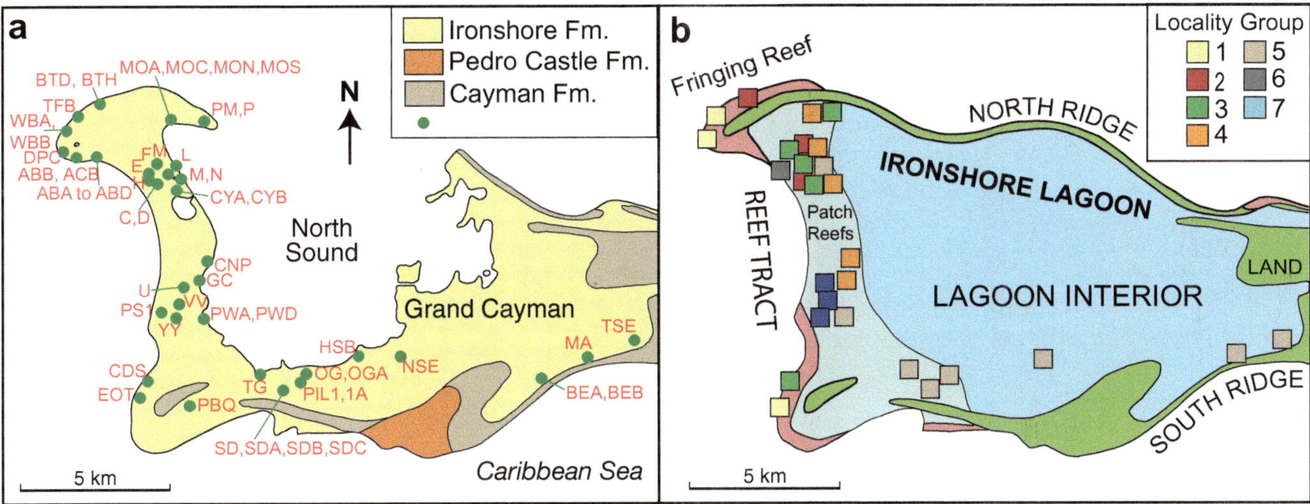

Fig. 7.23 Bivalve and gastropod collections from the Ironshore Formation, Grand Cayman. **a** Localities where bivalves and gastropods were collected from the Ironshore Formation. **b** Paleogeographic architecture of western part of Grand Cayman (see Fig. 7.18) showing distribution of Locality Groups 1–7 that are defined by their bivalve and gastropod assemblages (based on Cerridwen and Jones 1991, their Figs. 1 and 9)

7.10 Trace Elements

The limestones of the Ironshore Formation are characterized by variable concentrations of Mn, Fe, Na, and Sr (Figs. 7.25, 7.26 and 7.27). The Mn content of the corals and limestone matrices in the Ironshore Formation is relatively constant throughout Units A–F, with little difference in the mean values that are generally <20 ppm (Fig. 7.25a). Although the Fe concentrations are more variable, there is no discernable pattern between Units A–F (Fig. 7.25b). Some of the corals from Ironshore Formation also contain up to 700 ppm Fe (Fig. 7.25b). Samples from unit B and C show the largest range of values whereas the highest average valve is from Unit B (Fig. 7.25b).

Na concentrations are higher in the corals (up to 8353 ppm, based on 105 samples) than their surrounding matrices (up to 2244 ppm, based on 84 samples) (Fig. 7.26). Similarly, the average value from the corals (5236 ppm) is higher than the average value from the surrounding matrices (2244 ppm) (Fig. 7.26).

The average Sr content of the limestones in Units A–F of the Ironshore Formation ranges from 2743 ppm in Member B to 5553 ppm in Member E (Fig. 7.27). All of these values are higher than those found in the limestones and dolostones of the Bluff Group. Although the overall range in the Sr content of units A–F are similar, the mean values are lowest in units A and B and highest in units E and F (Fig. 7.27). Based on all samples from the formation, it is apparent that there is high correlation between the Sr content and the percentage of aragonite in the limestones (Fig. 7.27). Although the Sr content of the limestones and corals increases with the percentage of aragonite, there is no consistency from unit to unit within the formation (Fig. 7.27). The high Sr content of the limestones relative to the limestones in the Bluff Group largely reflects the fact that aragonite can, relative to calcite, more readily accommodate Sr in its lattice.

7.11 Diagenesis in the Ironshore Formation

The fact that limestones in the Ironshore Formation range from friable to well-lithified limestones reflects the stratigraphic and geographic variations in the diagenetic processes that have acted on these rocks. Dolomitization is not an issue given that dolomite has only been found in trace amounts in Unit A in the Rogers Wreck Point area. Diagenesis in the Ironshore Formation can be considered from the perspective of changes that involved alteration of the skeletal material and the processes that led to the transformation of the original sediments to their semi-lithified and lithified equivalents. Most of the information regarding diagenesis in the Ironshore Formation comes from the limestones and fossils in Unit D simply because that is the most widespread unit on each of the Cayman Islands.

The degree of diagenesis is highly variable with some aragonitic fossils displaying no evidence of alteration

Table 7.2 Bivalve (black) and gastropod (blue) faunal groups and assemblages in the Ironshore Formation on Grand Cayman as defined by Cerridwen (1989) and Cerridwen and Jones (1991)

Faunal Group I	**Faunal Group II**
Association I	**Association V**
Americardia guppyi	*Arca imbricate*
Linga pensylvanica	*Chama macerophylla*
Chione cancellata	*Barbatia domingensis*
Codakia costata	*Barbatia cancellaria*
Pitar fulminatus	*Chama congregata*
Ervilia concentrica	*Pseudochama radians variegata*
Crassinella martinicensis	*Chama* sp.
Chione paphia	*Gregariella coralliophaga*
Cerithium eburneum	Unidentified Chamidae
Cerithium lutosum	*Diplodonta punctata*
Association II	*Dendostrea frons*
Leporimetis intastriata	*Anachis hotessieriana*
Tellina similis	*Hipponix antiquatus*
Tellina radiata	*Diodora listeri*
Oliva reticularis	*Transennella gerrardi*
Polinices lacteus	**Association VI**
Bulla striata	*Lithopoma tectum*
Tricolia thalassicola	*Patelloidea pustilata*
Association III	*Columbella mercatoria*
Americardia media	*Cypraea* sp. B
Periglypta listeri	*Latirus carinfer*
Anadara floridana	*Lucapina suffusa*
Codakia orbicularis	*Cyphoma gobbosum*
Argopecten nucleus	*Leucozonia nassa leucozonalis*
Anodontia alba	*Cymatium pileare*
Tellina sybaritica	**Association VII**
Gouldia cerina	*Barbatia candida*
Codakia orbiculata	*Diodora minuta*
C. pectinella	
Ceritium eburneum algicola	
Divaricella quadrisulcata	
Tellina candeana	
Strigella mirabilis	
Laevicardium laevigatum	
Modulus modulus	
Semele proficua	
Zebina browniana	
Association IV	
Atys sp.	
Nassarius albus	

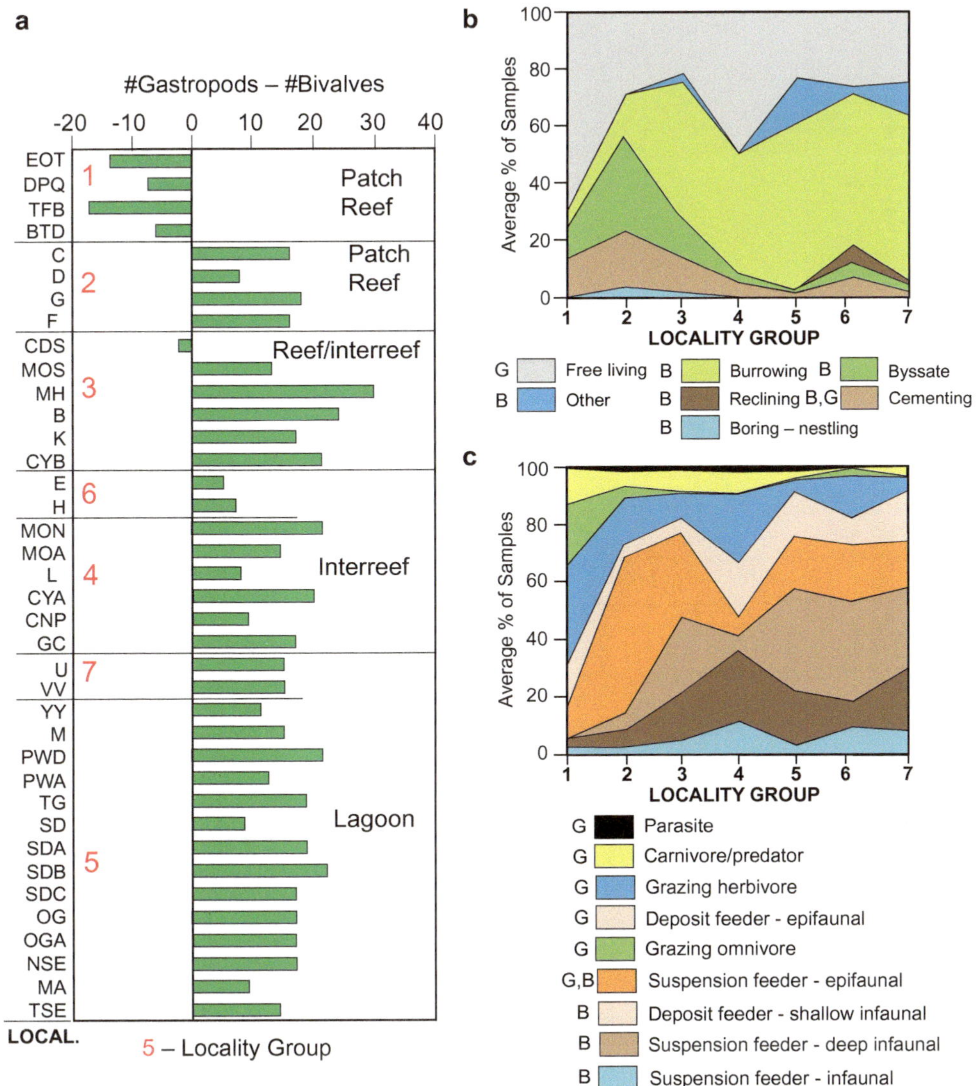

Fig. 7.24 Bivalve and gastropod assemblages from the Ironshore Formation, Grand Cayman. **a** Graph showing that the number of bivalve species exceeds the number of gastropod species at all localities except for EOT, DPQ, and TFB. **b, c** Life mode (**a**) and feeding mode (**c**) of bivalves from locality groups 1–7. Modified from Cerridwen and Jones (1991, their Figs. 13b and 14b, respectively)

whereas others have been totally calcified. Similarly, some limestones have been extensively cemented with calcite whereas others have not.

7.11.1 Alteration of Skeletal Material

Skeletal material in the Ironshore Formation, which is very common, ranges from complete, virtually intact colonial corals that are still in growth position to aragonite fossils that have undergone variable degrees of dissolution and/or neomorphism to calcite. Patterns of skeletal diagenesis are particularly evident in the large *Strombus gigas* (conch) shells and the corals that are so common in the formation.

Strombus Gigas Shells

Large *Strombus gigas* shells (15–30 cm long, columella and lips up to 2 cm thick), common throughout Unit D, clearly illustrate the variable nature of diagenesis (Fig. 7.28) given that these shells now range from 100% aragonite to 100% calcite (Rehman 1992; Rehman et al. 1994). Extensive boring by sponges, evident in many of these shells, are

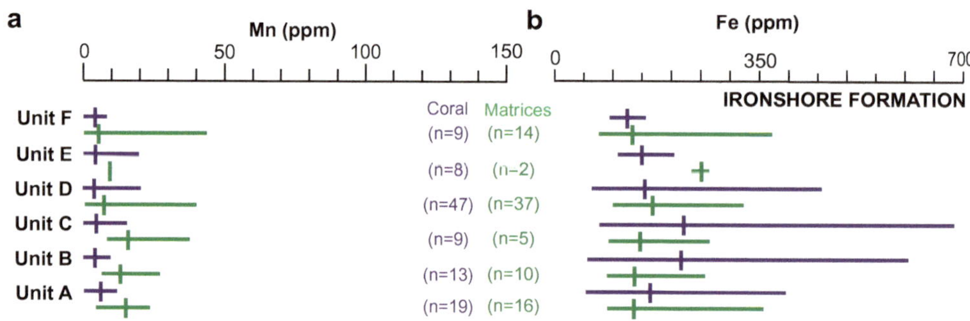

Fig. 7.25 Comparison of (**a**) Mn, and (**b**) Fe content (range and mean) in corals and limestone matrices in units A to F of the Ironshore Formation on Grand Cayman

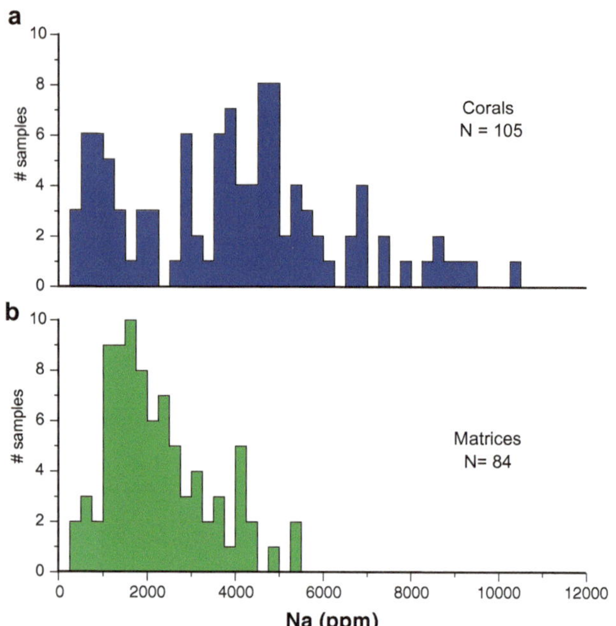

Fig. 7.26 Comparison of Na content in corals and matrices in Units A to E of the Ironshore Formation on Grand Cayman

commonly partly filled with sediment, spar calcite cement, peloids, and/or fiber calcite crystals. Diagenetic calcite is typically found along the outer edge of the shell where sponge borings are most common. Relict needle-shaped aragonite crystals are commonly evident in the layers that have been extensively replaced by blocky calcite crystals.

The $\delta^{18}O$ values for the calcite ranges from −7.10 to −4.90‰ whereas the range is from −3.84 to −1.37‰ for the aragonite (Fig. 7.29). On average, the $\delta^{18}O$ values of the calcite is 3‰ more negative than the aragonite values. The calcite $\delta^{13}C$ values range from 0.20 to 5.25‰ whereas the $\delta^{18}O$ values of the aragonite range from 0.46 to 4.54‰ (Rehman et al. 1994). The fact that the percentage of calcite in these shells ranges from 0 to 100% indicates that this facet of diagenesis was not controlled by the original shell structure and composition. The permeability of the host limestones may have played an important role in the diagenetic processes because shells in the low-permeability mudstones and wackestones underwent the least alteration, whereas those in the more permeable grainstones and packstones, found around the patch reefs and reefs, typically

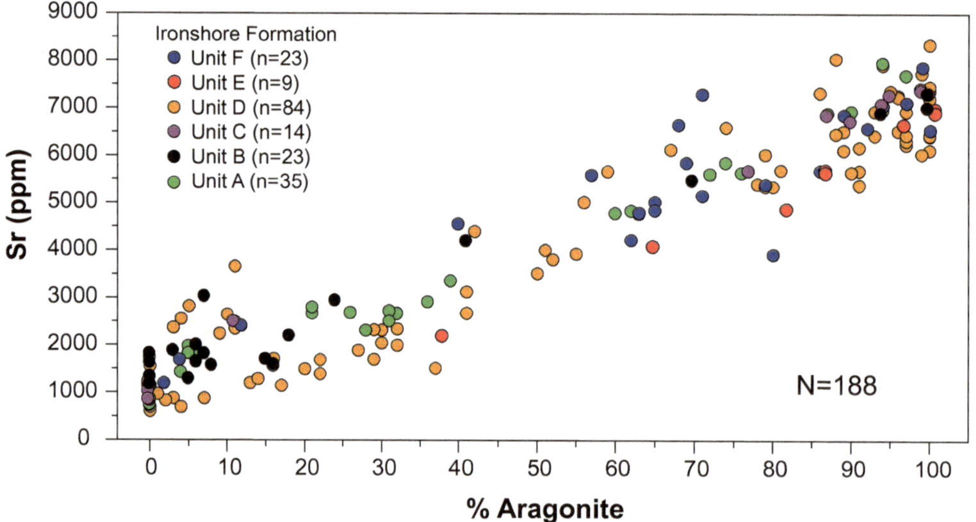

Fig. 7.27 Correlation between Sr and percent aragonite in limestones from Units A to F of the Ironshore Formation on Grand Cayman. Based on 188 analyses

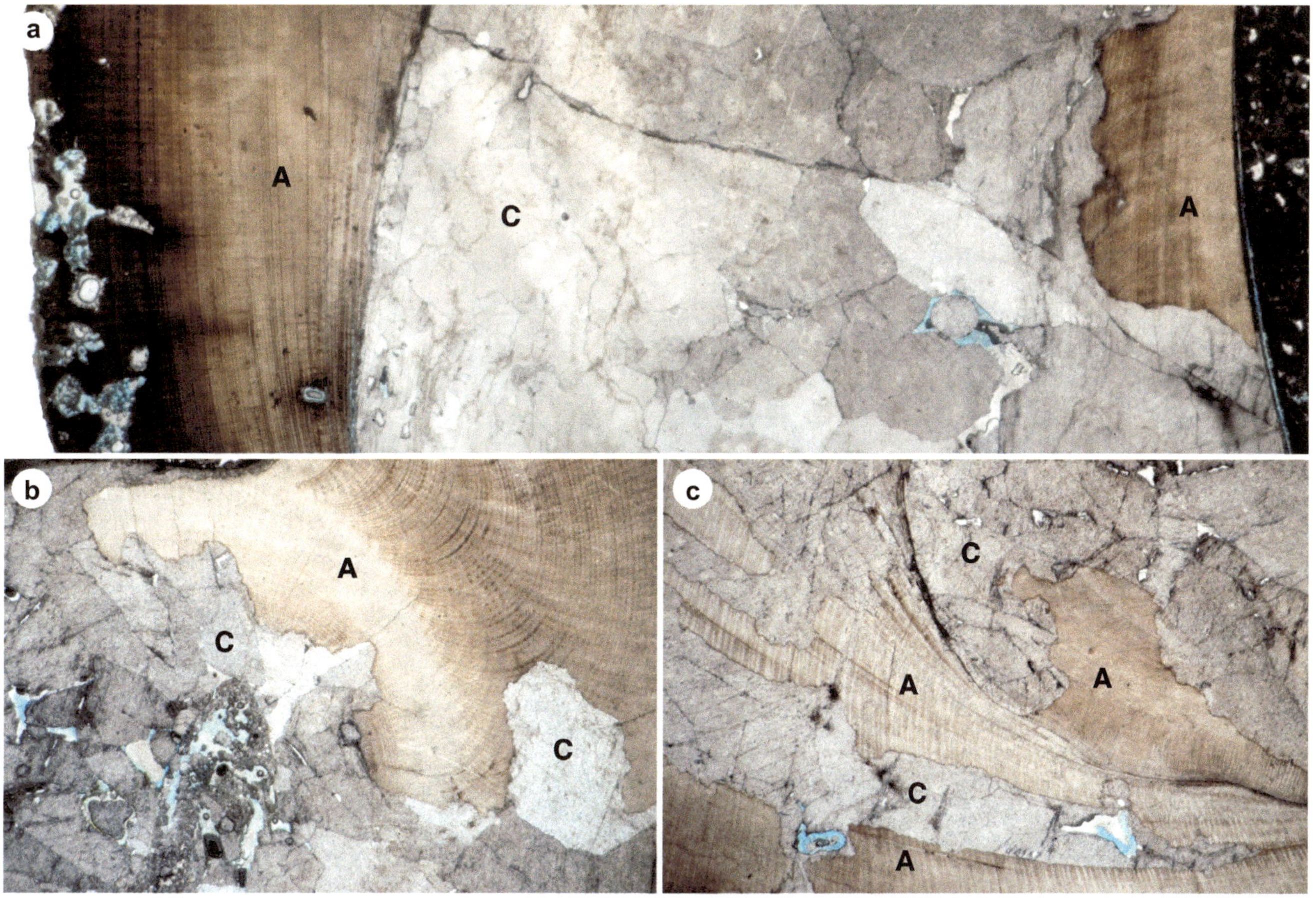

Fig. 7.28 Thin section photomicrographs showing neomorphic alteration in three different *Strombus gigas* shells from unit D of the Ironshore Formation, Grand Cayman. A = aragonite, C = calcite. **a** Central part of shell replaced by coarsely crystalline calcite. **b, c** Irregular patches of calcite replacing original aragonite shell

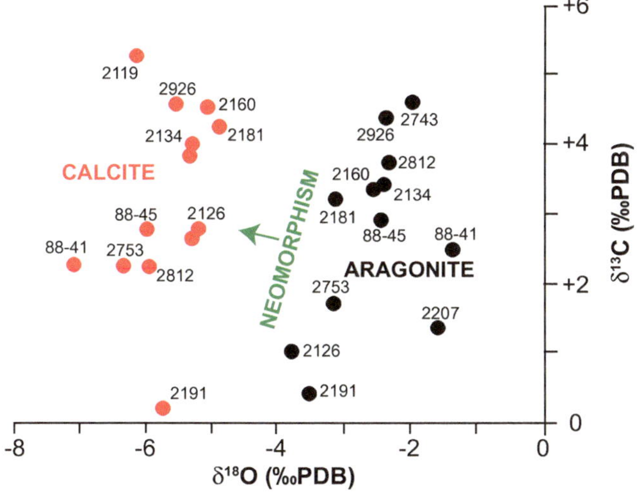

Fig. 7.29 Comparison of $\delta^{13}C$ versus $\delta^{18}O$ for aragonite and neomorphic calcite from *Strombus gigas* from Unit D of the Ironshore Formation on Grand Cayman. Numbers indicate matched calcite and aragonite samples from same shell. Modified from Rehman et al. (1994, their Fig. 8)

underwent more alteration to calcite. Rehman et al. (1994) also pointed out that the complex networks of sponge borings in some of the conch shells played a critical role in the process because they greatly increased the surface areas that could react with the meteoric waters.

Corals

Corals are present in units A, B, C, D, E, and F in the Ironshore Formation of the Cayman Islands. Their aragonitic skeletons were prone to diagenetic alteration that may have been mediated by many different processes while the coral was alive and following its death and burial in sediment. Although many corals from the Ironshore Formation seem to be perfectly preserved, others have clearly been modified through dissolution of their aragonitic skeletons, alteration of the aragonite skeleton to calcite, and/or precipitation of aragonite and/or calcite cements in the open skeletal voids (Li and Jones 2014b; Booker et al. 2020). Potential diagenetic changes to these skeletons include (1) boring by various organisms (e.g., sponges, worms, boring bivalves),

(2) neomorphic transformation to calcite, and/or (3) dissolution of the aragonite as groundwaters filtered through the porous coral skeletons.

Potentially, various geochemical parameters, such as the stable isotopes, can be used to determine the paleoclimate that existed when the corals were alive. These analyses, however, are only valid if it can be demonstrated that the coral skeletons are in virtually pristine condition and have not experienced any diagenetic changes that may have altered the original geochemistry of the skeletal aragonite. In some cases, calcification of the coral skeleton is obvious and negates its use for paleoclimate interpretations. Li (2014) and Li and Jones (2013) examined this issue by comparing *Montastraea annularis* and *Acropora palmata* from the Ironshore Formation at Rogers Wreck Point and the wells drilled offshore from the west coast (Fig. 7.1a). They found that there was no stratigraphic or geographic pattern to the mineralogical changes. Some of the corals were still formed entirely of aragonite whereas up to 90% of the aragonitic skeleton in other specimens had been replaced by calcite. Although some of these corals were in pristine condition, others had secondary aragonite cement, internal sediments, calcite cement, and/or evidence of aragonite dissolution. Analysis of these corals showed that some had been subjected to various combinations of (1) syndepositional precipitation of acicular aragonite cement on their internal surfaces, (2) bioerosion by fungi, algae, and/or bacteria, (3) deposition of marine calcite and/or aragonite sediments in their internal voids, (4) precipitation of calcite and/or aragonite in their internal voids, (5) calcification of the aragonite skeletons, and/or (6) random dissolution of the aragonite skeletons. Li and Jones (2013) showed that as *M. annularis* skeletons formed of their original aragonite had $\delta^{18}O_{VPDB}$ values of 4.96–2.97‰ and $\delta^{13}C_{VPDB}$ values of −2.37 to +0.94‰ whereas those formed of calcite had $\delta^{18}O$ values of −6.80 to −4.05‰ and $\delta^{13}C$ values of −8.39 to −4.10‰ (Fig. 7.30). They also demonstrated that there was a progressive change in these values as the percentage of calcite increased. A similar pattern was identified in the *A. palmata* from unit D with a specimen formed of 100% aragonite yielding a $\delta^{18}O_{VPDB}$ value of −1.05‰ and a $\delta^{13}C_{VPDB}$ value of +4.70‰ whereas a specimen formed of 100% calcite gave $\delta^{18}O$ value of −5.44‰ and a $\delta^{13}C$ value of −8.04‰. In general, the $\delta^{18}O_{VPDB}$ value of aragonitic *A. palmata* from units E and F were more positive than those from the lower units (Fig. 7.30).

A cross plot of the $\delta^{18}O_{VPDB}$ and $\delta^{13}C$ values for all of the *M. annularis* and *A. palmata* from Units A–F from Rogers Wreck Point and offshore George Town allows delineation of Group I (units E and F), Group II (Unit D from Rogers Wreck Point), and Group III (Unit D from offshore George Town) even though they all plot along the same trend line (Fig. 7.30). Units A to F of the Ironshore Formation formed during different highstands, whereas the unconformities that separate them are the record of the intervening lowstands. Based on data from other Caribbean islands, Li and Jones (2013) suggested, with caution, that the exposure times following the deposition of units A, B, C, D, E, and F was probably 52,000, 67,000, 65,000, 8000, 8000, and 64,000 years, respectively. Irrespective of the nuances associated with these estimates, it is not known if the lowstand that followed each highstand was low enough to place the older units in the vadose zone. It is, however, difficult to precisely align the $\delta^{18}O_{VPDB}$ and $\delta^{13}C$ values with a specific lowstand.

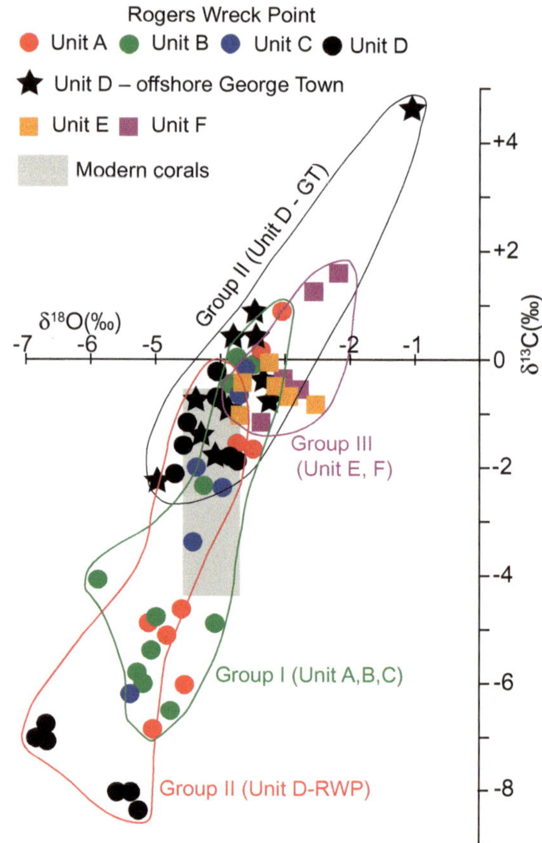

Fig. 7.30 Comparison of $\delta^{13}C$ versus $\delta^{18}O$ for corals from Units A, B, C, D, E, and F of the Ironshore Formation from Rogers Wreck Point and offshore George Town. The field of values for modern *Montastraea* and *Acropora* is shown for comparison (based on data from Fairbanks and Matthews 1978; Fairbanks and Dodge 1979; Leder et al. 1996; Smith 2006). Note diagenetic groups I, II, and III are aligned along a common trend. Modified from Li and Jones (2013, their Fig. 10c)

Potentially, the $\delta^{18}O$ in the aragonitic coral skeletons can be used to determine the surface seawater temperature (SST) that existed during the period when they grew.

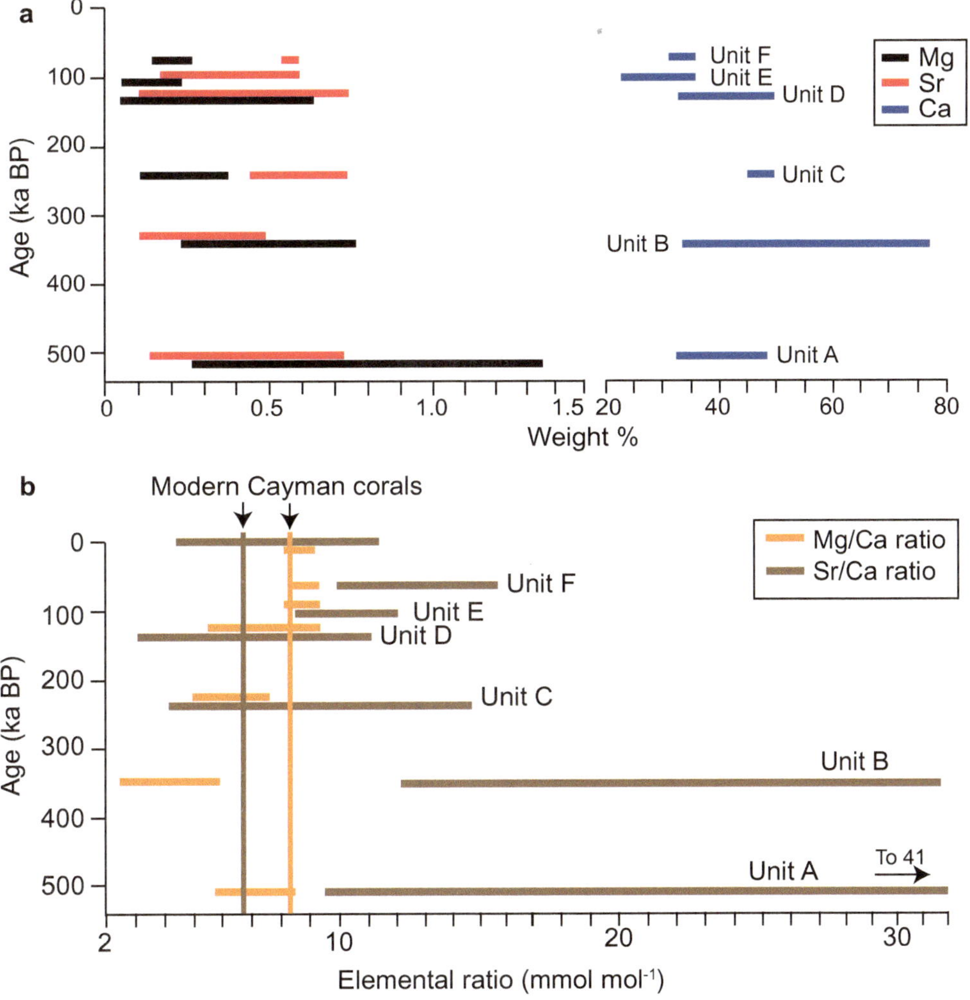

Fig. 7.31 Comparison of (**a**) weight percent of Mg, Sr, and Ca, and (**b**) Mg/Ca and Sr/Ca ratios in corals from Units A, B, C, D, E, and F of the Ironshore Formation and modern corals on Grand Cayman. Modified from Booker et al. (2020, their Fig. 5)

Although an extremely valuable paleoclimatic proxy, its function and accuracy demand that the skeleton of the coral has not been diagenetically altered in any way. The fact that many of the corals in the Ironshore Formation have been extensively altered (Li and Jones 2013, 2014b; Li 2014) has rendered them unsuitable for determination of the SST that existed when the original sediments of the Ironshore Formation were being deposited. Nevertheless, some of the corals from the Ironshore Formation are exceptionally well-preserved and appear to be ideal candidates for determination of the SST. Booker (2020) and Booker et al. (2020), however, cautioned that even subtle chemical changes can occur in corals with little or no physical change being evident in the skeletons. With respect to the corals from the Ironshore Formation, Booker et al. (2020), demonstrated that (1) although the coral skeletons from Units A and C had retained their primary skeletons, their Mg/Ca and Sr/Ca ratios had been changed, (2) corals from Unit B had been extensively recrystallized, and (3) the corals in units D, E, and F have retained their original skeletons and isotopic/elemental compositions. Graphs comparing the Mg, Sr, and Ca content and the Mg/Ca and Sr/Ca (Fig. 7.30) clearly show that the corals from units A and B have undergone substantial diagenetic changes whereas those from units C, D, E, and F have undergone relatively little change in these respects. Booker et al. (2020) suggested that the elemental and isotopic compositions of the corals in units A and C were altered in seawater that had been diluted by meteoric water in a semi-open diagenetic system. Despite this, the corals from these units show little physical evidence of diagenetic change. Corals in Unit B experienced pervasively recrystallization in an open meteoric diagenetic system that affected their chemical and physical characteristics. In contrast the corals from units D, E, and F show little evidence of alteration. Collectively, their research suggested that only those corals that are formed of >95% aragonite, show no evidence of cementation, and have Mg/Ca ratios <12.0 mmol mol^{-1}, Sr/Ca ratios >8 mmol mol^{-1}, $\delta^{18}O$ values $\geq$5.7‰, and $\delta^{13}C$ values $\geq$3‰ should be used to determine sea surface temperatures. Calculations based on the stable isotopes obtained from corals in units D, E, and F indicate that SST at the times when those corals were growing were similar to those of the modern ocean waters around Grand Cayman.

7.11.2 Calcretes

Calcareous crusts (up to 6 cm thick), which include laminar and non-laminar calcretes, commonly cover the exposed surfaces of the Ironshore Formation (Li and Jones 2014a), especially in areas where the formation is covered or was covered by peats that developed in the mangrove swamps (Fig. 7.32). These hard, brown, thinly laminated crusts, up to 6 cm thick, contrast sharply with the underlying soft white to off-white limestones. On Grand Cayman, these crusts are found on top of Member D that is well-exposed in the areas around Morgan's Harbour, Tarpon Springs, and Roger's Wreck Point. Locally, the crusts are overlain by black soil with a vegetation cover dominated by black mangrove (*Avicennia germinans*), white mangrove (*Laguncularia racemose)* and pickleweed (*Batis maritima*). These crusts are typically characterized by an irregular surface with numerous dish-like hollows (Fig. 7.32). Spencer (1981) suggested that these crusts slow the rate of weathering of the underlying limestones.

Typically, the succession comprises (1) unaltered limestone, (2) a transition zone, if present, up to 6 cm thick, and (3) laminated calcrete, to 6 cm thick (Li and Jones 2014a). The transition zone is mottled with irregular-shaped brown patches (up to 2.5 cm in diameter) that contrast sharply with the white host limestones. The limestone in this zone is typically formed of micrite that contains altered ooids, minor amounts of quartz, and voids left from the decay of plant roots that are commonly lined with Fe–Mn oxide–hydroxides. Laminated crusts are found near Morgan's Harbour, whereas non-laminated crusts are common in the areas around Roger's Wreck Point and Tarpon Springs. The non-laminated crusts are formed of low Mg calcite with trace amounts of quartz in some samples. The boundary between the crusts and the underlying bedrock is typically sharp and commonly truncates skeletal fragments in the underlying limestone. The calcareous crusts are formed of micrite, calcified roots, peloids, needle fibre crystals, alveolar septal structures, calcified filaments, and extracellular polymeric substances, microborings, and various types of void fillings (Li and Jones 2014a).

For the calcretes, the $\delta^{18}O$ which varies from −5.22‰ to +0.16‰ is positively correlated with the $\delta^{13}C$ values that vary from −11.75‰ to −5.43‰ (Fig. 7.33). There are minor differences between the samples from Morgans Harbour, Tarpon Springs, and Rogers Wreck Point (Li and Jones 2014a). Based on information derived from the equation of Gillson et al. (2004), Li and Jones (2014a) suggested that C_3 plants formed 81–89% of the vegetation in the Rogers Wreck Point area. In contrast, the crusts at Tarpon Springs and Morgan's Harbour appear to have formed when C_3 plants dominated the vegetation when the lower laminae developed (Group B—Fig. 7.33) but only 60–66% of the vegetation when the upper laminae formed (Group A—Fig. 7.33). That change may reflect a change in climatic conditions from cool, wet conditions to hotter and drier conditions.

7.11.3 Rhizoliths

Although not found in all exposures, rhizoliths are obvious features in some exposures of the Ironshore Formation on Cayman Brac (Figs. 7.13c, d and 7.16e). At the Water Authority site (Fig. 7.13c, d), rhizoliths that are at least 0.5 m long and 3–25 mm in diameter with complex downward branching patterns, are found in poorly lithified, fine-grained skeletal sands that locally contain small shell-fragments and small corals. Many of the rhizoliths are characterized by four orders of branching. Cross-sections (Fig. 7.13d) through the main parts of the rhizoliths show that they include up to four zones.

- Zone 1: open tube, 2–3 mm in diameter, commonly filled with a living root.
- Zone 2: up to 5 mm thick, formed of concentrically laminated light to medium brown, dense, micrite that contains scattered sand grains and/or skeletal fragments that are commonly partly altered. Porosity is low apart from scattered tubules that may have been formed by rootlets.
- Zone 3: this zone, 1–3 mm thick and light cream to white in colour, is transitionary from the dense micrite in zone 2

Fig. 7.32 Calcrete crusts developed on top of the Ironshore Formation (Member D) in area south of Morgan's Harbour on the western peninsula, Grand Cayman. **a** General view of calcrete crust with grasses growing in depressions where soil has collected. **b** Surface of hard, well indurated calcrete crust with characteristic brown colour. **c** Vertical cross-section through calcite crust and underlying limestones of the Ironshore Formation

to the poorly cemented sands in zone 4. It is formed of tightly packed grains that are commonly highlighted by a thin brown coating. Porosity is low.

- Zone 4: this outermost zone, characterized by high porosity (up to 45%), is formed of partly cemented skeletal sand that locally contains bivalve and coral fragments that are up to 1 cm long. Grains with a thin brown coating are similar to those in Zone 3.

Cements found in the rhizoliths include grain-coating needle mats, isolated calcite rhombs, calcite rhomb chains, calcified filaments, and calcified spherical bodies (Jones and

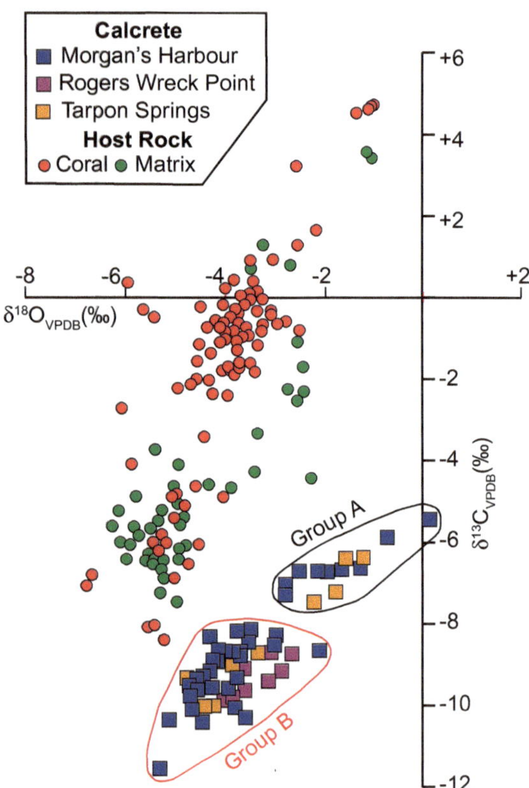

Fig. 7.33 Comparison of $\delta^{13}C$ versus $\delta^{18}O$ for calcrete crusts from Morgan's Harbour, Rogers Wreck Point, and Tarpon Springs relative to corals and limestones of the Ironshore Formation. Modified from Li and Jones (2014a, their Fig. 13)

Ng 1988). In contrast, the grainstones around the rhizoliths are friable because cements are restricted to scattered isolated rhomb chains, calcified rhombs, calcified spherical bodies, and calcified filaments (Jones and Ng 1988).

As noted by Jones and Ng (1988), the penetration of friable carbonate sediments by plant roots have a significant impact on local diagenesis because the roots act as conduits that allow rainwater to easily penetrate deep into the substrate, and the microbiota that is associated with the roots commonly mediates calcite precipitation. The cements in the rhizoliths formed through organic and inorganic processes, with variations in crystal morphologies being due to microscale variations in the physical and chemical conditions around the plant roots.

7.11.4 Roots and Caliche

On the east end of Cayman Brac, many of the bushes and trees are rooted directly into hard, indurated bedrock that appears to have developed as a caliche (Jones 1988). Some of the larger trees have roots that penetrate to depths of at least 4 m. Some of the holes around the larger roots are lined with laminated calcrete crusts that are up to 0.5 cm thick (Fig. 7.34). Rootlet borings are commonly associated with the larger holes. These borings provided avenues by which (1) algae, fungi, and bacteria could invade and interact with the subsurface rocks, and (2) rainwater could flow into the subsurface strata.

Collectively, plant root systems can lead to the destruction of lithified substrates while at the same time creating avenues by which rainwater can penetrate deep into the rocks and once there, mediate further diagenetic changes to the limestones.

7.11.5 Roots and Peloids

Although rhizoliths are rare on Grand Cayman and Little Cayman, there is evidence that plant roots have played a role in the diagenesis of the host limestones in the Ironshore Formation on Grand Cayman (Squair 1988; Jones and Squair 1989). In the area around Salt Creek on the western peninsula of Grand Cayman, the surface of the Ironshore Formation is characterized by three vegetation zones that are, from the shoreline inland, (1) Zone A, 6–8 m wide, characterized by abundant lichen, scattered succulents, and rare dwarf shrubs, (2) Zone B, ~2 m wide, that comprises densely packed shrubs, succulents, herbs, and creepers, and (3) Zone C that includes dense, tall, mainly mangrove trees (Squair 1988).

The oolitic limestones around Salt Creek are penetrated by numerous post-lithification borings that were probably produced by algae, fungi, roots, root hairs, bacteria, and actinomycetes (Squair 1988; Jones and Squair 1989). In those limestones, identification of the formative organism is commonly possible because the borings still contain partly decayed plant roots and/or root hairs. Nevertheless, it is impossible to trace these borings to their parent plant. Many borings contain spherical to elliptical peloids, to 80 μm long, that are formed of randomly oriented calcite plates formed of Ca and traces of Mg, bound together by organic material. Algae, fungi, and bacteria are commonly associated with the peloids and may be responsible for the borings that are evident in the peloids. Such peloids are not found in the bedrock around the borings. Available evidence suggests that the peloids may be fecal pellets that were produced by mites that tunneled into the central parts of the roots (Squair 1988; Jones and Squair 1989).

Fig. 7.34 Ironshore Formation exposed in disused quarry on Cayman Brac, northwest of airport terminal. **a** General view of outcrop showing lower recessive friable unit overlain by prominent, well-lithified unit. Note cavities created by action of plant roots at base of the upper unit (arrows). **b** Enlarged view of junction between friable and prominent units showing cavities created by plant roots. **c** Close up of basal part of cavity create by plant roots. Note highly irregular margins and plant roots

7.12 Conclusions

The Ironshore Formation on the Cayman Islands unconformably overlies the Pedro Castle Formation or the Cayman Formation. It is either overlain by mangrove peats or exposed at the surface along the coastlines of the islands. Analyses of this formation, based on detailed examinations of cores and surface exposures on each of the Cayman Islands, has yielded the following conclusions.

- The Ironshore Formation is divided into units A, B, C, D, E, and F that each formed during successive highstands that have been dated at ~400 ka, 345 ka, ~229 ka, ~125 ka, 95–110 ka, and 73–87 ka, respectively.
- Unit D is widely exposed across the islands, whereas units A, B, and C are known only from cores obtained from the northeast corner of Grand Cayman; Unit E has only been found in cores taken offshore from the west coast of Grand Cayman, just north of George Town; and Unit F is known from the offshore cores and scattered localities on the western peninsula of Grand Cayman.
- Limestones in the Ironshore Formation are formed of aragonite and calcite. The percentage of aragonite generally decreases from the youngest to oldest units.
- The limestones of the Ironshore Formation contain an abundant, diverse biota that is formed primarily of corals, bivalves, and gastropods, along with various other organisms. Fossil preservation is generally very good, with many organisms being preserved in life position and some displaying elements of their original colours.
- Many of the corals and other organisms still retain their original aragonite skeletons.
- Exposed surfaces of the Ironshore Formation are commonly capped by well-developed calcrete crusts.
- Poorly lithified skeletal sands and oolitic sands characterize the upper units of the Ironshore Formation in some areas. At various localities on Little Cayman and Cayman Brac, well-developed rhizoliths are evident in these deposits.

- Diagenesis of the limestones in the Ironshore Formation typically involved precipitation of calcite and aragonite cements, and neomorphic replacement of the aragonite skeletons by calcite. Only minor amounts (<5%) of dolomite are present in the oldest limestones of Unit A.
- The stable isotopes are generally negative, with only a few samples yielding positive $\delta^{18}O$ and $\delta^{13}C$ values.
- The limestones of the Ironshore Formation are characterized by low concentrations of Mn and Fe, variable concentrations of Na, and Sr contents that correlate with the percentage of aragonite in the limestones.

References

Booker S, Jones B, Li L (2020) Diagenesis in Pleistocene (80 to 500 ka) corals from the Ironshore Formation: implications for paleoclimate reconstruction. Sediment Geol 399:1–16

Booker SD (2020) Variation in climatic conditions from the Cayman Islands through stable isotope and element analyses from coral and sediment cores. PhD thesis, University of Alberta, Edmonton

Brunt MA, Giglioli MEC, Mather JD, Piper DJW, Richards HG (1973) The Pleistocene rocks of the Cayman Islands. Geol Mag 110: 209–221

Cerridwen SA, Jones B (1991) Distribution of bivalves and gastropods in the Pleistocene Ironshore Formation, Grand Cayman, British West Indies. Caribb J Sci 27:97–116

Cerridwen SA (1989) Paleoecology of Pleistocene mollusca from the Ironshore Formation, Grand Cayman, B. W. I. MSc thesis, University of Alberta

Coyne MK, Jones B, Ford D (2007) Highstands during Marine Isotope Stage 5e: evidence from the Ironshore Formation of Grand Cayman, British West Indies. Quat Sci Rev 26:536–559

Coyne MK (2003) Transgressive-regressive cycles in the Ironshore Formation, Grand Cayman, British West Indies. MSc thesis, University of Alberta, Edmonton

Donovan SK, Jones B, Harper DAT (2016) Neogene echinoids from the Cayman Islands, British West Indies. Geol J 51:864–879

Fairbanks RG, Dodge RE (1979) Annual periodicity of the $^{18}O/^{16}O$ and $^{13}C/^{12}C$ ratios in the coral *Montastrea annularis*. Geochim Cosmochim Acta 43:1009–1020

Fairbanks R, Matthews RK (1978) The marine oxygen isotope record in Pleistocene corals, Barbados, West Indies. Quat Res 10:181–196

Gillson L, Waldron S, Willis KJ (2004) Interpretation of soil $\delta^{13}C$ as an indicator of vegetation change in African savannas. J Veg Sci 15:339–350

Hills D, Jones B (2000) Peyssonnelid rhodoliths from the Late Pleistocene Ironshore Formation, Grand Cayman, British West Indies. Palaios 15:212–224

Hills D (1997) Rhodolite development in the modern and Pleistocene of Grand Cayman. MSc thesis, Alberta, Edmonton, Canada

Hunter IG, Jones B (1988) Corals and paleogeography of the Pleistocene Ironshore Formation on Grand Cayman, B. W. I. In: Proceedings of the Sixth International Coral Reef Symposium. Townsville, Australia, pp 431–435

Hunter IG, Jones B (1996) Coral associations of the Pleistocene Ironshore Formation, Grand Cayman. Coral Reefs 15:249–267

Hunter IG (1994) Modern and ancient coral associations of the Cayman Islands. PhD thesis, University of Alberta, Edmonton

Imbrie J, Hays JD, Martinson DG, Mcintyre A, Morley JJ, Pisias NG, Prell WL, Shackleton NJ (1984) The orbital theory of Pleistocene climate: support from a revised chronology of the marine $\delta^{18}O$ record. In: Berger A, Imbrie J, Hays J, Kuhla G, Saltzman B (eds) Milankovitch and climate part 1: NATO ASI series C. Reidel Publishing Co, Dordrecht, The Netherlands, pp 269–305

Jones B (1988) The influence of plants and micro-organisms on diagenesis in caliche: example from the Pleistocene Ironshore Formation on Cayman Brac, British West Indies. Bull Can Pet Geol 36:191–201

Jones B (1990) Tunicate spicules and their syntaxial overgrowths: examples from the Pleistocene Ironshore Formation, Grand Cayman, British West Indies. Can J Earth Sci 27:525–532

Jones B (2019) Diagenetic processes associated with unconformities in carbonate successions on isolated oceanic islands: case study of the Pliocene to Pleistocene sequence, Little Cayman, British West Indies. Sediment Geol 386:9–30

Jones B, Goodbody QH (1984) Biological factors in the formation of quiet water ooids. Bull Can Pet Geol 32:190–199

Jones B, Hunter IG (1990) Pleistocene paleogeography and sea levels on the Cayman Islands, British West Indies. Coral Reefs 9:81–91

Jones B, Hunter IG (1991) Corals to rhodolites to microbialites—a community replacement sequence indicative of regressive conditions. Palaios 6:54–66

Jones B, Ng KC (1988) The structure and diagenesis of rhizoliths from Cayman Brac, British West Indies. J Sediment Petrol 58:457–467

Jones B, Pemberton SG (1988b) *Lithophaga* borings and their influence on the diagenesis of corals in the Pleistocene Ironshore Formation of Grand Cayman Island, British West Indies. Palaios 3:3–21

Jones B, Pemberton SG (1989) Sedimentology and ichnology of a Pleistocene unconformity-bounded, shallowing-upward carbonate sequence: the Ironshore Formation, Salt Creek, Grand Cayman. Palaios 4:343–355

Jones B, Squair CA (1989) Formation of peloids in plant rootlets, Grand Cayman, British West Indies. J Sediment Petrol 59:1002–1007

Jones B, Zheng E, Li L (2018) Growth and development of notch speleothems from Cayman Brac, British West Indies: response to variable climatic conditions over the last 125,000 years. Sediment Geol 373:210–227

Jones B, Pemberton SG (1988a) Bioerosion of corals by *Lithophaga*: example from the Pleistocene Ironshore Formation of Grand Cayman, B.W.I. In: Proceedings of the Sixth International Coral Reef Symposium 3, pp 437–440

Leder JJ, Swart PK, Szmant AM, Dodge RE (1996) The origin of variation in the isotopic record of scleractinian corals I. Oxygen. Geochim Cosmochim Acta 60:2857–2870

Li R, Jones B (2013) Temporal and spatial variations in the diagenetic fabrics and stable isotopes of Pleistocene corals from the Ironshore Formation of Grand Cayman, British West Indies. Sediment Geol 286–287:58–72

Li R, Jones B (2014a) Calcareous crusts on exposed Pleistocene limestones: a case study from Grand Cayman, British West Indies. Sediment Geol 299:88–105

Li R, Jones B (2014b) Evaluation of carbonate diagenesis: a comparative study of minor elements, trace elements, and rare-earth elements (REE+Y) between Pleistocene corals and matrices from Grand Cayman, British West Indies. Sediment Geol 314:31–46

Li R (2014) The Pleistocene Ironshore Formation, Grand Cayman: diagenetic response to sea level change. PhD thesis, University of Alberta, Edmonton

Matley CA (1926) The geology of the Cayman Islands (British West Indies) and their relation to the Bartlett Trough. Q J Geol Soc Lond 82:352–387

Metz R (2011) Pleistocene trace fossils in the Ironshore Formation, Little Cayman, British West Indies. Cent Eur J Geosci 3:71–76

Rehder HA (1962) The Pleistocene mollusks of Grand Cayman Island, with notes on the geology of the island. J Paleontol 36:583–585

Rehman J, Jones B, Hagan TH, Coniglio M (1994) The influence of sponge borings on aragonite-to-calcite inversion in Late Pleistocene *Strombus gigas* from Grand Cayman, British West Indies. J Sediment Res A64:174–179

Rehman J (1992) Diagenetic alteration of *Strombus gigas*, *Siderastrea siderea*, and *Montastrea annularis* from the Pleistocene Ironshore Formation of Grand Cayman. MSc thesis, Alberta, Edmonton

Richards HG (1955) The geological history of the Cayman Islands. Notulae Naturae 284:1–11

Shourie A (1993) Depositional architecture of the Late Pleistocene Ironshore Formation, Grand Cayman, British West Indies. MSc thesis, University of Alberta, Edmonton

Smith JM (2006) Geochemical signatures in the coral *Montastrea*: modern and mid-Holocene perspectives. PhD thesis, University of South Florida, Tampa, USA

Spencer T (1981) Micro-topographic change on calcarenites, Grand Cayman Island, East Indies. Earth Surf Proc Land 6:85–94

Squair CA (1988) Surface karst on Grand Cayman Island, British West Indies. MSc thesis, University of Alberta

Stoddart DR (1980) Geology and geomorphology of Little Cayman. Atoll Res Bull 241:11–16

Vézina JL, Jones B, Ford DC (1999) Sea-level highstands over the last 500,000 years: evidence from the Ironshore Formation on Grand Cayman, British West Indies. J Sediment Res 69:317–327

Vézina JL (1997) Stratigraphy and sedimentology of the Pleistocene Ironshore Formation at Rogers Wreck Point, Grand Cayman. MSc thesis, University of Alberta, Edmonton

Warthin AS (1959) Ironshore in some West Indian islands. Trans N Y Acad Sci Ser 2 21:649–652

Woodroffe CD (1982) Geomorphology and development of mangrove swamps, Grand Cayman, West Indies. Bull Mar Sci 32:381–398

Woodroffe CD (1988) Vertical movement of isolated oceanic islands at plate margins. Z Geomorphol 69:17–37

Woodroffe CD, Stoddart DR, Harmon RS, Spencer T (1983) Coastal morphology and Late Quaternary history, Cayman Islands, West Indies. Quat Res 19:64–84

Zheng E (2017) Environmental controls on alternating aragonite-calcite laminations in notch-speleothems from Cayman Brac, Cayman Islands, British West Indies. MSc thesis, Alberta, Edmonton

8 Modern Carbonate Environments

Abstract

Each of the Cayman Islands are, in reality, the summits of submarine mountains that are parts of the Cayman Ridge and surrounded by deep oceanic waters. The seaward margins of the marine shelves around each island are defined by the 20 m isobath, which is the edge of the drop-off into deep water. As a result, the shelves around the islands are generally no more than 500 m wide. Lagoons, formed where fringing reefs have grown from headland to headland are common on Grand Cayman and Little Cayman, but rare on Cayman Brac. The shelves and lagoons are inhabited by luxuriant biotas that are typically dominated by taxonomically-diverse arrays of corals, bivalves, gastropods, and foraminifera along with *Thalassia*, and many different types of algae. The carbonate sands and muds in these depositional systems are formed largely of skeletal grains. Burrowing, commonly by *Callianassa*, is widespread in most of the lagoons. Although infrequent, storms and hurricanes have a major impact on these depositional systems by moving large amounts of sediment onshore or offshore from the lagoons and open shelves. The lagoons, characterized by a diverse array of facies, typically have extensive *Thalassia* banks in the quieter, near shore environments, and sands and patch reefs in the outer areas where energy levels are higher. Detailed mapping of the facies, based on image analyses of air photographs from various years, shows that the facies distributions in the lagoons are constantly evolving, commonly with the *Thalassia* banks expanding laterally over the bordering sand and mud facies. Beachrock is found along some of the beaches. Collectively, the modern depositional systems of the Cayman Islands offer perfect examples of modern carbonate systems and the processes that have dictated their evolution.

8.1 Introduction

Today, each of the Cayman Islands have similar modern depositional systems where carbonate sediments are actively forming and accumulating. Each island is, in reality, the peak of a submarine mountain that forms part of the Cayman Ridge (Fig. 2.1), bounded to the south by the deep waters (>6,000 m) of the Cayman Trench and to the north by the deep waters of the Yucatan Basin (up to 4,500 m). To the east and west of each island, the seafloor is formed of parts of the Cayman Ridge and therefore shallower than in the basins to the south and north. As a result, the shelves around each island (Figs. 8.1, 8.2, 8.3), with their seaward margin defined by the 20 m isobath (approximate edge of drop-off into deep water), are typically ~500 m wide (Logan 1994). Exceptions are found on the east end of Little Cayman where the shelf is 1,700 m wide, and the west of Cayman Brac where it is 1,300 m wide (Logan 1994). Similar dimensions are also apparent around Grand Cayman. The positions of the "drop-offs" into deeper water, which are near vertical walls, are critical because they define the maximum possible widths of coastal platforms around each island on which carbonate sediments can form and accumulate. If sea level drops, then the platform width will decrease because the position of the outer margin is essentially static.

The Cayman Islands are located in the central part of the Caribbean Basin where they enjoy a low amplitude diurnal and semidiurnal tidal regime, a moderately strong unidirectional trade wind and wave field, strong oceanic currents, and a sheltered position relative to high-latitude storm swell (Roberts 1994). The tropical climate of the islands is characterized by a wet season (May to November) and a dry season (December to April) with air temperatures from 11.2 to 36.5 °C with a monthly average of 24.7 to 28.4 °C (Turner et al. 2013). Between 1887 and 1987, hurricanes

B. Jones, *Geology of the Cayman Islands*,
https://doi.org/10.1007/978-3-031-08230-6_8

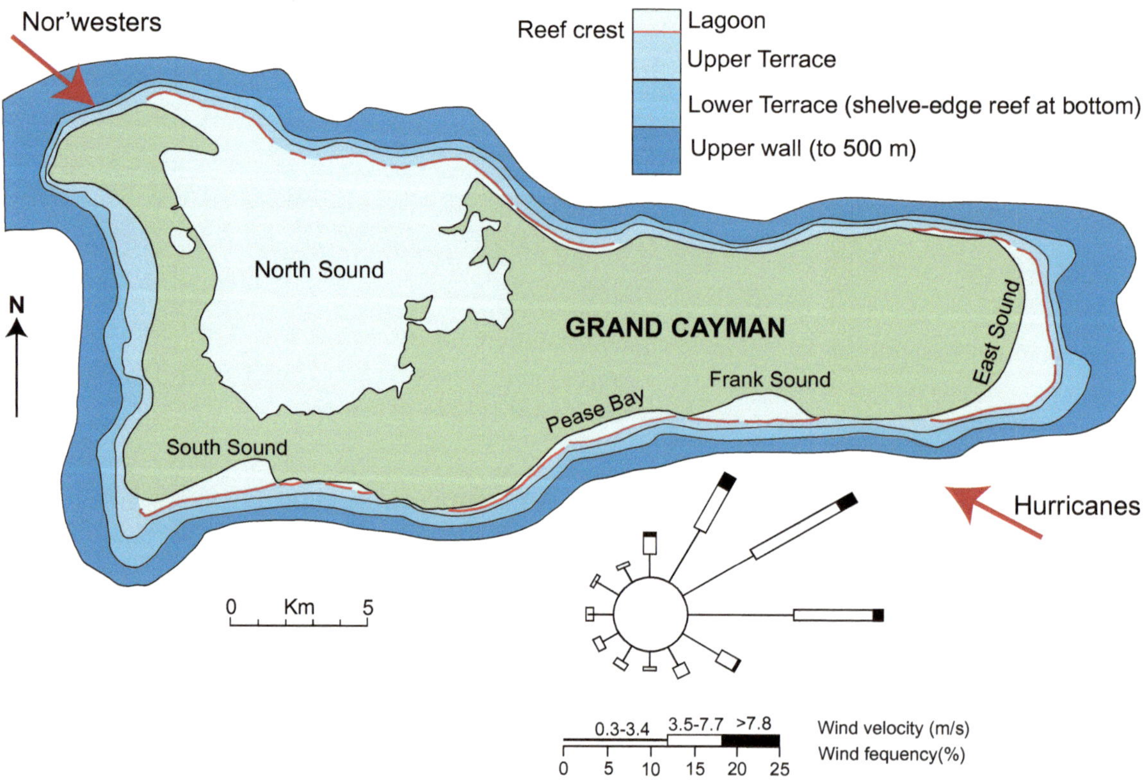

Fig. 8.1 Map showing lagoons and shelves around Grand Cayman and drop-off at the "wall". The shelf is divided into the upper and lower terraces. Grand Cayman is subject to hurricanes that typically come from the southeast and Nor'westers that come from the northwest. Prevailing winds from the east. From Blanchon and Jones (1997, their Fig. 8; reproduced and modified with permission from the Coastal Education and Research Foundation Inc.)

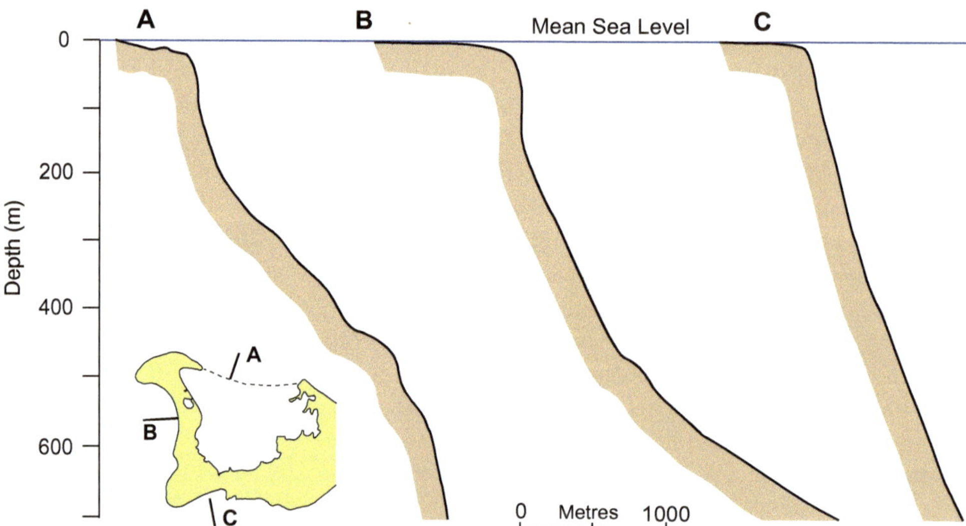

Fig. 8.2 Profiles of coastal areas on the western part of Grand Cayman showing narrow shelves and steep drop-offs to ocean depths. Modified from Roberts (1994, his Fig. 5.6)

passed over the islands, on average, once every 2.7 years (Tompkins 2005). Most hurricanes originate in the Atlantic Ocean and travel westward across the Cayman Islands. In 1932, Hurricane Cuba, a category 5 hurricane with wind speeds up to 320 km/hour, waves up to 16 m high, and a storm surge up to 10 m high devastated Cayman Brac (Sauer 1982; Fenner 1993). In recent years, Hurricanes Gilbert (1989) and Ivan (2004) passed over the islands and severely damaged the *Acropora* populations and moved sand and soft corals from the shallower reefs. Although short-lived, hurricanes can have a devastating effect on the shallow lagoons around the Cayman Islands, and commonly cause severe

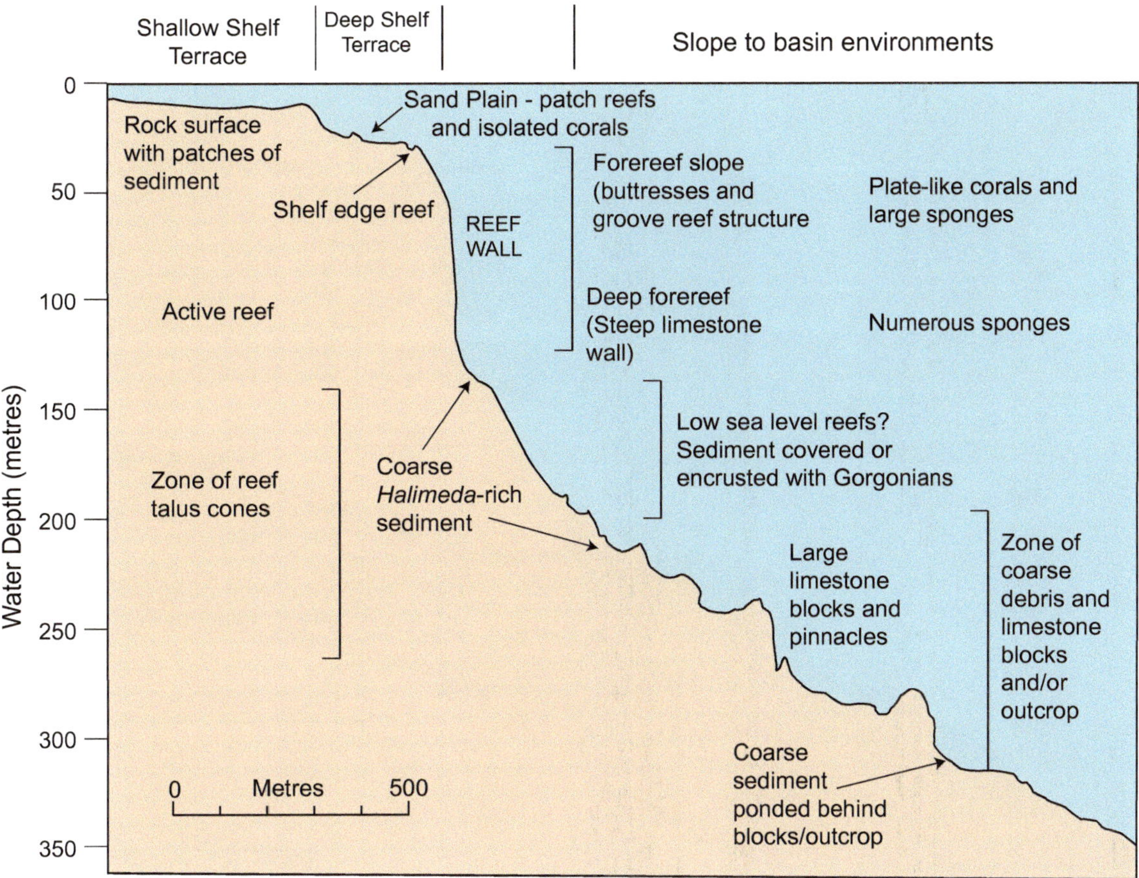

Fig. 8.3 Schematic diagram showing main features of the shelves, wall, and slope to basin environments. Modified from Roberts (1994, his Fig. 5.4)

damage to the reefs and move vast quantities of lagoonal sediment on land or out into the open ocean (Blanchon and Jones 1997). During winter months severe "Nor-Westers" from the northwest may severely impact the western part of Grand Cayman, which is technically the lee side of the island.

The maximum tidal range recorded on the Cayman Islands is 1 m, whereas the average range is 26 cm (Burton 1994). There is a seasonal variation in the tide levels with the highest tides being between July and September. For carbonate sedimentation, this restricted tidal range has a minimal effect on coastal deposits. The intertidal zone is generally restricted to narrow zones of beach sands with the best developed being the "Seven Mile Beach" on the western leeward side of Grand Cayman. In many areas, however, the intertidal zone is barely detectable. Supratidal areas are essentially non-existent and there are no examples of supratidal flats and/or evaporite deposits on any of the Cayman Islands.

The narrow shelves around each island are characterized by two gently sloping terraces, with the shallow, upper terrace that slopes from the fringing reef or shoreline to a depth of 8–10 m and the deep, lower terrace at a depth of 15–20 m (Logan 1994; Roberts 1994). As noted by Roberts (1977), Woodroffe et al. (1983), and Logan (1994), the similarity in the development and depth of these terraces on each of the islands (Fig. 8.4) suggests that they have been stable since the last interglacial period.

Detailed knowledge of the reefs and lagoons around Grand Cayman first came from Roberts (1971a, 1974, 1976, 1977, 1983), Roberts et al. (1975), Rigby and Roberts (1976, their Plates 24 to 26), Swain and Hull (1976), and Raymont et al. (1976). Fenner (1993) later documented some of the reefs around Little Cayman. Information regarding the facies and reefs of Cayman Brac and Little Cayman remained sparse until Logan (1994) produced detailed facies maps and described the depositional systems of those islands. Further details regarding the depositional systems around Grand Cayman come from the analyses by Tongpenyai (1989) and Tongpenyai and Jones (1991) who used image analyses to show how the facies in various lagoons around Grand Cayman had changed with time. Beanish (2000), Beanish and Jones (2002), and Beanish et al. (2002) focused on South Sound and examined the factors that controlled facies

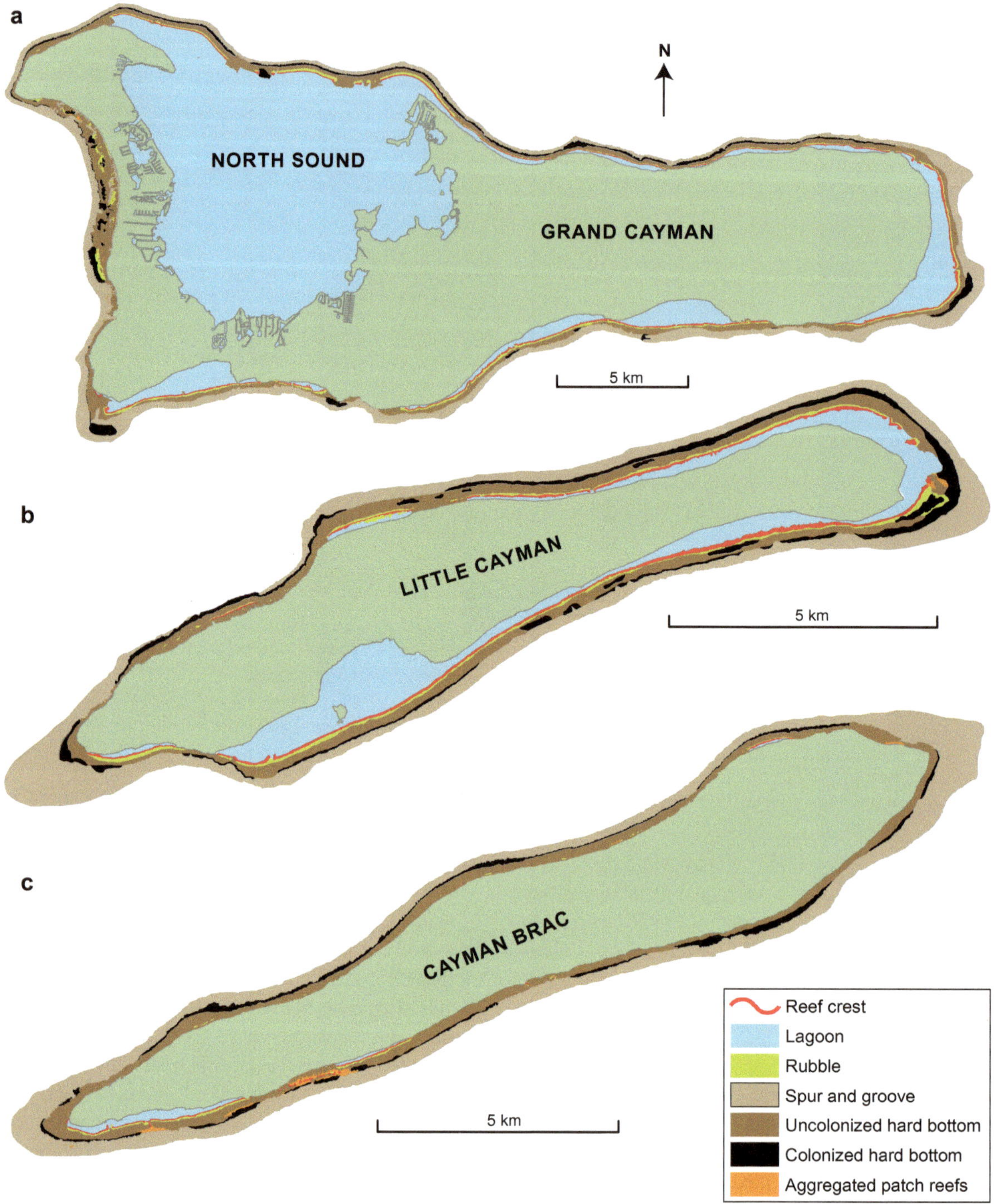

Fig. 8.4 Simplified maps of **a** Grand Cayman, **b** Little Cayman, and **c** Cayman Brac showing zonation of the shelves around each island. Benthic habitat maps provided courtesy of the Department of Environment, Cayman Islands

changes through time. MacKinnon (2000) and MacKinnon and Jones (2001) provided detailed information on the sedimentology of North Sound and Kalbfleisch (1995) and Kalbfleisch and Jones (1998) documented the sedimentological characteristics of Frank Sound. Li (1997), Li and Jones (1997), and Li et al. (1997, 1998) documented the foraminifera assemblages that inhabited the lagoons, whereas Corlett (2005) and Corlett and Jones (2007) examined the role that *Thalassia* plays in lagoonal sedimentation, and Blanchon (1995), Blanchon and Jones (1997), and Blanchon et al. (2002) provided detailed information on the evolution of the forereef areas around Grand Cayman. Collectively, the information obtained from these studies has provided a clear overview of the complex array of factors that control sedimentation in the idyllic marine habitats around the Cayman Islands.

The white carbonate sediments found on the seafloor around the Cayman Islands are formed of various proportions of aragonite, low-Mg calcite, and high-Mg calcite that are controlled by the source of the sediment. As in most carbonate systems throughout the world (e.g., James and Jones 2015), the sediments found in these environments are derived from many different sources, including the skeletons of animals (e.g., corals, bivalves) and plants (e.g., algae), direct precipitation from the seawater, epibiont organisms that attach themselves to various substrates (e.g., blades of *Thalassia*), bioerosion, physical breakdown of substrates during storms, and/or storm deposits.

8.2 The Coastal Shelves of the Cayman Islands

The shelf around Grand Cayman (Fig. 8.4a), which is typically <1 km wide and extends from the coast to the vertical shelf edge escarpment (Blanchon and Jones 1995; Blanchon 1995), is characterized by facies belts that parallel the coastline (Rigby and Roberts 1976). The upper terrace, with its edge defined by the 10 m isobath, is mostly a barren rocky pavement that is cut by erosional furrows (Fig. 8.5) that reflect ongoing erosion activity (Blanchon and Jones 1995). The mid-shelf scarp, commonly buried under modern sediments, is evident on the leeward shelf where a wave-cut notch is found at a depth of 18.5 m (Fig. 8.6). The lower terrace (12 to 40 m) extends from the mid-shelf scarp to the shelf edge. Blanchon and Jones (1995) suggested that the lower terrace formed between 11 and 7 ka when sea-level was slowly rising before being drowned by a rapid 5 m rise in sea level at ~7 ka and a subsequent slow rise in sea level to its present-day level.

Many descriptors have been applied to the different geomorphic elements of the coastal shelves around each of the Cayman Islands (e.g., Rigby and Roberts 1976; Logan 1994, 2013; Blanchon 1995). Herein, the terms proposed by Turner et al. (2013) are used because this is the most recently proposed scheme and it employs terms that are relatively easy to apply and use.

- **Aggregate reef**: Hard coral cover >70% with soft corals and sponges.
- **Spur and Groove**: Usually on seaward edge of reef, close to the escarpment, hard coral cover with high vertical relief and sand channels.
- **Individual patch reefs**: Isolated coral formations, hard corals but with some soft corals and sponges.
- **Aggregated patch reefs**: Aggregated coral colonies with >70% substrate coverage; hard corals dominate but soft corals and sponges may be present.
- **Reef rubble**: Unstable coral rubble, usually on landward side of reefs, commonly covered by algae.
- **Reef Crest**: Semi-emergent or emergent part of reef.
- **Sand plains**: Expanses of uncolonized sands between shallow and deep terrace reefs.
- **Colonized hard bottoms**: Areas with 10–170% cover: features include low-relief pavement or rubble, low-relief

Fig. 8.5 General view of linear ridges and furrows, at 90° to coastline, in the Ridge and Furrow Zone as defined by Blanchon and Jones (1995, their Fig. 7b). Ridge is ~3 m wide, water depth ~5 m. North coast of Grand Cayman, just east of Cayman Kai. Photograph courtesy of Paul Blanchon

Fig. 8.6 Submerged wave cut notch at depth of 18.5 m offshore from northwest coast (Hepps Pipeline) of Grand Cayman. Photograph courtesy of Croy McCoy, Department of Environment, Cayman Islands

rock and sand, colonized by algae, soft corals, and sparse hard corals that partly obscure the substrate.

- **Unconsolidated hardbottom**: Pavement, commonly dominated by hard corals with soft corals and sponges <10%.
- **Wall**: Near vertical to vertical slope form shelf-margin to great depths with coral and sponge coverage to depths of 120 m.

8.3 Corals and Coral Reefs

The ocean waters around each of the Cayman Islands are famous for their diverse coral faunas and the stunning reefs that they construct. Corals are found as isolated colonies, in patch reefs of varying size throughout the lagoons around the islands, and in the fringing reefs and shelf-edge reefs. These complex ecosystems are highly sensitive to environmental changes. Their distribution is controlled by a complex array of environmental parameters with many aspects still being open to debate.

8.3.1 Modern Corals of the Cayman Islands

Roberts (1994) listed 61 species of Anthrozoa and Hydrozoa found in the reefs around the Cayman Islands, whereas Fenner (1993) listed 40 species (and three forms) of corals and five species of hydrocorals from Little Cayman and 39 species of corals and three species of hydrocorals from Cayman Brac. On Cayman Brac, Fenner (1993) noted the presence of five species of small corals in the lagoon with *Siderastrea radians* being the most common, 10 species on the reef crest with *Porites porites* being the most common, and 14 species on the shallow terrace with *Diploria strigosa, Poroites asteroides, S. siderea*, and *Montastrea annularis* dominating. On the crests of the spurs, massive *M. annularis* dominate with large *Acropora palmata* on the middle and shoreward crests.

Hunter (1994) listed 44 species of hermatypic corals (i.e., those with zooxanthellae) that he found in the waters around Grand Cayman. Comparison of his list of corals with those for Little Cayman and Cayman Brac, as given by Fenner (1993), shows only a few differences between the islands (Fig. 8.7). Some caution must be exercised with respect to the minor differences in the recorded species because it is not clear if they are real differences or simply due to taxonomic issues. Estimates of the overall abundance of these species on the three islands are similar even though Hunter (1994, his Fig. 3.2) and Fenner (1993, his Table 1) used slightly different scales for assessing the abundances of each coral (Fig. 8.7).

On Grand Cayman and Cayman Brac much of the coral growth in water <8 m deep was removed during Hurricane Gilbert in 1988 (Blanchon and Jones 1997; Riegl 2001). Thus, the *Acropora palmata* zone as described in the 1970s and 1980s by Rigby and Roberts (1976) and Logan (1994) no longer exists because of the impact of hurricanes. Manfrino et al. (2003) and Turner et al. (2013) drew attention to the fact that many corals on the three islands are prone to the yellow band, white plague, and black band diseases. Over a five year period, Manfrino et al. (2013) noted that although bleaching and disease led to a decrease in coral coverage around Little Cayman, a full recovery took place over the following seven years.

Coral bleaching (Fig. 8.8), first noted in Cayman waters in 1983, also occurred in 1987, 1991, 1994, 1998, 2003, and 2005 (Turner et al. 2013). The El Niño event in 1997–1998, caused local sea surface temperature to rise above 30°C between August 9th and September 3rd, and affected ~90% of the corals in the waters around Grand Cayman. A similar event in 2009, which impacted waters to a depth of 460 m, also caused elevated sea water temperatures that were further exacerbated by the absence of cloud cover and calm weather over the same time period. The resultant mass bleaching of

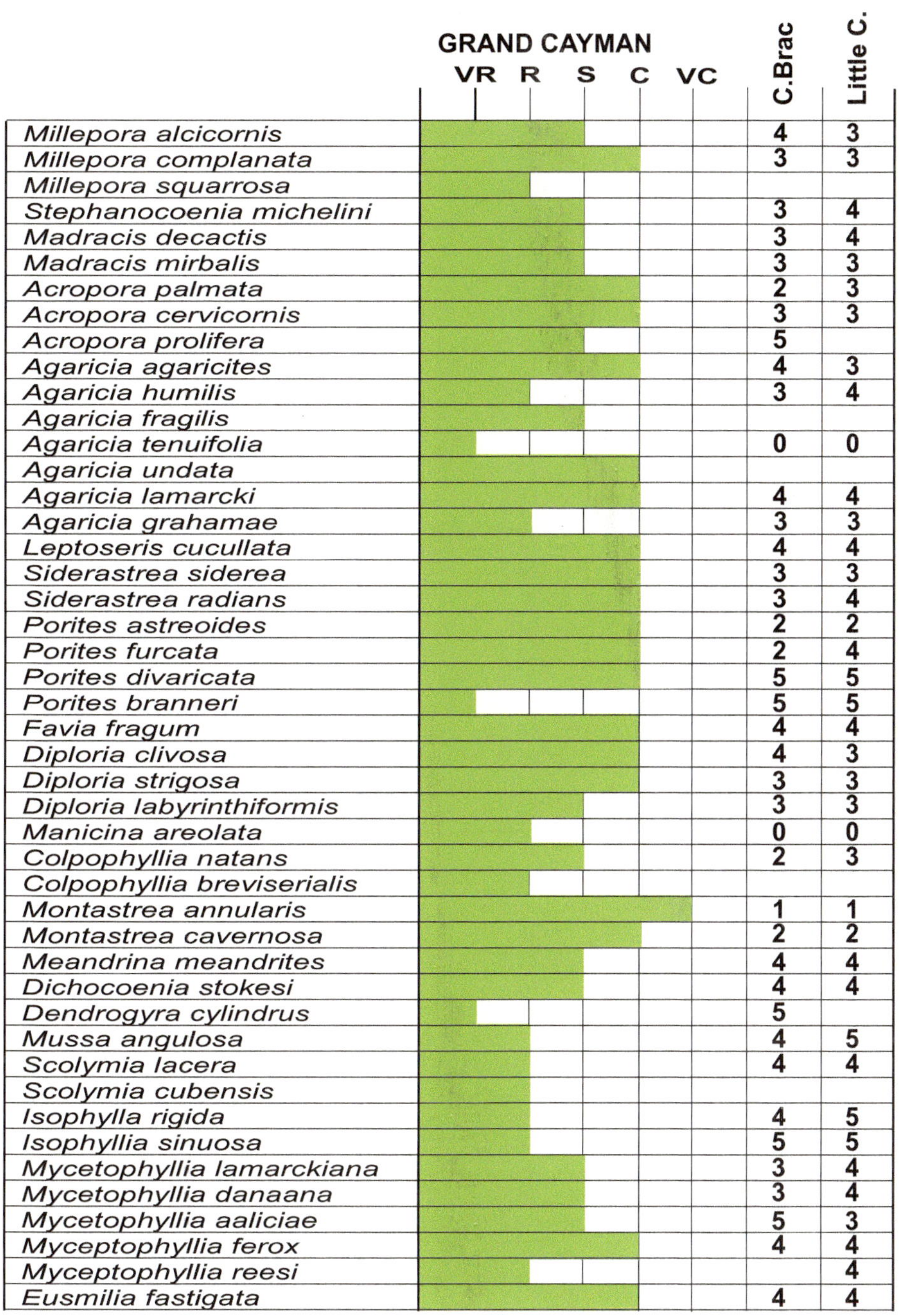

Species	GRAND CAYMAN (VR R S C VC)	C.Brac	Little C.
Millepora alcicornis	S	4	3
Millepora complanata	C	3	3
Millepora squarrosa	R		
Stephanocoenia michelini	S	3	4
Madracis decactis	S	3	4
Madracis mirbalis	S	3	3
Acropora palmata	C	2	3
Acropora cervicornis	C	3	3
Acropora prolifera	S	5	
Agaricia agaricites	C	4	3
Agaricia humilis	R	3	4
Agaricia fragilis	S		
Agaricia tenuifolia	VR	0	0
Agaricia undata	C		
Agaricia lamarcki	C	4	4
Agaricia grahamae	R	3	3
Leptoseris cucullata	C	4	4
Siderastrea siderea	C	3	3
Siderastrea radians	C	3	4
Porites astreoides	C	2	2
Porites furcata	C	2	4
Porites divaricata	C	5	5
Porites branneri	VR	5	5
Favia fragum	C	4	4
Diploria clivosa	C	4	3
Diploria strigosa	C	3	3
Diploria labyrinthiformis	S	3	3
Manicina areolata	R	0	0
Colpophyllia natans	S	2	3
Colpophyllia breviserialis	R		
Montastrea annularis	VC	1	1
Montastrea cavernosa	C	2	2
Meandrina meandrites	S	4	4
Dichocoenia stokesi	S	4	4
Dendrogyra cylindrus	VR	5	
Mussa angulosa	R	4	5
Scolymia lacera	R	4	4
Scolymia cubensis	R		
Isophylla rigida	R	4	5
Isophyllia sinuosa	R	5	5
Mycetophyllia lamarckiana	S	3	4
Mycetophyllia danaana	S	3	4
Mycetophyllia aaliciae	S	5	3
Myceptophyllia ferox	C	4	4
Myceptophyllia reesi	R		4
Eusmilia fastigata	C	4	4

Fig. 8.7 List of corals identified in modern reefs of the Cayman Islands with abundances as defined by Hunter (1994, his Fig. 3.2) and Fenner (1993, his Table 1)

corals was particularly severe around Grand Cayman with corals on the deep reefs rapidly expelling their zooxanthellae, turning white, and remaining like that until conditions improved (Turner et al., 2013). Van Hooidonk et al. (2012) showed that the species most susceptible to bleaching around Little Cayman were *Siderastrea siderea, Montastrea annularis,* and *M. favrolata*, whereas *Porites porites, P. asteroides*, and *M. cavernosa* were the least affected.

Hunter (1994), divided the corals that he recorded from around Grand Cayman into 15 Coral Associations (Fig. 8.9). Each association, named for its dominant genera and/or species, represents a recurring group of species, with low to high species diversities, that typically lived in environments that are characterized by different ranges of light and turbulence (Fig. 8.9). The distribution of the associations in the backreef areas is partly related to the type of coastline

Fig. 8.8 Example of coral bleaching in *Millepora complanata* from Grand Cayman. **a** Coral with white patches that formed during the early stages of coral bleaching. **b** Photograph of same coral as shown in panel A, but now completely bleached white. Photographs courtesy of the Department of Environment, Cayman Islands

Fig. 8.9 Coral associations found in the ocean waters around the Cayman Islands. Modified from Hunter (1994, his Fig. 3.6)

CORAL ASSOCIATION	WATER DEPTH	LIGHT	TURBULENCE	DIVERSITY (0–30)
divaricata	< 5 m	High	Low	
annularis—porites	< 5 m	High	Low	
Diploria clivosa	< 7 m	High	Moderate	
porites—agaricites	< 5 m	High	Moderate	
annularis—agaricites	< 5 m	High	Moderate	
Massive coral head	< 5 m	High	Moderate	
Acropora cervicornis	< 5 m	High	Moderate	
annularis—rigida	< 5 m	High	Mod. to High	
Acropora palmata	< 2 m	High	Very High	
Millepora complanata	< 2 m	High	Very High	
cervicornis—mirabilites	3—20 m	Moderate	Moderate	
Lepioseris - Mycetophyllia	10—20 m	Low	Low	
Stephenocoenia— Madracis	20—30 m	Mod. to Low	Mod. to Low	
Agaricia lamarcki	25—50 m	Low	Very Low	
Agaricia undulata	40—60 m	Very Low	Very Low	

(Fig. 8.10). This approach to understanding the distribution of the modern coral assemblages was taken in order to provide a basis for understanding the factors that may have controlled the distribution of corals in the Pleistocene Ironshore Formation as well as those in the older rocks of the Bluff Group.

8.3.2 Shelf-Edge Reef of Grand Cayman

The perimeter of the shelf around Grand Cayman is characterized by an impressive Shelf-Edged Reef that has grown over the outer 200 m of the terrace and extends to a depth of ~75 m (Blanchon 1995; Blanchon and Jones 1997). This spectacular reef, rising to depths of 12 m below mean sea level, is formed of buttresses, up to 100 m long and 10 m wide, that are separated by steep-sided canyons that are commonly known as "spur and groove structures" (Figs. 8.11, 8.12). Each buttress, oriented at 90° to the shelf edge, consists of a wall, a crown, and a spur. The near-vertical walls, up to 40 m high, may overhang in their upper part and the grooves between adjacent buttresses are in shadow for most of the day. The broad rounded crown rises into water that is 15–20 m deep. The spur, which may be up to 300 m long, starts from the crown and extends shoreward into the lower terrace. The canyons between neighbouring buttresses are floored with skeletal sands and cobble-sized sediment (Fig. 8.13).

Blanchon et al. (1997) demonstrated that the fringing-reef complexes around Grand Cayman have cores that are formed of large, abraded coral clasts that probably originated from hurricane activity. Between storms these clasts are stabilized by coralline-algal crusts that provide an ideal substrate for new coral growth and, eventually, development of the fringing reefs. They also suggested that this mechanism was responsible for the variations in reef architecture,

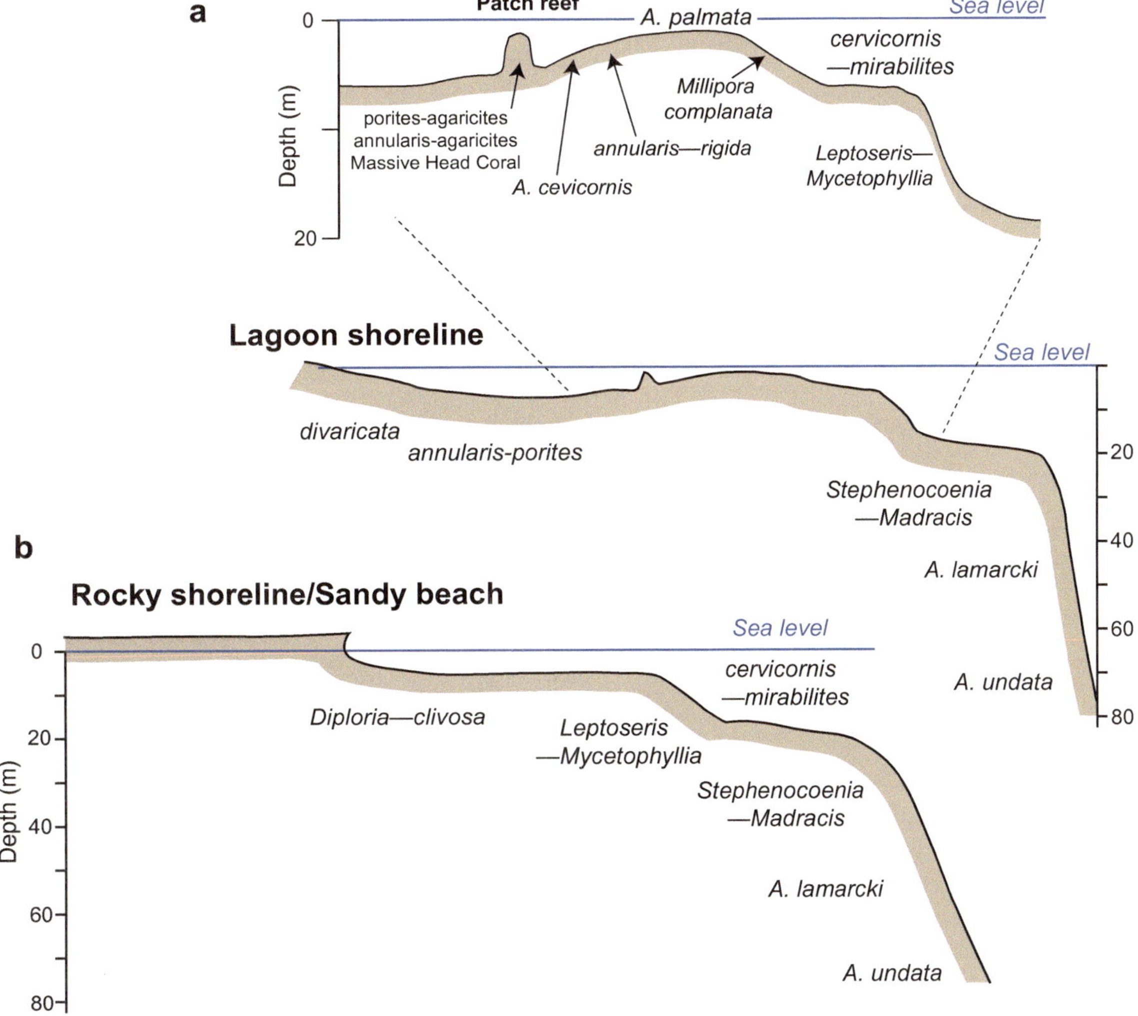

Fig. 8.10 Distribution of coral associations in oceanic waters of the Cayman Islands associated with **a** lagoonal shoreline, and **b** rocky shoreline/sandy beach shoreline. Modified from Hunter (1994, his Figs. 3.7 and 3.8)

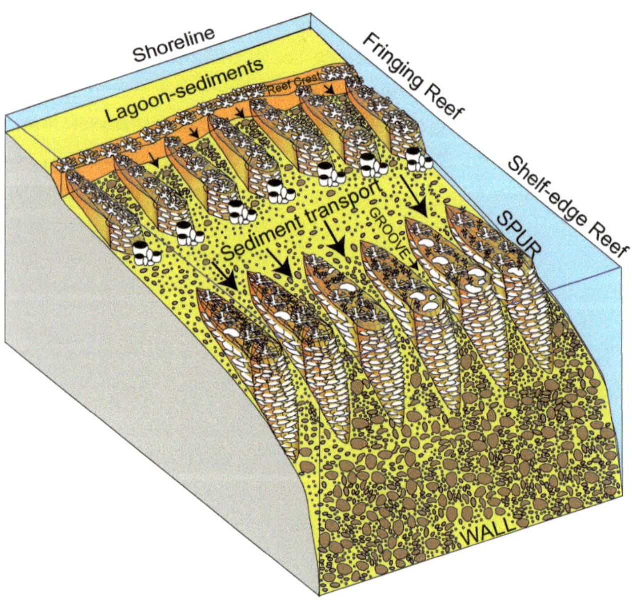

Fig. 8.11 Schematic model showing spur and groove structures that characterize the perimeter of the shelf around Grand Cayman. Sediment transport is funneled through the grooves to the edge of the shelf

the locations where reef development took place, and the uniform location of reefs in these depositional systems.

The Shelf-Edge Reefs are home for a diverse array of animals and plants that include stony corals, sponges, soft corals, gorgonians, fleshy algae, and calcareous algae (Blanchon and Jones 1997). This includes 44 species of corals with 33 that are found only in these reefs (Hunter 1994). The architecture of these reefs changes with the orientation of the coastline (Blanchon and Jones 1997). Thus, shelf-edge reefs found along the exposed windward margins of the island have the deepest and highest buttresses with the most extensive coral coverage. In contrast, the shelf-edge reef on the leeward western margin of Grand Cayman is characterized by shallow, low-amplitude buttresses, little spur development, and numerous branching corals. In protected areas, the canyons are absent and the buttresses merge to form a continuous reef.

Blanchon and Jones (1995) examined the origin of the two seaward-sloping terraces around Grand Cayman that are separated by the mid-shelf scarp. Where exposed to wind, the upper terrace is the site of luxurious coral growth, whereas on the sheltered parts of the coastline, the upper terrace is a barren rocky pavement characterized by erosional features that probably formed during the major storms that frequently impact the island. The mid-shelf scarp is typically partly to completely buried by sediment. In some areas along the shelf on the western leeward side of the island, the scarp is well exposed with a wave-cut notch at a depth of 18.5 m bsl being clearly evident (Fig. 8.6). The lower terrace complex (12–40 msl), which extends from the mid-shelf scarp to the shelf edge, is covered with sediment and reefs that are up to 40 m thick in some areas. This complex package of sediments developed when sea level was at a lower level. Blanchon and Jones (1995) suggested that the lower terrace and mid-shelf scarp developed between 11 and 7 ka when sea level was rising slowly before being suddenly drowned at ~7 ka. Following that, sea-level rose gradually to its present level.

Fig. 8.12 View of landward part of spur and groove structure, seaward of the fringing reef at South Sound. Note coral growth on the spur (right side) and sand and debris in the open groove (left side). Water is ~6 m deep in the groove

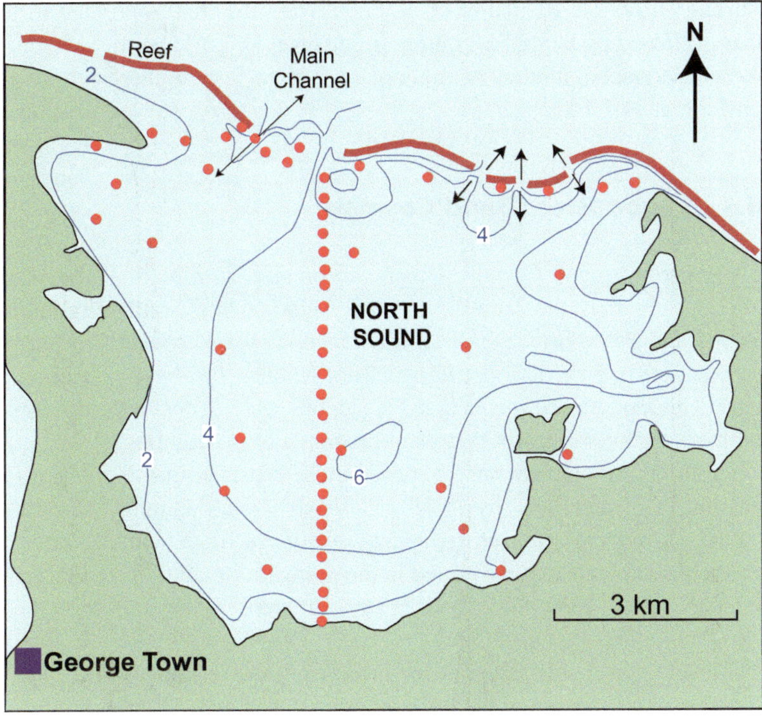

Fig. 8.13 Map of North Sound showing water depths based on Roberts (1971b) and MacKinnon and Jones (2001, their Fig. 1c) and location of cores and surface sediment samples used by MacKinnon and Jones (2001). Contour intervals in metres

Blanchon and Jones (1997) proposed the hurricane-control hypothesis to explain the architecture and development of the shelf-edge reefs around Grand Cayman. They argued that variations in the height and approach directions of hurricane waves influenced growth of the reefs by the following processes.

- Controlling the coral zonation on the buttresses with robust domal forms dominating on the crowns where there is maximum turbulence and branching corals growing on the sides of the spurs.
- Variations in the architecture of the self-edge reefs and the angle at which the hurricane waves approach with those sections that face directly into the waves having the pruned detritus swept back onto the shelf to produce large buttresses and high spur frequencies, whereas reefs affected by obliquely approaching waves have the detritus swept off the shelf with the result that spur development is limited.
- Variations between reefs on different sections of the coastline are controlled by differences in the intensities of buttress pruning and sand flushing during hurricanes.
- The vertical growth of self-edge reefs may be controlled, at least in part, by the frequency of pruning by hurricanes.

The historical development of reefs around the Cayman Islands was controlled largely by sea level, which may in turn, have been controlled by (1) vertical movements of the islands, and/or (2) eustatic sea level changes. On the northwest coast of Grand Cayman, the presence of a wave-cut notch at 18.5 m below modern sea level (Fig. 8.6) offers mute testimony to a time when sea level was, relative to today, much lower. Blanchon et al. (2002), based on cores taken close to a submerged sea-cliff at a depth of 21 m on the east coast of Grand Cayman, showed that a reef formed largely of *Acropora palmata* had once flourished there. The cores penetrated a lower *Palmata*-rudstone Facies and an upper Mixed Framestone Facies. The *Palmata*-rudstone Facies was formed of clasts, derived from branches of *A. palmata*, held in a cemented skeletal grainstone. A layer below that facies is formed of photophilic crustose coralline algae, foraminifera, and vermetid gastropods that were overlain by a cryptic assemblage of sclerosponges, foraminifera and serpulid worms. The uppermost Mixed Framestone Facies, found only locally, includes a sequence of deposits attributed to repeated cycles of destruction, bioerosion, and recolonization akin to that associated with repeated hurricane pruning of corals. Blanchon et al. (2002) argued that the basal *Palmata*-rudstone Facies was formed when a hurricane destroyed a well-established, viable *Acropora* reef. U-Th dating of the near-surface corals from that facies indicates that it developed 8.9 to 8.1 ka, when there was a breakwater reef crest at 21 m bsl, which is consistent with –19 m shoreline that had previously been proposed for Grand Cayman between 8.1 and 7.6 ka (Blanchon and Jones 1995). Data from Grand Cayman and other localities throughout the Caribbean indicate that the older reefs died out over a period of ~160 years, starting ~7.6 ka. Blanchon et al. (2002) argued that was then followed by a rapid 6 m jump in sea level ~7.5 ka and the subsequent initiation of modern reefs 4 to 9 m higher upslope at ~7.4 ka.

8.4 Lagoons of Grand Cayman

The lagoons around Grand Cayman vary in size from North Sound, which is the largest, to small lagoons like South Sound and Pease Bay (Fig. 8.1). Most lagoons were created as coral reefs grew and developed from headland to headland. The modern sediments in these lagoons were deposited on hard, rocky substrates that were produced by weathering when the last major lowstand exposed those coastal areas to erosion. Characterized by "life", these dynamic lagoons include a diverse array of sediments, a diverse biota with animals and plants that are rooted in the sediment or attached to the rocky substrates, and many different types of free-swimming organisms.

Developing facies schemes for these lagoonal sediments is a challenge because of the different approaches that can be used for delineating and naming the surface and subsurface facies. The surface facies have typically been delineated and named based on the plants and animals that live there and the sediment types. For example, the *Thalassia Facies* is based on the *Thalassia* (seagrass known locally as "turtle grass"), that covers large areas of the lagoons, and is easy to recognize while in the water and on air photographs. Recognition of this facies in subsurface sediments, however, becomes more difficult once the *Thalassia* and associated plants die without leaving any direct evidence of their presence. Accordingly, a dual system has generally been applied to these lagoons—one that deals with the surface sediments/biota and the other that deals with the subsurface sediments. In most cases, attempts have been made to genetically link the subsurface sediments to their surface counterparts.

Although extensive research has been completed on the lagoons around Grand Cayman, there are some that have received scant attention—including those found along the north coast of the island. Relatively little is known about the lagoons on Little Cayman. On Cayman Brac, there are no well-defined lagoons because of the paucity of off-shore reefs (Logan 1994).

8.4.1 North Sound

North Sound, which is ~9 km (N to S) and up to 10 km wide (E-W) is the largest lagoon on any of the Cayman Islands (Fig. 8.13). Originally, mangrove swamps were present around most of North Sound, but development has led to the loss of extensive tracts of mangrove tracts on the western peninsula. Although the water in North Sound is typically <2 m deep near the shoreline, it is up to 6 m in the central part of the lagoon with channels through the reef being up to 4 m deep (MacKinnon 2000; MacKinnon and Jones 2001). The water has an average temperature of ~28 °C with normal marine salinity near the reef where good water circulation exists (Rigby and Roberts 1976). In the lagoon interior, where water circulation is poorer, the salinity can be up to 42‰ (Rigby and Roberts 1976).

Information on the facies, sediment thickness (Fig. 8.14), and other attributes of North Sound first came from the work of Roberts (1971a) and Rigby and Roberts (1976). Roberts (1971a) divided the depositional environments into the Reef Shoal Environment, the Grass Plain Environment and Shore Zone Environment (Fig. 8.15). Building on that early work, North Sound was the focus of detailed sedimentological analysis by Mackinnon (2000), MacKinnon and Jones (2001), Booker (2020), and Booker and Jones (2020). Collectively, the information from all of these studies means that the age and development of the sediments in North Sound are now well understood.

MacKinnon (2000) and MacKinnon and Jones (2001) delineated seven facies in North Sound that were based on consideration of the surface and subsurface sediments that were evident in the numerous soft-sediment cores collected from the lagoon (Fig. 8.16). These facies were assigned to the Fresh-Brackish Water Association and the Marine Facies Association. The former includes the Composite Grain Facies, the Gastropod Facies, and the Bivalve Facies, whereas the latter includes the *Halimeda* Facies, the Bivalve-*Halimeda* Facies, and the Foraminifera-Bivalve Facies. The relationship of these facies to the underlying bedrock and to each other was determined from the sediment cores that came from North Sound (Fig. 8.16).

The Fresh-Brackish Water Association

The *Composite Grain Facies* is a well-sorted, fine-grained, light beige to white sediment formed of composite grains that consist of biofragments derived from bivalves, gastropods, and foraminifera held in a micrite matrix. In the central part of North Sound, this facies rests on the bedrock.

The *Gastropod Facies,* formed of small *Amnicola forsthi* (average 2 mm long, 1.5 mm diameter) embedded in organic material, is found in lenses in the Mangrove Peat Facies or resting directly on the bedrock.

The *Bivalve Facies* is characterized by disarticulated, fragmented, and articulated *Mytilopsis domingensis* that are embedded in organic material. The shells lack the characteristic dark-brown periostracum of the living shells. This facies rests on the bedrock or above the Mangrove Peat Facies. It is locally overlain by the *Halimeda* Facies and, in some areas, the Mangrove Peat Facies.

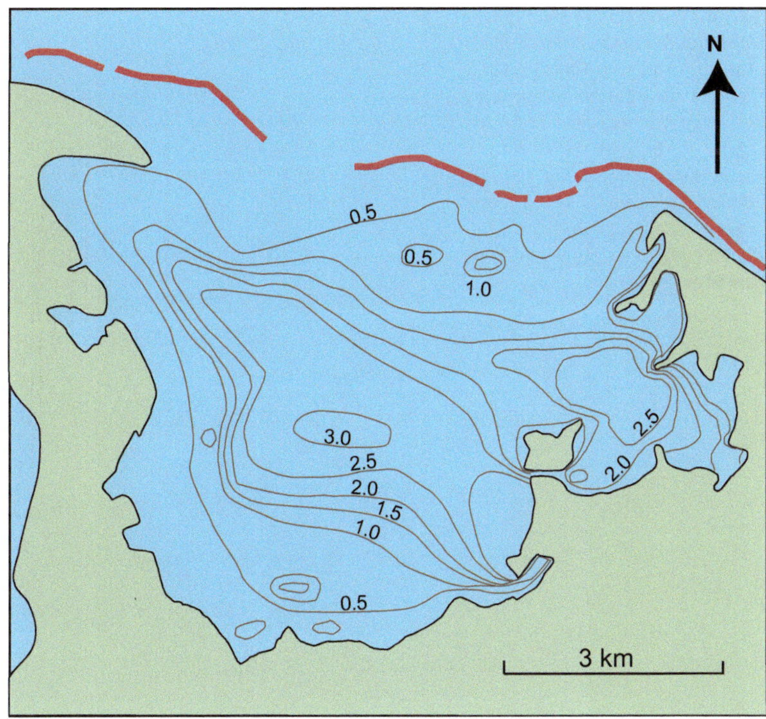

Fig. 8.14 Map of North Sound showing thickness of sediments on the lagoon floor. Modified from MacKinnon and Jones (2001, their Fig. 1d)

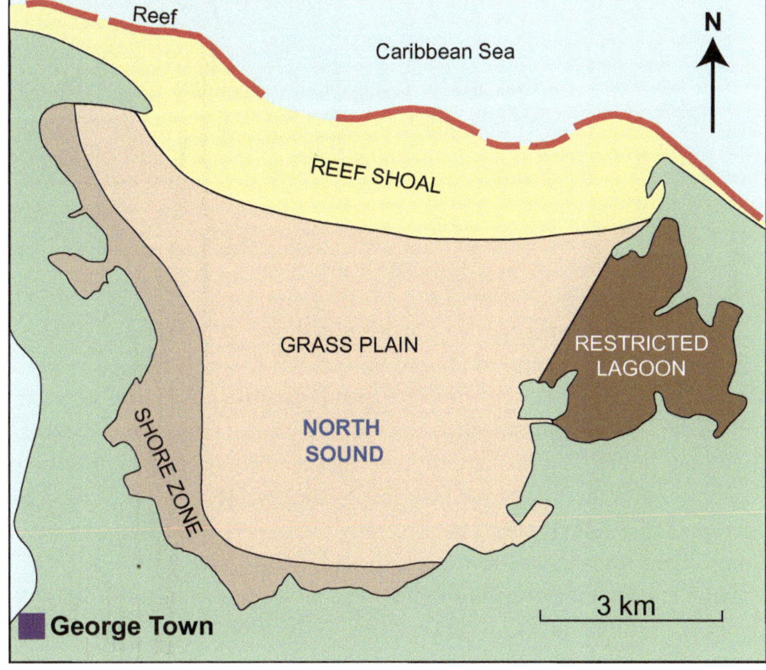

Fig. 8.15 Map of North Sound showing main depositional zones. Modified from Roberts (1971b, his Fig. 3) and Mackinnon and Jones (2001, their Fig. 2a)

The Marine Facies Association

The *Mangrove Peat Facies* comprises a black to dark brown organic rich sediment that contains wood fragments, up to 1.5 cm long, that probably came from mangrove trees. Well-preserved articulated and disarticulated *Chione cancellate* are common. The lower part of this facies commonly contains small (<3 mm) fragments of indeterminate bivalve shells. Locally, bioturbation has resulted in some mixing with the overlying *Halimeda* Facies. Although commonly resting on the bedrock, it locally lies on top of the Bivalve Facies or the Composite Grain Facies. In most areas it is overlain by the *Halimeda* Facies but locally the Bivalve or Gastropod Facies lie on it.

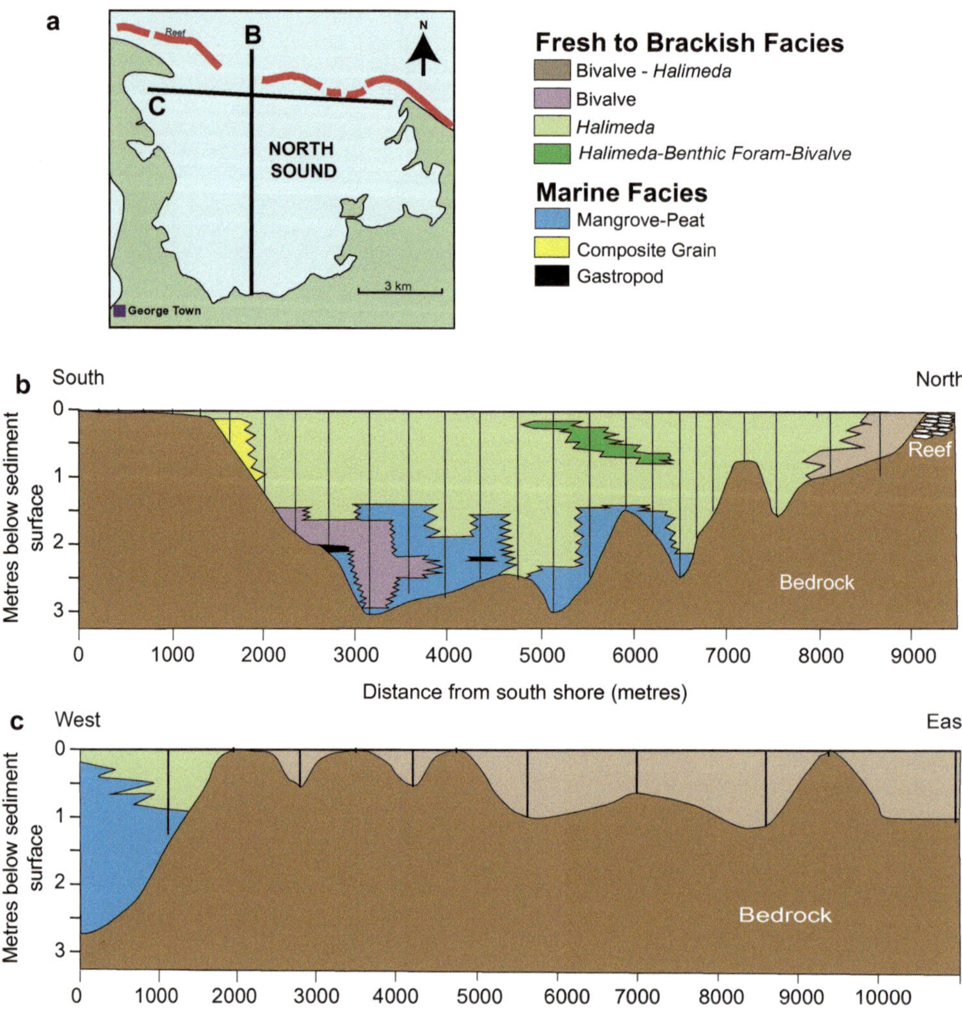

Fig. 8.16 Vertical and lateral facies distribution in North Sound. **a** Locations of transects B and C. **b** Facies distribution along transect B based on cores collected along north–south transect. **c** Facies distribution along transect C based on cores collected along east–west transect. Modified from MacKinnon and Jones (2001, their Figs. 3, 4)

The *Halimeda Facies* is characterized by numerous *Halimeda* plates, scattered bivalves (up to 4 cm long), benthic foraminifera, gastropods, coral fragments, and composite grains. Micritization of the *Halimeda* plates is common. In the central part of the lagoon, the *Halimeda* Facies overlies the Mangrove Peat Facies, but locally it rests on top of the Bivalve Facies. Laterally, it passes into the Composite Grain Facies to the south and the Bivalve-*Halimeda* Facies to the north (Fig. 8.16). The sediments in these facies are similar to those that are presently accumulating beneath the *Thalassia* banks in North Sound.

The *Halimeda–Benthic Foraminifera–Bivalve Facies* is formed of articulate and disarticulate bivalves and gastropods held in a sand matrix with grains derived largely from *Halimeda* (extensively micritized), benthic foraminifera, bivalves, composite grains, gastropods, red algae, corals, and echinoids. This facies has only been found in the central part of North Sound where it is encased by the *Halimeda* Facies.

The *Bivalve–Halimeda Facies* is characterized by moderately to well sorted, coarse-grained skeletal allochems, derived largely from *Halimeda* along with fewer red algae, benthic foraminifera, and echinoids with most grains being partly micritized. The bivalve shells and *Halimeda* plates are typically fragmented. This facies forms a belt, ~1 km wide, on the landward side of the reef that defines the northern margin of North Sound.

Rare Earth Elements

Booker (2020) and Booker and Jones (2020) examined the distribution of rare earth elements (REE and Y) in three of the soft-sediment cores from North Sound (Fig. 8.17). The concentrations are linked to the facies.

- *Bivalve Facies*: This facies contains 0.1 to 5.6 ppm REE +Y, but lacks detectable amounts of Tb, Tm, and Lu. Eight samples lacked detectable Eu, Ho, or Yb, and five samples lacked detectable amounts of Er.

- *Lower Peat Facies*: All samples from this facies contained all of the rare earth elements (totals of 0.1 to 14.3 ppm), and exhibit negative Eu and positive Gd anomalies.
- *Upper Peat Facies*: This facies contains 0.1 to 4.5 ppm REE+Y with concentrations increasing towards the surface. The REE+Y profiles are characterized by negative Eu and positive Gd anomalies. None of the samples contains detectable quantities of Tm and Lu, and one sample lacked Tb.
- *Halimeda Facies*: This facies contained 0.1 to 3.8 ppm REE+Y with all samples lacking detectable amounts of Eu, Tb, Ho, Er, Tm, Yb, and Lu. Eu and Gd anomalies were not evident in any of the samples.

Booker and Jones (2020) suggested that the REE and Y may have theoretically originated from (1) weathering of insoluble material from the underlying bedrock, (2) fluvial detrital material, (3) marine sourced biogenic or colloidal materials, (4) wind-blown volcanic ash, and/or (5) wind-blown aerosols from distant regions. They argued that the REE and Y did not come from the underlying limestone bedrock and the lack of rivers on the island argues against the possibility of fluvial influx. The REE–Y profiles from these sediments are different from the seawater profile from Misteriosa Bank, which suggests that the North Sound REE+Y was not sourced directly from the seawater. Of the remaining two possibilities, the most likely possibility is that the REE and Y was brought into the area as wind-blown dust that originated from the Sahara Desert. This notion is supported by the fact that most of the REE in the terra rossa found on Grand Cayman probably came from that source (Jones 2021).

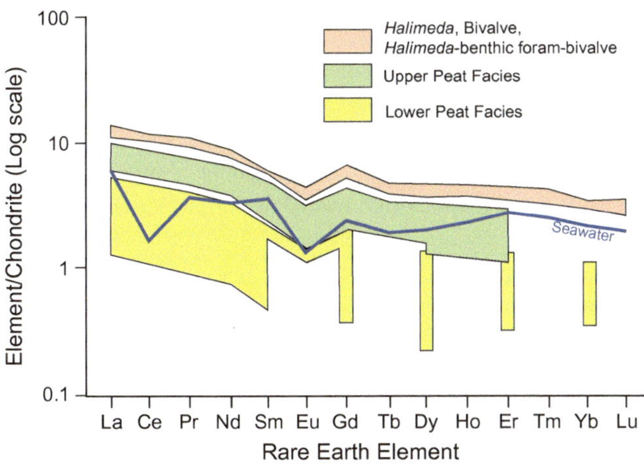

Fig. 8.17 Rare earth element distribution (Chondrite normalized) in cores B10, B15, and B16 from North Sound relative to facies identified by MacKinnon (2000) and MacKinnon and Jones (2001). Seawater concentrations are from the Mysteriosa Bank ($\sim$400 km west of Grand Cayman, 10 m water depth), multiplied by a factor of 10^6—data from Osborne et al. (2015). Modified from Booker and Jones (2020, their Fig. 4a)

Age

MacKinnon (2000) and MacKinnon and Jones (2001) reported C^{14} dates for six samples that ranged from 5060 $\pm$ 80 years to 1584 $\pm$ 40 years BP. These dates, along with some additional C^{14} dates that Booker (2020) and Booker and Jones (2020) obtained from the cores, led to calibration of all the ages using CALIB 7.10. This produced ages from $\sim$4000 BCE to 500 BCE with an uncertainty of $\pm$ 200 years (Booker and Jones 2020). Uncertainties with these dates reflect analytical uncertainty, the choice of the ^{14}C marine reservoir correction factor that is poorly known for the Caribbean, uncertainties in the calibration curve, and assumptions used in the BACON age-depth curve. Nevertheless, these dates do provide a time framework for sediment deposition.

Facies Architecture

As noted by MacKinnon and Jones (2001), the age-versus-depth positions of the samples in North Sound are very similar to the minimum sea-level curve that Fairbanks (1989) developed for Barbados. Deposition started $\sim$5,000 years BP when a freshwater swamp developed in the central part of North Sound with the growth and gradual expansion of mangrove swamps. This fresh to brackish water regime, with deposition of facies of the Fresh-Brackish Water Association, continued for $\sim$1,000 years. With the development and growth of the reef along the northern margin of North Sound, circulation patterns changed as the lagoon was effectively cut off from the open ocean. As a result of that and the continued rise in sea level, sediments belonging to the Marine Facies Association accumulated in North Sound. Over the last 4,200 years, sediments have accumulated at a rate of $\sim$85 cm/1000 years.

8.4.2 South Sound

South Sound, located on the southwest corner of Grand Cayman is 5 km long (E-W) and narrows from a maximum width of 1.1 km at its eastern end to 275 m at its western end (Fig. 8.18). It is bounded by land to the north and east, and a barrier reef forms its southern margin. The northeast margin is fringed by a red mangrove swamp. The only passages to the open sea are through a man-made channel through the reef and a natural opening at its west end. Maximum water depths of 2 to 3 m are found along the central axis of the lagoon (Fig. 8.18a). The maximum sediment thickness

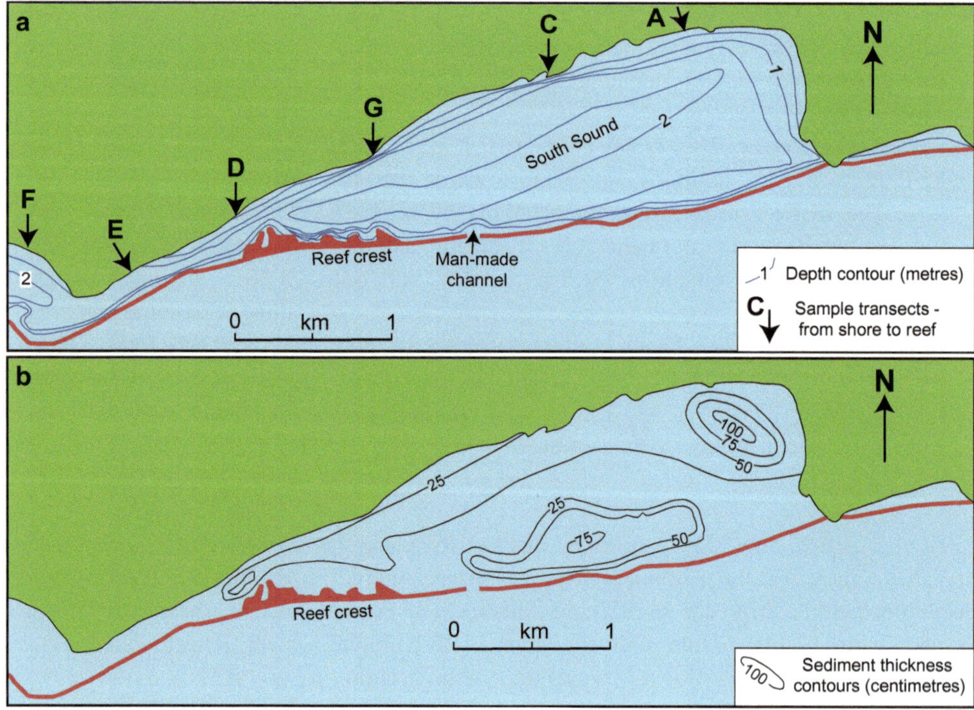

Fig. 8.18 South Sound **a** water depth and **b** sediment thickness. Arrows labelled A, C, D, E, F, and G indicate positions of transects that were used to map the facies distribution in South Sound. From Beanish and Jones (2002, their Figs. 1 and 3: reproduced and modified with permission from the Coastal Education and Research Foundation Inc.)

of ~1 m occurs in the northeast corner of the South Sound where water circulation is poor (Fig. 8.18b). Over the rest of the lagoon, sediment thicknesses of <75 cm are the norm.

Beanish (2000), Beanish and Jones (2002), and Beanish et al. (2002) described various aspects of the sedimentology of South Sound based on (1) assessment of the water depth, sediment thickness, substrate type, biota, biota density and sedimentary textures along six shore to reef transects, (2) sediment samples collected at set intervals along each transect, (3) sediment cores, and (4) digital image analyses of air photographs that were used to map the facies and assess their changes over time.

Facies

Facies identified on the basis of their resident biota, sediment components, grain size, and location include the widespread *Thalassia* facies, Sand Facies, Rock Bottom Facies and the restricted Rubble Facies, Coral Head Facies, and Brown Algae Facies (Fig. 8.19).

The *Thalassia Facies*, with its landward margin along the shoreline, is up to 380 m wide. Isolated patches of *Thalassia* are present beside the Rubble Facies that is close to the reef crest (Fig. 8.19). Based on the number of plants per square metre, colour tone on air photographs, and biotic composition, this facies was divided into the Sparse, Medium, Dense, and Very Dense *Thalassia* Facies, with the Medium *Thalassia* Facies being the most common. Sediment associated with the *Thalassia* facies is typically formed of mollusc fragments, *Halimeda* plates, planktonic and benthic foraminifera, red algae, and coral fragments. Many fragments are partly micritized. Locally, *Porites porites*, *Siderastrea radians*, *Favia fragrum*, and sponges are present. Sea cucumbers, echinoids, hydroids, parrotfish, and gastropods (including *Strombus gigas* and *S. costatus*) are found in and around the *Thalassia* banks.

The *Sand Facies*, found in the central part of the lagoon (Fig. 8.19), is sparsely vegetated in some areas. The rippled sands are poorly sorted medium sand (average grains size of 1.8ϕ) with grains derived from molluscs, *Halimeda*, foraminifera, red algae, and corals.

The *Rock Bottom Facies* is found in the channel at the west end of the sound (Fig. 8.19), where the currents are strongest, and near Prospect Point where waves crash over the reef crest. Poorly sorted lithoclasts, up to 50 cm long, are common. Locally, green and brown algae are attached to the substrate. Corals, including *Siderastrea radians, S. siderea, Diploria, Acropora cervicornis,* and *A. palmata* are common.

The *Rubble Facies*, a 60–350 m wide belt landward of the reef crest (Fig. 8.19), is formed of pebble- to cobble-sized clasts that rest on the rocky substrate. Associated sediments are coarse grained. Most clasts are derived from *A. palmata* with others coming from *Montastrea annularis, Diploria, Siderastrea, and Millepora*. Patch reefs, individual corals, gorgonians, and algae are common. *Homotrema rubrum* commonly encrust the undersides of the clasts.

The *Coral Head Facies* is characterized by individual coral heads and coral colonies. Local patch reefs, 1.2 to

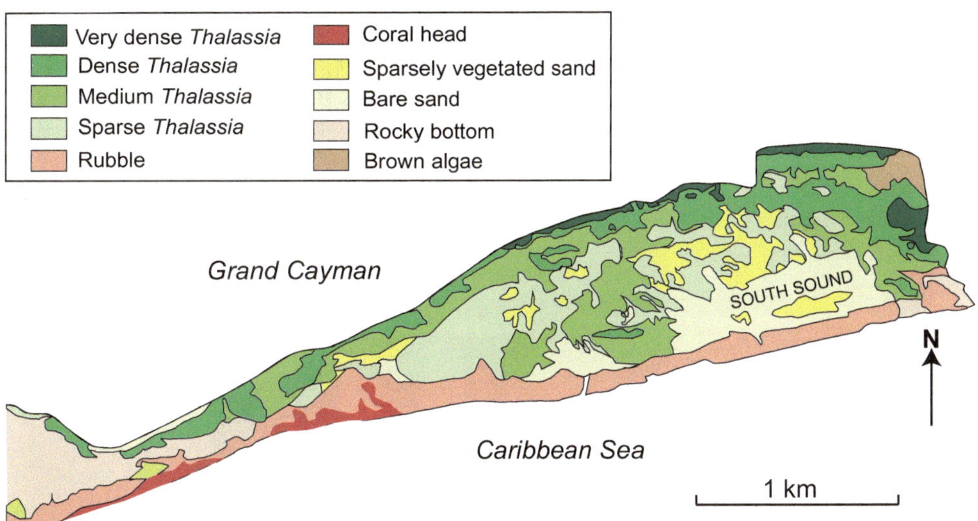

Fig. 8.19 Facies map for South Sound based on air photograph interpretations and mapping of lagoon along transects A, C, D, E, F, and G (see Fig. 8.18). From Beanish and Jones (2002, their Fig. 4: reproduced and modified with permission from the Coastal Education and Research Foundation Inc.)

1.8 m high, include a diverse array of corals and various encrusting coralline algae. The diversity and abundance of corals decreases landward.

The *Brown Algae Facies*, found in the northeast corner of the lagoon (Fig. 8.19), is characterized by bare sand, *Callianassa* mounds, and algae. The poorly sorted sands include organic material derived from the red mangroves that grow along the shoreline. Brown algae coats the surface of the sand.

Age of Sediments

Sediment cores from the northeast corner of South Sound show that the surface facies have existed for long periods of time. In cores 1 and 6, the basal Peat Facies are overlain by the *Thalassia* Facies, whereas in cores 2, 7, 8, and 9 the Peat Facies is overlain by the Brown Algae Facies (Fig. 8.20). A ^{14}C date from a crocodile vertebra in core #7 (26 cm below surface) yielded an age of 1710 ± 60 years whereas ^{14}C dates from three bivalve shells in core #8, at depths of 9.8 cm, 33 cm, and 62 cm yielded dates of 350 ± 70 years, 440 ± 70 years, and 590 ± 70 years, respectively. Beanish and Jones (2002) suggested that the crocodile bone had probably been reworked from an older deposit (Fig. 8.21).

Facies Changes Through Time

Tongpenyai (1989) and Tongpenyai and Jones (1991) used air photographs taken in 1971 (black and white, scale 1:12,500), 1979 (colour, scale 1:50,000) and 1985 (colour, scale 1:10,000) to ascertain the changes in facies distribution in South Sound over a 24 year period. They verified facies identification derived from those photographs by comparison with maps produced by Rigby and Roberts (1976) and fieldwork. They showed that between 1971 and 1985 the area covered by *Thalassia* increased from 38.2 to 45.1% of the lagoon, while the area of sand decreased (Table 8.1). They pointed out that the increase in the *Thalassia* was important because it promoted sedimentation by providing substrates where epibionts live, by baffling currents and promoting deposition, and stabilizing the sediment through its extensive root system.

With the advent of more sophisticated image analysis systems and better developed statistical methods, Beanish (2000) and Beanish et al. (2002) examined the temporal changes in the facies of South Sound by comparing the facies, as mapped from a colour set of air photographs taken in 1992 (scale of 1:10,000) and the black and white set of aerial photographs taken in 1971 (scale of 1:12,500). Recognition of the facies in the 1971 black and white photographs was achieved by combining known information with the facies map that Rigby and Roberts (1976) produced from those photographs.

Beanish et al. (2002), based on various statistical techniques, showed that the area and distribution of facies in South Sound had changed between 1971 and 1992. The area covered by the *Thalassia* Facies increased by 17.3% with most of that taking place in the central part of the lagoon as the *Thalassia* invaded the sand areas. *Thalassia* also expanded into the Brown Algae Facies. In the west, the *Thalassia* Facies increased by becoming established in sand that had covered the Rock Bottom Facies. The only other facies that increased in coverage, by 5.8%, was the Coral Head Facies.

Sedimentary Processes

Sedimentation in South Sound, as noted by Roberts (1983, 1994) is controlled primarily by the action of storms on the sediments that form under everyday conditions. Most sediment in South Sound forms through the breakdown of the various animals and plants that dwell there, bioturbation, and micritization.

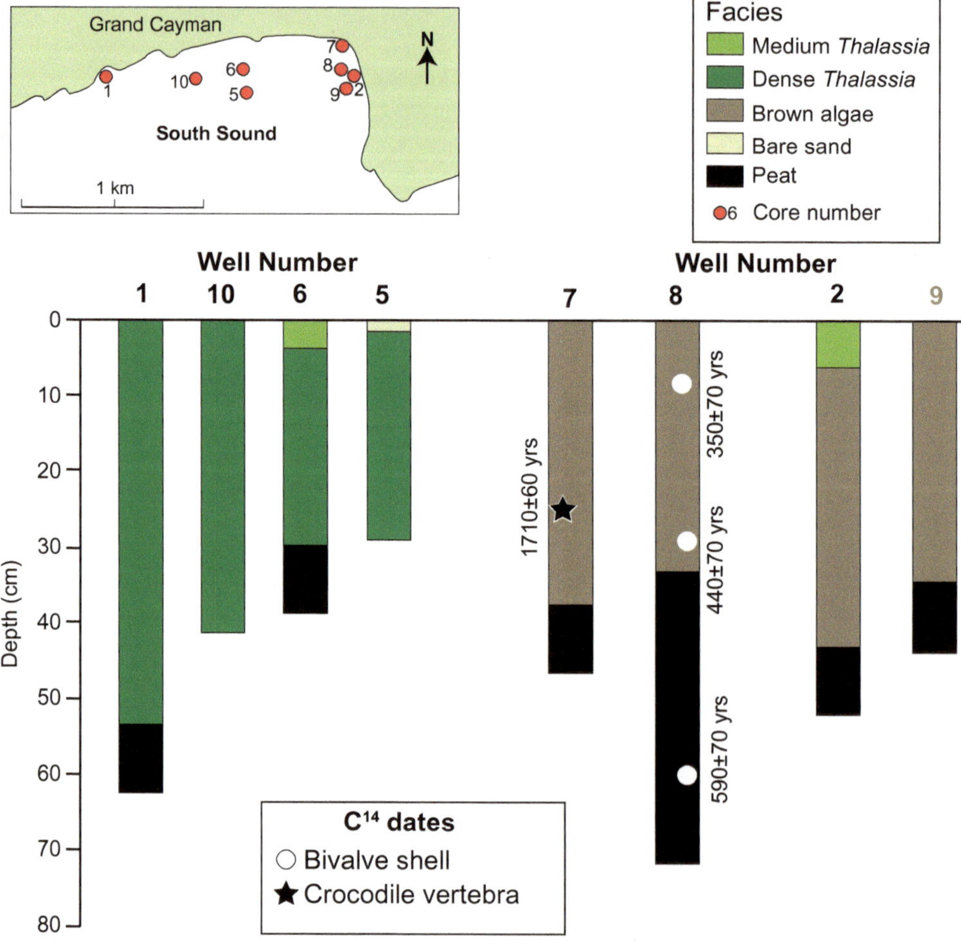

Fig. 8.20 Vertial facies succession with C^{14} ages from northeast corner of South Sound. From Beanish and Jones (2002, their Fig. 8: reproduced and modified with permission from the Coastal Education and Research Foundation Inc.)

South Sound, located on the windward south coast of Grand Cayman, receives wind-generated waves that approach from the SSE and break over the reef and thereby generate currents in the lagoon. Roberts et al. (1975) and Suhayda and Roberts (1977) showed that the wave surge produces currents in the lagoon with a velocity of 10 cm/sec that are at a low angle to the shoreline. Currents that are 6 cm/sec at the east end of the lagoon increase up to 45 cm/sec as they are funneled through the narrow western exit. In South Sound, the thickest sediment is found where the currents are the weakest. Accordingly, sediments are up to 1.2 m thick at the east end, 25–50 cm thick in the central part, and <5 cm thick at the west end. Facies distribution are also a reflection of current strength with the Very Dense and Dense *Thalassia* Facies being found in bands near the shoreline where currents are the weakest.

Hurricanes generate huge waves that become higher and more violent as they break over the reef crest and thereby increase current strengths in the lagoon. Although aperiodic, these storms have a significant impact on the sediments in South Sound. Coral rubble is generated as the waves break over the corals in the reef. Sand transportation is greatly enhanced and sediment is rapidly washed out of the west end of the lagoon (Roberts 1983, 1994).

8.4.3 Frank Sound and Pease Bay

Frank Sound, located on the central part of the south coast of Grand Cayman, is a large semi-circular shaped lagoon with a maximum E-W length of 4.5 km and N-S width of ~1 km (Fig. 8.22). In contrast, Pease Bay that is located west of Frank Sound is ~4 km long with a maximum width of 0.5 km (Fig. 8.23). The water in Frank Sound is up to 4 m deep but mostly <2.5 m deep (Fig. 8.22a). The sediment is up to 3.0 m thick but <2.0 m thick over much of the lagoon (Fig. 8.22b). The thickest sediments are found landward of the reef crest on either side of the channel. In Pease Bay, the water is mostly <2.0 m deep (Fig. 8.23b). The sediment in Pease Bay is up to 3.0 m thick, but <2.0 m thick over much of the lagoon.

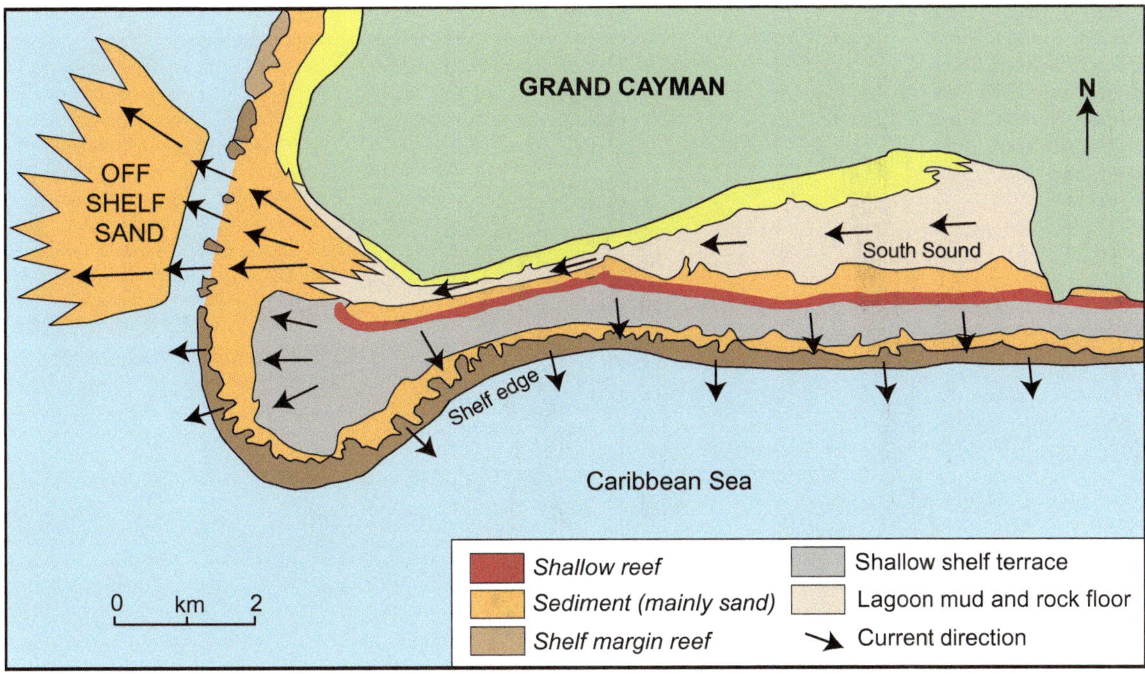

Fig. 8.21 Model of sediment transport out of South Sound mediated by currents that cross the reef at the east end of the Sound and then flow along the axis of the lagoon before leaving at the west end. Sediment from the lagoon is washed across and then off the shelf at the southwest corner of the island. Modified from Roberts (1994, his Fig. 5.34)

Table 8.1 Percentage areas covered by each facies in South Sound in 1971, 1979, and 1985 as determined by Tongpenyai and Jones (1991, Table 2)

Facies	1971	1979	1985
High density *Thalassia*	8.3	8.7	12.5
Low density *Thalasssia*	29.9	30.3	32.6
Transitional	6.9	4.4	5.0
Sand	36.0	38.1	30.3
Corals	7.0	5.7	5.7
Brown algae	11.9	12.8	13.9

Facies

Large areas of the substrates (~98%) in Frank Sound and Pease Bay are covered by the *Thalassia* Facies, Rubble and Knob Facies, the Bare Sand Facies, and Sand Facies (Fig. 8.24). Collectively, the Bare Rock Facies, Coral Knolls Facies, and Shoreline areas form <2% of the lagoon floors (Kalbfleisch 1995; Kalbfleisch and Jones 1998). Many of these facies are similar to those in South Sound.

The *Thalassia* Facies is characterized by *Thalassia testudinum,* numerous green algae, and red and brown algae. Sea urchins, green turtles, and parrotfish graze in these banks. Gastropods (including *Strombus gigas, S. costatus, Cerithium litteratum*, *Fasciolaris tulipa*) are common. Corals are limited to scattered *Porites divaricata*, *Siderastrea radians*, and *Favia fragum*. *Callianassa* mounds are found throughout this facies. The fine to medium grained sand in the substrate, with a shoreward decrease in grain size, is formed of corals fragments, foraminifera, and mollusc fragments along with numerous completely micritized grains (up to 70%). Oncoids, up to 3 cm long, like those described by Jones and Goodbody (1985), are found locally.

The *Rubble and Knob* Facies, which is a shoreward sloping mass of coral rubble forms an 80 to 120 m wide band on the landward side of the reef crest (Fig. 8.24). Waves breaking on the reef crest are transformed into shoreward moving currents with the strongest being just landward of the reef crest. The area is inhabited by various algae, encrusting *Porolithon*, and some green algae. Near the reef crest, scattered *Acropora palmata* are present. The coral diversity decreases landward from a maximum of ~25 species. Gorgonians, sponges, bryozoans, and hydrocorals live among the corals. The rubble is extensively eroded by gastropods, fish, and echinoids. The sediment is typically bimodal and poorly sorted.

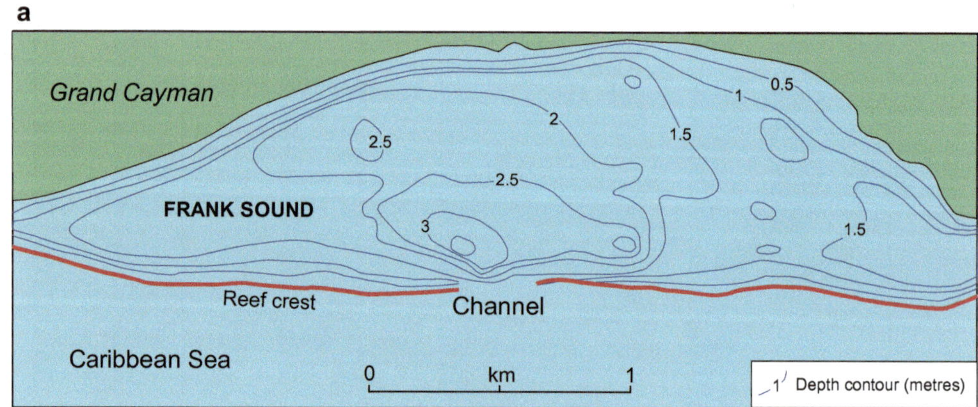

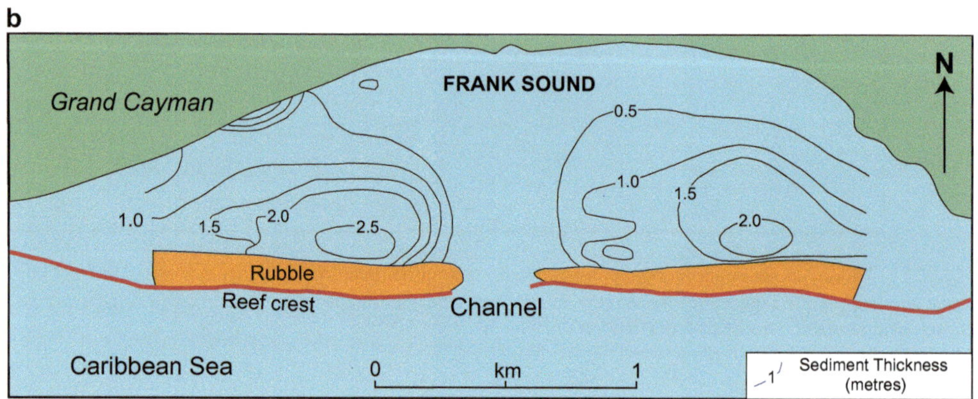

Fig. 8.22 Maps of Frank Sound (Fig. 8.1) showing **a** water depths and **b** sediment thicknesses. From Kalbfleisch and Jones (1998, their Figs. 4a and 5b: reproduced and modified with permission from the Coastal Education and Research Foundation Inc.)

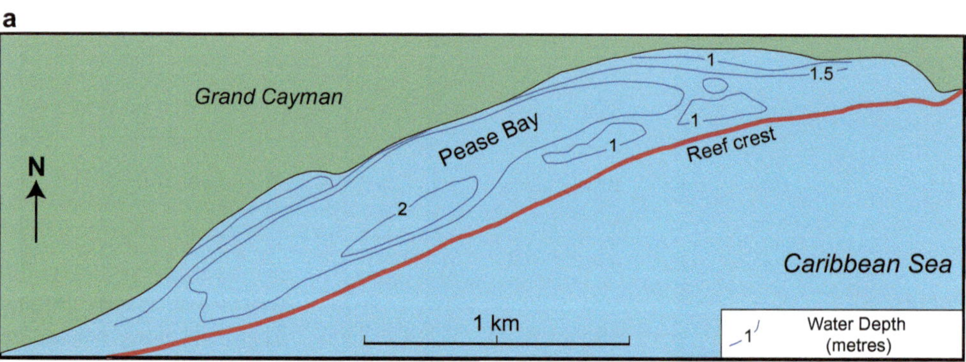

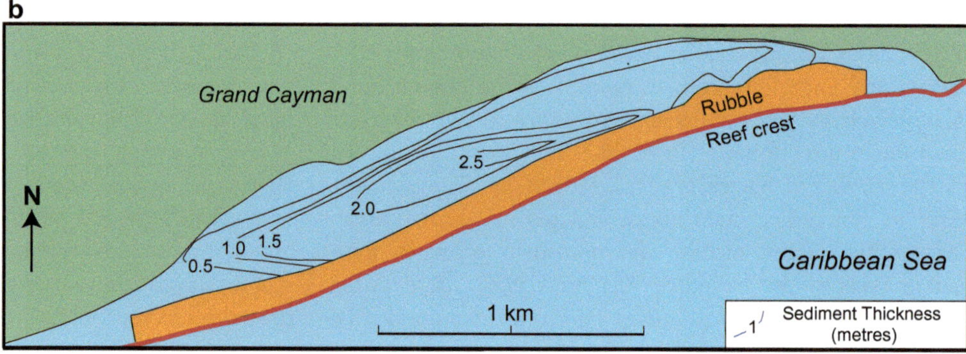

Fig. 8.23 Maps of Pease Bay (Fig. 8.1) showing **a** water depths and **b** sediment thicknesses. From Kalbfleisch and Jones (1998, their Figs. 4B and 5D: reproduced and modified with permission from the Coastal Education and Research Foundation Inc.)

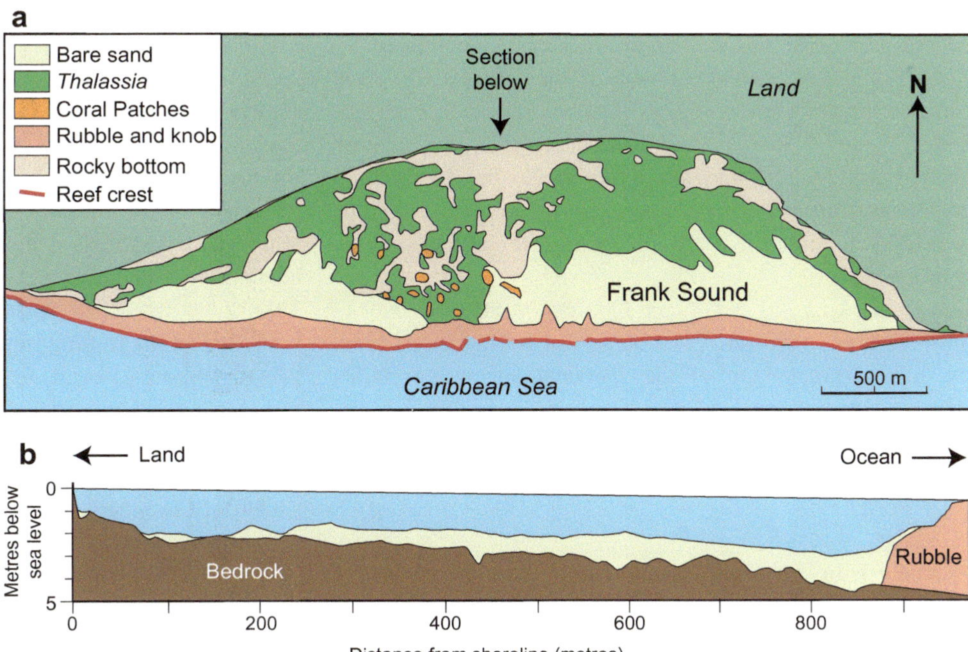

Fig. 8.24 **a** Facies map for Frank Sound. Arrow indicates location of cross-section shown in panel B. Modified from Li et al. (1998, their Fig. 1d). **b** North–South cross-section across Frank Sound showing relationship of sediment thickness to underlying bedrock. Cross-section from Kalbfleisch and Jones (1998, their Fig. 7b: reproduced and modified with permission from the Coastal Education and Research Foundation Inc.)

The *Bare Rock* Facies, found nearshore on the lagoon flanks and near the channel, experience strong currents during storms that quickly move any sediment that may have settled there. In some areas, shallow sediment pockets may support poorly rooted *Thalassia* and green algae. Various corals are present locally.

The *Bare Sand* Facies, is a 200–300 m wide belt on the landward side of the rubble zone, where the water is 1.5 to 2.0 m deep (Fig. 8.24). Asymmetrical ripples are evident locally. Only scattered *Thalasssia* and green algae live in these areas. *Callinassa* mounds are found near the boundary with the *Thalassia* facies. Circular depressions in these sands were made by stingrays (*Dasatis americana*). These fine- to medium-grained sands are formed of skeletal grains derived from corals, foraminifera, and molluscs that have been micritized to varying degrees. A core through these sediments (134 cm long) consists of clean pink sand with coral fragments, foraminifera (including *Homotrema rubrum*) mollusc, and *Halimeda*. Layers at 40 cm and 50–60 cm below the surface are formed of pebbles.

Patch reefs, up to 30 m in diameter and 3 m high are found throughout the lagoon with most rooted on rocky substrates. They are most common around the channel where there is constant current activity. The largest reefs are dominated by *Montastrea annularis* and *Siderastrea siderea* along with fewer *M. cavernosa, Diploria strigosa*, *Porites porites, P. asteroides, Agaricia agaricites*, and a few other species. Coral diversity is highest in the offshore patch reefs.

The narrow, steep beaches around Frank Sound and Pease Bay stretch ~5 m on each side of mean sea-level. Bedrock exposures are locally common. The beach sands are medium to coarse grained and coral rubble can be found where the beach is close to the reef. These beaches are inhabited by various organisms, including crabs. Rocky exposures are bored and encrusted.

Facies Changes Through Time

Based on image analyses of various sets of air photographs, Kalbfleisch and Jones (1998) showed that *Thalassia* coverage between 1985 and 1992 expanded by 5% of the lagoon area—an assessment that agreed with the conclusions obtained by Tongpenyai (1989) and Tongpenyai and Jones (1991). This increase in *Thalassia* coverage led to a decrease in the area of the Bare Sand Facies.

Foraminifera in Frank Sound

Li (1997) and Li et al. (1998) analyzed the foraminifera populations in surface and subsurface samples from Frank Sound with the view of finding "tracer species" that could provide information regarding sediment transportation. The foraminifera in the surface samples including living and dead specimens whereas the subsurface samples contained only dead specimens. Using cluster analysis, Li et al. (1998) delineated Group I that was dominated by lagoonal species and Group II that included numerous forereef species. Group I was divided into three assemblages.

Assemblage I-1: *Archaias angulatus–Cymbaloporetta squamosa–Asterigerina carinata* Assemblage, with a high species diversity and numerous epiphytic forms, found in the *Thalassia* and Sand Facies and inner part of the Bare Sand Facies near the channel.

Assemblage I-2: *Archaias angulatus–Valvulina oviedoiana–Discorbis mira* Assemblage that is dominated by abrasion-resistant species, found in the Bare Rock Zone north of the channel.

Assemblage I-3: *Archaias angulatus–Quinqueloculina agglutinans–Discorbis rosea* Assemblage that has a high percentage (up to 77%) of the lagoonal species *Archaias angulatus*, is found on the Bare Rock Facies and around the coral knobs.

Group II: *Archaias angulatus–Cymbaloporetta squamosa–Amsphistegina gibbosa* Assemblage that is characterized by the lowest percentage of the lagoonal species *Archaias angulatus* but numerous forereef species. It is found in the Rubble and Knob Facies and the seaward part of the Bare Sand Zone and locally in the inner part of the *Thalassia* and Sand Facies.

Analysis of subsurface samples from a transect across the eastern part of Frank Sound (line X–Y, Fig. 8.25a) found Assemblage I-1 near the surface in the *Thalassia* and Sand Facies but extended landward and seaward in the subsurface (Fig. 8.25b). Assemblage I-2 found near the surface in core CB is replaced landward by Group II and seaward by Assemblage I-1 (Fig. 8.25b). Group II is found in the bottom and top parts of core CF and the upper part of Core CA.

The notion of tracer species is based on a particular species being common in the samples, and that the preferred habitat of that species is known. Among the 117 species of foraminifera found in samples from Frank Sound (Li 1997), only the lagoonal *Archaias angulata* and the forereef *Amphistegina gibbosa, Asterigerina carinata, and Discorbis rosea* met the criteria of "tracer species" (Li et al. 1998). The forereef species varies from 1–5% in the *Thalassia* Sand Facies to $\geq 10\%$ in the Bare Sand Facies. Landward from the reef crest land, the number of forereef species generally decreases. Some samples from nearshore settings contained abnormally high numbers of forereef species such that analyses showed that they belong to Group II (Li et al. 1998).

Based on the foraminifera assemblages and tracer species, Li (1997) and Li et al. (1998) showed that (1) the lagoon sediments are a mixture of sediments that originated in the forereef and in the lagoon, (2) in some areas of the lagoon there are elevated numbers of forereef foraminifera, (3) the subsurface sediments record an ongoing history of storm and inter-storm deposition, and (4) bioturbation mixes the lagoonal sediments during the inter-storm periods.

8.4.4 East Sound

Situated on windward side of Grand Cayman, East Sound is a large lagoon with a bounding reef that embraces the entire east coast (Fig. 8.26). East Sound, which is 200 m to 1.6 km wide, can be divided into the inner and outer zones (Fig. 7.26). The inner zone has a thin sediment layer (<2 m) that is generally covered by *Thalassia*. In contrast, the outer zone is a sand apron that passes seaward into the rubble zone

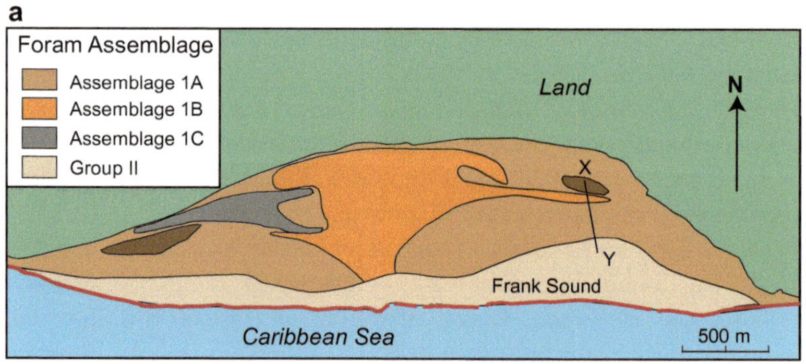

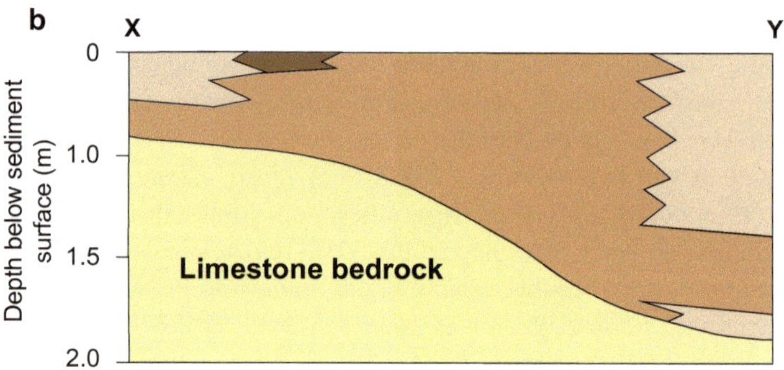

Fig. 8.25 Distribution of foraminifera assemblages 1A, 1B, 1C, and Group II (see text for definition and description of each assemblage) in **a** surface sediment samples, and **b** in subsurface sediment samples from transect X to Y (see panel A for location) in Frank Sound. Modified from Li et al. (1998, their Fig. 4)

that lies landward of the reef that forms the lagoon boundary. On the open shelf, seaward of the reef, little sediment accumulates on the upper terrace because of the high energy conditions associated with the onshore winds. In contrast, the sediment on the lower terrace is up to 10 m thick (Blanchon 1995).

The rubble zone behind the reef is formed of large pieces of rounded corals that were probably deposited during hurricanes and severe storms that impacted the reef (Fig. 8.27a, b). These storms were also probably responsible for the large overturned *Acropora palmata* that are found in some areas (Fig. 8.27c). The sandy substrates in some of the deeper, more protected parts of East Sound are characterized by numerous mounds formed by the burrowing shrimp *Callianassa* (Fig. 8.27d) or thickets of calcareous algae, including *Halimeda* (Fig. 8.27e).

Using the notion of foraminifera tracer species, Li et al. (1997) first used 50,000 foraminifera belonging to 150 species from 5 transects across East Sound to determine their distribution. Based on that, they identified *Archaias angulatus, Amphistegina gibbosa, Asterigenia carinata*, and *Discorbis rosea* as the best "tracer species". The living specimens of these species have distinct ecological preferences with (1) *A. angulatus* being most common in the *Thalassia* banks but significantly reduced in numbers on the sands of the inner and outer lagoon and rare in the forereef, (2) *A. gibbosa* and *A. carinata* being most common on the rubble in the forereef area, and (3) *D. rosea* being most common on the upper terrace.

In general, the distribution of *A. angulatus* is inversely proportional to that of the forereef species (Fig. 8.28). Forereef species are found landward of the reef crest such that the $\leq 5\%$ contour of the forereef species coincides with the $\geq 40\%$ contour of the lagoonal species, with that boundary also being the boundary that separates the outer lagoon from the inner lagoon (Fig. 8.28).

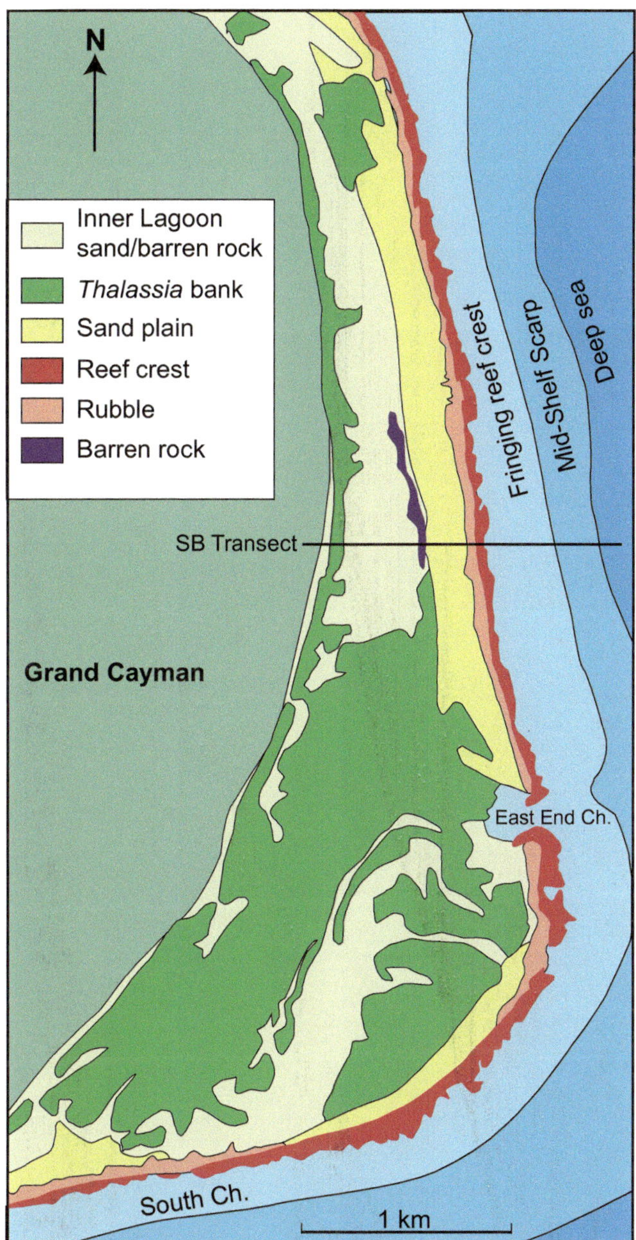

Fig. 8.26 Facies map of East Sound. Modified from Li et al. (1997, their Fig. 1d). Note position of the Sand Bluff Traverse where the foraminifera samples were collected

8.5 Bioerosion

Bioerosion, which plays an important role in the production of sediment in carbonate depositional systems like those around the Cayman Islands, involves a variety of organisms including parrotfish (Bruggermann et al. 1996), urchins (Bak 1994), boring sponges (Acker and Risk 1985; Rose and Risk 1985) and various macro- and micro-endolithic taxa (Hutchings 2011). Boring sponges are particularly important because their infestation of corals can lead to the production of significant amounts of sediment and weaken their attachment sites to the extent that the corals become detached from the substrate during severe storms and hurricanes. Ghiold et al. (1994) listed 23 common species of sponges that are important components of the biota found in the waters around the Cayman Islands. Among these species are boring sponges, including *Cliona*, that generate large amounts of sediment through their boring activities.

Boring sponges are important because they bore into any substrates, including living corals and exposed rock surfaces. Acker and Risk (1985), who focused on the boring sponge *Cliona caribbaea* found offshore from the southeast corner of Grand Cayman, showed that the sponges excavated corals and hard rocky substrates to a depth of ~1 cm and

Fig. 8.27 East Sound. **a** General view of rubble zone exposed during low sea level, area to south of East Channel (Fig. 8.26). **b** Underwater view of rubble zone on landward side of reef. **c** Overturned *Acropora palmata* colony, close to rubble zone. **d** Seafloor in East Sound, water ~6 m deep, with numerous coalescing mounds created by the burrowing shrimp *Callianassa*. Hammer handle is 0.3 m long. **e** Sandy substrate with numerous *Halimeda*

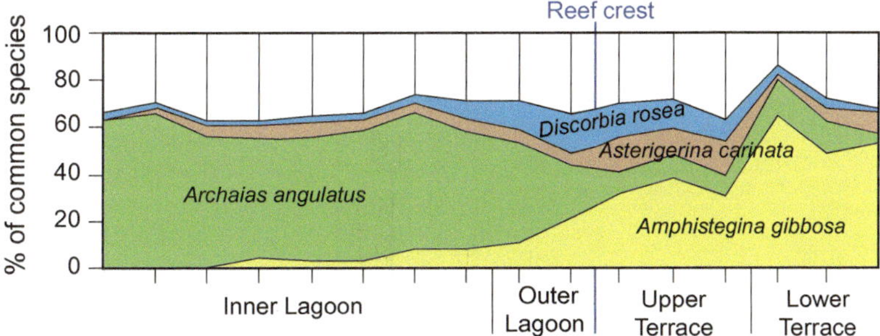

Fig. 8.28 Cross-section across East Sound (Sand Bluff Traverse—Fig. 7.26) showing distribution of tracer foraminifera. Based on samples collected from soft-sediment cores (vertical black lines). Modified from Li et al. (1997, their Fig. 3)

produced up to 8 $kg/m^2/yr^{-1}$ of silt-sized chips that was an important contribution to the surrounding sediments. The scalloped surfaces of these borings (Fig. 8.29a, b) reflects the fact that the sponges dissolve the aragonite or calcite along the edges of their tissues that they insert into the substrate (Fig. 8.29b). Once produced, the chips are then removed and eventually expelled into the surrounding ocean. The sponge borings are easily recognized by the scalloped surfaces and the silt-sized multifaceted chips that reflect the boring mechanism (Fig. 8.29c, d). It is important to note that sponges, like *Cliona*, will bore into any hard, calcareous substrate that is available. Jones and Goodbody (1984), for example, showed that beachrock from Bodden Town on the south coast of Grand Cayman had been extensively altered by borings that had been produced by *Cliona*.

Rose and Risk (1985) demonstrated that in the area around the Turtle Farm on the northeast corner of Grand Cayman, there was high sponge biomass that they attributed to the increased amount of untreated fecal sewerage found in that area. They argued that that was responsible for the high numbers of *Cliona delitrix* that infested the *Montastrea cavernosa* found there and, hence, the high amounts of silt-sized sediment that was produced. They argued that the sponge borings had removed 53% of the *M. cavernosa* and 47% of the *Porites astreoides* skeletons. Murphy et al. (2016), who focused attention on the boring activity of *Cliona tenuis* and *Siphonodictyon brevitubulatum* found on the west and south coasts of Grand Cayman, showed that the mean bioerosion rates ranged from 0.036 to 0.172 kg $CaCO_3$ m^{-2} yr^{-1}, and noted that these were similar to the rates reported from other parts of the world. Although estimates of the amount of sediment generated by sponge borings varies from study to study, it is readily apparent that the sponges do a vast amount of damage to hard substrates with that damage being further accentuated as storms and hurricanes that uproot the corals that have had their basal attachment areas weakened by the borings.

8.6 Temporal Changes in Surface Seawater Temperatures

Corals grow as the living coral mediates aragonite precipitation beneath its outer layer of living tissue. The yearly product is a laminae couplet that comprises a light and dark band couplet on X-ray or computerized tomography images (Fig. 8.30). As the aragonite is being precipitated it also incorporates various trace elements (e.g., Sr, Mg) and the ^{18}O and ^{13}C stable isotopes with the quantities/magnitudes of these elements being largely controlled by water temperature. These geochemical proxies can therefore be used to determine the water temperature at the time of coral growth. Although full details of the complex array of analytical techniques used to determine the paleotemperatures are given in Booker et al. (2019), they basically involved (1) mapping of the growth bands using high density computerized tomography (CT) scans produced on an Aquilion ONE helical CT scanner (InnoTech Alberta, Edmonton), and/or X-rays generated on a portable SY-21-100P X-ray machine (Fig. 8.30), (2) application of ImageJ to these scans to produce a map of the grey level values which were then analyzed by ERDAS Imagine to produce a greyscale curve that highlights the growth lines, (3) C^{14} and U/Th dating to determine the age of the oldest part of each coral, (4) determination of the Mg, Ca, and Sr concentrations in each growth layer, and (5) determination of $\delta^{18}O$ and $\delta^{13}C$ from those layers that contained >98% aragonite. Data from any coral band that showed evidence of diagenetic alteration were not used in the subsequent calculations of the paleotemperatures (Booker et al. 2019, 2020).

Water samples from Magic Reef yielded $\delta^{18}O$ values of +0.75 ± 0.18‰ for surface water and +0.79 ± 0.06‰ for water at a depth of 18.3 m. Eighty samples collected from corals from that reef yielded concentrations of 277,656 to 415,544 ppm Ca, 5672 to 7878 ppm Sr, and 1035 to

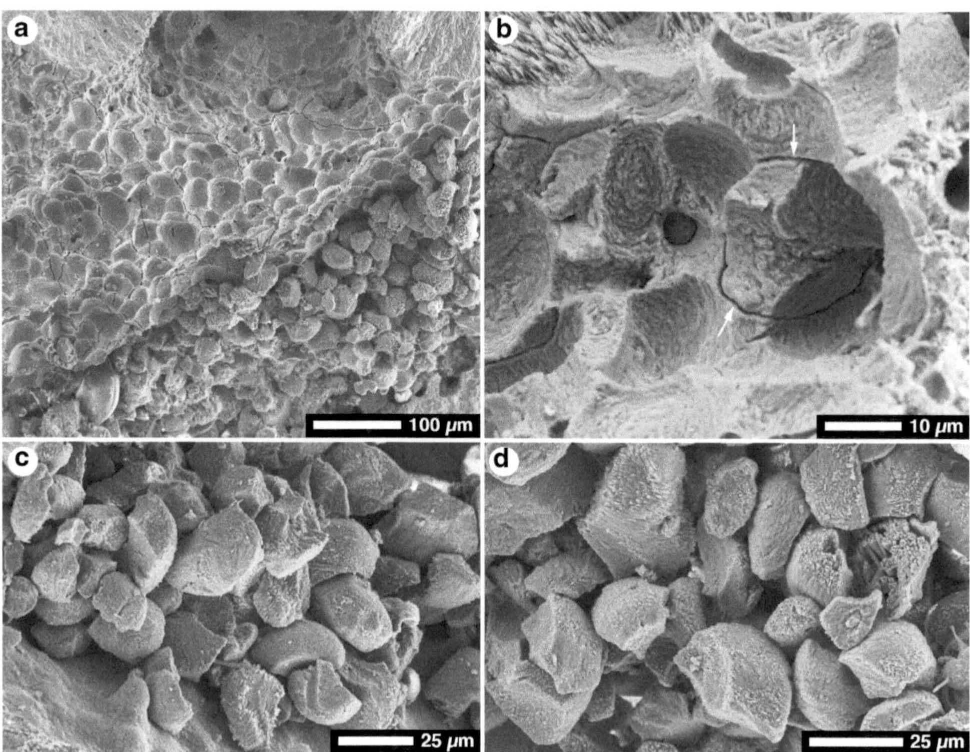

Fig. 8.29 SEM photomicrographs of sponge borings and chips, probably produced by *Cliona*, in modern *Orbicella annularis* from Magic Reef, which is located off the southwest corner of Grand Cayman. **a** Scalloped surface of boring (upper left corner) with chips produced by the boring sponge (lower right corner). **b** Enlarged view of scalloped surface of boring with one chip that is still in place being outlined by narrow gap (arrows) that was produced by the sponge as it excavated the substrate. **c**, **d** Examples of chips produced by boring sponge—note curved facets of the grains. Images courtesy of Dr. Simone Booker

2504 ppm Mg with Sr/Ca ratios of 8.3 to 9.6 mmol/mol. Although numerous equations have been developed for calculating paleotemperatures from the geochemical proxies in corals, Booker et al. (2019) used the equation developed by Saenger et al. (2008) because it had been developed from *Montastrea* that grew in conditions similar to those found around the Cayman Islands. The temperatures of 23.3 to 32.3 °C (average 28.4 ± 0.1 °C) derived from that equation are in accord with the seawater temperatures measured around Magic Reef.

By integrating data from various corals from Grand Cayman, Booker et al. (2019) produced a summary of the changes in the subsurface seawater temperatures that have occurred in the Caribbean Sea around the Cayman islands between 1815 and today (Fig. 8.31). Although these data apply to corals that grew between ~1474 and 2014 CE, it should be noted that no data are available for the period between ~1512 and ~1814 because of the lack of corals from that time period, possibly because of adverse growth conditions. Booker et al. (2019) recognized the following fluctuations in the calculated seawater temperatures (T_{cal}) around the Cayman Islands.

- *Cool Period 1 (~1474 to 1512CE)*: T_{cal} of 21 to 29°C, with average of 25.1°C.
- *Cool Period 2 (~1815 to 1924 CE)*: T_{cal} of 18 to 31°C, with average of 23.9°C; including two cool intervals that were separated by a warm period.
- *Warm Period 1 (~1924 to 2006CE)*: T_{cal} of 23 to 34°C, with average of 28.3°C; including two warm intervals and two cool periods.
- *Mild Period (~2006 to 2014)*: T_{cal} of 25 to 33°C, with average of 27.5°C.

Coral bleaching, which took place around Grand Cayman in 1982, 187–1988, 1990, 1995, 1998, and 2000 (Bruckner 2010) occurs when high seawater temperatures cause the corals to expel their resident zooxanthellae that leads to reduced vertical growth of the coral and in some cases ^{18}O enrichment. These issues influence the growth banding and calculated water temperatures.

The warm, cool, and mild temperature periods delineated from the temperatures calculated from the $\delta^{18}O$ in the corals are consistent with similar temperature regimes recognized

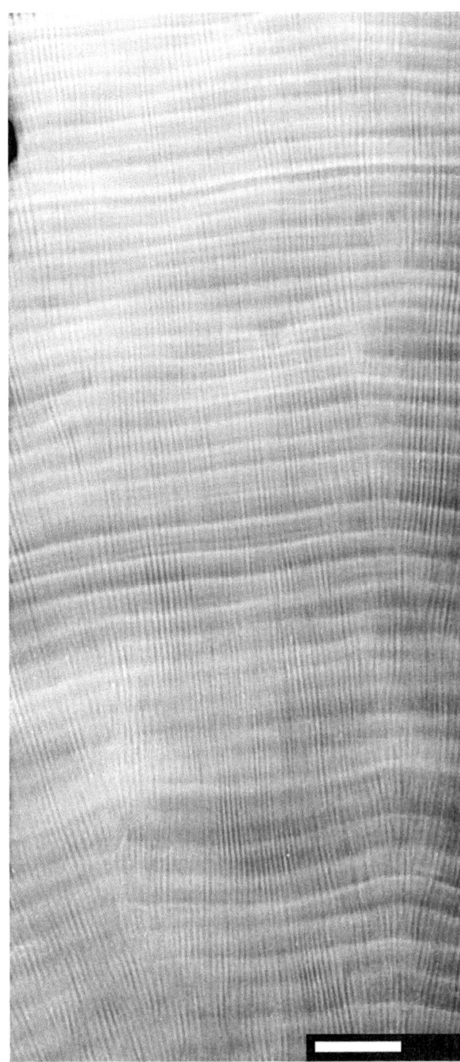

Fig. 8.30 X-ray scan of slab cut from *Orbicella annularis* from Grey's Wall, Grand Cayman highlighting growth band with each dark band and light band couplet representing one year of coral growth. Image courtesy of Dr. Simone Booker

from other Caribbean locations (see Booker et al. 2019, their Fig. 15). Minor variations between these different areas probably reflect local environmental conditions that may have impacted coral growth and the $\delta^{18}O$ signals in the corals on which the temperature calculations are based.

8.7 The Critical Role of *Thalassia* in the Lagoons of the Cayman Islands

Thalassia testudinium, or "turtle grass" as it is known locally, plays numerous critical roles in the lagoons around each of the Cayman Islands. Critically, *Thalassia* baffles currents and thereby induces sediment deposition, binds sediment on the seafloor (Fig. 8.32a, b), and its leaves provide ideal habitats for numerous epiphytic organisms (Corlett 2005; Corlett and Jones 2007). In the lagoons around Grand Cayman, *Thalassia* covers ~60% of the seafloor in North Sound (MacKinnon and Jones 2001), 51% in South Sound (Beanish et al. 2002), 19.8% in Pease Bay and 31.8% in Frank Sound (Kalbfleisch and Jones 1998), and 38.3 and 14.9% in the south and north parts of East Sound, respectively (Tongpenyai and Jones 1991). In those settings, *Thalassia* forms banks along the landward margins of the lagoons, except for North Sound where it is found in the interior of the lagoon.

Thalassia typically has 3 to 5 leaves that are up to 1 m long (typically <0.5 m) and 4–12 mm wide with new shoots being generated from the rhizomes that form dense, tangled mats 5 to 10 cm below the sediment surface (Fig. 8.32b). It commonly forms dense colonies such that the sandy substrate is barely visible (Fig. 8.32c–e). The leaves baffle local currents and thereby induce sedimentation, and the roots stabilize the seafloor and prevent erosion even under strong currents. Oncoids (Fig. 8.32d) are commonly associated with the *Thalassia* banks (Jones and Goodbody 1985). Other organisms commonly grow among the *Thalassia* and/or feed on the epiphytes that live on the leaves. Echinoderms ("sea urchins") are common among the *Thalassia* (Fig. 8.32f) because the leaves provide protection and the organisms on the leaves are a potential food source.

The leaves of *Thalassia* are commonly white because of the diverse, dense array of epiphytic organisms that encrust their surfaces (Fig. 8.32c, e). Based on samples of *Thalassia* leaves and sediment from around the base of the plant collected from South Sound, Pease Bay, Frank Sound, East Sound, and North Sound, Corlett (2005) and Corlett and Jones (2007) showed that the epiphytic community included three species of coralline algae, 72 species of foraminifera, and 61 species of diatoms along with lesser numbers of sponges, gastropods, ostracods, and coccoliths with ~85% of these organisms having calcareous or siliceous skeletons. Non-skeletal organisms include dinoflagellates, brown algae, and worms. These organisms collectively form a tripartite community succession that comprises (1) a basal epiphytic layer, formed of the diatom *Cocconeis*, that coats the leaf surface, (2) a middle layer formed largely of the red algae *Hydrolithon farinosum*, except in North Sound where the leaves are coated with brown algae rather than red algae, and (3) the upper epiphytic layer that is formed of a diverse array of foraminifera, gastropods, bivalves, coccoliths, diatoms, and calcareous and siliceous sponge spicules.

Although the epiphytic communities found on the *Thalassia* leaves include at least 136 species, only 20 of those species were found in the sediment around the plants. Among the 136 species of foraminifera found in the lagoon sediments (Li 1997; Li and Jones 1997) and 72 species on

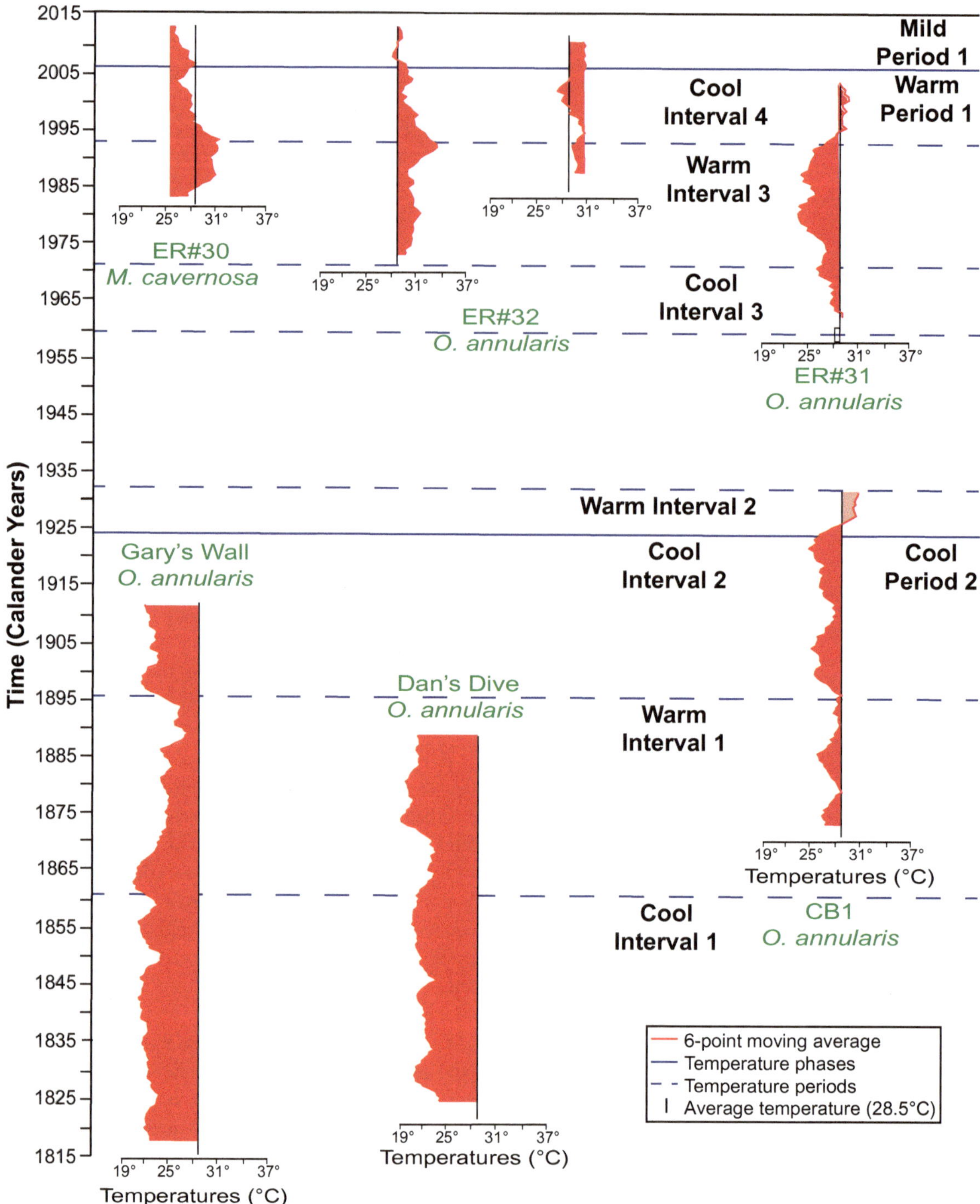

Fig. 8.31 Temperature trends based on temperatures calculated from $\delta^{16}O$ values determined from various species of corals from different localities around Grand Cayman. Modified from Booker et al. (2019, their Fig. 14)

Fig. 8.32 *Thalassia* in lagoons around Grand Cayman. **a**, **b** Steep cut on landward edge of *Thalassia* bank that was probably caused by erosion associated with a storm. Prospect Point Lagoon. **b** Close up view of *Thalassia* bank shown in panel A, showing dense array of intertwined roots. **c**–**d** Examples of dense arrays of *Thalassia* leaves, many white in colour because of dense coatings formed of various epibionts. Note oncoid between leaves in panel D. Prospect Point Lagoon. **e** Echinoid among *Thalassia* leaves. North Sound

the *Thalassia* leaves, only 9 species are common to both. Corlett and Jones (2007) suggested that this apparent discrepancy may be an artifact created by the different methods that were used in each study because Li (1997) only examined those foraminifera that were larger than 0.125 mm. Although careful examination of some of the fine-grained sediment from some of the lagoons did reveal another 14 species of foraminifera that were also present on the *Thalassia* leaves, none of them formed a significant part of the sediment.

One of the factors typically associated with *Thalassia* is the generation and trapping of sediment in the lagoons. In the Cayman lagoons the lack of finer sediment can be probably be attributed to the storms/hurricanes that

Fig. 8.33 South coast of Cayman Brac showing ridge formed largely of large coral heads that probably formed during Hurricane Cuba in 1932. **a** General view of boulder ridge formed of large coral heads. Ridge in foreground is ~2 m high. **b** Close-up of large, rounded coral heads that form the boulder ridge shown in panel A

periodically impact the lagoons and are known to remove and wash out to sea much of the finer sediment (Li et al. 1997, 1998; Kalbfleisch and Jones 1998; Beanish and Jones 2002).

8.8 Impact of Storms on the Coastal Areas of the Cayman Islands

On a typical day, the lagoons around the Cayman Islands are idyllic settings with clear, warm, tropical waters with minimum tidal activity and minimal wave activity. In these serene settings, the resident biota thrives, and sediment is produced and accumulates. Periodically, these peaceful settings are transformed into short-lived periods of terror by severe storms or hurricanes that generate high winds and huge onshore waves. Over the last 270 years, the Cayman Islands have been affected by at least 39 hurricanes (Blanchon and Jones 1997). The tranquility of the lagoons is quickly lost as the high winds, rain, and massive sea waves start coming ashore. Historical records, for example, noted Hurricane Cuba (category 5), in 1932, with wind speeds up to 320 km/hour, 16 m waves, and storm surge of up to 10 m, and eyewitness accounts described large boulders and ships being thrown onshore (Sauer 1982; Fenner 1993). A ridge of large coral heads (up to 2 m long) that formed on the south coast of the island (Fig. 8.33) during that hurricane was documented and photographed by The Oxford Expedition in 1934.

Hurricane Ivan in 2010 (September 7 and 8), with winds of 240 km/hour (gusts to 355 km/hour) affected each of the Cayman Islands with its eye passing only 33 km SW of Grand Cayman. The storm surge in North Sound was 3 m high. That hurricane clearly demonstrated the power and sedimentological importance of hurricanes by transporting vast quantities of sediment out of the lagoons and depositing onshore and burying many of the coastal roads (Fig. 8.34).

Around the Cayman Islands, the effects of a storm on the lagoonal sediments are apparent from (1) storm boulders, (2) boulder ramparts, and (3) large scale movement of lagoonal sediments.

8.8.1 Storm Boulders

Along the south coast of Grand Cayman, fringing reefs are absent from some stretches of the coastline, including the areas offshore from Great Pedro Point and Blowholes (Fig. 8.35). These sections of the coastline face the open ocean and commonly receive the full power of the storms and hurricanes that come in from the southeast quadrant. It is not a coincidence that these two stretches of rocky coastline are characterized by large boulders that are located up to 100 m inland from the coastline (Figs. 8.35, 8.36).

The coastal terrace at the south end of High Rock Drive (Fig. 8.35), bounded at its seaward margin by vertical cliffs that rise ~4 m above sea level, is formed of massive dolostones of the Cayman Formation. The terrace between the shoreline and the coastal road, 100 to 120 m wide, is divided into zones A and B that are separated by a break in slope that is ~6 m above sea level (Fig. 8.35) that may have formed with the +6 m highstand during the Pleistocene, ~125,000 years ago. Zone A, 30 to 45 m wide, is characterized by phytokarst with up to 1 m relief, whereas Zone B, located between Zone A and the road, is characterized by a more subdued phytokarst (Fig. 8.35). The

Fig. 8.34 Photograph of coastal area on east end of Grand Cayman after Hurricane Ivan (September 9 and 10, 2010). Note damage to trees, and road that is completely covered with sand that was transported onshore from East Sound and the narrow beach. Sand on road was up to 0.5 m thick. Photograph courtesy of the Water Authority, Cayman Islands

exposed dolostones are cut by numerous joints (Rigby and Roberts 1976), many of which are filled with flowstone, breccia, and/or lithified terra rossa.

Strewn across this area are 51 boulders (Jones and Hunter 1992), formed of dolostone from the Cayman Formation with 50 being located in Zone B (Fig. 8.36). Most have a flat base, flat sides, and phytokarst on their upper surfaces. Although the largest boulder is 5.5 m long, 2.8 m wide, and 1.5 m high (Fig. 8.36b, c), most are less than half that size (Fig. 8.36d). Some have a flat basal surface that then curves upwards in a shape reminiscent of a wave-cut notch (Fig. 8.36c). Today, such boulders are always oriented so that the notch points to the east. Twenty-six boulders have their long axis-oriented E-W, 18 are oriented N-S, and the rest display no preferred orientation because they are equidimensional. The boulders commonly occur in small clusters with many resting against each other (Fig. 8.36).

Some boulders have been bored by sponges and/or *Lithophaga* and/or encrusted by *Homotrema* that are clearly not part of the original dolostone. One block has numerous, very well preserved *Astrangia solitaria* growing on its surface (Fig. 8.36d, e). Originally, Jones and Hunter (1992) used uncalibrated ^{14}C dating of these corals to suggest that they grew 660 ± 50 years BP with an age of 1662 AD when calibrated against the Dendro calibration curve. Based on that age, they suggested that the event that transported the boulders onshore occurred ~330 years ago. A reassessment of that age, using Calib version 8.2 with the calibration curve MARINE20 and a marine reservoir correction of –28 per mil, produced an age of 1803, which is substantially younger than the date first proposed by Jones and Hunter (1992). It is possible that one of the major hurricanes that affected the Cayman Islands in 1793, 1812, and 1837 and 1838 may have been responsible for the onshore movement of these boulders.

In the Great Pedro Point area, vertical sea cliffs, carved in dolostones of the Cayman Formation, are ~10 m high with terraces that rise to ~15 m asl over a distance of 70 m. The box-like boulders on those terraces, which are commonly concentrated in clusters, like those in the Blowholes area, have smooth sides with phytokarst on their upper surface. If present, the wave-cut notch on these boulders always points to the east, like those at the Blowholes. The exact age when these boulders formed cannot be determined.

The coastal areas near Blowholes and at Great Pedro Point are prone to high waves because there are no offshore reefs to dampen the onshore waves (Fig. 8.35). During hurricanes, these areas receive the full force of the massive storm waves. The encrusting corals and boring organisms on some boulders show that they had, at one time, been submerged by ocean waters in an off-shore locality. Dating suggests that these boulders were carried onshore ~220 years ago. It appears, therefore that some of the large boulders (up to 2.0 × 2.0 × 1.3 m with an estimated weight of 2 tons) were moved 15 to 18 m vertically and 50 to 60 m horizontally. Following Hurricane Ivan in September 2004, it became evident that some of the boulders in the Blowholes area had been moved. The boulder with the corals encrusting its upper surface (Fig. 8.36d, e), for example, had been overturned so that the corals are now on the underside. Other boulders had been moved laterally by at least 5–10 m and in some cases, gouges on the bedrock provided evidence of that movement. There is no evidence that any of these boulders had been moved by human activity.

Jones and Hunter (1992) suggested that the boulders on the south coast of Grand Cayman were moved onshore by hurricane waves or a tsunami. Stoddart (1980, his Plate 21) reported the presence of large storm boulders (1.2 × 1.5 11.5 m and 1.8 × 1.5 × 11.5 m) on the south coast of Little Cayman but provided no detailed information on them.

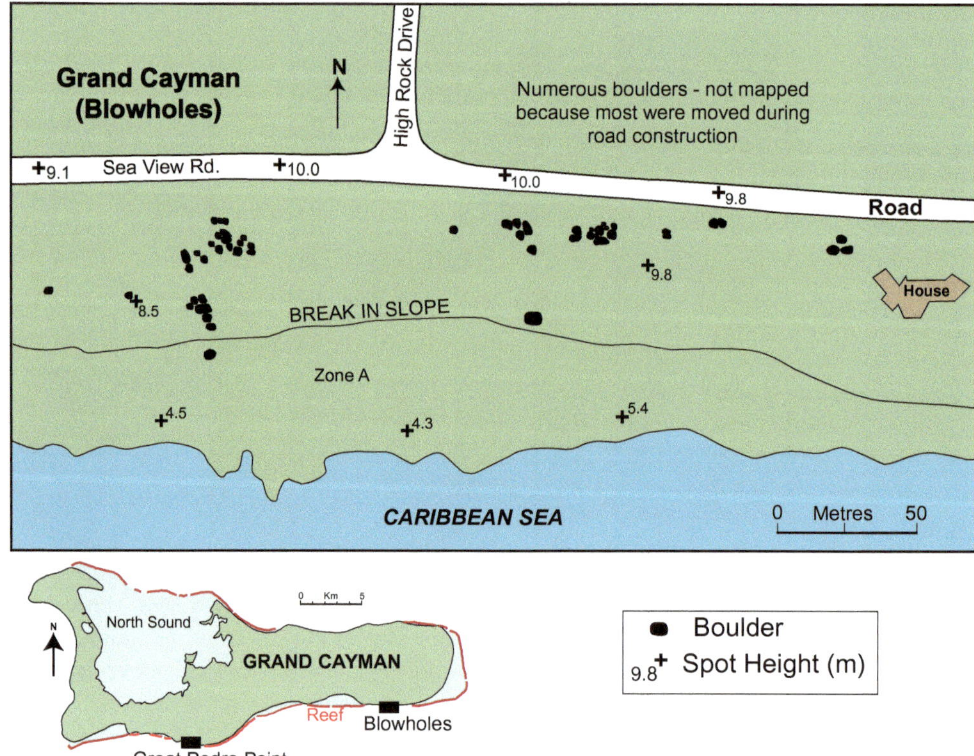

Fig. 8.35 Map of south coast, east Grand Cayman south of High Rock Drive, showing distribution of boulders with most being in Zone B that is landward of Zone A. Zones A and B are separated by a break in slope (see Fig. 7.34). From Jones and Hunter (1992, their Fig. 2: reproduced and modified with permission from the Coastal Education and Research Foundation Inc). Inset map of Grand Cayman (lower left corner) shows location of Blowholes and Great Pedro Point where large boulders are found on the rocky coastlines

8.8.2 Boulder Ramparts

Doran (1954) described coarse hurricane-built boulder ramparts that are found on the east and southeastern coasts of Grand Cayman. Later, Rigby and Roberts (1976, their Fig. 13) documented the ramparts found around Breakers, Cottage Point, and Old Isaacs on the southeast coast, Rogers Wreck Point, Great Bluff, and Little Bluff on the northeast coast, and Boddentown and Great Pedro Point on the south central coast. The coral heads and associated fragments, up to 1 m in diameter, are typically *Siderastrea, Montastrea*, and *Diploria* along with some *Dichocoenia, Isophyllastera, and Meandrina*. *Acropora* is rare in these deposits. In many cases, the corals are imbricated. Given that many of these corals show little evidence of damage, Rigby and Roberts (1976) suggested that the transportation time from their growth habitat to the onshore ridge was minimal. They argued that most of the coral heads that were washed onshore during the 1931 and 1932 hurricanes originated from the shallow shelf terrace in water that is 7 to 15 m deep. Today, it is difficult to find evidence of these boulder ramparts on Grand Cayman because they were extensively quarried with the coral heads being used for road material and building. There are numerous garden walls on Grand Cayman that have been built using the coral heads from these storm deposits.

8.8.3 Storm Impact on Lagoonal Sediments

The fate of lagoonal sediments during major storms is largely a function of the direction of the onshore waves and wind relative to the geometry of the lagoon. Specific examples that illustrate these points include South Sound, Frank Sound, and East Sound. South Sound is perhaps unique because it has a natural opening on its western leeward margin (Fig. 8.21). Wind driven waves from the east break over the reef at a slight angle and generate currents that then flow down the length of the lagoon (Rigby and Roberts 1976; Roberts and Suhayda 1983). With the westward narrowing of the lagoon, the currents accelerate and hence carry more and more sediment through the natural opening at the west end where the sediment is then deposited on the shelf off the southwest corner of Grand Cayman (Fig. 8.21).

For Frank Sound and East Sound there are no natural outlets other than the channels through the reef crests

Fig. 8.36 Coastal area west of the Blowholes area on the south coast of Grand Cayman (see Fig. 8.34), where there is no off-shore reef. **a** General view of area showing exposed dolostones of the Cayman Formation. The break in slope is in the middle of the photograph, just seaward of the left-hand side of the house in the background. **b**, **c** Largest boulder in the area, with "wave-cut notch" shape on right side (east). **d** Boulder formed of dolostones of the Cayman Formation with encrusting corals. **e** Close up of encrusting corals on boulder shown in panel D. These corals were used to obtain the C^{14} date

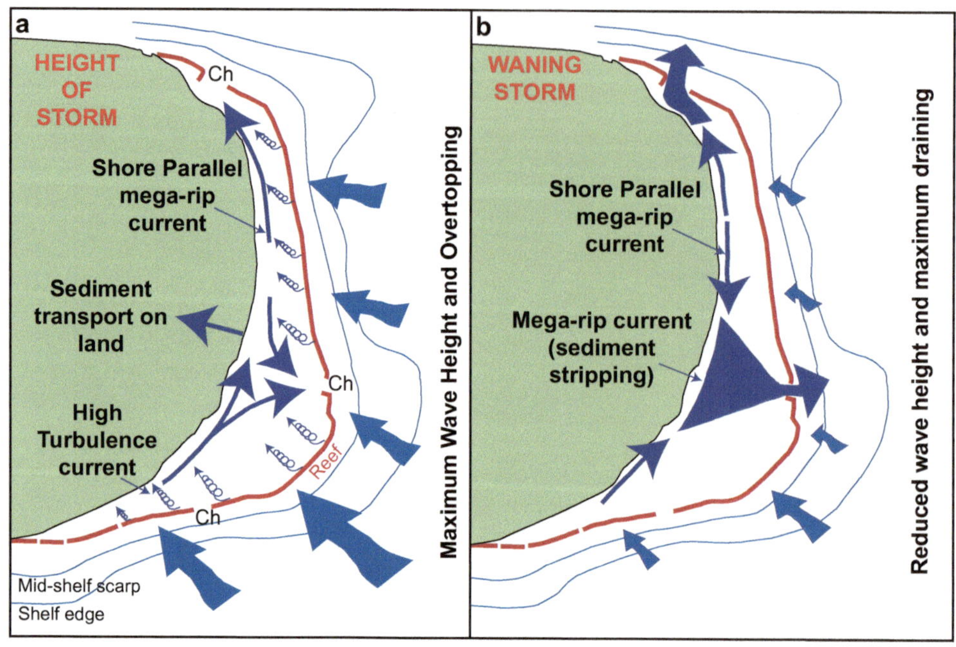

Fig. 8.37 Model of sedimentary processes associated with hurricane action on lagoons around Grand Cayman—although based on East Sound, this model can be applied to most other lagoons on Grand Cayman with the exception of South Sound and North Sound. **a** Current directions setup during height of storm, driven by powerful onshore winds. **b** Current regime associated with waning stage of storm as water flows out of the lagoon back to the open ocean. Modified from Li et al. (1997, their Fig. 7a, b)

(Figs. 8.22). As the hurricanes gain force, waves crash over the reef crest and commonly deposit coral heads and other boulders on the landward side of the reefs (Figs. 8.24a, 8.26, 8.27a–b). The force of the winds then piles the water onshore with rip currents commonly being generated with flow directions being determined by the wind direction and morphology of the lagoon (Fig. 8.37). As the storms abate, the water that was piled against the shoreline then starts to flow out of the lagoon with the direction of the rip currents being the reverse of those formed during the height of the storm (Fig. 8.37). The power of the currents generated during the height of the storm and the waning stages means that they are capable of eroding sediments from the lagoon floors and transporting them in all directions. Thus, forereef sediments may be transported into the lagoons during the initial stages of the storms, but as the storms wane, lagoonal sediment may be transported out of the lagoons and into the forereef areas. This has been amply demonstrated by Li (1997), Li and Jones (1997), and Li et al. (1998) who used tracer species of foraminifera to show how the lagoonal sediments had been moved around during the major storms and hurricanes (Figs. 8.25, 8.28). Corlett (2005) and Corlett and Jones (2007) also showed that much of the fine-grained sediment produced from the epiphytes that lived on the *Thalassia* leaves was removed from the lagoons during these storms.

8.9 Vermetid Buildups

Buildups formed by vermetid gastropods (mainly *Dendropoma*) are found at various locations around Grand Cayman, including Bodden Town on the south coast and Great Bluff estates on the northeast coast (Jones and Hunter 1995). The vermetid shells, which are tightly coiled to loosely sinuous to almost straight are at least 1.5 cm long, with inside diameters up to 1.3 mm and an average shell thickness of 0.3 mm. These shells collectively form large honeycombed masses that encrust hard substrates that exist where rocky shorelines are exposed to wave action (Fig. 8.38). The vermetid gastropods are commonly encrusted by foraminifera, coralline algae, and cyanobacteria. Spaces between the shells are filled with carbonate sand and silt that probably originated from the seafloor. C^{14} dating of the gastropod shells indicates that they died off during the period of 1625 to 1740 AD (Jones and Hunter 1995).

At Bodden Town, the vermetids encrust seaward dipping sheets of beachrock where the water is 0.3 to 0.5 m deep. There, the buildups have suffered extensive bioerosion due to the activity of echinoids, sponges, worms, bivalves, algae, and fungi. Today, many of the vacated vermetid tubes are inhabited by polychaete annelid worms. At Great Bluff Estates on the northeast coast of Grand Cayman, vermetid

Fig. 8.38 Vertical cross-section through vermetid buildup that developed on the beachrock at Bodden Town. White carbonate sand/silt fills the spaces between the gastropod shells. The red colouration evident in some areas reflects the presence of *Homotrema rubum*

buildups have formed a lower ledge and an upper ledge that gave C^{14} dates of 318 ± 90 years BP and 278 ± 90 tears BP, respectively (Jones and Hunter 1995).

Today, *Dendropoma* typically thrive in <1 m zone that is centered on mean sea level where there is a rocky substrate and exposure to onshore waves. The reason why the *Dendropoma* at Bodden Town and Great Estates are not alive today may be related to the fact that those locations suffered a reduction in wave energy as the offshore reefs developed and greatly reduced the onshore waves.

8.10 Beachrock

Moore (1973) first documented and described beachrock from Grand Cayman by noting its presence in the West Bay area, various localities on the west coast, the southwest corner, Spotts Bay, Bodden Town, northeast corner, and near Old Man Village. As with beachrock found elsewhere in the world, its origins still remain open to debate.

When exposed, beachrock forms flat to arcuate, seaward dipping sheets that generally parallel the slope of the nearby sandy beaches (Fig. 8.39a–c). The beachrock is formed of lithified sand that is of the same composition as nearby unconsolidated beach sand (Fig. 8.39d). Beachrock commonly includes foreign items such as the glass fragments that Jones and Goodbody (1982) reported from the beachrock near West Bay (Fig. 8.39e). Although typically well-lithified, beachrock is prone to erosion by tidal action and/or storms (Fig. 8.39f). The extensive, well-indurated beachrock near West Bay evident in February 1981 had vanished by February, 1982, presumably due to storm activity. Although beachrock appeared to be forming at the same site in 1984, it also subsequently disappeared. Beachrock is transitory in nature. Once exposed to the atmosphere, beachrock is subject to subaerial weathering that is commonly aided by the action of boring and rasping organisms such as chitons (Fig. 8.39g).

Moore (1973) demonstrated that the sand in the Cayman beachrock was cemented by (1) acicular crusts aragonite crystals that were 20 μm long and 1 to 5 μm wide, (2) micrite crusts that locally contain peloids and/or organic matter that coated individual grains and locally filled intergranular pore spaces, and (3) clear, bladed to equant crusts, 30 to 40 μm thick, formed of high Mg-calcite. Groundwater sampling suggested that the beachrock formed under the influence of mixed marine and meteoric waters. Moore (1973) also argued that there may be a biological influence on some of the precipitation.

The surfaces of exposed beachrock are typically weathered with evidence of both physical and biological processes being operative. Exposed surfaces on the beachrock at the southwest corner of Grand Cayman are characterized by erosional runnels that are perpendicular to the strike of the dipping sheets of beachrock (Fig. 8.39c) and erosional pockets occupied by chitons (Fig. 8.39g) and gastropods are common. Jones and Goodbody (1984) demonstrated that some of the Cayman beachrock was being actively eroded by the combined efforts of pelecypods, polychaets, annelid worms, sipunculid worms, spionid worms, algae, sponges, and fungi(?). Such information indicates that beachrock goes through cycles of formation (probably in the subsurface) that are followed by periods of destruction that are mediated by biological and physical weathering.

The origin of beachrock is open to debate because it is not clear why it forms along some stretches of a beach but is absent from neighbouring stretches even though conditions appear to be the same in both zones. For Grand Cayman, there is no obvious explanation why it has formed at the localities noted by Moore (1973) but not elsewhere even though the coastal settings look identical. As yet, this question has not been completely resolved.

8.11 Conclusions

The shallow waters around each of the Cayman Islands are famous for their numerous reefs and the diverse biota that inhabit their warm, clear waters. Numerous studies of these tranquil environments have provided valuable insights into their ecology, sediments and reefs. Important conclusions derived from these studies include the following.

- The vertical to overhanging "drop-offs" to the abyssal depths around each of the islands define the offshore limit of the "shallow-water" depositional environments.

Fig. 8.39 Beachrock on Grand Cayman. **a**, **b** General view, southwest corner of Grand Cayman, showing seaward dipping accurate sheets of beachrock. **c** Beachrock, southwest corner of Grand Cayman showing exposed eroded surfaces on seaward dipping sheets of beachrock. **d** View of lithified carbonate sands in beachrock, west coast of Grand Cayman, near West Bay. **e** Piece of glass, bored by microbes, cemented into beachrock, west coast of Grand Cayman, near West Bay. **f** Exposed sheets and slabs of beachrock after unconsolidated sand was removed by storm action. Beachrock disappeared within the following year, presumably by erosion associated with storms. West coast of Grand Cayman, near West Bay. **g** Example of chitons in hole on exposed surface of beachrock, southwest corner of Grand Cayman. Chitons erode the beachrock as they graze on the surface of the beachrock

- Today, the shelf areas around each of the islands are relatively narrow and characterized by depositional belts that generally parallel the coastlines.
- Spur and groove structures are commonly well-developed seaward of the fringing reefs.
- Lagoons formed where coral reefs have extended from headland to headland and thereby separated them from the open ocean.
- The clear, warm ocean waters around the islands provide ideal conditions for the growth of a diverse biota that includes numerous species of corals, bivalves, gastropods, and extensive tracts covered with *Thalassia* and various types of algae.
- Mangrove swamps commonly fringe the lagoons.
- Fringing and patch reefs, formed of diverse arrays of corals, are common around each of the Cayman Islands.
- Although the lagoons and shelves around each of the Cayman Islands are generally pictures of tranquility, storms and hurricanes are periodic severe events that can severely damage the reefs and/or move vast quantities of sediment around.
- Sedimentation in North Sound, which is the largest lagoon on any of the islands, began ~5,000 yrs BP with the early formed fresh-brackish water ponds eventually being replaced by marine environments.
- The lagoons of the Cayman Islands include a diverse array of facies that largely reflect water depths and distance from the shoreline. *Thalassia* banks are common in the shoreward areas, whereas sands are common in the middle and outer parts of the lagoons. Major storms commonly move a lot of the sediment onshore or out of the lagoons.
- Beachrock is found along some of the sandy beaches on the west coast of Grand Cayman and the shorelines of some of the lagoons.

References

Acker KL, Risk MJ (1985) Substrate destruction and sediment production by the boring sponge *Cliona caribbea* on Grand Cayman Island. J Sediment Petrol 55:705–711

Bak R (1994) Sea urchin bioeroeion on coral reefs: place in the carbonate budget and relevant variables. Coral Reefs 13:99–103

Beanish JMR (2000) Sedimentology of a current-dominated lagoon: case study of South Sound, Grand Cayman, B.W.I. (M.Sc. thesis), University of Alberta, Edmonton

Beanish JMR, Jones B (2002) Dynamic carbonate sedimentation in a shallow coastal lagoon: case study of South Sound, Grand Cayman, British West Indies. J Coastal Res 18:254–266

Beanish JMR, Sanchez-Azofeifa A, Jones B (2002) Application of image analysis for mapping of sedimentary facies in a shallow lagoon: case study, South Sound, Grand Cayman, British West Indies. Int J Remote Sens 23:2877–2890

Blanchon P, Jones B (1995) Marine-planation terraces on the shelf around Grand Cayman: a result of stepped Holocene sea-level rise. J Coastal Res 11:1–33

Blanchon P, Jones B (1997) Hurricane control on shelf-edge architecture around Grand Cayman. Sedimentology 44:479–506

Blanchon P, Jones B, Ford DC (2002) Discovery of a submerged relic reef and shoreline off Grand Cayman: further evidence for an early Holocene jump in sea level. Sed Geol 147:253–270

Blanchon P, Jones B, Kalbfleisch W (1997) Anatomy of a fringing reef around Grand Cayman: storm rubble, not coral framework. J Sediment Res 67:1–16

Blanchon PA (1995) Controls on modern reef development around Grand Cayman. (Ph.D. thesis), University of Alberta, Edmonton

Booker S (2020) Variation in climatic conditions from the Cayman Islands through stable isotope and element analyses from coral and sediment cores. (Ph.D. thesis), University of Alberta, Edmonton

Booker S, Jones B (2020) A 6000-year record of environmental change from Grand Cayman, British West Indies. Sediment Geol 409:105779

Booker S, Jones B, Chacko T, Li L (2019) Insights into sea surface temperatures from the Cayman Islands from corals over the last ~540 years. Sed Geol 389:218–240

Booker S, Jones B, Li L (2020) Diagenesis in Pleistocene (80 to 500 ka) corals from the Ironshore Formation: implications for paleoclimate reconstruction. Sed Geol 399:1–16

Bruckner A (2010) Coral Islands coral reef health and resilience assessment. https://www.livingoceansfoundation.org/publication/cauman-islands-coral-reefhealth-and-resilience-assessments

Bruggermann JH, van Kessel AM, van Rooij JM, Breeman AM (1996) Bioerosion and sediment ingestion by the Caribbean parrotfish *Scarus vetula* and *Sparisoma viride*: implications of fish size, feeding mode and habitat use. Mar Ecol Prog Ser 134:59–71

Burton FJ (1994) Climate and tides of the Cayman Islands. In: Brunt MA, Davies JE (eds) The Cayman Islands: natural history and biogeography. Kluwer Academic Publishers, Dordrecht-Boston-London, Monographiae Biologicae, pp 51–60

Corlett HJ (2005) Epiphyte growth and community structure on *Thalassia testudinum*: A case study from Grand Cayman. (M.Sc. thesis), University of Alberta, Edmonton

Corlett HJ, Jones B (2007) Epiphyte communities on *Thalassia testudinum* from Grand Cayman, British West Indies: their composition, structure, and contribution to lagoonal sediments. Sed Geol 194:245–262

Doran E (1954) Landforms of Grand Cayman Island, British West Indies. Tex J Sci 6:360–377

Fairbanks RG (1989) A 17,000-year glacio-eustatic sea level record: influence of glacial melting rates on the Younger Dryas event and deep-ocean circulation. Nature 342:637–642

Fenner DP (1993) Some reefs and corals of Roatan (Honduras), Cayman Brac, and Little Cayman. Atoll Res Bull 388:111–130

Ghiold J, Roundtree GA, Smith SH (1994) Common sponges of the Cayman Islands. In: MA Brunt, JE Davies (eds) The Cayman Islands: Natural History and Biogeography. Kluwer Academic Publishers

Hunter IG (1994) Modern and ancient coral associations of the Cayman Islands. (Ph.D. thesis), University of Alberta, Edmonton

Hutchings P (2011) Bioerosion. In: D. Hopley (Ed.), Encyclopedia of Modern Coral Reefs. Springer, pp 139–156

James NP, Jones B (2015) Origin of Carbonate Sedimentary Rocks. Wiley, Oxford, England

Jones B (2021) Formation, dispersal and accumulation of terra rossa on the Cayman Islands. Sedimentology 68:1–45

Jones B, Goodbody QH (1982) The geological significance of endolithic algae in glass. Can J Earth Sci 19:671–678

Jones B, Goodbody QH (1984) Biological alteration of beachrock on Grand Cayman Island, British West Indies. Bull Can Pet Geol 32:201–215

Jones B, Goodbody QH (1985) Oncolites from a shallow lagoon, Grand Cayman Island. Bull Can Pet Geol 32:254–260

Jones B, Hunter IG (1992) Very large boulders on the coast of Grand Cayman: the effects of giant waves on rocky coastlines. J Coastal Res 8:763–774

Jones B, Hunter IG (1995) Vermetid buildups from Grand Cayman, British West Indies. J Coastal Res 11:973–983

Kalbfleisch WBC (1995) Sedimentology of frank sound and pease bay, two modern shallow-water hurricance-affected lagoons, Grand Cayman, British West Indies. (M.Sc. thesis), University of Alberta, Edmonton

Kalbfleisch WBC, Jones B (1998) Sedimentology of shallow, hurricane-affected lagoons: Grand Cayman, British West Indies. J Coastal Res 14:140–160

Li C (1997) Foraminifera: their distribution and utility in the interpretation of carbonate sedimentary processes around Grand Cayman, British West Indies. (Ph.D. thesis), University of Alberta, Edmonton

Li C, Jones B (1997) Comparison of foraminiferal assemblages in sediments on the windward and leeward shelves of Grand Cayman, British West Indies. Palaios 12:12–26

Li C, Jones B, Blanchon P (1997) Lagoon-shelf sediment exchange by storms—evidence from foraminiferal assemblages, east coast of Grand Cayman, British West Indies. J Sediment Res 67:17–25

Li C, Jones B, Kalbfleisch WBC (1998) Carbonate sediment transport pathways based on foraminifera: case study from Frank Sound, Grand Cayman, British West Indies. Sedimentology 45:109–120

Logan A (1994) Reefs and lagoons of Cayman Brac and Little Cayman. In: Brunt MA, Davies JE (eds) The Cayman Islands: natural history and biogeography. Kluwer Academic Publishers, Dordrecht-Boston-London, Monographiae Biologicae, pp 105–124

Logan A (2013) Coral reefs of the Cayman Islands. Coral Reefs of the United Kingdom Overseas Territories. Springer, Dordrecht, pp 61–68

MacKinnon L (2000) Sedimentology of North Sound, Grand Cayman, British West Indies. (M.Sc. thesis), University of Alberta, Edmonton

MacKinnon L, Jones B (2001) Sedimentological evolution of North Sound, Grand Cayman—a freshawater to marine carbonate succession driven by Holocene sea-level rise. J Sediment Res 71:568–580

Manfrino C, Jacoby CA, Camp E, Frazer TK (2013) A positive trajectory for corals at Little Cayman Island. PLoS ONE 8:1–9

Manfrino C, Riegl B, Hall JL, Graifman R (2003) Status of coral reefs on Little Cayman, Grand Cayman and Cayman Brac, British West Indies, in 1999 and 2000 (Part 1: Stony corals and algae). Atoll Res Bull 496:204–225

Moore CH (1973) Intertidal carbonate cementation, Grand Cayman, West Indies. J Sediment Petrol 43:591–602

Murphy GN, Perry CT, Chin P, McCoy C (2016) New approaches to quantifying bioerosion by endolithic sponge populations: applications to the coral reefs of Grand Cayman. Coral Reefs 35:1109–1121

Osborne AH, Haley BA, Hathorne EC, Plancherel Y, Frank M (2015) Rare earth element distribution in Caribbean seawater: continental input versus lateral transport of distinct REE compositions in subsurface water masses. Mar Chem 177:172–183

Raymont JEG, Lockwood APM, Hull LE, Swain G (1976) Cayman Islands natural resources study - Pt. IVB - Results of the investigations into the coral reefs and marine parks. Ministry of Overseas Development, pp C1-C18

Riegl B (2001) Inhibition of reef framework by frequent disturbence: examples from the Arabian Gulf, South Africa, and the Cayman Islands. Palaeog Palaeocl Palaeoecol 175:79–101

Rigby JK, Roberts HH (1976) Grand Cayman Island: geology, sediments and marine communities 4. Brigham Young University, Geology Studies, Special Publication no. 4, p 97

Roberts HH (1971a) Environments and organic communities of North Sound, Grand Cayman Island, B.W.I. Carib J Sci 2:67–79

Roberts HH (1971b) Mineralogical variation in lagoonal carbonates from North Sound, Grand Cayman Islands, B.W.I. Sed Geol 6: 201–213

Roberts HH (1974) Variability of reefs with regard to changes in wave power around an island. In, Proceedings of the Second Corral Reef Symposium, Australia. Great Barrier Reef Committee, Brisbane, pp 497–512

Roberts HH (1976) Carbonate sedimentation in a reef-enclosed lagoon. Geology, Sediments, and Marine Communities. Brigham Young University, Provo, Utah, North Sound, Grand Cayman Island, Grand Cayman Island, pp 97–122

Roberts HH (1977) Field guidebook to the reefs and geology of Grand Cayman Island, B.W.I. In, Third International Symposium on Coral Reefs, p 41

Roberts HH (1983) Shelf margin reef morphology: a clue to major off-shore sediment transport routes, Grand Cayman Island, West Indies. Atoll Res Bull 263:1–11

Roberts HH (1994) Reefs and lagoons of Grand Cayman. In: Brunt MA, Davies JE (eds) The Cayman Islands: Natural history and biogeography. Kluwer Academic Publishers, Netherlands, pp 75–104

Roberts HH, Murray SP, Suhayda JN (1975) Physical processes in a fringing reef system. J Mar Res 33:233–260

Roberts HH, Suhayda JN (1983) Wave-current interactions on a shallow reef (Nicaragua, Central America). Coral Reefs 1:209–214

Rose SR, Risk JM (1985) Increase in *Cliona delitrix* infestation of *Montastrea cavernosa* heads on the organically polluted portion of the Grand Cayman fringing reef. Mar Ecol 6:345–363

Saenger C, Cohen AL, Oppo DW, Hubbard D (2008) Interpreting sea surface temperature from strontium/calcium ratios in *Montastrea* corals: link with growth rate and implications for proxy reconstructions. Paleoceanography 23:1–11

Sauer JD (1982) Cayman Islands seashore vegetation: a study in compartive biogeography. University of California publications in geography, 25. University of California Press, London, England, p 161

Stoddart DR (1980) Geology and geomorphology of Little Cayman. Atoll Res Bull 241:11–16

Suhayda JN, Roberts HH (1977) Wave action and sediment transport on fringing reefs. In, Proceedings of the 3rd International Coral Reef Symposium. Miami. Rosenthiel School of Marine and Atmospheric Science, University of Miami, pp. 65–70

Swain GWJ, Hull LE (1976) Biological investigations in a tropical lagoon, Grand Cayman, British West Indies. Cooperative Investigations of the Caribbean and Adjacent Region II (Symposium on progress in marine research in the Caribbean and adjacent regions), 449–467

Tompkins EL (2005) Planning for climate change in small islands: insights from national hurricane preparedness in the Cayman Islands. Global Environ Change Part A 15:139–149

Tongpenyai B (1989) An assessment of the use of image analysis in carbonate sedimentology. (Ph.D. thesis), University of Alberta, Edmonton

Tongpenyai B, Jones B (1991) Application of image analysis for delineating modern carbonate facies changes through time: grand Cayman, western Caribbean Sea. Mar Geol 96:85–101

Turner JR, McCoy C, Cottam M, Olynik J, Austin T, Blumenthal J, Bothwell J, Burton FJ, Bush P, Chin P, Dubock O, Godbeer D, Gibb J, Hurlston L, Johnson BJ, Logan A, Parsons G, Ebanks-Petrie G (2013) Biology and ecology of the coral reefs of the Cayman Islands. In: Sheppard CRC (ed) Coral Reefs of the United Kingdom overseas terrritories, coral Reefs of the World 4. Springer Science +Business Media, Dordrecht, pp 69–88

van Hooidonk RJDPM, Moye J, Brandt ME, Hendee JC, McCoy C, Manfrino C (2012) Coral bleaching at Little Cayman, Cayman Islands, 2009. Estuarine, Coast Shelf Sci 106:80–84

Woodroffe CD, Stoddart DR, Harmon RS, Spencer T (1983) Coastal morphology and Late Quaternary history, Cayman Islands, West Indies. Quatern Res 19:64–84

9 Karst on the Cayman Islands

Abstract

Karst developed on the Cayman Islands whenever the islands were exposed to the atmosphere. Modern karst, developed largely on the hard limestones and dolostones of the Cayman Formation and Pedro Castle Formation, is characterized by dark grey to black phytokarst surfaces that are reflective of various types of microbes, including bacteria, fungi, and cyanobacteria, that actively sculpted the topographically complex phytokarst surfaces. Surface karst features include sinkholes of various sizes and solution-widened joints, whereas the subsurface features include complex cave systems and tunnels of all sizes. Many caves are adorned with complex arrays of speleothems (stalactites, stalagmites, flowstone) that are formed of aragonite and calcite and, in some cases, minor amounts of dolomite. Oncoids ("cave pearls") are found in some of the cave pools. Locally, karst development commonly enhanced the porosity and permeability of the limestones and dolostones. The formational boundaries in the Oligocene to Pleistocene succession are paleokarst surfaces that are commonly characterized by significant karstic relief with complex topographies at all scales. Karst features are well-developed in the hard limestones and dolostones of the Cayman Formation and Pedro Castle Formation but poorly developed in the "soft" limestones of the Ironshore Formation. Paleokarst features include sinkholes of various dimensions, solution-widening fractures and joints, and complex cave systems. Fortuitous cross-sections through some of these paleocaves during quarrying show that many of them are filled with complex successions that include terra rossa and oceanic sediments that were washed into the cave systems. These deposits commonly buried the earlier formed speleothems.

9.1 Introduction

Today, the Cayman Islands enjoys a sub-humid tropical climate with distinct seasonal variations in rainfall such that the year can be divided into the "wet" season (May to November) and the "dry season" (December to April) (Burton 1994). Air temperatures on Grand Cayman range from 11.2° to 36.5 °C with daily temperatures that peak at 28.4 °C in July and 24.8 °C in February. Between 1920 and 1965, the average rainfall registered at the Owens Roberts International Airport on Grand Cayman was 1,740 mm (Ng and Jones 1995). The rainfall on Grand Cayman decreases from west to east with the highest amounts being around George Town (Ng and Jones 1995, their Fig. 4). These climatic conditions, combined with the lush vegetation that covers large parts of the islands are ideal for karst development. Apart from saline ponds found in coastal areas on Grand Cayman (e.g., Malportas Pond, Colliers Pond, Meagre Bay Pond) there are no surface lakes or rivers on the Cayman Islands. Rainwater quickly disappears from the surface of the islands as it drains into the land via the myriad arrays of fractures and solution-widened joints that characterized the bedrock. Collectively the conditions on each of the Cayman Islands are ideal for the development of surface and sub-surface karst features (Fig. 9.1). It is also evident that karst formation was associated with the unconformities

B. Jones, *Geology of the Cayman Islands*,
https://doi.org/10.1007/978-3-031-08230-6_9

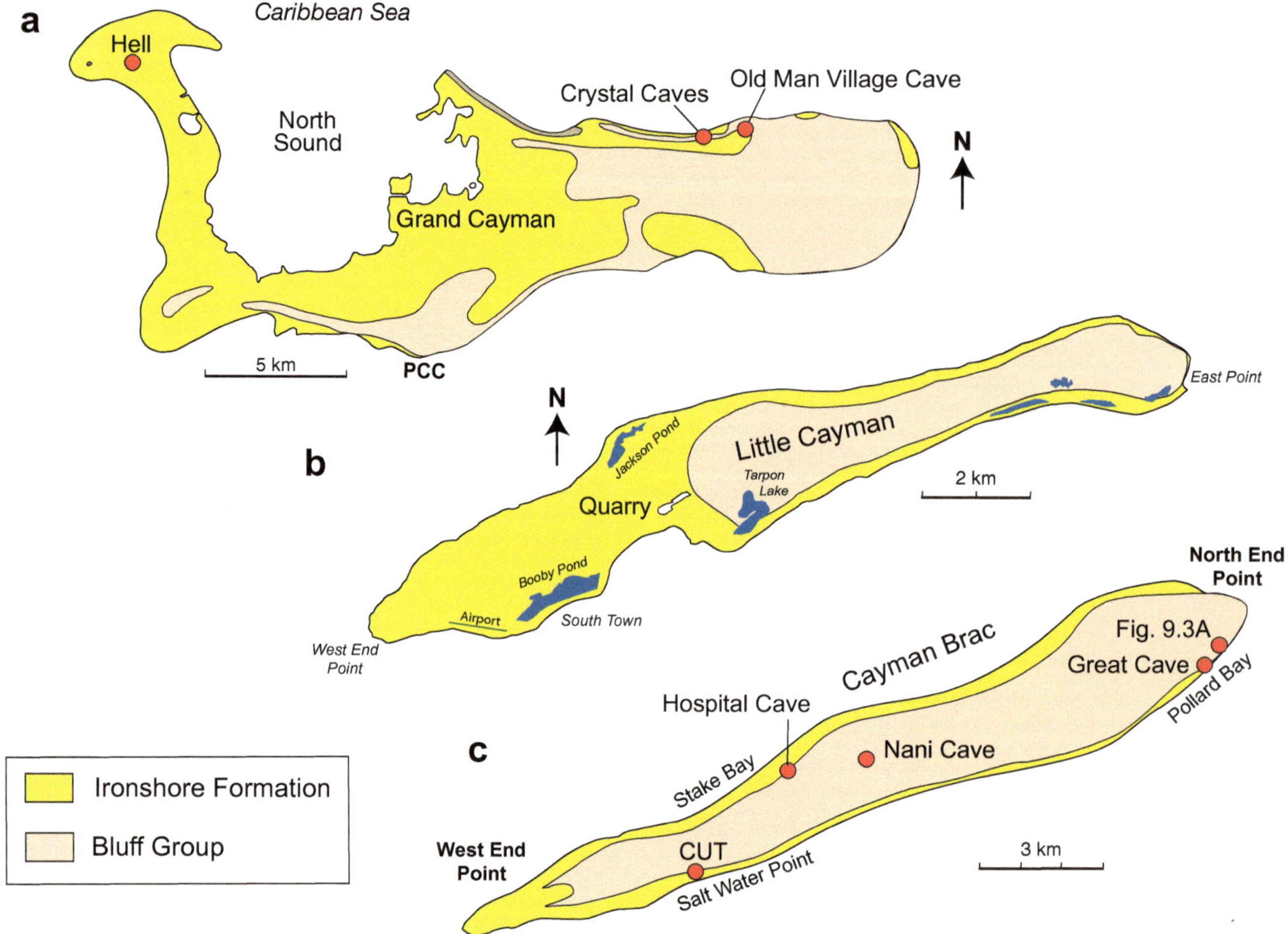

Fig. 9.1 Maps of **a** Grand Cayman, **b** Little Cayman, and **c** Cayman Brac showing locations of caves and karst features

that now define formational boundaries. Today, ancient caves that are filled with various combinations of speleothems and internal sediments are common (Jones 1992d, 2016).

9.2 Modern Karst of the Cayman Islands

9.2.1 Phytokarst

Folk et al. (1973, p. 2351), based on exposures at Hell on Grand Cayman (Figs. 9.1a and 9.2a, b) defined phytokarst as a "…landform produced by rock solution in which boring plant filaments are the main agent of destruction, and the major morphological features are determined by the peculiar nature of this mode of attack". They specifically noted that their described processes were restricted to carbonate rocks that had been attacked by endolithic algal filaments, and argued that it was a landform that probably develops under humid tropical climates. At Hell, there is a small inlier of the Cayman Formation (Bluff Limestone of Folk et al. 1973) that has a weathered surface characterized by a by highly irregular pinnacles with razor-sharp edges (Fig. 9.2a, b). There, the pinnacles are typically less than 2 m high, which is common for most weathered surfaces that have developed on the dolostones of the Cayman Formation. Along the coastline to the east of Pedro Castle, however, there are scattered pinnacles that are up to 4 m high. The development of the phytokarst and the height of the pinnacles seem to be related to the "hardness" of the bedrock. On coastal exposures to the east of Pedro Castle, for example, the boundary between the Pedro Castle Formation and the Cayman Formation can be mapped by the change in the relief on the phytokarst as it is more subdued on the "soft" limestones and dolostones of the Pedro Castle Formation than the "hard" dolostones of the Cayman Formation. Weathering of the poorly consolidated limestones of the Ironshore Formation is even more subdued than that associated with the Pedro Castle Formation. Jones (2000) highlighted the fact that microbes mediate a wide array of destructive and constructive processes that contribute to the karstic landscape that is developed on these rocks.

Fig. 9.2 **a** General view of phytokarst developed on hard dolostones of the Cayman Formation at Hell, Grand Cayman. Most pinnacles in the foreground are ~0.5 m high whereas some of the peaks in the background are up to 0.75 m high. **b** Close up of phytokarst pinnacles at Hell showing sharp edges and concavities on surfaces of ornate pinnacles that are up to 0.5 m high. **c** Large tree rooted in dolostones of the Cayman Formation, east end of Grand Cayman. Tree trunk is ~1 m in diameter. **d** Sinkhole developed in dolostones of the Cayman Formation, central part of Cayman Brac. The sinkhole is ~0.5 m wide at the top and ~10 m deep

Well-developed phytokarst on the dolostones of the Cayman Formation is characterized by concavities that are typically <5 cm in diameter and separated from each other by razor-sharp ridges (Fig. 9.2b). Squair (1988) and Jones and Smith (1988) agreed with Folk et al. (1973) regarding the notion that this style of weathering was mediated by the action of the microorganisms that lived on the surface of the rock. Locally, concavities on opposite sides of thin parts of the rock pinnacles have merged to produce holes through the rock. The dark grey to black phytokarst surfaces contrast sharply with the white dolostone on which it has developed. As noted by Folk et al. (1973), these weathered surfaces are covered with various microorganisms that have bored into the surfaces of the dolostones. Spencer (1981), however, disagreed with this notion and followed the terminology of Trudgill (1976) by arguing that the weathering was due to "solutional disintegration". Spencer (1985b) suggested that this took place because selective dissolution removed the high-Mg calcite cement that was present in the rocks. Spencer (1981, 1985a, b) argued that microbes were not involved in weathering of the rocks because the limestones at many localities were bluff coloured. Jones (1989b), however, clearly demonstrated that the organic coating on dolostones were formed largely of fungal(?) sporangia and spores and fewer algae and bacteria, whereas on limestones the coating is formed largely of epilithic and endolithic filamentous microbes. Viles (2001) argued that landscapes, like the phytokarst on the Cayman Islands are multi-scale features. Specifically, Viles and Spencer (1986) argued that structural controls, including the location of joints, influenced the overall pattern of phytokarst development.

A wide variety of calcite crystal morphologies are commonly associated with the microorganisms that inhabit phytokarst surfaces. Jones (1987) demonstrated that calcite cements that cemented dolostone clasts in a karst breccia on Grand Cayman had subsequently been altered by algae and/or fungi that produced (1) borings, (2) micrite envelops, (3) spiky calcite, (4) almond-shaped etch pits, (5) smooth crystal faces to surfaces with a blocky topography, and (6) etching of many crystals. Collectively, this demonstrated that infestations of microbes were capable of mediating both constructive and destructive processes and thereby played a significant role in vadose diagenesis. Jones and Kahle (1986) showed that complex multi-branched dendrite crystals developed around algal filaments that were found on cavity walls in a breccia that filled sinkholes in dolostones of the Cayman Formation found on the southeast corner of Grand Cayman. Later, Jones and Kahle (1993, their Fig. 2), based on samples from the Ironshore Formation showed that crystals associated with these microbes ranged from fibre crystals (hexagonal, 2-lobed, 4-lobed, 6-lobed, ribbon, rhomb chains) to complex dendritic crystals that involved numerous levels of branching. They also demonstrated that subsequent diagenesis commonly led to the modification of those crystals by various destructive and constructive processes. The destructive processes included physical breakdown, dissolution, and/or micritization, whereas epitaxial overgrowths commonly led to the original crystals being completely disguised. Similarly, Jones and Peng (2014), showed that biogenic coatings developed around lithoclasts in a breccia from a sinkhole in the Cayman Formation on the southeast coast of Grand Cayman, involved amorphous calcium carbonate, complex dendrite crystals, acicular crystals, triradiate crystals, and skeletal rhombs that had developed on the surfaces of the atmophytic cyanobacterium *Scytonema julianum.* They argued that this microbe had exerted a direct influence over the various precipitates that now coated its surface.

Many dolomite crystals in samples from the Cayman Formation are severely etched with rhomb-shaped etch pits, holes, one or more of their crystal faces absent, and partly leached cores. Jones (1989b) suggested that such weathering was mediated by surface-reaction-controlled kinetic processes that were operative within the organic coatings. In comparison, the limestones of the Ironshore Formation seem to weather at a much faster rate and therefore lack evidence of such slow breakdown.

Today, phytokarst development is common on the exposed surfaces of the Cayman Formation, Pedro Castle Formation, and the Ironshore Formation. The unconformities that constitute the boundaries between the formations on the Cayman Islands, however, generally have locally smooth, relatively flat surfaces that are devoid of intricate weathering structures such as those generated through phytokarst development. The only exception to this was found in the northeast corner of the quarry on Little Cayman (Fig. 9.1b) where the limestones of the Ironshore Formation unconformably overlie the limestones and dolostones of the Pedro Castle Formation (Fig. 9.3). There, parts of the unconformity are characterized by a well-developed phytokarst surface with irregular-shaped pinnacles, up to 2 m high and typically <25 cm in cross-sectional width, that had developed on the limestones/dolostones of the Pedro Castle Formation (Fig. 9.3). With the transgression that led to deposition of the Ironshore Formation (Member D), the rugged phytokarst surface was buried under the Pleistocene fossiliferous limestones (Fig. 9.3). Although beautifully exposed in 2018, the quarry walls that best showed the cross-section through these deposits were destroyed by renewed quarrying activity in 2019.

9.2.2 Caves and Speleothems on Cayman Brac

Gilleland (1998), using a broad definition of a cave, noted that there are hundreds of caves on Cayman Brac and

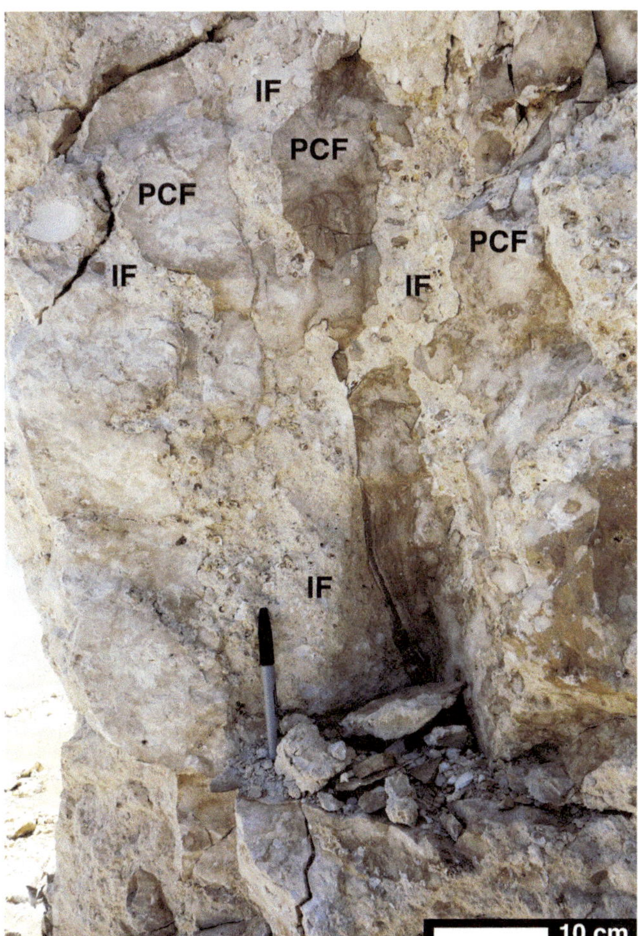

Fig. 9.3 View of middle part of wall in quarry on Little Cayman showing fossiliferous limestones of the Ironshore Formation (IF) filling phytokarst topography that had developed on the underlying Pedro Castle Formation (PCF)

divided them into (1) inland cockpits, (2) cliff face caves, (3) joint fractures, and (4) classic sinkholes. He noted that the cliff face caves are found on the north and south sides of the elevated core of the island with many being ~10 m below the top of the cliffs (~40 m asl on the east end of the island). Tarhule-Lips (1999), after noting that more caves are evident in cliff faces on the south coast than on the north side (Fig. 9.4a), divided them into the (1) notch caves, with entrances 1–2 m above the wave-cut notch that is evident at 6 m asl in the cliff-face around much of the island, and (2) "upper caves" that are apparent in the cliff faces at random elevations above the notch.

On Cayman Brac, modern caves are common in the Brac Formation and Cayman Formation, rare in the Pedro Castle Formation, and absent from the Ironshore Formation. Their rarity in the Pedro Castle Formation may be misleading because that formation is not widely exposed and there is only a limited thickness of the succession exposed on the west end of the island. The caves on Cayman Brac are 5–450 m long with entrances 1–23 m above the wave–cut notch (Tarhule-Lips and Ford 1998b) that is 6 m above sea level. Tarhule-Lips and Ford (1998a) suggested that many of these caves formed through condensation corrosion rather than from flooding of the cave system. This development style occurs in closed systems where surface streams are absent and the fresh water needed for bedrock dissolution only comes from condensation or rainwater that has infiltrated into the cave.

Caves in the Brac Formation

Numerous caves are visible in exposures of the Brac Formation along the southeast coast of Cayman Brac (Figs. 9.1, 9.4). These irregular-shaped caves are unusual because they typically have flat roofs that coincide with the Brac Unconformity (Fig. 9.4a, b). Great Cave, located in the Brac Formation at the east end of the road on the south side of the island, is decorated with various speleothems, including stalactites that hang from the flat cave roof that is the Brac Unconformity (Fig. 9.4b). Evidence of water flow into the cave is only apparent after periods of heavy rainfall when droplets of water appear on the tips of the stalactites (Fig. 9.4c). In one branch of the cave, coated grains up to 10 cm long cover the floor (Fig. 9.4d). These coated grains probably formed when water flowed freely through the cave.

Caves in the Cayman Formation

Most caves on Cayman Brac are housed in the dolostones of the Cayman Formation. Perhaps the best known cave is Peters Cave (also known as the Hurricane Cave), which is a large cave, 29 m high above sea level (asl) that was a place of refuge for local inhabitants when major hurricanes passed close to or over the island.

Tarhule-Lips (1999) described the notch caves, like those evident near road level in the cliff faces opposite the hospital (Figs. 9.1c, 9.5a–c), as being characterized by one or two large rooms close to the entrance that are up to 4 m high. Smaller passages, rooms and recesses that radiate from them all end abruptly. Most of these caves are devoid of speleothems and today, there is little evidence of water seeping into them (Fig. 9.5a–c). The situation in the past was obviously different because the cave walls commonly include vertical cross-sections through old tunnels that were partly to totally filled with layered internal sediments and/or calcite precipitates (Fig. 9.5b, c). Tarhule-Lips (1999) suggested that notch caves did not form at the same time as the notch because (1) the caves are 1–2 m above the notch, (2) the shape of main-chambers suggest that they formed in the rock without connection to the cliff face and only later became exposed as cliff retreated and intersected them, and (3) in some caves (e.g., Bats Cave), U-series dating of stalagmites

Fig. 9.4 Caves on Cayman Brac. **a** Caves in upper part of the Brac Formation with roofs formed by the Brac Unconformity (white arrows). In this area, the Brac Unconformity is ~12 m asl. South coast (Pollard Bay), east end of Cayman Brac. **b** Large stalactite (~1 m long) hanging from roof of the Great Cave where the base of the Cayman Formation forms the flat cave roof. **c** Groundwater bead hanging from tip of stalactite (~5 mm diameter) in Great Cave—such drips are only evident in the days following heavy rainfalls. **d** Cave pisoliths in old pool on floor of Great Cave, Cayman Brac

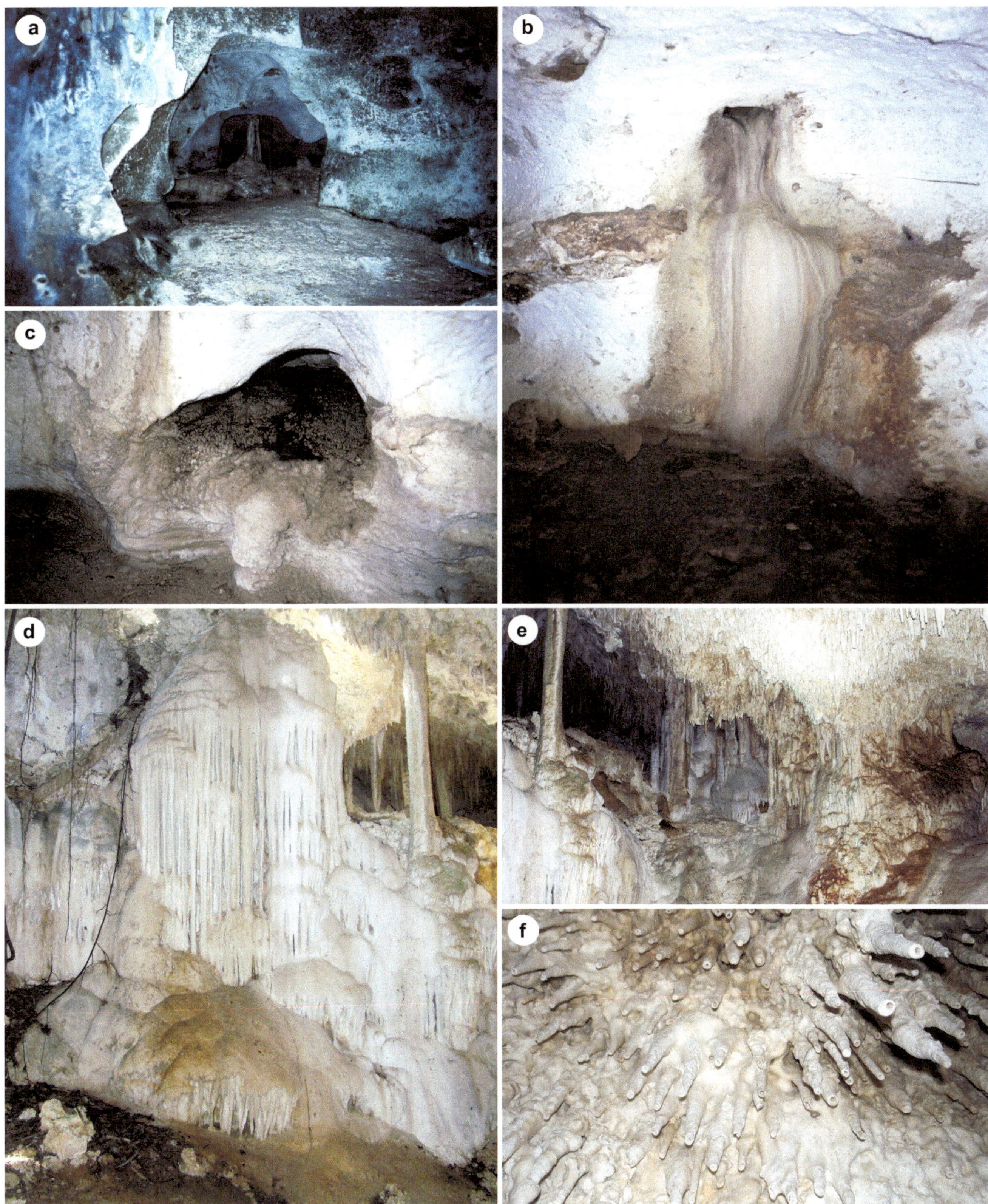

Fig. 9.5 Caves on Cayman Brac. **a** General view of cave, north coast of Cayman Brac, opposite hospital. Note general lack of speleothems and bare walls. Chamber is ~4 m high. **b**, **c** Different parts of cave shown in panel **a**, showing tunnels with speleothem precipitates on floor and extending into main cave. Tunnel in **b** is ~1 m wide; tunnel in **c** is ~2 m wide. **d–e** General views of speleothems in Nani Cave, which is located on top of the bluff on Cayman Brac, to the west of Bluff Road. Panel **d** is ~10 m high, panel **e** is ~5 m high at maximum; panel F, broken end of stalactites (upper right) are ~2 cm diameter

and stalactites indicates that their growth commenced ~200 ka, long before the formation of the notch that developed ~125,000 years ago. The areas around the entrances of the notch caves are commonly characterize by upward tapering bell holes, 0.25–1.3 m in diameter and 0.2–5.7 m high, that rise vertically upwards from the cave ceilings (Tarhule-Lips and Ford 1998b). Tarhule-Lips and Ford (1998b) argued that these features formed in a subaerial setting through the combined effects of microbiological activity and condensation corrosion.

Tarhule-Lips (1999) showed that the upper caves are more varied than the notch caves and divided them into the (1) vertical joint (e.g., First Cay Cave), (2) low angle fissure (e.g., Peters cave), and (3) flank margin (e.g., Tibbet Turn cave) types and suggested they are all phreatic caves. Today, the temperatures in these caves are relatively steady at 25.5–26.5 °C with relative humidity of 100%.

Cave Speleothems

According to Tarhule-Lips (1993, her Fig. 5.8), the cave speleothems on Cayman Brac, up to 500,00 years old, developed through episodic growth periods that were separated by hiatuses that commonly lasted for long periods of time. She suggested that over the last 125,000 years, growth decreased relative to that between 125,000 and 500,000 years ago and Tarhule-Lips and Ford (1996) argued that speleothem growth ceased between 75 and 20 Ka. Tarhule-Lips (1999) divided the speleothems into the "old" and "young" groups for before and after ~120,000 years ago. Most of the young speleothems are found in the notch caves, whereas the old group come from the "upper caves" that are located at various heights above the wave-cut notch.

Nani Cave, a recently discovered cave on top of the bluff (Fig. 9.1c), has provided additional perspectives about the caves and their speleothems (Jones 2011b). This cave is highly decorated with numerous, densely packed stalactites, scattered stalagmites, and areas of flowstone that cover the cave floor (Fig. 9.5d–f). Dark, organic-rich laminae in many of the stalactites developed as biofilms that are characterized by a diverse biota of actinomycetes, calcified filaments, and extracellular polymeric substances (Jones 2011b). Associated with these biofilms are various types of calcite crystals that grew in the extracellular polymeric substances, and in some areas, etching of the substrate. Many speleothems in this cave have a reddish hue and a thin layer of terra rossa covers part of the cave floor (Fig. 9.5d, e). Many stalactites are highly ornate. The stalactites in this cave are formed of aragonite and calcite along with minor amounts of Ca-rich dolomite (identified from XRD analysis), gypsum, and Mg-Si needles (Jones 2010b). The dolomite is divided into the blocky, filamentous mat, crust, and "beehive" types, whereas the gypsum is evident as sheet and tabular types. Jones (2001) demonstrated that microbes play an important role in the genesis of many cave deposits through a variety of constructive and destructive processes. The constructive processes may include microbe calcification, trapping and binding of detrital particles, and precipitation of new crystals. In contrast, the destructive processes may include boring and/or etching of substrates. Jones (2010b) recognized six microbe morphotypes (M1 to M6) based on the size of the hyphae, branching, and various other physical criteria. Microbes, including filamentous forms, spores (probably actinomycetes) and associated exopolysaccharides (EPS) are unevenly distributed throughout the speleothems. The close association between the dolomite, Mg–Si needles and the EPS implies that the precipitates may be genetically linked to the EPS. Dolomite precipitation in caves has been attributed to many different factors including Mg poisoning, high Fe and Mn contents in the water, Sr concentrations, temperature, humidity levels, sulfate poisoning, saturation levels and precipitation rates, CO_2 content of cave air, substrate type, and composition of bedrock (see Jones 2010b for appropriate references for all of these possibilities). The close association between the dolomite and the microbes in Nani Cave led Jones (2010b) to suggest that they may be genetically related.

Tarhule-Lips (1999) demonstrated that the speleothems from caves on Cayman Brac developed through alternating periods of growth and periods of little or no growth. She noted, however, that the growth hiatuses in the speleothems occurred at different times and that not all speleothems stopped growing at the same time. Speleothem growth and growth cessation may, in some cases, be related to climate conditions within an individual cave as opposed to external climate changes. Tarhule-Lips (1999) also demonstrated that the growth rates of the speleothems vary from cave to cave and that such variance was also apparent in individual speleothems.

Jones (2009b), based on samples from the Old Man Village cave, showed that active growth of speleothems is commonly replaced by corrosion that leads to the development of solution-widened boundaries between adjacent crystals, microcavities, and etched crystals. Parts of the microbial mats were replaced and/or coated with calcium phosphate. Mineralization of the microbes, which may be actinomycetes, must have been rapid because the eight morphotypes of hyphae and nine morphotypes of spores are superbly preserved.

Notch Speleothems

Numerous notch speleothems, including stalactites, stalagmites, and columns are found in the wave-cut notch (Fig. 9.6) that is evident in the cliff faces on the western part of Cayman Brac (Jones 2010c; Jones et al. 2018). Transverse cross-sections through speleothems from locality CUT, located on the south coast of the island, show that they are

Fig. 9.6 View of speleothems associated with wave-cut notch that is cut into the vertical cliff face on the south side of Cayman Brac, close to locality CUT

formed of discrete packages of laminae that are separated by erosional discontinuities (Fig. 9.7). U/Th dating of the speleothems supports the notion that these discontinuities represent substantial periods of time when speleothem growth ceased.

The notch speleothems are formed of small and large crystals of aragonite (Fig. 9.8) that are intercalated with crystalline calcite (columnar, fibers, grain-coating mats) and minor amounts of halite, dolomite, gypsum, and a Mg-Si precipitate (kerolite?). Laminae formed of aragonite are characterized by bundles of radiating crystals with crystals up to 3–8 µm wide and 300 µm long (Fig. 9.8b, c). The calcite laminae are formed of compact columnar crystals that are 80–100 µm wide and 200–300 µm long (Fig. 9.8a, b, d), columnar open crystals that are ~120 µm wide and 200–300 µm long, elongate columnar crystals that are 100–150 µm wide and 0.6–1.5 mm long, mosaics with crystals 200–300 µm long, and microspar with crystals ~300 µm long. The relationship between calcite and aragonite is complex and variable at all scales. Dissolution discontinuities commonly cross-cut and truncate the laminae. The green hue on the surfaces of these speleothems attests to colonization by microbial mats that are characterized by a diverse biota of filamentous and coccoid microbes. Microbes and/or microborings are evident in some laminae (Fig. 9.8e).

The well-defined laminae in the speleothems are variable in thickness up to a maximum of 2 mm with individual laminae prone to lateral thinning and swelling. Individual laminae may be formed entirely of aragonite, entirely of calcite, or a mixture of calcite and aragonite. Growth hiatuses in these speleothems are evident from changes in colour, mineralogy, crystal fabrics, truncation of laminae, and/or microbial borings.

U/Th dating of the notch speleothems shows that the major growth phases were between ~73,000–59,000 years ago, 49,000–46,000 years ago, and ~4,000–1,000 years ago (Zheng 2017; Jones et al. 2018). These periods of active speleothem growth were interrupted by extended periods when growth was minimal and erosion was the norm. The growth hiatuses are evident from changes in colour, mineralogy, crystal fabrics, and truncation of the crystals, presence of bored surfaces, and significant differences in ages of the laminae as determined by U/Th dating. In the cross-section through one of the stalactites there is evidence of three major hiatuses with each sector also being characterized by minor hiatuses.

9.2.3 Caves and Speleothems on Grand Cayman

On Grand Cayman, known caves are restricted to the north-central part of the island where there is an elevated ridge of dolostones (Cayman Formation), up to ~10 m asl, that parallels the north coastline of the island. These include the informally named Old Man Village cave (Jones 2009a), located to the east of Old Man Village and the "Crystal Cave" system (Fig. 9.1a) that includes dozens of caves of various sizes, located to the west of Old Man Village. This system was opened to tourists in 2016. The largest cave in the Crystal Cave system includes an underground freshwater lake (Fig. 9.9a) that is, in reality, the near-surface expression of the freshwater North Side Water Lens that is located in

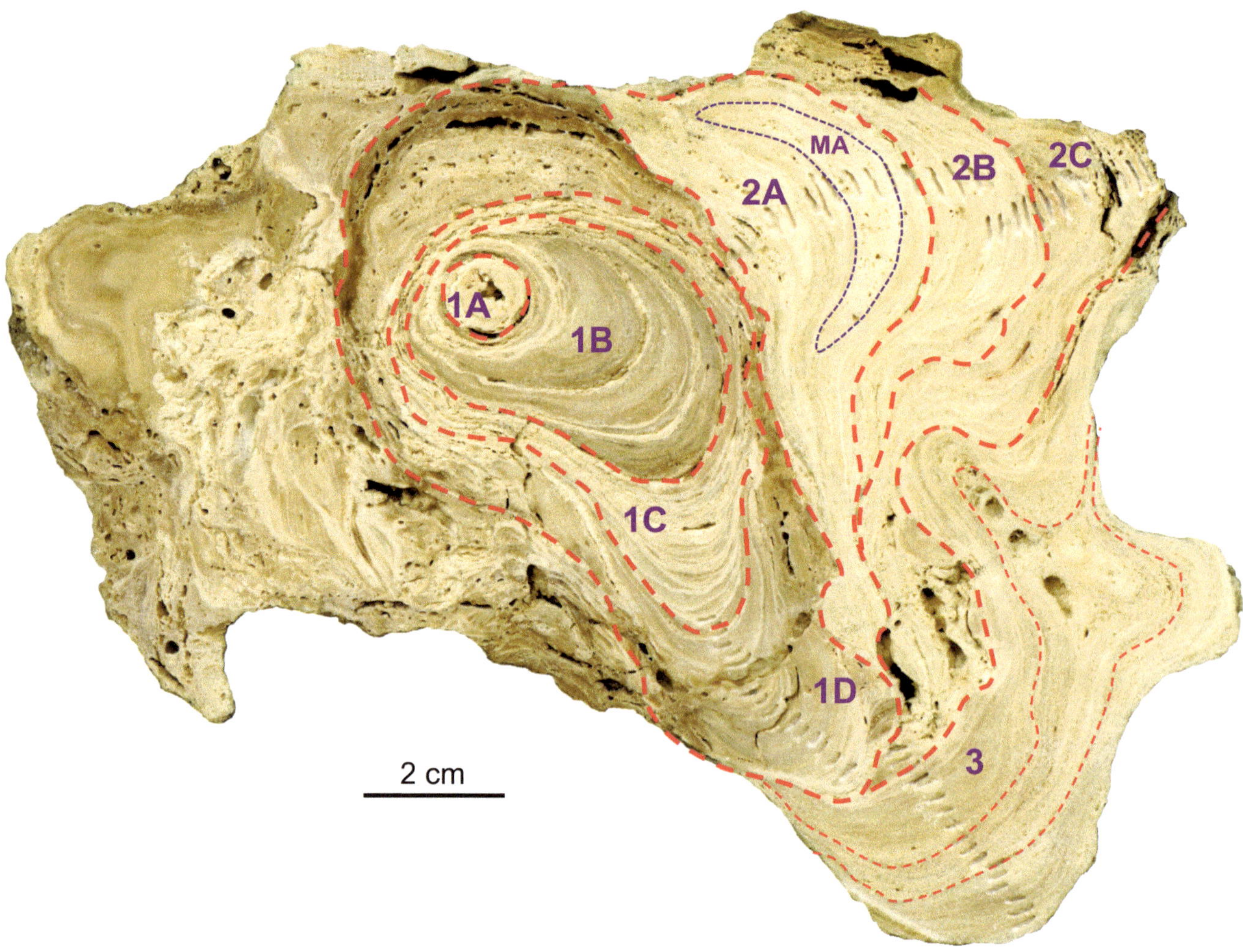

Fig. 9.7 Transverse cross-section through stalactite from locality CUT showing laminae packages 1, 2, and 3 separated by growth hiatuses (red dashed lines). Minor hiatuses are also evident within each of the main packages of laminae. From Jones et al. (2018, their Fig. 4a)

that part of Grand Cayman. The caves in this system are decorated with vast numbers of highly ornate stalactites, stalagmites, and columns that are of variable sizes (Fig. 9.9b, c). Many of the speleothems are characterized by diverse arrays of microbes that appear to have mediated both destructive and constructive processes (Jones 2010a). Collectively, the speleothems in those caves attest to the vast volumes of water that must, at one time, have flowed into them.

The Twilight Zone

The "twilight zone" of a cave, which is the entrance area of a cave, is characterized by a progressive decrease in light levels, and represents the transition from the outer, well-illuminated environment to the darkness of the cave interior. The walls of the twilight zone commonly have a distinct greenish hue (Fig. 9.10b) that is a reflection of the diverse microbial biota that thrives in these reduced light conditions (Cox 1977; Cox and Marchant 1977). Based largely on samples collected from Old Man Village Cave on Grand Cayman, and to a lesser extent on samples from Bats Cave and the Hospital Cave on Cayman Brac, Jones (1995, his Figs. 2 and 3) demonstrated that the diverse microbial biotas that thrive on the cave walls in the twilight zones were responsible for various microscale processes. The walls are coated with a 100–200 µm thick biofilm that is formed of large amounts of mucus that protects the bacteria, various types of filamentous microbes, and different types of spores and sporangia. The dolomite and calcite substrates underneath these biofilms are extensively etched. Constructive processes operative within the thin biofilms included calcification of the microbes, trapping and binding of detrital grains, and some precipitation of calcite, dolomite, gypsum, halite and sylvite that probably formed from through evaporation of seawater. It appears, however, that the destructive processes outpace the constructive processes.

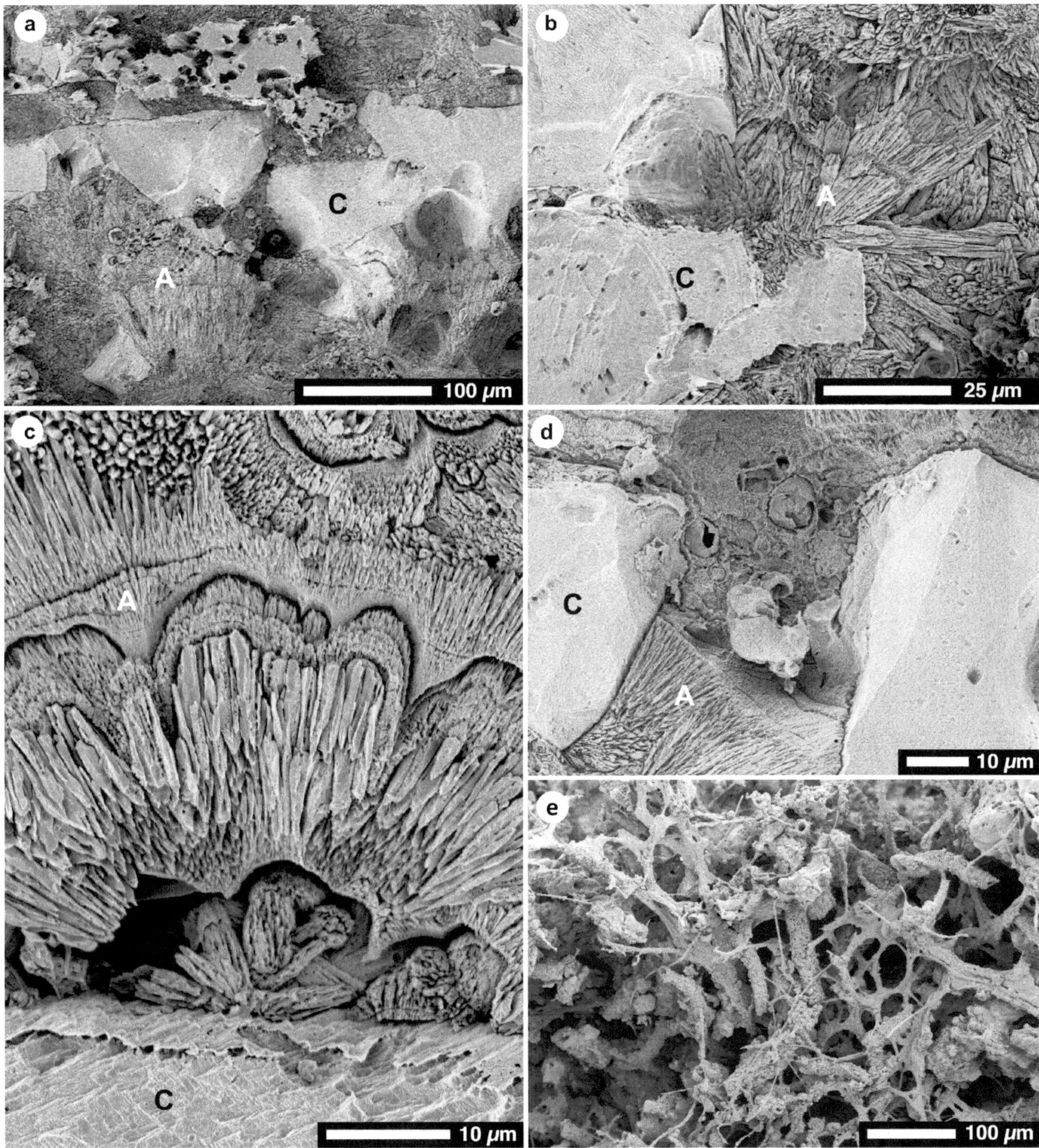

Fig. 9.8 SEM photomicrographs showing microscale relationships between aragonite and calcite in stalactite from locality CUT on Grand Cayman. A = aragonite; C = calcite. **a** Interlaminated aragonite and calcite. **b** Aragonite crystals radiating from surface of a calcite crystal. **c** Succession of aragonite laminae with each one being formed of crystals of different sizes. **d** Aragonite growing between calcite crystals. Note spherical microbes in upper middle part of image. **e** Network of calcified filaments in calcitic laminae

Fig. 9.9 Crystal Caves, north coast of Grand Cayman. **a** General view of cave with large underground freshwater lake (North Side Water Lens). **b** Stalactite cluster, stalactite on left side of image is ~1.5 m long. **c** Small alcove on side of larger cavern with densely packed arrays of stalagmites and stalactites. Floor covered with thin layer of terra rossa. Stalagmite in right-centre position is ~1 m long

Fig. 9.10 Comparison of cave surfaces in twilight zone (near entrance) and the dark, interior, Old Man Village Cave (Fig. 9.1a). **a** Entrance to cave amid lush tropical vegetation and phytokarst surface on exposed dolostones of the Cayman Formation. **b** Twilight zone close to cave entrance where walls are coated with a thin green microbial mat that thrives in low light level. **c**, **d** Stalactites **c** and stalagmites **d** in the dark, interior of the cave (no light) characterized by clean surfaces that are devoid of the green mats evident in the entrance zone. Red colouration is due to coating of terra rossa dust that has been washed into the cave following major rainstorms

Stalactites and Stalagmites

Many caves in the Crystal Cave system are decorated with visually stunning arrays of stalactites and stalagmites of variable sizes and shapes (Fig. 9.9). In some areas they are widely spaced whereas in other areas, they are densely crowded together. Many of these speleothems are characterized by complex surface decorations that commonly have a reddish hue due to the terra rossa that has been brought into the cave with the water from which the speleothems are precipitated.

Jones and Motyka (1987), showed that some of the stalactites from the Old Man Village Cave and a small old, unnamed and filled cave in the Cayman Formation near Breakers (south coast of Grand Cayman) contained laminae and bulbous masses of micrite. One stalactite from Breakers, for example, had an open, central soda straw that was surrounded by core, up to 1.7 cm in diameter, formed of radiating outward expanding, internally laminated columns. The core tapered distally in accord with the external stalactite shape. The sequence of laminae in each column is the same. The micrite cores of these stalactites were encased by concentric crystalline calcite laminae.

In the cores of the stalactites, the micrite laminae commonly contain black ovate bodies (up to 0.3 mm long) and rod-shaped bodies that are probably bacteria. Calcified filaments, commonly >2 mm long, were also present locally. Jones and Motyka (1987) suggested that the micrite may be detrital grains transported from elsewhere, inorganic precipitates formed from the cave waters, organic precipitates, and/or sparmicritization. They argued that much of the micrite may have formed through the breakdown of calcified algal filaments.

Cave Pearls

The Old Man Village cave and one cave in the Crystal Cave system include shallow pools (now dry) that contain numerous spherical to discoid coated grains (Fig. 9.11). Although colloquially known as "cave pearls", they are really cave pisoids or cave pisoliths that formed as calcite was progressively precipitated around various types of nuclei. The cave pools are situated on terraced mounds that are up to 2 m in diameter and typically ~0.5 m high (Fig. 9.11a, b). The surfaces of the mounds are characterized by rimstone dams (up to 2 cm high) that encase shallow pools that are typically up to ~6 cm long, ~3 cm wide, and <2 cm deep (Fig. 9.1b–d). The pool on top of the mound is up to 2 m in diameter and up to 6 cm deep. Water is typically fed into the pool via a channel that commonly extends back into the bedrock. Today, none of these pools contain water and there is no evidence of recent water flow from the feeder conduits. Many of these pools contain numerous coated grains, termed cave pisoliths by Jones and MacDonald (1989), that are highly variable in terms of their size, shape, external appearance, type of nucleus, and the type and morphology of their cortical laminae (Fig. 9.11c–g). These pisoliths, with smooth or reniform outer surfaces, are subspherical, lensoid, discoid, or roller in shape. About 30% of the pisoliths developed around small bone fragments that were derived from birds (Fig. 9.11e), or bats and 5–10% of them formed around small, lithified masses of terra rossa. Some of the larger pisoliths, up to 10 cm long, are shaped like a dog biscuit, whereas other elongate forms have a bulbous mass at one end (Fig. 9.11e). These shapes reflect the shapes of the bones that form their nucleii. The concentric, pseudo-concentric, or mamillated cortical laminae are 0.02–0.04 mm thick (Fig. 9.11f) and commonly thinner on their bottom than their sides and top. Many laminae are characterized by a reddish hue (Fig. 9.11g) that reflects the amount of terra rossa that has been included in the laminae. Jones and MacDonald (1989) also showed that various types of calcified and non-calcified filamentous microbes and spherical to subspherical microbes are present in the cortical laminae. The laminae are formed of micrite, dendrite crystals (commonly filled-in), and trigonal crystals. The micrite laminae may have formed through the calcification of filamentous microbes, whereas the dendritic and trigonal crystals probably formed from waters with different levels of saturation. These cave pisoliths are different from those found in one of the unnamed caves in the Crystal Cave system (Fig. 9.11b–d) and the Great Cave on Cayman Brac (Fig. 9.4c–d). Some pisoliths, like those from the Great Cave, appear to have formed without the microbial activity being involved.

Calcite Versus Aragonite Precipitation

Primary calcite and aragonite precipitates are important components of the notch speleothems, where they are present in alternating laminae or in the same laminae with the calcite grading laterally in aragonite and/or calcite growing around the aragonite (Fig. 9.8). Similar alternations between aragonite and calcite are also found in cave speleothems on Grand Cayman and Cayman Brac. The factors that control the precipitation of these two $CaCO_3$ polymorphs are poorly understood and many different theories have been proposed for their precipitation in caves. Jones (2017) noted that aragonite precipitation in caves has been variously attributed to temperature, supersaturation levels with respect to calcite and aragonite, CO_2 concentrations, nucleation issues related to the substrate mineralogy, high Zn concentrations, drip water that controlled the rate of CO_2 degassing, and/or the Mg/Ca ratio of the drip waters. In some cases, the precipitation of aragonite and calcite has been attributed to seasonal changes that influenced the water saturation states with respect to calcite and aragonite.

The drip waters in the modern caves and shallow groundwater on Cayman Brac vary from saturated to undersaturated with respect to dolomite, calcite, and aragonite (Ng 1990; Tarhule-Lips 1993). The chemistry of the groundwater housed in the dolostones of the Cayman Formation (to a depth of 20 m below surface) on Cayman Brac varies from place to place. Ground water from the eastern part of Cayman Brac had Mg concentrations of 38–167 mg/l and Ca concentrations of 37–336 mg/l (Ng 1990, Table III.6). Mg/Ca ratios range from 0.7:1 to 0.8:1 for water from the perched water lens to 1.6:1 and 0.8:1 for samples from water collected from other wells. Rainwater on Grand Cayman had Mg concentrations of 0–7 mg/l and Ca concentrations of 8–20.4 mg/l with Mg/Ca ratios of 0.06:1–0.2:1 (Ng 1990, his Table III.2). It must, however, be remembered that there is presently very little active precipitation in the caves and notch speleothems. On Cayman Brac, for example, water dripping from the tips of stalactites (Fig. 9.4c) is only evident for a few days following periods of heavy rainfall. Thus, the premise that the modern geochemical parameters of the groundwater can be used to explain the distribution of calcite and aragonite in the speleothems must be treated with some caution.

Comparison of Notch Speleothems and Cave Speleothems

The $\delta^{18}O$ values for the cave speleothems on Cayman Brac range from −1.16 to −7.04‰, whereas the $\delta^{13}C$ values vary from −0.40 to −12.13‰ (Tarhule-Lips and Ford 1998b; Tarhule-Lips 1999). These values include the "old group" with average $\delta^{18}O$ values of -5.43 ± 1.01‰ VPDB,

Fig. 9.11 Cave pisoliths from caves on Grand Cayman. Panels **a–d** from unnamed cave in the Crystal Cave system; **e**, **f** from Old Man Village cave (Fig. 9.10). **a** General view of stalactites and stalagmites with red terra rossa on floor that has been transported into the cave. **b** Terraced mound with shallow rimstone pools that contain numerous cave pisoliths. **c**, **d** Views of different cave pools showing variations in size and shape of densely packed cave pisoliths. **e–g** Cross-sections through cave pisoliths showing variations in growth development and structures in the cortices. The nucleus in the pisolith in **e** is bone that may have come from a bird or a rat

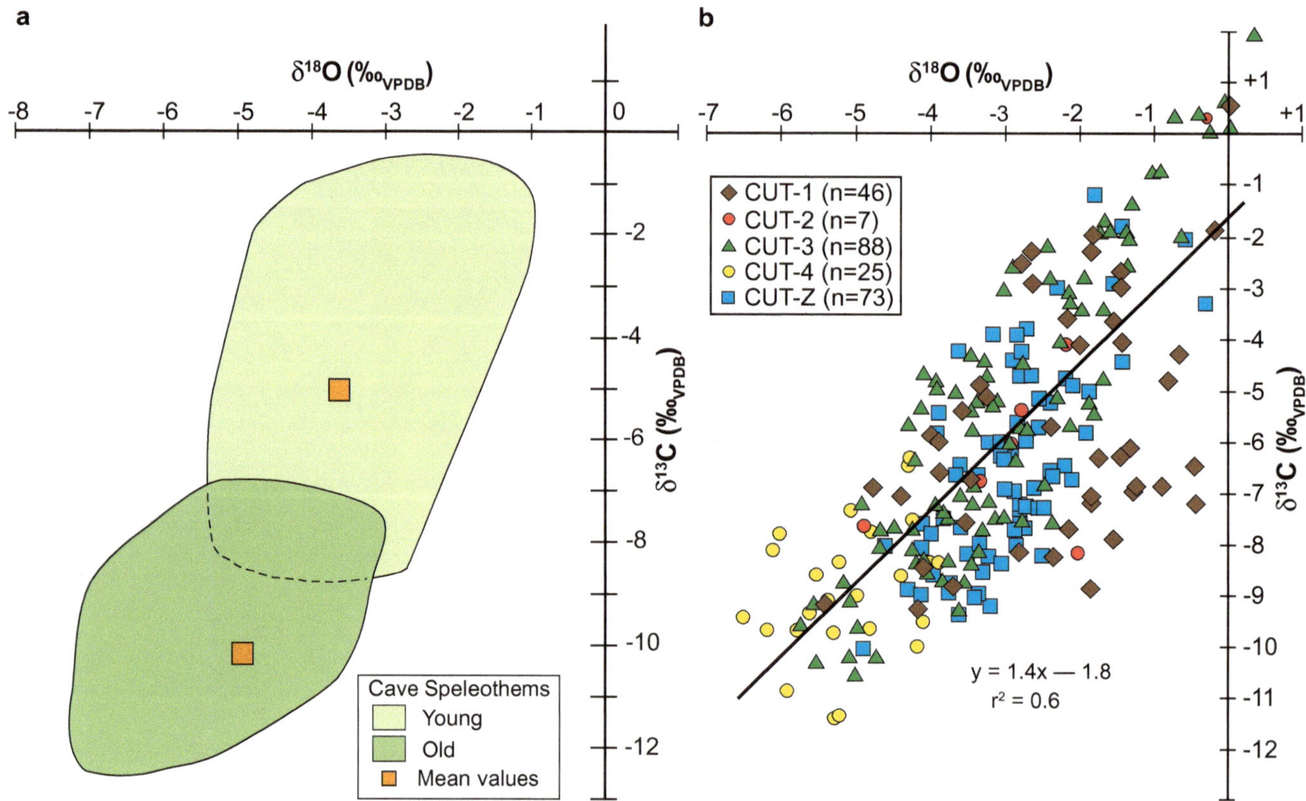

Fig. 9.12 Stable isotopes for cave and notch speleothems on Cayman Brac. **a** Comparison of stable isotopes for "old" and "young" cave speleothems. Adapted from Tarhule-Lips (1999, her Fig. 3.14). **b** Comparison of stable isotope values for five different notch speleothems from locality CUT. Modified from Jones et al. (2018, their Fig. 6a). Note similarity to trend derived from the cave speleothems shown in panel **a**

whereas the average $\delta^{13}C$ for the young group is −3.42 ± 0.84‰ VPDB. For the $\delta^{13}C$, the equivalent values are −10.17 ± 1.13‰ and −4.94 ± 2.40‰ VPDB, respectively (Fig. 9).

The $\delta^{18}O$ values for the notch speleothems (based on 285 analyses) range from −6.5 to +0.3‰ with an average of −3.1‰ whereas the $\delta^{13}C$ values range from −10.5 to +1.9‰ with an average of −6.1‰. The values obtained from the notch speleothems overlap with both the young and old groups of the cave speleothems (Fig. 9.12).

Precipitation of the calcite and aragonite in speleothems, irrespective of whether they develop in caves or the wave-cut notch, is ultimately controlled by the rainwater that seeps down through the bedrock to the site of precipitation. Today, rainwater on Grand Cayman has $\delta^{18}O_{SMOW}$ values that range from −1.6 to −7.3‰ with the heavier values coming from low-intensity rainfalls (Ng 1990). Freshwater from perched water lens on Cayman Brac has $\delta^{18}O_{SMOW}$ values of −5.2 to −5.5‰ (Ng 1990). Following a period of heavy rain in June, 2018, seven samples of drip water collected from the tips of four stalactites in Peters Cave, and three in Great Cave (Fig. 9.4c), yielded $\delta^{18}O_{SMOW}$ values of −5.17 to −3.43‰. The isotope values in the speleothems should reflect those of the rainwater and any changes that took place as the water seeped down through the soils and bedrock above the caves/notch. Theoretically, where the bedrock and soils above a cave are thin, minimal changes will happen to the rainwater as it seeps down into the cave.

9.3 Vertebrate Fossils

Fossil vertebrates reported from Cayman Islands have generally been found in (1) sediments that have accumulated in caves and fissures, and (2) organic sediments (peat) that are found in cow wells or mangrove swamps that now cover the karst surface (Morgan 2017). Morgan et al. (2019) described the history of vertebrate discoveries on the Cayman Islands since they were first noted by Lord Moyne in 1938. Morgan (1994) noted 31 known vertebrate deposits and other localities have been discovered since then (e.g., Morgan and Albury 2013; Harvey et al. 2019). Morgan and Patton (1979) were the first to report the presence of the crocodiles on the islands when they discovered a tooth in Crab Cave, which is located near East End on Grand Cayman. Since then, fossils found in the cave and fissure sediments and peat deposits include crocodiles (*Crocodylus rhombifer*), frogs, lizards, birds, and rats with many species no longer being found on the islands (Steadman and Morgan 1985; Morgan and Albury 2013; Morgan 2017). Crocodiles from two sites on Grand Cayman yielded radiocarbon dates of 860 ± 50 yr BP and 375 ± 60 yr BP (Morgan et al. 1993).

The fossil evidence indicates that crocodiles were once common on the Cayman Islands as has been noted in many historical records (Morgan et al. 2019). Although there have been rare reports of the American alligator found in the waters around Grand Cayman, living Cuban crocodiles disappeared from the islands in the 1800s (Morgan 2017). Extensive vertebrate extinctions throughout the West Indies since the late Pleistocene have generally been attributed to climate change, rising sea levels, and/or the arrival of human colonists (Morgan et al. 2019).

9.4 Paleokarst on the Cayman Islands

As demonstrated by Jones and Hunter (1994), Jones (1994b), and Liang and Jones (2015a), the unconformities that form the boundaries between the formations in the Bluff Group and the boundary between the Bluff Group and the overlying Ironshore Formation are, in reality, karst surfaces that developed during periods when the islands were subaerially exposed. Throughout the Cayman Formation and Pedro Castle Formation there are also numerous examples of subsurface karst features that developed in association with those paleokarst surfaces.

9.4.1 Paleokarst Surfaces

Knowledge of the paleokarst surfaces is restricted, to varying degrees, by available exposures and data from wells that have drilled on the islands. Accordingly, the Cayman Unconformity that defines the upper boundary of the Cayman Formation is well known, whereas the Brac Unconformity that defines the upper boundary of the Brac Formation and the Pedro Castle Unconformity that delineates the upper boundary of the Pedro Castle Formation are less well known.

The Brac Unconformity

In photographs of the sheer cliff faces on the east end of Cayman Brac, it seems that the Brac Unconformity may be as high as 30 m above sea level (Figs. 3.2, 9.4a). Tracking of that boundary in the north- and south-facing cliffs to the west shows that height of the unconformity declines westward before finally disappearing below ground level. In well EOR#1, which is only ~1.7 km to the west of the eastern end of the island, the base of the Ironshore Formation is 30 m below sea level (bsl). Although the exact subsurface relationship between the Ironshore Formation and the underlying Brac Formation is not known, it does show that there is at least 60 m of relief on that boundary. Similarly, on the south coast, there is clear evidence of the Brac Unconformity becoming progressively deeper to the west (Fig. 3.1). In well CRQ#1, the boundary is 65.5 m bsl, whereas in wells CUT#1 and SQW#1 the boundary must be >80 m bsl as it was not reached when drilling stopped at the maximum depth possible with the rig that was being used. The real dip on the Brac Unconformity is difficult to precisely define because some of the apparent dips determined from the available wells also encompass the tectonic tilting of the island. Although the angle of tectonic tilting is unknown with certainty, available evidence indicates that it is probably ~1° to the west. Irrespective of the precise angle of tilting, it is apparent that there is considerable relief on the Brac Unconformity.

The relief associated with the Brac Unconformity on Grand Cayman is difficult to establish in detail because that surface is deeply buried and has only been located in a few wells. Based on available information, McCormick (2019) and McCormick and Jones (2021) showed that the Brac Unconformity has a subdued relief on the western part of the island, with only a gentle dip to the north, from 72 m bsl in well CUC#4 (near George Town) to 83 m in well TW#2 (near West Bay). In contrast, there is a greater increase in depth to the east as it is found at a depth of 123 m in LV#2 and 129 m in RTR#1. Based on these numbers, it appears that there is at least 57 m of erosional relief on this unconformity. Well NSC#1/2, located in the east central part of Grand Cayman, that was drilled to a depth of 244 m is excluded from this assessment because the position of the Brac Unconformity in that well has yet to be located with confidence.

The Cayman Unconformity

The Cayman Unconformity, which is best known from Grand Cayman, is the upper boundary of the Cayman Formation. As demonstrated by Jones and Hunter (1994) and Liang and Jones (2015a), this is a complex unconformity that is characterized by significant topography and one that is still evolving today. On the western part of Grand Cayman, the unconformity is largely buried beneath the Pedro Castle Formation, and in some areas, the Ironshore Formation (Fig. 9.13). Accordingly, the unconformity on that part of the island is a record of the topography that existed prior to it being buried by the sediments of the Pedro Castle Formation. On the eastern part of the island, however, the situation is more complicated. If it is assumed that the eastern area was once buried by the Pedro Castle Formation and perhaps in part by the Ironshore Formation, then those formations have since been removed by erosion so that the

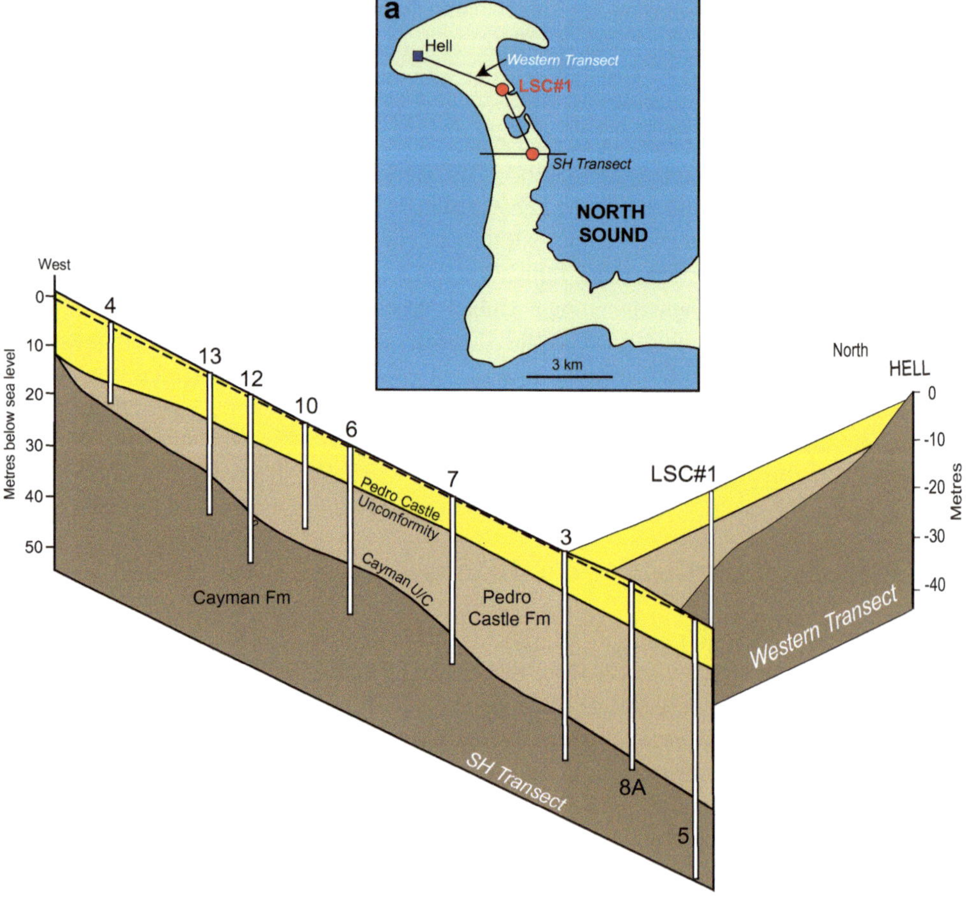

Fig. 9.13 East–west and north–south cross-sections showing stratigraphic relationships between the Cayman Formation (dark brown), Pedro Castle Formation (light brown), and Ironshore Formation (yellow). Note topography on the Cayman Unconformity. Based on information from wells drilled at various locations

rocks of the underlying Cayman Formation became exposed once again and are now being modified as modern weathering affects the area. Similarly, it is the dolostones of the Cayman Formation that form the uplifted core of Cayman Brac and are still being modified by modern weathering. Jones and Hunter (1994) originally modelled the surface of the Cayman Unconformity by combining well information from the western part of the island and modern topographic data for the eastern part of the island as determined from the 1988 topographic maps (published by the Cayman Islands Government) and spot heights taken from the 1:2500 maps that had been determined from photogrammetric analysis of air photographs. Liang and Jones (2015a) subsequently updated those models by using digital elevation models (based on digital terrain data supplied by the Land and Survey Department of the Cayman Islands), air photo interpretations, and field observations. They included models of the Cayman Unconformity for Grand Cayman (Fig. 4.9a) and Cayman Brac. The lack of well information for Little Cayman precluded the development of an equivalent map for that island.

On Grand Cayman, the main features apparent from modelling the topography of the Cayman Unconformity (Fig. 9.14a) include the following.

(1) On the western part of the island, a large basin developed where North Sound is now located. Between the floor of that basin and Pedro Castle, there was a least 40 m of relief on the unconformity. That may be a minimum figure because the true depth of the unconformity beneath North Sound is unknown (Fig. 9.14a).
(2) In Pedro Castle Quarry the unconformity is up to 10 m asl, whereas in well LV#2 it is ~30 m bsl, and the top of the Mountain at an elevation of 16 m asl, is formed of dolostones that belong to the Cayman Formation. Between LV#2 and the Mountain, the relief is ~46 m.
(3) An elevated ridge on the eastern part of the island that parallels the coastline and rises to ~8 m asl. Inland of that ridge, the surface of the Cayman Formation is commonly a few metres below or above sea level (Fig. 4.9a) and commonly covered by the Ironshore Formation.

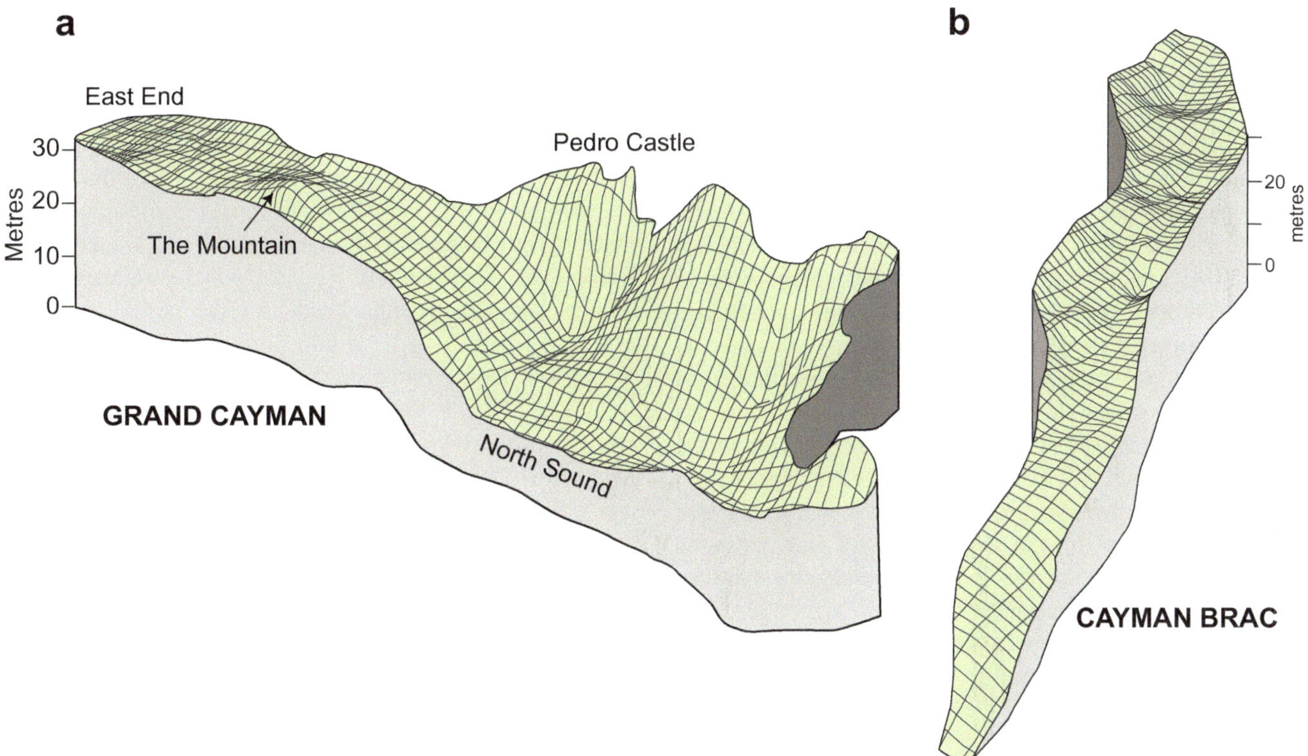

Fig. 9.14 Three-dimensional model of Cayman Unconformity on **a** Grand Cayman and **b** Cayman Brac based on all available well data and modern-day surfaces where the Cayman Formation is exposed. Modified from Liang and Jones (2015a, their Figs. 3 and 5, respectively)

These examples clearly illustrate the relief that characterizes the topography that exists on the Cayman Unconformity on Grand Cayman (Fig. 9.14a). On the eastern part of the island, where the Cayman Formation is widely exposed at the surface, modern weathering continues to deflate that surface and reduce the relief that is apparent from some of the comparisons noted above. Irrespective of these nuances, the topography that existed on the Cayman Unconformity on Grand Cayman is far greater than that of the modern-day topography.

On Cayman Brac, the situation is more complicated than on Grand Cayman because the island was uplifted and tilted to the west following formation of the Cayman Unconformity (Fig. 9.14b). As a result, the original surface of the unconformity has been subject to considerable weathering ever since it was uplifted and now has a topography that is probably quite different from the original one. At the west end of Cayman Brac, the Cayman Unconformity in SWQ is generally ~3 m asl and displays minimal relief. In well BW#1 the Cayman Unconformity is ~20 m lower than in the quarry, even though they are only 1.3 km apart, and clearly illustrates that there was significant topography on the surface. As demonstrated by Liang and Jones (2015a, their Figs. 6 and 10), there is evidence of peripheral ridges along the north and south coasts at the east end of the island that are akin to those found on Grand Cayman.

The Pedro Castle Unconformity

Given the restricted distribution of the Pedro Castle Formation on Grand Cayman, Cayman Brac, and Little Cayman, relatively little is known about the Pedro Castle Unconformity that defines its upper boundary. Surface exposures of the Pedro Castle Formation are relatively rare and few exposures show the limestones of the Ironshore Formation on top of the upper boundary. Evidence from drilling on the western peninsula of Grand Cayman shows that the Pedro Castle Unconformity is characterized by a relatively flat surface that contrasts sharply with the topography of the Cayman Unconformity (Fig. 9.13).

On Little Cayman, the Pedro Castle Unconformity used to be evident in the weathered quarry walls where the fossiliferous limestones of the Ironshore Formation filled the rugged phytokarst topography that had developed on the limestones of the Pedro Castle Formation (Fig. 9.3). On Cayman Brac the boundary between the Pedro Castle Formation and the overlying Ironshore Formation is not exposed.

9.5 Subsurface Paleokarst Features

Exposures of the limestones and dolostones of the Bluff Group commonly include clear evidence of various paleokarst features such as caves, sinkholes, and solution widened joints that are open or filled with speleothems and/or various types of precipitates and/or sediments (Smith 1987; Jones and Smith 1988; Jones 1989a, 1992d; Liang and Jones 2014; Liang 2015). In effect, the solution widened joints and other avenues of egress became the channels that allowed rainwater and, after major storms, seawater to transport various types of sediment into the subsurface. As a result, cavities of all types, sizes, and shapes became the receptacles for deposits such as flowstone, marine sediments, terra rossa, various types of breccia, and caymanite. Plants with roots that penetrated the hard-rocky substrates also played an important role in these processes and generated features such as the rootcretes that line the cavities that they created in the rocky substrates (Alonso-Zarza and Jones 1996). The diversity of these precipitates and deposits are best illustrated by using specific examples found on Grand Cayman and Cayman Brac.

9.5.1 Caymanite

Caymanite (Fig. 9.15), the local name for the rock used for making jewelry, is a banded, multicoloured (various shades of white, red, black) deposit that is formed of very finely crystalline dolomite (Lockhart 1986; Jones 1992a). It is a cavity-filling deposit with laminae that commonly appear to be dipping at steep angles in a variety of directions. It was this aspect that was particularly problematic in explaining the origin of caymanite. Resolution of this problem came through the fortuitous discovery of exposures in the outcrops along the coastline near Pedro Castle that showed that the caymanite was commonly deposited within cavities in cone-shaped masses, up to 90 cm in diameter, 40 cm high, with sides dipping at 35–38° (Fig. 9.15a, b) with sediment being fed-in from above the top of the cone (Jones 1992a). If the space in the cavity allowed, the cones have a circular base and are formed internally of laminae that have the appearance of stacked cones and surfaces that dip in all directions away from the central feed point (Fig. 9.15a, b). The situation is more complicated where the feeder channel is located so that the sides of the chamber interfere with deposition of the sediment. Caymanite formed and developed in any type of cavity, including caves and fossil-mouldic cavities (Fig. 9.15c).

Caymanite is formed of mudstones, wackestones, packstones and grainstones that locally may contain foraminifera, red algae, gastropods, bivalves, and detrital grains of microcrystalline dolomite. Sediment structures, evident in some samples, include planar laminations, graded bedding, desiccation cracks, and geopetal fabrics. The colours reflect the presence of Mn, Fe, Al, Ni, Ti, P, and K with the colours generally attributable to the presence of the Mn and/or Fe (Table 9.1). The original deposits, which originated from offshore lagoons, swamps, brackish water ponds, and/or the terra rossa, may have been transported on land during hurricanes and then washed down into the cavities along with the water. The $\delta^{18}O$ and $\delta^{13}C$ values of the caymanite overlap with those of the dolostones in which the caymanite is hosted (Jones 1992a, his Fig. 9).

Available evidence indicates that the caymanite was originally formed of calcite and was transformed in dolomite at the same time as the surrounding bedrock. One specimen of caymanite found in the Brac Formation on Cayman Brac was still formed of limestone.

9.5.2 Paleo-Caves

Quarrying operations on Grand Cayman and Cayman Brac have, over the last 30 years or so, produced clean cross-sections through filled caves in the dolostones of the Cayman Formation and Pedro Castle Formation. Unfortunately, most of those outcrops were destroyed as quarrying continued. Excavation of the southwest corner of Pedro Castle Quarry (Jones 1992d) revealed a filled cave, about 22 m long with a maximum height of 4.5 m, with its floor $\sim$9 m below the Cayman Unconformity (Fig. 9.16a). The cave was filled with dolomitized cross-bedded skeletal grainstones and mudstones, terra rossa, and flowstone (Fig. 9.16a–c). Although subsequent quarrying operations destroyed the filled cave, they did reveal a filled tunnel that extended from the west end of the filled cave and led to the base of sinkhole that was $\sim$12 m wide (in cross-section) and extended upwards to the Cayman Unconformity. The fact that sediments of the Pedro Castle Formation lay on top of the filled sinkhole showed that its formation must have been associated with the development of the unconformity. The grainstones and mudstones that filled the sinkhole are distinctly different from the bedded carbonates on either side of the sinkhole. Those sediments, however, were dolomitized, seemingly at the same time as the surrounding bedrock of the Cayman Formation.

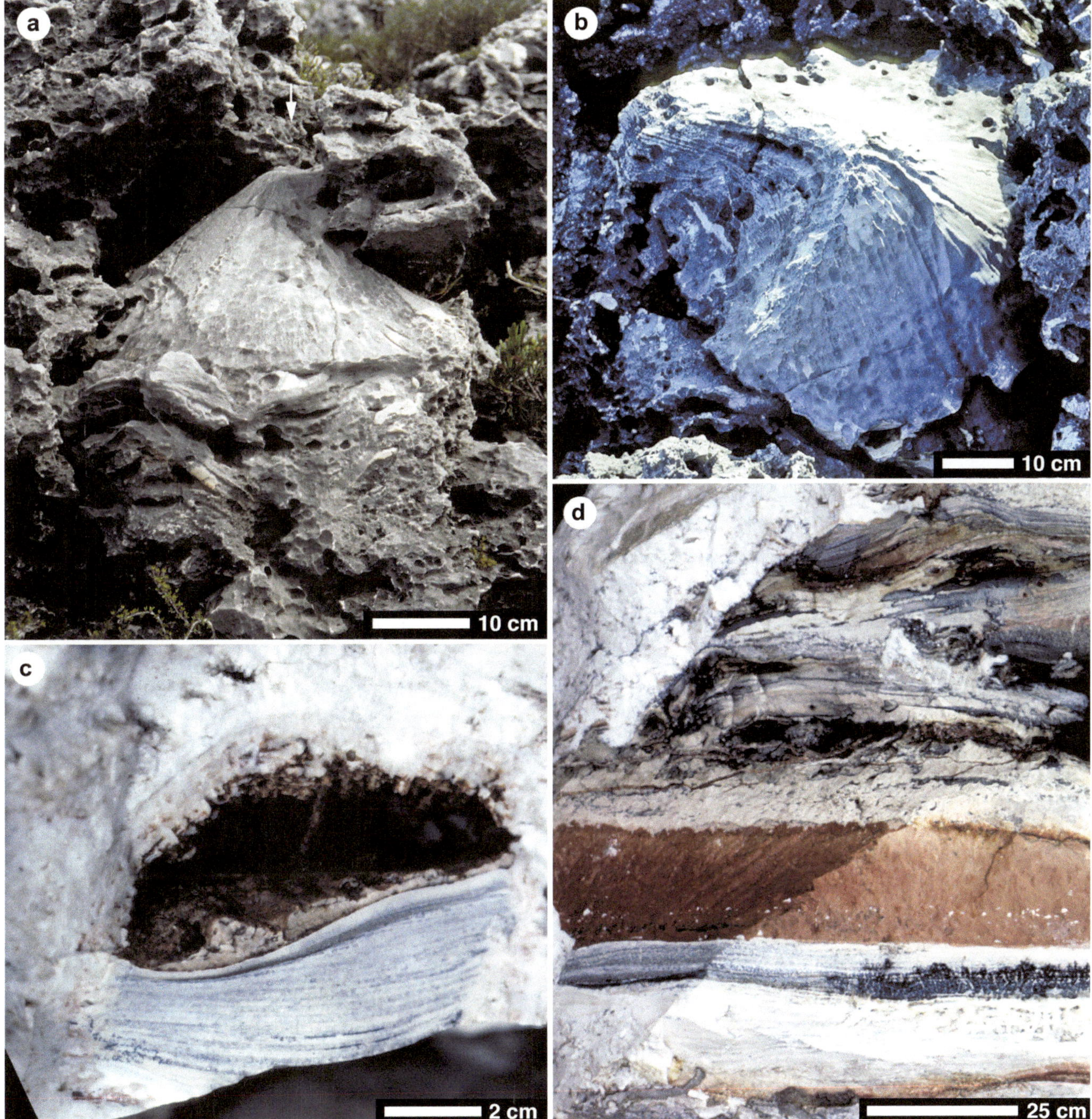

Fig. 9.15 **a** Cone formed of caymanite in cavity in Cayman Formation, south coast of Grand Cayman near Pedro Castle. White arrow indicates point where sediment was fed into the cavity. **b** View of top of cone-shaped deposit of caymanite in cavity in Cayman Formation, south coast of Grand Cayman near Pedro Castle. **c** Leached colonial coral in Cayman Formation partly filled with thinly laminated caymanite. High Rock Quarry, east end of Grand Cayman. **d** Laminated, multicoloured caymanite filling cavity in Cayman Formation, Queen's Highway, northeast Grand Cayman

On Cayman Brac, exposures through the Cayman Formation near the Water Authority site (WA on Fig. 9.1c) revealed a vertical cross-section through a cave that was completely filled with calcitic stalactites, stalagmites, and flowstone (Fig. 9.16d). Another small cave of a similar size, found at locality CUT (Fig. 9.1c), was partly filled with interbedded terra rossa and flowstone that had filled the cave around small stalactites (Fig. 9.16e). The upper part of the

Table 9.1 Elemental compositions of caymanite samples collected from Pedro Castle Quarry. Values are the average of two samples for each colour. Data from Lockhart (1986, Table 1). L.O.I. = Loss on ignition

Constituent	White	Black	Red–Orange
L.O.I	47.52	45.83	46.80
CaO	32.23	32.15	33.21
MgO	20.09	18.02	18.30
MnO	<0.01	3.06	0.09
Fe_2O_3	0.02	0.25	0.99
Na_2O	0.07	0.07	0.08
K_2O	<0.01	0.04	0.05
Al_2O_3	0.02	0.06	0.25
P_2O_3	0.0	0.02	0.0
CuO	Trace	Trace	Trace

cave, which was not completely filled, was decorated with numerous stalactites. Unfortunately, both of these filled caves have since been destroyed.

In the large active quarry on the west end of Cayman Brac, a filled cave complex was revealed in the east wall that spanned the upper part of the Cayman Formation and the lower part of the overlying Pedro Castle Formation (Fig. 9.17). The lower cave (>8 m long, up to 2.5 m high) in the Cayman Formation was connected to the upper cave (>23 m long, up to 2.5 m high) that was in the Pedro Castle Formation via a small-diameter tunnel (Jones 2016). The cave was filled with caymanite that was formed of laminated varicoloured dolomitized mudstones and grainstones that contained foraminifera and red algae. The upper cave was filled with dolostones (with foraminifera and red algae grains), calcareous mudstone, terra rossa, gastropod coquina, oncoids, and speleothems. U/Th dating showed that the flowstones in the lower part of the cave were >500,000 years old, whereas flowstones higher in the cave yielded an age of ~21,000 years. Based on a detailed analysis of these complex successions (Fig. 9.18), Jones (2016) suggested that the lower cave and its deposits formed in conjunction with the development of the Cayman Unconformity and before deposition of the sediments that now form the Pedro Castle Formation. The upper cave formed after the deposition and lithification of the sediments that now form the Pedro Castle Formation. Jones (2016) suggested that development of the caves and the sediments in the upper cave provided a record of a transition from marine to non-marine deposition was controlled by eustatic sea-level fluctuations and possibly, the westward tilting of Cayman Brac that took place during the Late Pliocene. Unfortunately, subsequent quarrying in 2016 and 2017 destroyed the outcrops in which these filled caves were once evident.

9.5.3 Sinkholes, Joints, and Other Cavities

Sinkholes and open, solution-widened joints, and other cavities are common features in the dolostones of the Cayman Formation and Pedro Castle Formation that are commonly filled with a wide variety of deposits (Fig. 9.19). The variance in the size of these cavities reflects their different modes of development. Many of the smaller cavities developed due to the preferential dissolution of the aragonitic skeletons of corals, bivalves, and gastropods. All cavities, irrespective of their size, become places where speleothems and internal sediments may be deposited. The variable types of deposits in some cavities reflect ever-changing sources from where the sediment and precipitates originated.

Sinkholes and other cavities that open at the surface are commonly filled with various combinations of rootcrete (Fig. 9.19a–c), breccias, loose limestone and dolostone lithoclasts, and white, red, and orange limestones, coated grains, and/or terra rossa (Liang and Jones 2015b). Many of the trees that grow on the Cayman Islands seem to root directly into the bedrock where soil has accumulated in small fractures (Fig. 9.19a). As the trees grow, their roots enlarge the cracks in which they are growing. These roots and their associated microbes commonly play an active role in the diagenetic alteration of the bedrock (Jones 1994a) by actively mediating dissolution of the bedrock and/or precipitation of calcite. Many of these root cavities are lined with laminated calcrete crusts (also known as rootcrete—Jones 1992b; Alonso-Zarza and Jones 2007) that are up to 8 cm thick (9.19B, C). As demonstrated by Alonso-Zarza and Jones (2007) these crusts include a corrosion zone in the host dolostone, an alteration zone, and an accretionary zone. The complex accretionary zone may involve various

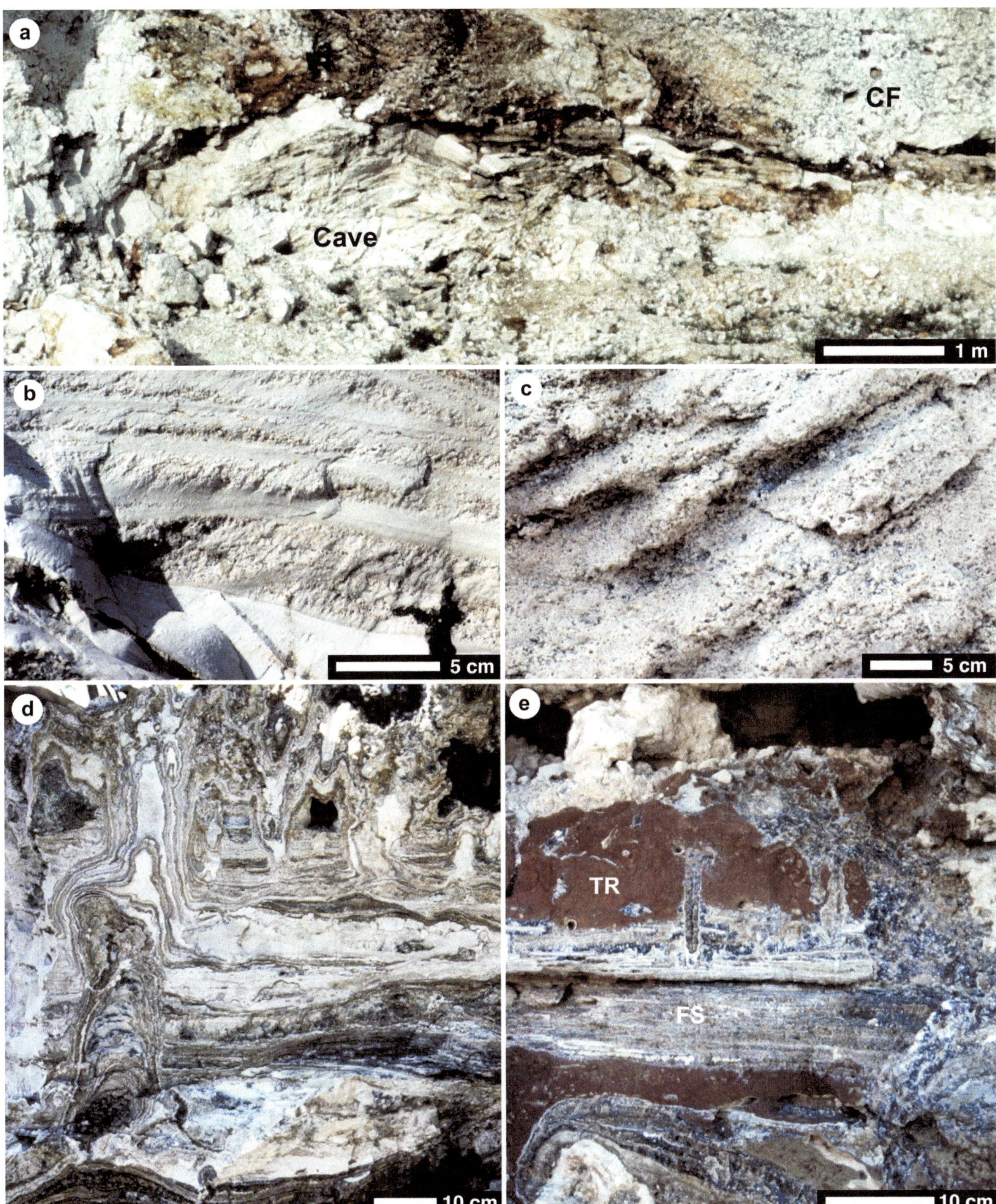

Fig. 9.16 Filled caves. **a** Vertical cross-section through large cave in Cayman Formation in Pedro Castle Quarry (southwest corner) filled with a variety of deposits, including terra rossa, dolomitized grainstones and mudstones, and caymanite. Quarry face was destroyed by continued quarrying in 1998–99. **b** Interlaminated dolomitized mudstones and grainstones in filled cave shown in panel **a**. **c** Cross-bedded dolomitized grainstones in filled cave shown in panel **a**. **d** Vertical cross-section through filled cave in Cayman Formation, west end of Cayman Brac. Note stalagmites, stalactites, flowstone and internal sediments. **e** Cave in upper part of Cayman Formation, near locality CUT, Cayman Brac, partly filled with stalactites, stalagmites, lithified terra rossa (TR), and flowstone (FS). Cave was destroyed by continued quarrying

Fig. 9.17 Filled cave system in upper part of the Cayman Formation and lower part of the Pedro Castle Formation exposed in east wall of quarry on west end of Cayman Brac. Wall was subsequently destroyed by quarrying operations. Caves were filled with complex succession of rock types as shown in Fig. 9.18

combinations of laminar/nodular crusts, dolomite crystals and micritic septae, alveolar septal structures, micrite, peloidal laminae, microspar, dense micrite, and pisolithic crusts. Exogenetic and endogenetic internal sediments that are found in many these cavities formed through many different processes (Jones and Kahle 1995). Microborings, spores, and needle fiber calcite are present in some examples. Laminae in the rootcretes contain numerous different types of filaments and calcified spores (Liang and Jones 2015b). Jones and Kahle (1985) demonstrated that breccias found in the sinkholes have undergone various diagenetic processes that were largely governed by the lichen, algae, and fungi that colonized the surfaces of the lithoclasts. Such processes included the fixing of Mn, trapping sediment to form microstromatolites, and/or inducing the precipitation of aragonite. These processes played a key role in the lithification of the breccias.

Morphological variable but distinct black Mn precipitates, which may be formed largely of birnessite, are common in the karst terrains of the Cayman Islands (Jones 1992c). There is no recognizable pattern to the distribution of these precipitates, which can be found in stalactites, karst breccias, caymanite, terrestrial oncoids, and root calcites, and coating fractures in the bedrock. The Mn, with trace amounts Al, Si, Fe, K, Ti, Na, Mg, and/or Ca, probably originated from the terra rossa soils that are common on the islands. Biogenic processes may have been involved in fixing of some of the Mn-rich precipitates.

Following rainfall, fine-grained "soil" can be washed down any fractures that are present in the bedrock before being deposited in any subterranean cavity. In some sinkholes, the white dolostones of the Cayman Formation are commonly coated with thinly laminated, black, Mn-rich or red crusts that may be up to 1 cm thick (Fig. 9.19d). As noted by Liang and Jones (2015b), the black laminae typically contain Mn and Fe whereas the red laminae contain Al, K, Fe, and Si. The black laminae, which are formed largely of micrite and microspar, include minor quantities of Mn precipitates, Fe precipitates, chlorite, feldspar, quartz, and possibly zeolites.

Some sinkholes contain numerous terrestrial oncoids that are up 5.5 mm in diameter, that generally lack a nucleus and are typically characterized by vague concentric laminations (Jones 1991). These coated grains are typically formed of

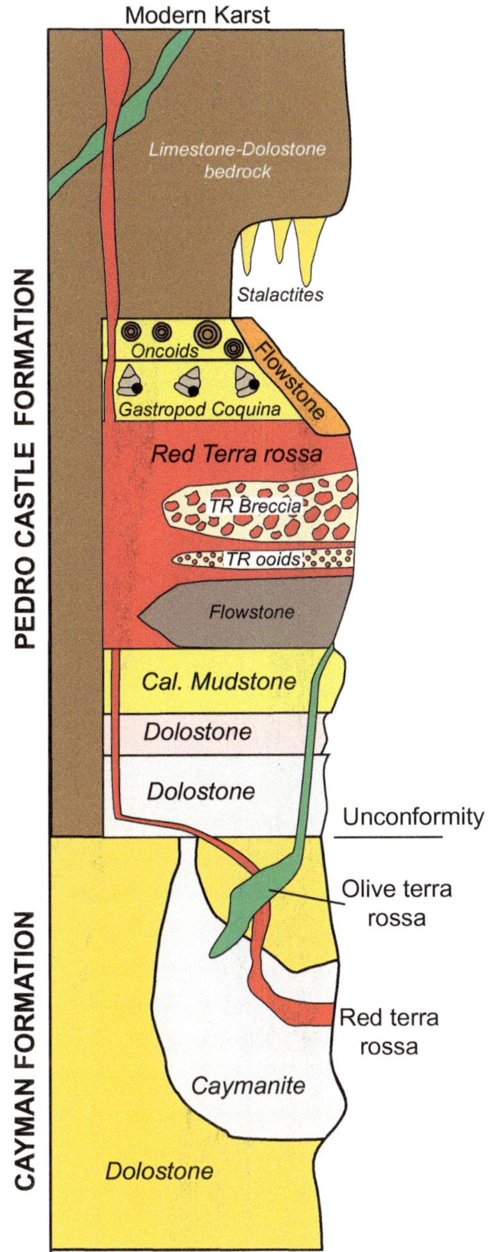

Fig. 9.18 Schematic stratigraphic succession through strata shown in Fig. 9.17. Note distribution of caves and cave-fills in the Cayman Formation and overlying Pedro Castle Formation. Adapted from Jones (2016, his Fig. 24)

detrital micrite, clays, and dolomite along with calcified filaments and spores and, in some cases, insect fragments. The diverse microbial biotas include fungi, algae, and/cyanobacteria along with the mucus mats that they generated. These microbes appear to have played a formative role in the genesis of these grains through their calcification and trapping and binding of detrital grains. The identification of such grains is important because they indicate proximity to an exposure surface.

Many sinkholes are filled with terra rossa that contain angular lithoclasts, <5 cm long, that are of unknown origin because they are formed of lithologies that are unknown in any of the surrounding bedrock successions. In some cases, the terra rossa includes terrestrial oncoids (Fig. 9.20), up to 8.5 cm long, that have a nucleus of white, finely crystalline dolostone nucleus that is encased by a light to tan cortex which is formed of micrite with vaguely defined laminations (Jones 2011a). Although the surfaces are infested with a rich diverse microbiota, the inner layers are characterized by relatively few preserved microbes. Nevertheless, available evidence suggests that the microbiota played an important role in the formation of these coated grains.

Solution-widened joints that are evident in some of the coastal areas are commonly filled with a wide variety of deposits, including flowstone that coats the walls of the joints (Fig. 9.19e), breccias with clasts formed a wide variety of lithologies, including coral heads, and other lithologies for which there are no obvious sources in the surrounding bedrock. Storms commonly transport sediments and skeletal material onshore, where they commonly come to rest in the open joints. Even small cavities in the bedrock are commonly occluded by various combinations of caymanite, grainstones, and/or calcite cements (Fig. 9.19f).

9.6 Conclusions

The geological development of the Cayman Islands is, in its simplest sense, an alternation between sediment deposition when sea levels are high and karst development when sea levels are low. With specific reference to the karst development, the following conclusions are important.

- During sea-level lowstands, the exposed carbonates limestones and dolostones are subject to karst development that affect both the surface exposures and the subterranean strata.
- Surface karst processes led to the development of phytokarst that is characterized by its rugged topography with sharp pinnacles and ridges being common. Much of the phytokarst development involved microbes that mediated surface alteration of the exposed bedrock. Sinkholes are common in some areas.
- Subterranean weathering of the exposed carbonate strata commonly led to the development of large caves, tunnels, and solution-widened joints. Terraced pools developed in many of the caves.
- Speleothems are common in many caves adorned with vast arrays of flowstone, stalactites, and stalagmites. Cave pearls developed in some of the cave pools.

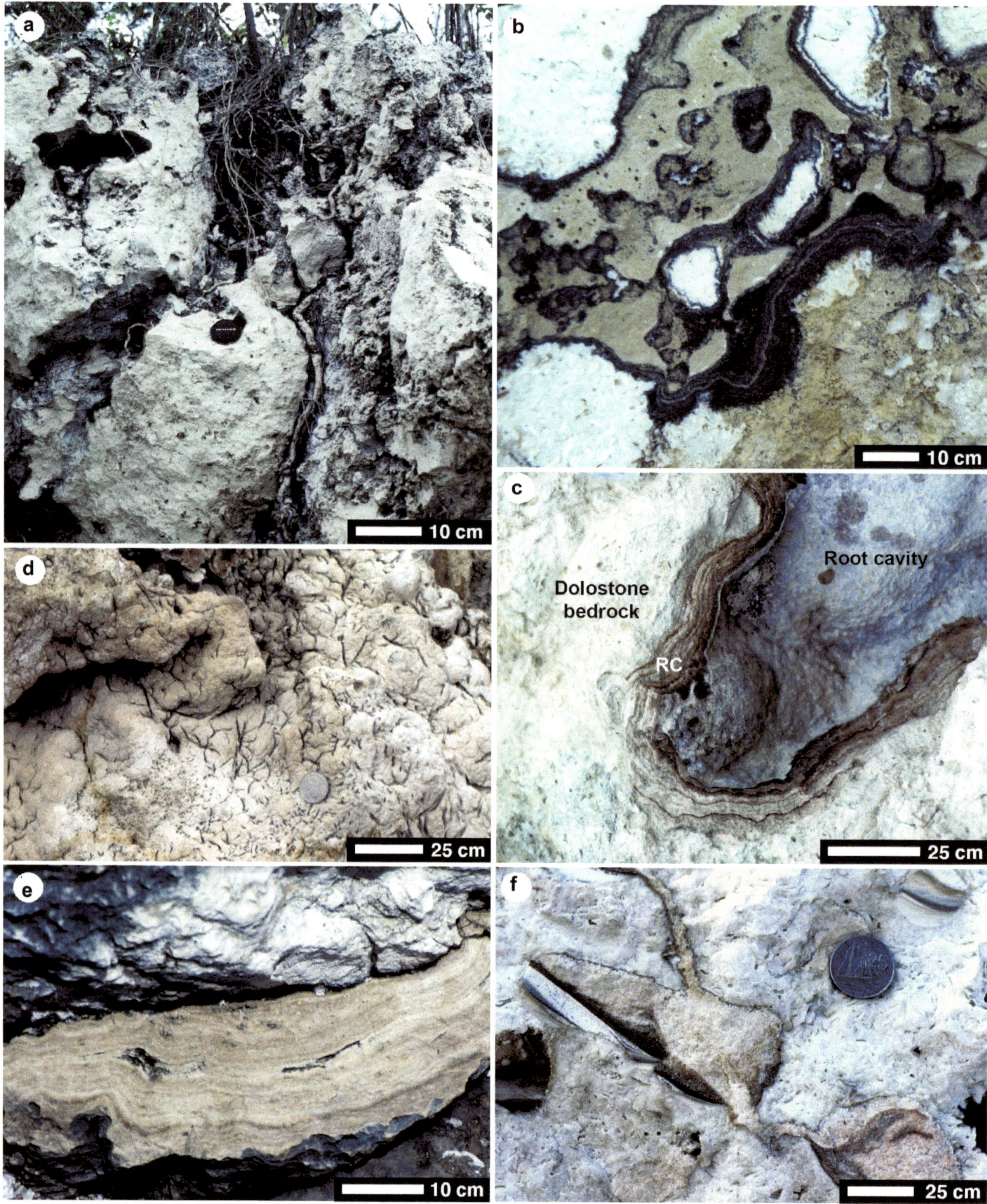

Fig. 9.19 Examples of plant roots and cavity-filling deposits in the Cayman Formation on Grand Cayman. **a** Tree roots penetrating dolostones of Cayman Formation, Queen's Highway, northeast Grand Cayman. **b** Rootcrete coating wall of cavity that was once filled with soil and tree roots. Cayman Formation, Queen's Highway, northeast Grand Cayman. **c** Interior face of rootcrete lining cavity in dolostones of the Cayman Formation, High Rock Quarry, east end of Grand Cayman. **d** Mn-rich laminated crusts coating surfaces of cavity in Cayman Formation. High Rock Quarry, east end of Grand Cayman. **e** Flowstone coating wall of joint in dolostones of the Cayman Formation, coastal exposures, southern end of High Rock Drive, south coast of Grand Cayman. **f** Small interconnected cavities in dolostones of Cayman Formation filled with caymanite and dolomitized grainstone, Queen's Highway, northeast Grand Cayman

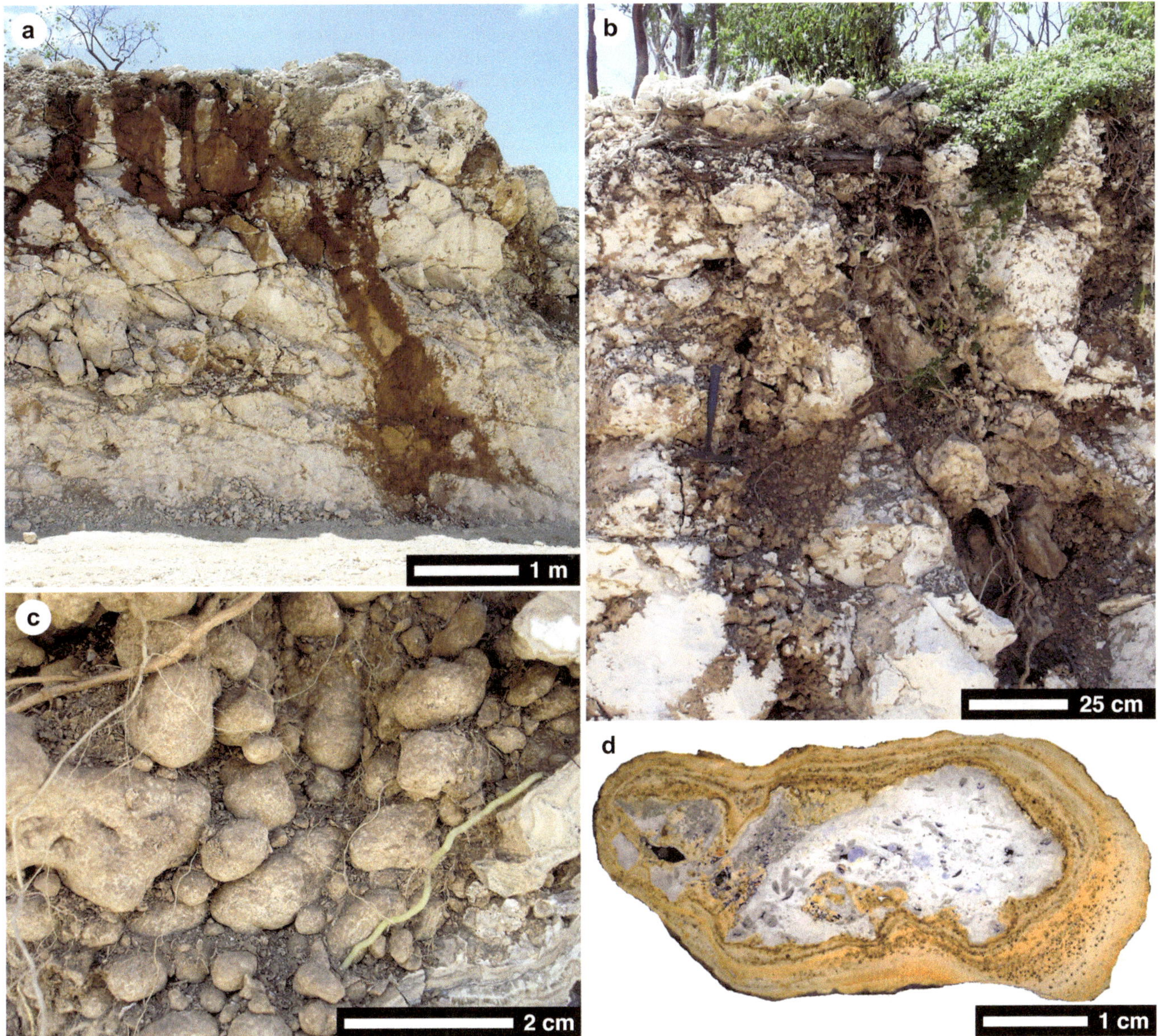

Fig. 9.20 Modern terra rossa and terrestrial oncoids in Cayman Formation, Anne Tatum Drive, Cayman Brac. **a** Fractures and sinkholes filled with red, poorly lithified terra rossa that contrasts sharply with the white dolostones of the Cayman Formation. **b** Small, solution-widened fractures cutting through Cayman Formation (to south of area shown in panel **a**), filled with unlithified terra rossa that contains terrestrial oncoids shown in panel **c**. **d** Cross-section through terrestrial oncoid showing large nucleus formed of dolstone from the Cayman Formation encased by concentric cortical laminae

- Many of the speleothems in the caves and adorning the wave-cut notches are formed of interlaminated calcite and aragonite. Microstromatolites with preserved microbes are common in many of these speleothems.
- The Brac Unconformity, Cayman Unconformity, and Pedro Castle Unconformity that define the upper boundaries of the respective formations are karstic surfaces that developed following deposition of the sediments in each formation. Complex topographies with high relief characterize each of these unconformities.
- Paleokarst features commonly developed in association with each karstic unconformity. As a result, many caves and paleocaves are filled or partly filled with complex arrays of speleothemic deposits that are, in some cases, interbedded with terra rossa and/or marine sediments that were washed into the caves.

References

Alonso-Zarza AM, Jones B (1996) Desarrollo de calcretas en cavidades cársticas y/o de raíes: Las calcretas Cuaternaries de la Isla Grand Cayman (Root development in root karstic cavities : Quaternary calcretes from Grand Cayman Island). Geogaceta 20:262–265

Alonso-Zarza AM, Jones B (2007) Root calcrete formation on quaternary karstic surfaces of Grand Cayman. Geol Acta 5:77–88

Burton FJ (1994) Climate and tides of the Cayman Islands. In: Brunt MA, Davies JE (eds) The Cayman Islands: natural history and biogeography. Monographiae Biologicae. Kluwer Academic Publishers, Dordrecht, pp 51–60

Cox G (1977) A 'living fossil' in the twilight zone: a cave-wall bacterium of unique ultrastructure. In: Ford TD (ed) Proceedings of the 7th international speleological congress, Sheffield. British Cave Research Association, Bridgewater, U. K., pp 155–156

Cox G, Marchant H (1977) Photosynthesis in the deep twilight zone: micro-organisms with extreme structural adaptations to low light. In: Ford TD (ed) Proceedings of the 7th international speleological congress, Sheffield. British Cave Research Association, Bridgewater, U. K., pp 131–133

Folk RL, Roberts HH, Moore CH (1973) Black phytokarst from Hell, Cayman Islands, British West Indies. Geol Soc Am Bull 84:2351–2360

Gilleland T (1998) The caves of Cayman Brac. Natl Speleol Soc News 56:202–207

Harvey VL, Egerton VM, Chamberlain AT, Manning PL, Sellers WI, Buckley M (2019) Interpreting the historical vertebrate biodiversity of Cayman Brac (Greater Antilles, Caribbean) through cologen fingerprinting. The Holocene 29:531–542

Jones B (1987) The alteration of sparry calcite crystals in a vadose setting, Grand Cayman Island. Can J Earth Sci 24:2292–2304

Jones B (1989a) Calcite rafts, peloids, and micrite in cave deposits from Cayman Brac, British West Indies. Can J Earth Sci 26:654–664

Jones B (1989b) The role of micro-organisms in phytokarst development on dolostones and limestones, Grand Cayman, British West Indies. Can J Earth Sci 26:2204–2213

Jones B (1991) Genesis of terrestrial oncoids, Cayman Islands, British West Indies. Can J Earth Sci 28:382–397

Jones B (1992a) Caymanite, a cavity-filling deposit in the Oligocene-Miocene Bluff Formation of the Cayman Islands. Can J Earth Sci 29:720–736

Jones B (1992b) Construction of spar calcite crystals around spores. J Sediment Petrol 62:1054–1057

Jones B (1992c) Manganese precipitates in the karst terrain of Grand Cayman, British West Indies. Can J Earth Sci 29:1125–1139

Jones B (1992d) Void-filling deposits in karst terrains of isolated oceanic islands: a case study from tertiary carbonates of the Cayman Islands. Sedimentology 39:857–876

Jones B (1994a) Diagenetic processes associated with plant roots and microorganisms in karst terrains of the Cayman Islands, British West Indies. In: Wolf KH, Chilingarin GV (eds) Diagenesis IV. Elsevier, Amsterdam, pp 425–475

Jones B (1994b) Geology of the Cayman Islands. In: Brunt MA, Davies JE (eds) The Cayman Islands: natural history and biogeography. Kluwer, Dordrecht, pp 13–49

Jones B (1995) Processes associated with microbial biofilms in the twilight zone of caves: examples from the Cayman Islands. J Sediment Res A65:552–560

Jones B (2000) Microbial sediments in tropical karst terrains: a model based on the Cayman Islands. In: Riding RE, Awramik SM (eds) Microbial sediments. Springer, Berlin, pp 171–178

Jones B (2001) Microbial activity in caves - a geological perspective. Geomicrobiol J 18:1–13

Jones B (2009a) Cave pearls – the integrated product of abiogenic and biogenic processes. J Sediment Res 79:689–710

Jones B (2009b) Mineralized actinomycetes and phosphates in speleothems from Grand Cayman, British West Indies. Sed Geol 219:302–317

Jones B (2010a) Microbes in caves: agents of calcite corrosion and precipitation. In: Pedley HM, Rogerson M (eds) Tufas and speleothems: unravelling the microbial and physical controls. Geological Society of London, vol 36, Special Publications, London, England, pp 7–30

Jones B (2010b) The preferential association of dolomite with microbes in stalactites from Cayman Brac, British West Indies. Sed Geol 226:94–109

Jones B (2010c) Speleothems in a wave-cut notch, Cayman Brac, British West Indies: the integrated product of subaerial precipitation, dissolution, and microbes. Sed Geol 232:15–34

Jones B (2011a) Biogenicity of terrestrial oncoids formed in soil pockets, Cayman Brac, British West Indies. Sed Geol 236:95–108

Jones B (2011b) Stalactite growth mediated by biofilms: example from Nani Cave, Cayman Brac, British West Indies. J Sediment Res 81:322–338

Jones B (2016) Cave fills in Miocene-Pliocene strata on Cayman Brac, British West Indies: implications for the geological evolution of an isolated oceanic island. Sed Geol 341:70–95

Jones B (2017) Review of calcium carbonate polymorph precipitation in spring systems. Sed Geol 353:64–75

Jones B, Hunter IG (1994) Messinian (Late Miocene) karst on Grand Cayman, British West Indies: an example of an erosional sequence boundary. J Sediment Res 64:531–541

Jones B, Kahle CF (1985) Lichen and algae: agents of biodiagenesis in karst breccia from Grand Cayman Island. Bull Can Pet Geol 33:446–461

Jones B, Kahle CF (1986) Dendritic calcite crystals formed by calcification of algal filaments in a vadose environment. J Sediment Petrol 56:217–227

Jones B, Kahle CF (1993) Morphology, relationship, and origin of fiber and dendrite calcite crystals. J Sediment Petrol 63:1018–1031

Jones B, Kahle CF (1995) Origin of endogenetic micrite in karst terrains: a case study from the Cayman Islands. J Sediment Res 65:283–293

Jones B, MacDonald RW (1989) Micro-organisms and crystal fabrics in cave pisoliths from Grand Cayman, British West Indies. J Sediment Petrol 59:387–396

Jones B, Motyka A (1987) Biogenic structures and micrite in stalactites from Grand Cayman Island, British West Indies. Can J Earth Sci 24:1402–1411

Jones B, Peng X (2014) Multiphase calcification associated with the atmophytic cyanobacterium *Scytonema julianum*. Sed Geol 313: 91–104

Jones B, Smith DS (1988) Open and filled karst features on the Cayman Islands: implications for the recognition of paleokarst. Can J Earth Sci 25:1277–1291

Jones B, Zheng E, Li L (2018) Growth and development of notch speleothems from Cayman Brac, British West Indies: response to variable climatic conditions over the last 125,000 years. Sed Geol 373:210–227

Liang T (2015) Evolution of erosional unconformities in the Cenozoic succession of the Cayman Islands. PhD thesis, University of Alberta, Edmonton

Liang T, Jones B (2014) Deciphering the impact of sea-level changes and tectonic movement on erosional sequence boundaries in carbonate successions: a case study from tertiary strata on Grand Cayman and Cayman Brac, British West Indies. Sed Geol 305: 17–34

Liang T, Jones B (2015a) Ongoing, long-term evolution of an unconformity that originated as a karstic surface in the Late Miocene: a case study from the Cayman Islands, British West Indies. Sed Geol 322:1–18

Liang T, Jones B (2015b) Petrographic and geochemical features of sinkhole-filling deposits associated with an erosional unconformity on Grand Cayman. Sed Geol 315:64–82

Lockhart EB (1986) Nature and genesis of caymanite in the Oligocene-Miocene Bluff Formation of Grand Cayman, British West Indies. M.Sc. thesis, University of Alberta, Edmonton

McCormick CA (2019) Eustatic and tectonic controls on the development of the stratigraphic architecture of the Cayman Islands, British West Indies, University of Alberta, Edmonton

McCormick CA, Jones B (2021) On the efficacy and limitations of isolated carbonate platforms as "oceanic dipsticks" to reconstruct subsidence histories, a case study from the Paleogene to Neogene strata on Grand Cayman and Cayman Brac, B.W.I. Mar Geol 436:106470

Morgan GS (1994) Late quaternary fossil vertebrates from the Cayman Islands. In: Brunt MA, Davies JE (eds) The Cayman Islands: natural history and biogeography. Kluwer Academic, Dordrecht, pp 465–508

Morgan GS (2017) Fossil vertebrates from the Cayman Islands. Flicker 29:1–6

Morgan GS, Albury NA (2013) The Cuban crocodile (*Crocodylus rhombifer*) from Late Quaternary fossil deposits in the Bahamas and Cayman Islands. Fla Mus Nat Hist Bull 52:161–236

Morgan GS, Patton TH (1979) On the occurrence of *Crocodylus* (Reptilia, Crocodilidae) in the Cayman Islands, British West Indies. J Herpetol 13:289–292

Morgan GS, MacPee RDE, Woods R, Turvey ST (2019) Late Quaternary fossil mammals from the Cayman Islands, West Indies. Bull Am Mus Nat Hist 428:1–19

Morgan SG, Franz R, Crombie RI (1993) The Cuban crocodile, *Crocodylus rhombifer*, from Late Quaternary fossil deposits on Grand Cayman. Caribb J Sci 29:3–4

Ng KC (1990) Diagenesis of the Oligocene-Miocene Bluff Formation of the Cayman Islands – a petrographic and hydrogeochemical approach. PhD thesis, University of Alberta, Edmonton

Ng KC, Jones B (1995) Hydrogeochemistry of Grand Cayman, British West Indies: implications for carbonate diagenetic studies. J Hydrol 164:193–216

Smith DS (1987) The genesis of speleothemic calcite deposits on Grand Cayman, British West Indies. M.Sc. thesis, University of Alberta, Edmonton

Spencer T (1981) Micro-topographic change on calcarenites, Grand Cayman Island, West Indies. Earth Surf Proc Land 6:85–94

Spencer T (1985a) Marine erosion rates and coastal morphology of reef limestones on Grand Cayman Island, West Indies. Coral Reefs 4:59–70

Spencer T (1985b) Weathering rates on a Caribbean reef limestone: results and implications. Mar Geol 69:195–201

Squair CA (1988) Surface karst on Grand Cayman Island, British West Indies. M.Sc. thesis, University of Alberta

Steadman DW, Morgan GS (1985) A new species of bullfinch (Aves: Emberizinae) from a Late Quaternary cave deposit on Cayman Brac, West Indies. Proc Biol Soc Wash 98:544–553

Tarhule-Lips RFA (1993) Speleogenesis on Cayman Brac, Cayman Islands, British West Indies. M.Sc. thesis, McMaster University

Tarhule-Lips RFA (1999) Karst processes on Cayman Brac, a small oceanic island. PhD thesis, McMaster University, Hamilton

Tarhule-Lips RFA, Ford DC (1996) Timing and causes of speleothem dissolution on Cayman Brac, Cayman Islands, BWI. In: Climate change: the Karst record, vol 2. Karst Waters Institute Special Publication, Bergen, Norway, pp 165–166

Tarhule-Lips RFA, Ford DC (1998a) Condensation corrosion in caves on Cayman Brac and Isla de Mona. J Cave Karst Stud 60:84–95

Tarhule-Lips RFJ, Ford DC (1998b) Morphometric studies of bell hole development on Cayman Brac. Cave Karst Sci 25:119–130

Trudgill ST (1976) The marine erosion of limestones on Aldabra Atoll, Indian Ocean. Z Für Geomophologie 26:164–200

Viles HA (2001) Scale issues in weathering studies. Geomorphology 41:63–72

Viles HA, Spencer T (1986) 'Phytokarst', blue-green algae and limestone weathering. In: Paterson K, Sweeting MM (eds) New directions in Karst, proceedings of the Anglo-French Karst symposium, pp 115–140

Zheng E (2017) Environmental controls on alternating aragonite-calcite laminations in notch-speleothems from Cayman Brac, Cayman Islands, British West Indies. (M.Sc. thesis), Alberta, Edmonton

Terra Rossa, Phosphate, and Mangrove Swamps

10

Abstract

Exposure of bedrock successions during periods of sea level lowstand is commonly accompanied by the development of terrestrial deposits that, in the case of the Cayman Islands, included terra rossa, phosphates, and mangrove swamp deposits. The red terra rossa, which contrasts with the white limestone and dolostone bedrock, is formed of various combinations of X-ray amorphous material, anatase, hematite, goethite, Mn precipitates, biofragments, glaebules, coated grains, and calcite cements. The composition and rare earth elements indicate that much of the terra rossa is formed of wind-blown material that originated from the Sahara Desert. Stratigraphic evidence indicates that the Saharan dust clouds have been depositing material on the Cayman Islands over the last 5 million years. The phosphate deposits, once found on each of the Cayman Islands, were extensively mined and exported during the late 1800s. The phosphates that remain on Little Cayman are mostly formed of hydroxylapatite with boehmite and crandallite being present locally. These deposits, formed largely of P_2O_5, CaO, Al_2O_5, Fe_2O_3, and MnO, are petrographically complex with various admixtures of lithoclasts, coated grains, fossils, phosphate rafts, and cements. The phosphate probably originated from seabird guano that was produced by large flocks of seabirds that nested in various areas on the islands. The mangrove swamp deposits, found over large areas on Grand Cayman and Little Cayman, are characterized by plastic mud and peat deposits that are locally up to 2 m thick. These deposits started to accumulate with a transgression that was initiated ~2,160 years ago.

10.1 Introduction

Terrestrial deposits on the Cayman Islands include terra rossa, phosphates, and peats. The terra rossa was examined by Jones (2021), the phosphates were recently described and investigated by Jones (2019b) and the peats that formed in the mangrove swamps were described by (Woodroffe et al. 1980; Woodroffe 1981, 1982, 1983). Each of these deposits appear to have formed and accumulated since the islands were last exposed following a eustatic drop in sea level. Depending on location, these deposits overlie the upper surfaces of the Cayman Formation, Pedro Castle Formation, or Ironshore Formation. Terra rossa is transported into the subsurface where it may accumulate in any cracks, pores, or caves.

10.2 Terra Rossa

"Terra rossa" is a term that was originally applied to red soil that, in the Mediterranean region, lies on limestone bedrock. It is found on islands throughout the Caribbean region and on some islands, such as Jamaica, it grades into bauxite (Comer 1974). Terra rossa is common on the Cayman Islands where its distinctive red colour contrasts sharply with the white to light cream coloured limestones and dolostones of the Cayman Formation, Pedro Castle Formation, and Ironshore Formation. These Caymanian deposits cannot be regarded as bauxite because they contain <50 wt% Al_2O_3, which is one of the criteria that Comer (1974) used to define bauxite on Jamaica.

Modern soils on the Cayman Islands, typically concentrated in low-lying areas and pockets in the karst terrains, have not been extensively studied. Baker (1974) recognized

B. Jones, *Geology of the Cayman Islands*,
https://doi.org/10.1007/978-3-031-08230-6_10

24 types of soil among the diverse range of dark grey to brown and red soils that have a pH of 6.1 to 8.4. He divided the soils into Group I that were found on the soft limestones of the Ironshore Formation and Group II that are found on the hard limestones and dolostones of the Bluff Group. Subsequently, Ahmad (1996) and Ahmad and Jones (1969) provided more comprehensive classifications of these soils and compared them with similar soils found on the Bahamas. For Ahmad (1996), agricultural uses of the soil were his primary focus with the major aim being the identification and grading of the land that was suitable for mechanized agriculture. Ahmad (1996, his Table 6) showed that gibbsite, boehmite, kaolinite, smectite, illite, quartz, hematite, and goethite were the main components in the soil samples that he collected from Grand Cayman and Cayman Brac. He used these analyses to demonstrate the similarity between the Cayman soils and the bauxite on Jamaica. Ahmad and Jones (1969) showed that the aluminous lateritic soils from Grand Cayman (one sample) and the Bahamas (three samples) were not compatible with the ore grade deposits found on Jamaica (Fig. 10.1).

10.2.1 Stratigraphic Distribution of Lithified Terra Rossa on the Cayman Islands

Terra rossa is common in the Cayman Formation and Pedro Castle Formation but rare in the Brac Formation and Ironshore Formation (Jones 2021). The apparent rarity in the Brac Formation, however, may reflect the fact that outcrops of that formation are restricted to the east end of Cayman Brac where it is only exposed in vertical cliff faces and is therefore far less extensive than outcrops of the other formations. Terra rossa is found in surface depressions, filling cracks and joints throughout the bedrock, and filling or partly filling subsurface cavities of all sizes and shapes (Fig. 10.2a, b). It commonly fills leached fossils (Fig. 10.2c) and is commonly part of the complex successions that fill caves (Fig. 9.16e) that are found in the carbonate bedrock (Jones 1992; Liang and Jones 2015). Surface depressions are commonly filled with loose, non-lithified terra rossa (Fig. 10.2d). In many cases, however, irregular masses of partly to fully lithified terra rossa are present (Fig. 10.2e). Non-lithified terra rossa is commonly washed into the subsurface where it accumulates on the cave

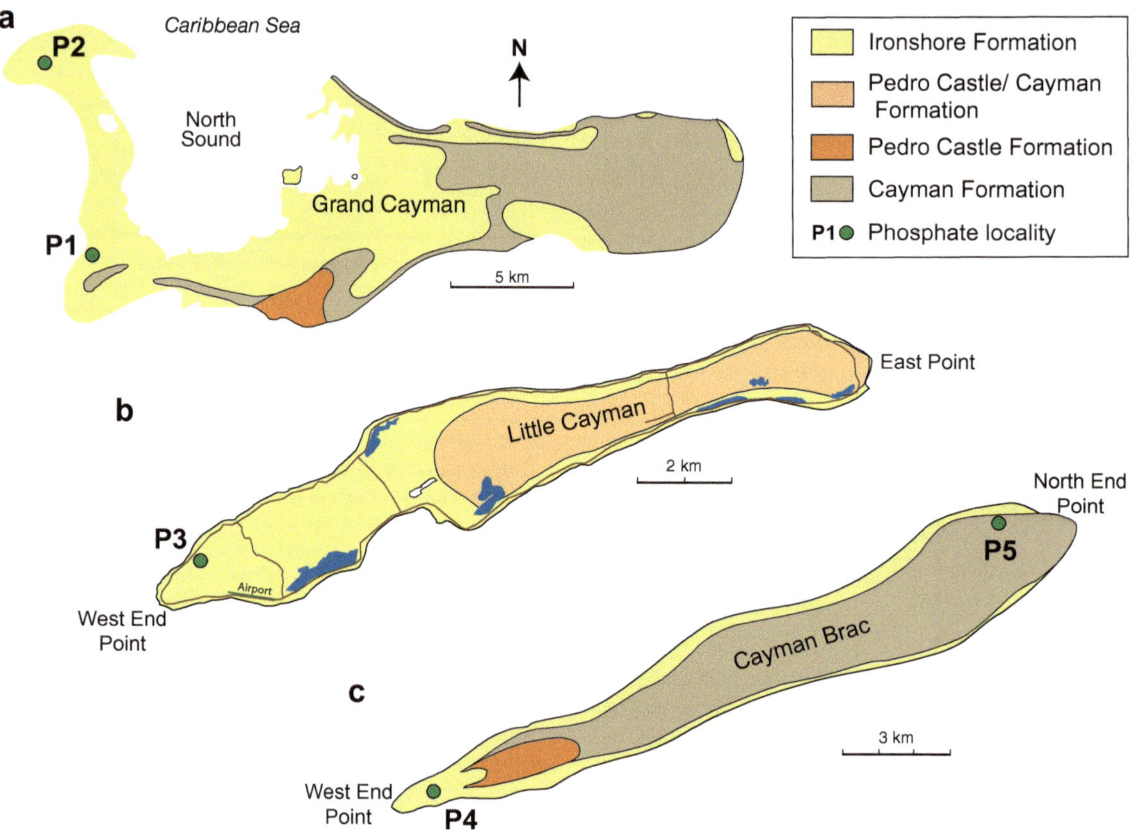

Fig. 10.1 Maps showing localities where phosphates are found. Phosphate locations from Matley (1926a, his Figs. 2, 5, and 8)

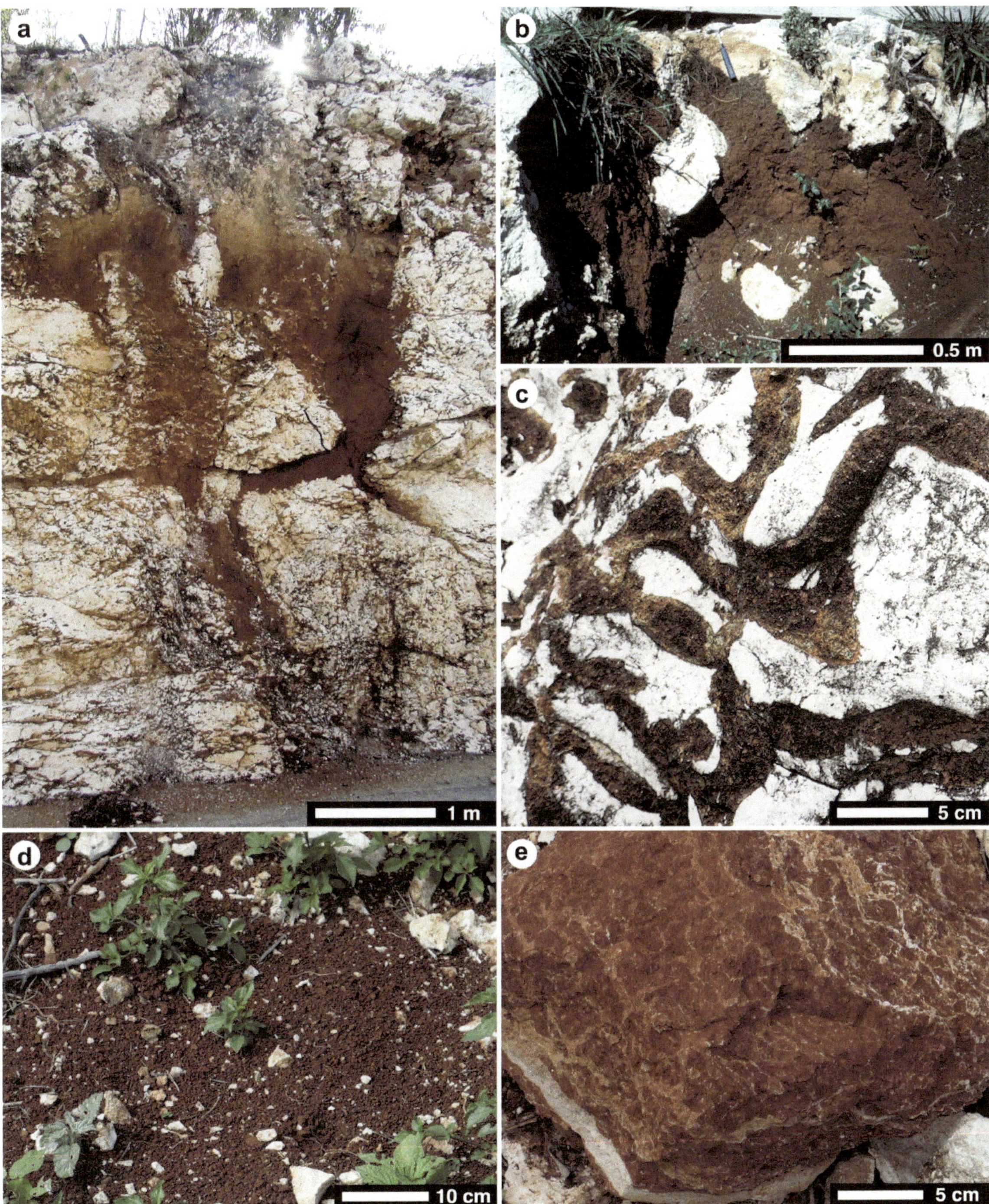

Fig. 10.2 Field occurrences of terra rossa on the Cayman Islands. **a** Unconsolidated terra rossa filling sinkholes and fractures in the Cayman Formation. Anne Tatum Drive, Cayman Brac. **b** Terra rossa filling large cavities in Cayman Formation, Grand Cayman, near Boddentown. **c** Terra rossa filling cavities formed by leaching of *Stylophora*, Cayman Formation, south end of Bluff Drive, Cayman Brac. **d** Loose, unlithified terra rossa resting on top of Cayman Formation, northeast coast of Grand Cayman, note plants growing in soil. **e** Lithified terra rossa with calcite-filled fractures, Cayman Formation, central eastern part of Grand Cayman, near High Rock Quarry

Fig. 10.3 Cave in Crystal Caves complex (Grand Cayman) showing red terra rossa coating floor and some of the speleothems

floors and/or coats the surfaces of speleothems (Fig. 10.3). Locally, lithoclasts formed of limestone/dolostone derived from the bedrock and speleothemic calcite are embedded in the terra rossa. Lithified terra rossa, commonly cemented by calcite, is found throughout the carbonate successions on the Cayman Islands.

10.2.2 Composition

Jones (2021) used data derived from X-ray diffraction analyses (normal and Rietveld methods), thin section analyses, scanning electron microscopy, and chemical analyses to demonstrate that terra rossa found on the Cayman Islands is mineralogically and chemically diverse. Kaolinite, halloysite, böehmite, gibbsite, and hematite are present in every sample and, depending on location, these minerals are associated with variable quantities of calcite, X-ray amorphous material, vermiculite, gibbsite, goethite, crandallite, apatite, hematite, anatase, and/or hydrotalcite.

Whole rock chemical analyses of 22 samples from the Cayman Formation, Pedro Castle Formation, and Ironshore Formation on the three islands showed that the terra rossa is formed largely of SiO_2 (up to 25 wt%), Al_2O_3 (4 to 41 wt%), and CaO (0.6 to 46 wt%) along with lesser amounts of MnO, MgO, Na_2O, K_2O, and Ti_2O. Loss on ignition was 16 to 41%. P_2O_5 (up 24 wt%) that is present in some samples from Grand Cayman and Cayman Brac probably originated from bird guano.

Petrographically, the terra rossa is formed of various combinations of clays and X-ray amorphous material, anatase, hematite, goethite, manganese precipitates, biofragments, glaebules, clay-coated grains, terra rossa ooids, and calcite cements (Jones 2021, his Figs. 5 to 9).

Clays, X-ray Amorphous Material, Anatase, Hematite, Goethite, and Manganese Precipitates

The very fine-grained clays and X-ray amorphous material form the dense homogeneous groundmasses of the terra rossa. The individual clay grains are so small (most <5 μm long) that it is impossible to see the individual grains in thin section and difficult to image on the SEM (Fig. 10.4). Similarly, the anatase, hematite, and goethite are difficult to identify in thin section, but can be recognized using backscatter images coupled with Energy Dispersive X-ray analyses on the SEM.

The Mn precipitates, easy to identify in hand samples and thin section because of their distinctive black colour (Fig. 10.4b), are found as dendritic arrays, diffuse irregular-shaped masses, or small grains dispersed throughout the clay groundmass.

Biofragments

Biofragments, up to 6 mm long, derived from thin-shelled bivalves and gastropods, are common in some samples. None of these shell fragments display any evidence of dissolution or alteration (Jones 2021, his Fig. 13).

Glaebules

The distribution of glaebules is highly variable between samples and within the same sample. The round to elongate glaebules, mostly <2 mm long, appear black and featureless when viewed with plane polarized light (Fig. 10.4a). Backscattered imaging on the SEM, however, shows that are they characterized by concentric laminations that are defined by variable average atomic weights (Fig. 10.4). Formed largely of Fe, Mn, Al, and Si, the concentric laminations are largely defined by the distribution of Fe (Fig. 5.10c, f). There is little to no similarity between glaebules, even those found close to each other commonly display markedly different zonation styles.

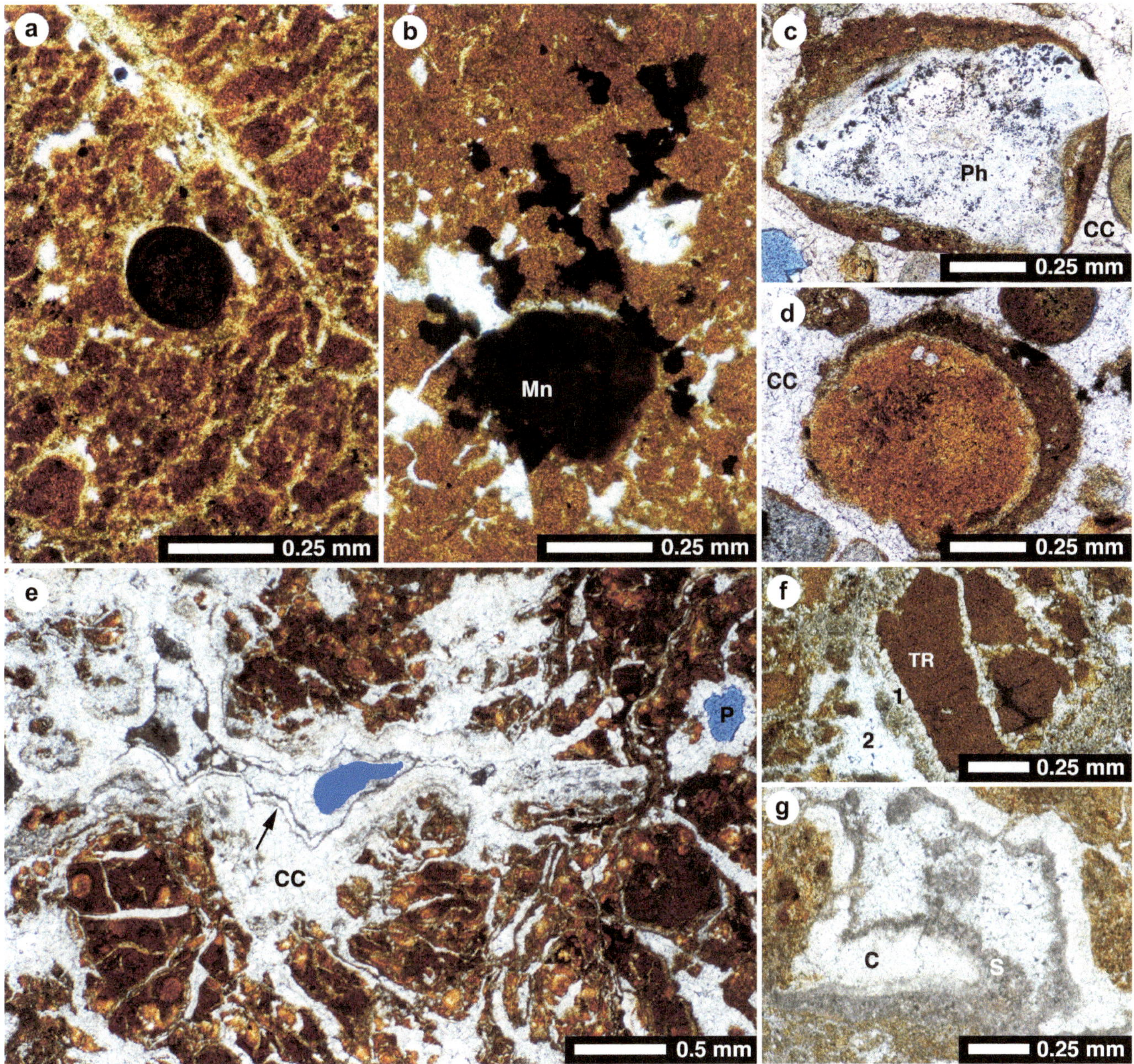

Fig. 10.4 Thin section photomicrographs of terra rossa. **a** Glaebule (central round structure) embedded in terra rossa groundmass that is cut by fracture that is partly filled with calcite cement. **b** Manganese (Mn) precipitates in terra rossa groundmass. Note cavities filled with calcite cement (white). **c** Phosphate grain (Ph) with terra rossa cortex and calcite cement (CC) around it. Anne Tatum Drive, Cayman Brac. **d** Terra rossa ooid surrounded by calcite cement (CC). **e** Terra rossa (red to brown) cut by numerous fractures filled with calcite cement (CC). Note thin layers of silt (arrow) interlayered with calcite cement. Blue = porosity. **f** Small mass of terra rossa (TR) cut by fractures filled with calcite cement. Note two generations (1 and 2) of cement filling cavity (lower left). **g** Cavity filled with multiple phases of calcite cement separated by thin silt layer (S)

Clay-Coated Grains and Terra Rossa Ooids

Coated grains are typically found mixed with terra rossa that has been deposited in subsurface cavities. The clay-coated grains, up to 5 mm long but typically <2 mm long, have a small grain of dolostone, phosphate, or clay nucleus that is encased by a cortex formed of vaguely laminated clay (Fig. 10.4c, d). Circumgranular cracks that are open or filled with calcite cement are apparent in some of these grains.

The terra rossa ooids, up to 2 mm in diameter, are formed largely of vaguely laminated terra rossa with some having a small nucleus formed of calcite and/or dolomite.

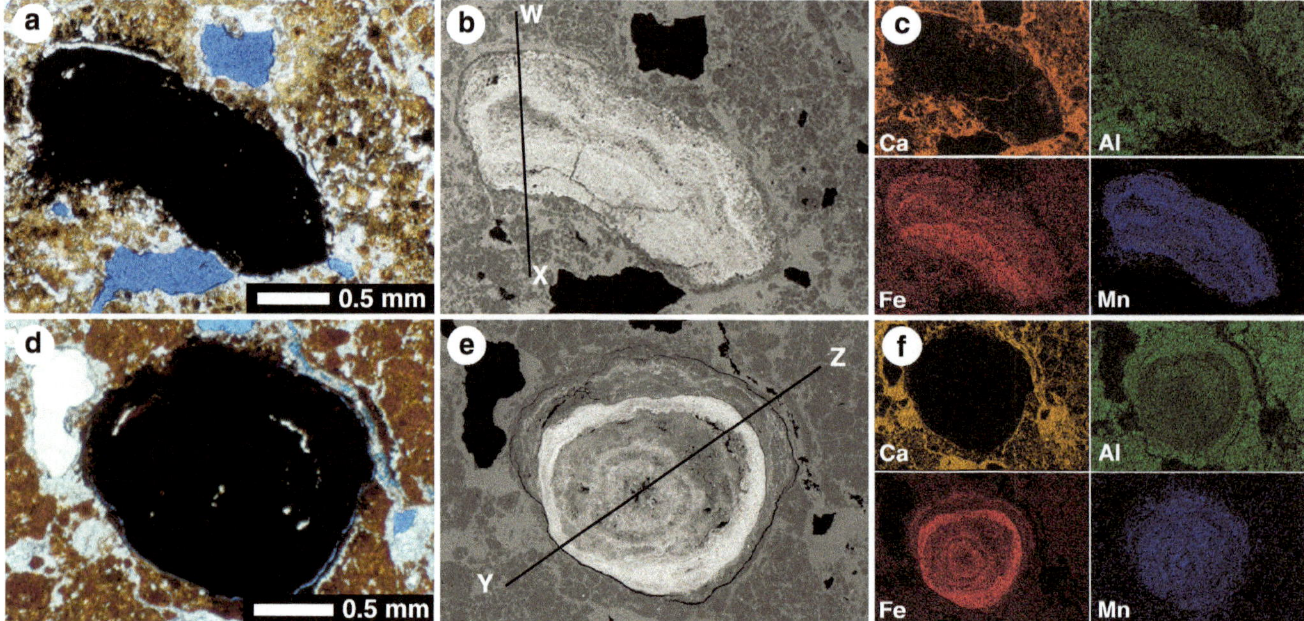

Fig. 10.5 Thin section photomicrographs (**a** and **d**), SEM backscattered images showing distribution of Ca, Al, Fe, and Mn (**c** and **f**) in glaebules. Samples from localities BW (**a–c**) and OW (**d–f**) (Fig. 10.1a). Panels B and E in backscatter mode showing variations in average atomic weights of different layers in the glaebules. Panels C and F show distribution of Ca, Al, Fe, and Mn in each glaebule

Fractures

Lithified terra rossa is commonly characterized by multiple generations of fractures that are commonly filled with multiple generations of calcite cement and layers of clay-rich sediment (Fig. 10.4e, g). Cavities that are inter-connected by these fractures are commonly filled in a similar manner. At least two generations of calcite cement are evident in some cavities (Fig. 10.4f). Complex patterns arise due to the fact that successive generations of these fractures commonly cross-cut the earlier formed and filled fractures (Fig. 10.5).

10.2.3 Stable Isotopes

Based on 106 samples of terra rossa, the $\delta^{18}O$ ranges from −17.8 to + 2.6‰ (average −5.2‰) and the $\delta^{13}C$ ranges from −13.9 to −0.3‰ (average −8.1‰) (Fig. 10.6a). There is no identifiable pattern to the isotope values in terms of the island from which the samples came, the host formation, or the location of the terra rossa relative to the surface (Jones 2021). Samples from one locality commonly display ranges that are the same or even greater than samples from other localities (Fig. 10.6a). The range of $\delta^{18}O$ values from the terra rossa is greater than that from the bedrock limestones/dolostones found on the three islands, whereas the range of $\delta^{13}C$ values are similar to the range derived from the dolostones in the Cayman Formation and Pedro Castle Formation.

10.2.4 Rare Earth Elements

In the terra rossa from the Cayman Islands, the ΣREE ranges from 29 to 648 ppm (average 251 ppm based on 106 samples) whereas the ΣREE+Y ranges from 39 to 826 ppm (average 313 ppm) (Fig. 10.6b). Herein, the REE values are normalized against the Chondrite Standard (suffix "CN") and plotted on a logarithmic scale against their atomic number (cf., McLennan 1989). All of the REE_{CN} profiles are characterized by (1) high heavy rare earth element (HREE), lower middle (MREE), and light (LREE) values with the La/Yb ratio between 4.9 to 15.8 (average 8.5), and (2) negative Eu anomalies with Eu*/Eu values of 0.56 to 2.7.

10.2.5 Origin of Terra Rossa on the Cayman Islands

Terra rossa found on various Caribbean islands, the Bahamas Islands, and Florida has generally been attributed to residual accumulations (e.g., Muhs et al. 2012) or aerosol dusts that came from various volcanoes in the Caribbean region, the Sahara Desert, or the Lower Mississippi Valley loess (e.g., Muhs et al. 2007). For Jamaica, Comer (1974) suggested that the terra rossa may have come from alluvial deposits that were older than the limestone bedrock on which it is now found.

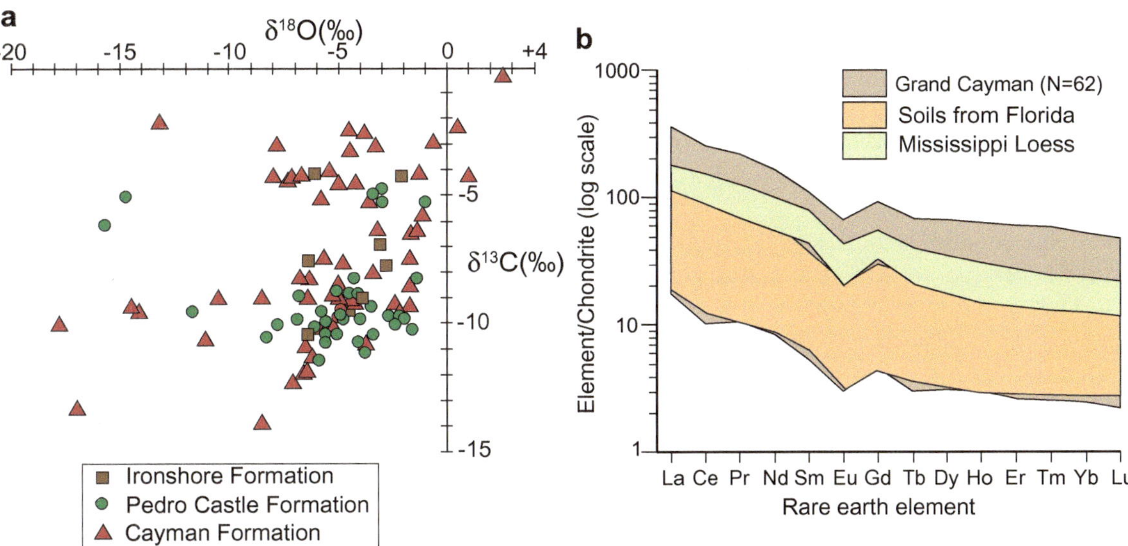

Fig. 10.6 **a** $\delta^{18}O$ versus $\delta^{13}C$ for samples of terra rossa from the Cayman Formation, Pedro Castle Formation, and Ironshore Formation on Grand Cayman, Cayman Brac, and Little Cayman. The stable isotope values are assumed to come from the calcite cements in the terra rossa (see Fig. 10.4). Modified from Jones (2021, his Fig. 23b). **b** Chondrite normalized REE values for terra rossa from Grand Cayman compared to REE values from soils from Florida and Mississippi Loess (Muhs et al. 2007, his Fig. 18a). Modified from Jones (2021, his Fig. 23a and c)

The residual model argues that the terra rossa is formed of insoluble materials that were once in the limestone/dolostone bedrock that has been lost to weathering and dissolution. Inherent to much of this discussion is the notion that the limestones lost to weathering had the same composition as the limestones that remain. It is, however, virtually impossible to demonstrate whether or not this was the case. Most of the limestones/dolostones found on the Caribbean islands contains relatively little insoluble materials. On the Cayman Islands, for example, the limestones and dolostones of the Brac Formation, Cayman Formation, Pedro Castle Formation, and Ironshore Formation are very pure—a fact that has been amply demonstrated by the innumerable XRD, thin section, SEM, and microprobe analyses of these samples that have routinely failed to find any detrital materials in them (e.g., Jones 1994; Zhao and Jones 2012a, b; Ren and Jones 2018). The fact that the ΣREE and ΣREE+Y values from these limestones and dolostones are typically <10 ppm also attests to the dearth of detrital materials in them (Jones 2021). This is consistent with the fact that these Cayman carbonate successions accumulated on small, isolated carbonate banks that are surrounded by deep oceanic waters and are far removed from any possible source of non-carbonate sediments.

The notion that dust from the Sahara Desert could be blown over the Atlantic Ocean was first suggested by Darwin (1845, 1846) after he observed dust settling on ships around the Cape Verde Islands in the Atlantic Ocean. Today, satellite images from the National Aeronautics and Space Administration (NASA) routinely tract large dust storms from the Sahara Desert that blow across the Atlantic Ocean and into the Caribbean region. Reddish skies during the summer months in the Cayman Islands signal the presence of the Saharan dust. Once the fine-grained dust is deposited on the islands, subsequent rainfalls quickly wash it into low-lying areas, and into the subsurface via the numerous open cracks and joints that characterize the bedrock of the islands. Dating the terra rossa on the Cayman Islands is very difficult because it typically lacks the minerals needed for radiometric dating. As noted by Jones (2021) dating can only be achieved relative to the bedrock in which it is found, or in rare situations where the terra rossa is interbedded between layers of rock that can be dated using radiometric methods. On Cayman Brac, for example, the interlayering of lithified terra rossa with flowstone in caves housed in the Pedro Castle Formation and Cayman Formation (Jones 2016) allowed such dating by virtue of U/Th dates obtained from the flowstone units. Such dating indicated that terra rossa had been periodically accumulating in that cave for more than 500,000 years.

Cross-sections through the upper part of the Cayman Formation and the basal part of the overlying Pedro Castle Formation in a quarry near Pedro Castle on Grand Cayman included a vertical section through a cave in the Cayman Formation that had been filled with marine sediments, terra rossa, and flowstone (Fig. 9.16a). Jones (1992) demonstrated that the cave was linked to a tunnel that then opened into the basal part of a large sinkhole that once opened at the erosional surface that is now represented by the unconformity that separates the Cayman Formation from the overlying Pedro Castle Formation. The field relationships clearly indicate that the sediments in the cave, including the terra rossa, were deposited in the cave before deposition of the marine sediments that now form the Pedro Castle Formation took place about 5 million years ago.

Dust storms that originate in the Sahara Desert are known to transport a vast array of minerals that originate from the weathered bedrock of the desert. Scheuvens et al. (2013), for example, showed that these constituents may include quartz, feldspar, carbonates (calcite, dolomite), sulphates (gypsum, anhydrite), halite, mica (muscovite, biotite), oxides (hematite, magnetite), and/or goethite along with other rarer minerals. The composition of the dust that reaches the Cayman Islands is controlled by (1) the source area in the Sahara from which the dust originated because not all areas have the same bedrock, and (2) the "travelled size"—a term coined by Goudie and Middleton (2001) to reflect that fact that as the wind strength decreases with distance from the source area the larger sediment grains will be progressively lost and deposited. By the time that the winds have reached the Caribbean region, the Saharan dust will be formed largely of grains that are <10 μm long (e.g., Stuut et al. 2005; Garrison et al. 2006). As shown by Jones (2021), the main characteristics of Cayman terra rossa are as follows.

- Most of the grains are <2 μm long apart from some of the rare grains formed of Fe and Ti that are up to 20 μm long. Clays are the dominant component.
- X-ray amorphous material is a common constituent of the Cayman terra rossa.
- No quartz grains were found and only minor amounts (<3%) were detected by XRD analysis in a few samples.
- No sulphides and minimal chlorides are present in the Cayman terra rossa.
- The lithified terra rossa on the Cayman Islands contain up to 78% calcite, which is thought to have formed largely through diagenetic processes.
- Phosphatic minerals that originated from local bird guano are present in some samples of Cayman terra rossa.

Assessing the origin of the Cayman terra rossa can be based on various parameters including the SiO_2/Al_2O_3 ratio. The SiO_2/Al_2O_3 molar ratio of Saharan dust is between 5.0 and 21.6 with an average of 8 to 10 that largely reflects the high content of quartz and/or SiO_2 that comes from diatomaceous beds that occur in some parts of the Sahara Desert (e.g., Moskowitz et al. 2016). That ratio rapidly decreases as the quartz grains are deposited. For the Cayman terra rossa, the SiO_2/Al_2O_3 ratio is 0.8 to 1.6 with an average of 1.1 and the modern soils have a ratio of 0.6 (Ahmad and Jones 1969). Other ratios, including the Ti/Th and Ti/Y are generally similar to those reported by Muhs et al. (1990) for terra rossa elsewhere in the Caribbean region.

The chondrite normalized REE profiles derived from the Cayman terra rossa are characterized by (1) high HREE values, (2) lower, relatively constant MREE and LREE values, and (3) negative Eu anomalies with Eu/Eu* between 0.56 and 0.71 (average 0.66). These profiles are comparable to those from terra rossa soils on Barbados, the Florida Keys, the Bahamas, and Jamaica (Muhs et al. 2007; Muhs and Budahn 2009).

The lithified terra rossa on the Cayman Islands is commonly characterized by numerous generations of calcite-filled fractures that are typically <2 mm wide (Fig. 10.4e–g). These fractures commonly link round to ovate cavities that are <2 mm long. In many cases, successive generations of calcite cement in the fractures and cavities are separated by thin layers of very-fine grained sediment (Fig. 10.4e, g). The soils on the Cayman Islands are subjected to repeated cycles of heavy rainfall that are followed by hot, cloud-free days that dry out the soils. As the soils dries and hardens, cracks become common. Calcite cements precipitate from the groundwaters that may then circulate through the soils. The stable isotopes derived from the Cayman terra rossa, which is assumed to come largely from the calcite cements, ranges from −17.8 to +2.6‰ (average −5.2‰) for the $\delta^{18}O$ and −13.9 to −0.3‰ (average −8.1‰) for the $\delta^{13}C$ (Jones 2021). For the Cayman Islands, Ng (1990) showed that the $\delta^{18}O$ of rainwater ranges from −7.3 to −1.6‰, whereas fresh groundwater is −4.8 to −4.4‰ (average −4.5‰), and brackish water is −4.5 to −3.9‰ (average −4.2‰). Tarhule-Lips (1999) gave a value of −7.4‰ for rainwater on Cayman Brac. The stable isotope values of the calcite in the terra rossa cannot be linked directly to the stable isotope values of the rainwater because, as noted by Cerling and Quade (1993), (1) as rainwater infiltrates into the soil it may mix with waters of different isotopic compositions, (2) the rainwater may be modified by evaporation and/or (3) the local flora may, during the growing season, modify the water by evapotranspiration.

The $\delta^{13}C$ values reflect differences between C_3 ($\delta^{13}C$ of −25 to −32‰, average −26‰) and C_4 ($\delta^{13}C$ of −10 to −14‰, average −12‰) vegetation (Cerling and Quade 1993). For notch speleothems on Cayman Brac, which form from groundwaters that permeate through the soils and rocks before reaching the notch, Jones et al. (2018) noted that the $\delta^{13}C$ can potentially be affected by (1) rainfall that controls the local flora, (2) temperature that controls isotopic equilibrium and biological activity, (3) C_3 or C_4 vegetation cover, (4) bedrock dissolution, and/or (5) CO_2 degassing. Potentially, these processes could also affect the calcite that is precipitated in terra rossa. Given the array of potential controls, it is difficult to exactly pinpoint the factor(s) that controlled the stable isotope values of the calcite cement in the terra rossa.

10.3 Phosphate Deposits

With the high worldwide demand for phosphate fertilizer in the late 1800s, large volumes of phosphate were produced from many islands throughout the oceans of the world. In the northern part of the Caribbean region, for example, phosphate was mined and exported from East Guano Cay and Cayos de los Jardinillos, the Swan Islands, Pedro Cays, south of Jamaica, Navassa Island, and Mona (Hutchinson 1950). Between 1884 and 1890, large quantities of phosphate were also exported from Grand Cayman, Cayman Brac, and Little Cayman (Hirst 1910), with much-needed revenue being provided for the people of the Cayman Islands (Fawcett 1888; Billymer 1946; Hutchinson 1950; Sauer 1982). According to Craton (2003), mining ceased in ~1890 because the Cayman phosphate became too expensive to produce following the discovery of large phosphate deposits in Florida.

Matley (1926a, b) recorded five main phosphate localities on the Cayman Islands (Fig. 10.1)—P1 and P2 on Grand Cayman (Matley 1926a, his Fig. 8), P3 on Little Cayman (Matley 1926a, his Fig. 5), and localities P4 and P5 on Cayman Brac (Matley 1926b, his Fig. 2). Although the approximate locations where the phosphate deposits on Grand Cayman are known, attempts to find remnants of the deposits have proven unsuccessful. Similarly, attempts to find the phosphate deposits at localities P4 and P5 on Cayman Brac have been unsuccessful, although it should be noted that detrital phosphate grains have been found in the terra rossa that is associated with the Cayman Formation in outcrops near locality P5. On the west coast of Little Cayman, locality P3 is located ~400 m ESE of Salt Water dock where a sign explains the nature of the deposit and the mining methods that were used (Fig. 10.7). Located near the sign is the original pen where the donkeys that pulled the mining carts from the deposit to the coast were kept. A path now leads to the phosphate site where pieces of the rail track are still evident (Fig. 10.8a). Given that large quantities of the phosphate deposits were removed while the site was being actively worked, only remnants of the phosphate deposits now remain. Today, the light tan to yellowish-red

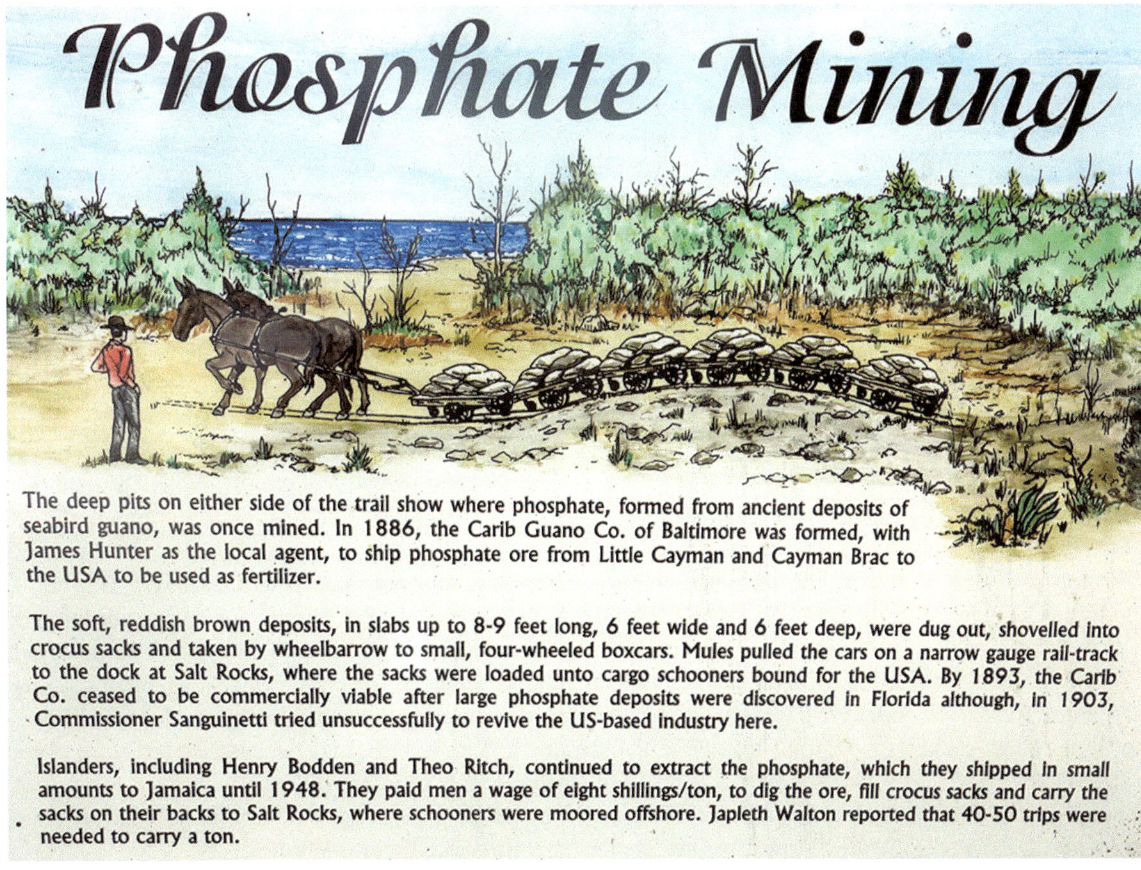

Fig. 10.7 Photograph of sign at Salt Rocks on east end of Little Cayman near the footpath that leads to area where phosphates were once mined. Provides information on the history of the mining and the methods used. Photograph used with permission of the Cayman Islands Government

Fig. 10.8 Photographs of phosphate mining area, west end of Little Cayman. **a** General photograph of path and area where mining took place. Note pieces of the original rail track (arrows) for carts that were once used to transport the ore to the coastline. **b** General view of phosphatic limestones in area where mining took place. **c** Thin section photomicrograph showing phosphatic material in mudstone matrix

lithified phosphatic limestone is found in depressions and sinkholes of the phytokarst that characterizes the upper surface of the Pedro Castle Formation. Although the phosphates are younger than the Pedro Castle Formation, their exact age cannot be determined.

10.3.1 Composition of Phosphatic Limestone

X-ray diffraction analyses show that the phosphatic limestones from Little Cayman are formed of calcite with variable quantities of hydroxylapatite [$Ca_5(PO_4)_3(OH)$] and, in some samples, boehmite and crandallite [$CaAl_3(PO_4)(PO_3OH)(OH_6)$]. Based on 13 analyses, Jones (2019b) showed that the phosphatic limestones contain CaO (36–46 wt%), P_2O_5 (6–24%), Al_2O_3 (1.6–4.8%), Fe_2O_3 (0.4–1.1%), Mg (0.2–1.1%) and < 1 wt% of MnO, Na_2O, and K_2O. Relative to the limestone of the Pedro Castle Formation, the phosphatic limestones contain more Al, P, Fe, Cu, Mn, Zn, and K.

10.3.2 Petrography of Phosphatic Limestones

The phosphatic limestones are formed of lithoclasts, detrital phosphatic grains, coated grains, composite coated grains, fossils, and phosphate rafts that are found in cavities in the limestone (Figs. 10.8c and 10.9). Thin section analyses show that the textures and compositions of the phosphate-rich rocks are highly variable, even at the micro-scale.

Lithoclasts and Detrital Phosphate Grains

Lithoclasts up to 8 × 4 × 2 cm in size are formed of phosphatic grains, coated grains, and phosphatized fossils that are bound together by thin, isopachous microlaminated phosphate cements. In some lithoclasts, the original components are difficult to recognize because the replacive phosphate has largely masked their identity. In many samples, detrital phosphate grains (up to 5 mm long) with variable textures, are intermixed with the lithoclasts.

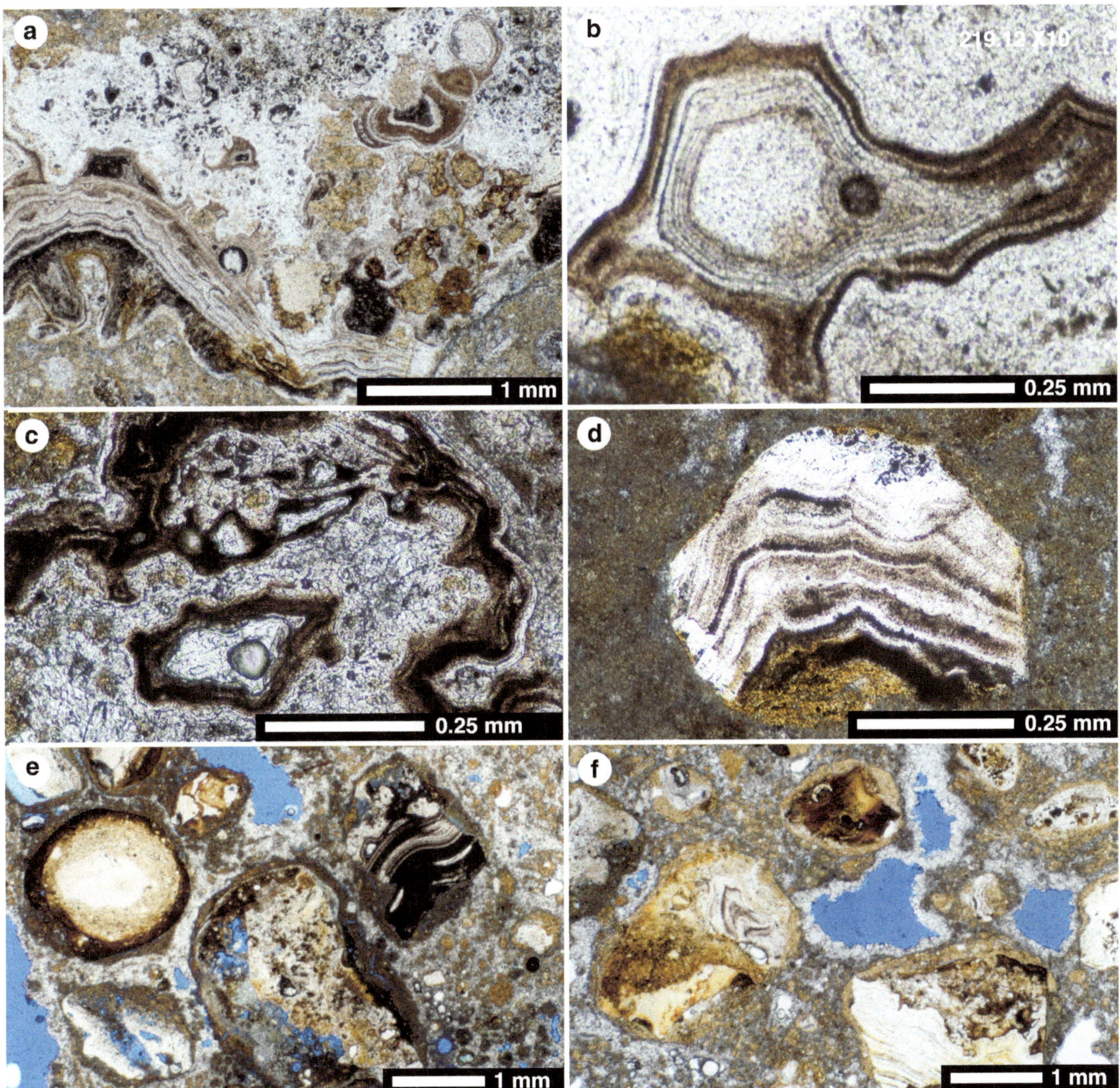

Fig. 10.9 Thin section photomicrographs of phosphates from Little Cayman. Blue = porosity. **a** General view of complex fabrics in phosphates. **b**, **c** Phosphate groundmass with banded phosphate cements filling a cavity. **d** Grain formed of banded phosphate that probably originated as a cement in a cavity. **e**, **f** Examples of phosphate grains in limestone matrices

Glaebules

These grains, up to 2 mm in diameter, have various types of nuclei (e.g., phosphate grains, phosphatized bioclasts, micrite) that are encased by 1 to 6 cortical laminae of various colours that are formed largely of Al with variable amounts of Fe and minor amounts of Ca. Coated grains like these, found in phosphate deposits throughout the world, have been variously labelled as ooids, pisolites, pellets, and many other terms (see Jones 2019b for full listing of names). Jones (2019b), however, argued that they should be termed 'glaebules' because they conform with the term glaebule, which Brewer and Sleeman (1964) originally defined as three-dimensional bodies that develop in soils and have greater concentrations of some constituents relative to the surrounding soil. Composite coated grains, up to 4 mm long, are formed of numerous smaller grains and/or glaebules that are encased by an outer cortex.

Phosphate Rafts

Phosphate rafts, up to 5 mm long and 0.5 mm thick, are found in some of the larger cavities that exist in some of the phosphatic limestone. Each raft has a paper-thin nucleus that is encased by successive thin layers of phosphate. The rafts are typically stacked one on top of the other with the cavities between being open or partly filled with phosphate and/or calcite cements.

Matrix

The micritic matrix of the phosphatic limestones contains minor amounts of microspar and small clusters of non-crystalline precipitates that are formed of P, Al, Si, and/or Fe. Locally, the micrite grades into areas formed of loosely packed peloids (<0.25 mm long).

Microbes

Small cavities in the limestones commonly contain branching septate filaments (~500 nm diameter, >100 μm long) and small clusters of spores(?).

Calcite Cement

Throughout the phosphatic limestones, there are variable amounts of calcite cement that are typically located in the small cavities between the allochems.

10.3.3 Stable Isotopes

Two samples from the Pedro Castle Formation yielded $\delta^{18}O$ values of −5.7 to −5.8‰ and $\delta^{13}C$ values of −3.7 to −5.2‰; two samples from the phosphatic soil yielded $\delta^{18}O$ values of −7.4 to −5.4‰ and $\delta^{13}C$ values of −8.5 to −7.8‰, and ten samples of phosphatic limestones yielded $\delta^{18}O$ values of −5.9 to −3.8‰ and $\delta^{13}C$ values of −11.5 to −6.6‰ (Fig. 10.10a). Although all of these values fall within the field of stable isotopes that came from the calcite cements in the terra rossa, it should be noted that the latter is characterized by broader ranges of $\delta^{18}O$ and $\delta^{13}C$ values.

10.3.4 Rare Earth Elements

The ΣREE+Y of the soil and phosphatic limestones ranges from 51 to 139 ppm with all of the REE being present (Fig. 10.10b). By comparison, Y, La, Ce, Nd, and Gd were below detection in the limestones of the underlying Pedro Castle Formation that yielded ΣREE+Y values of <2 ppm. The ΣREE+Y of the phosphatic limestones are also significantly higher than those reported from the Pedro Castle Formation and Ironshore Formation on Cayman Brac (Zhao and Jones 2013). The Chondrite normalized profiles of the

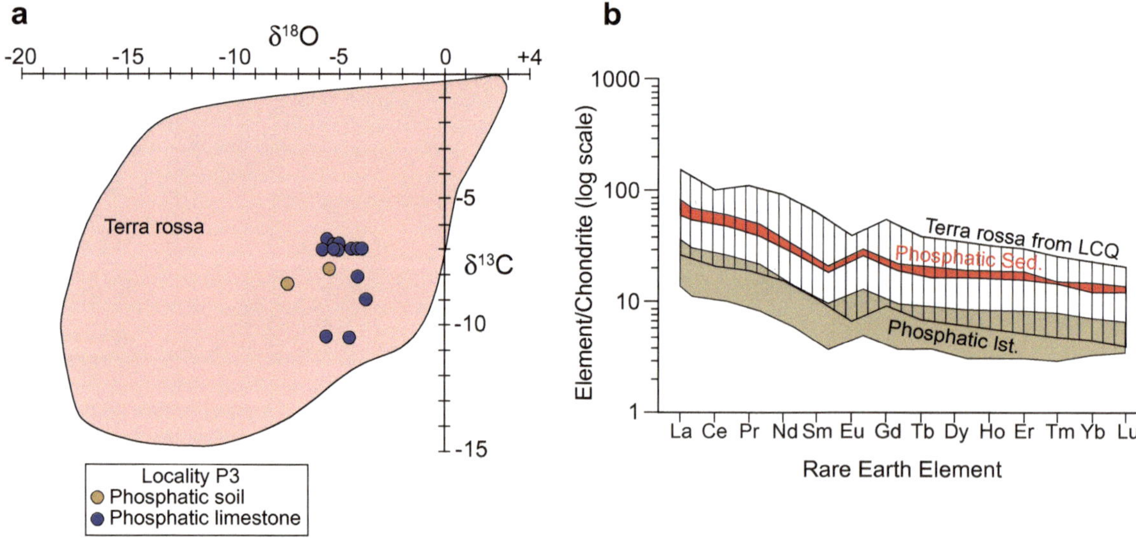

Fig. 10.10 **a** Comparison of stable isotopes from phosphatic soil and phosphatic limestone from locality P3, west end of Little Cayman (modified from Jones 2019a, b, his Fig. 21) with stable isotopes from terra rossa (modified from Jones 2021, his Fig. 23b). **b** Rare earth element profiles (chondrite normalized) for phosphatic sediments and phosphatic limestone from Little Cayman (modified from Jones 2019a, b, his Fig. 24) relative to the REE profiles for terra rossa from the nearby quarry on Little Cayman (modified from Jones, 2021, his Fig. 18)

ΣREE from the soil are similar to the ΣREE profiles that Jones (2019a) reported from terra rossa from the quarry on Little Cayman (Fig. 10.10b).

10.3.5 Origin of Phosphate Deposits

Phosphate formation on isolated oceanic islands has typically been attributed to P-rich fluids that originate from seabird guano (e.g., Trichet and Fikri 1997; Dawson and Smithers 2010). Alternate explanations include derivation from fecal material from animals other than seabirds (Piper et al. 1990), and/or dissolution of the bedrock (Rogers 1989). With seabird guano, decomposition leads to a decrease in the nitrogen and organic material while the P increases and forms apatite solutions that then percolate down through the underlying sediments (e.g., Stoddart and Scoffin 1983; Trichet and Fikri 1997).

Interpretation of the phosphate deposits on Little Cayman is complicated because mining removed most of the original deposits. Nevertheless, information derived from the remaining deposits indicates that the Cayman phosphatic deposits must have once been like the insular phosphates found elsewhere in the world. Many of their attributes are, for example, similar to those documented from the phosphate deposits found on Navassa Island, which is located ~30 km west of the southwest tip of Haiti (Miller et al. 2007).

The presence of the glaebules and phosphates in the Little Cayman deposits indicates that the original depositional environment was characterized by phosphate development with nearby sources of the Al and Fe that became incorporated in the cortical laminae that formed around the detrital phosphate grains (Jones 2019b). This could have evolved as phosphate deposits in coastal areas were broken up during major storms/hurricanes with the resulting grains becoming mixed with nearby terra rossa deposits. Ground waters that subsequently seeped through the terra rossa would have become enriched with Al and Fe and subsequently re-precipitated in the cortical laminae that developed around the detrital phosphate grains. Although the exact conditions that mediated this process are unknown, it may have been related to the alternating wet and dry seasons that characterize the climate of the Cayman Islands.

Jones (2019b, his Fig. 25) proposed a five stage model to explain the origin and development of the phosphatic deposits on Little Cayman.

Stage I: The depositional setting was probably a low-lying coastal area, possibly with some terra rossa located nearby, that was home for a large, nesting seabird community. The guano produced by these seabirds led to phosphatization of the underlying carbonate sediment.

Stage II: The phosphatic deposits were broken-up to produce lithoclasts and detrital grains, possibly by storm (hurricane) activity or during the early stages of a transgression.

Stage III: During this stage, the detrital phosphate grains were mixed with nearby terra rossa due to storm activity or were gradually covered with wind-blown dust from the Sahara Desert. Glaebules developed as cortical laminae developed around the detrital phosphate grains. The Al, Fe, and various trace elements involved in this process probably came from terra rossa that was present in the depositional system. Some of the terra rossa may have subsequently been removed during a storm or as a transgression flooded the area.

Stage IV: During this stage, the glaebules and phosphatic lithoclasts were buried under marine calcareous muds that contained scattered biofragments. This sediment may have been washed into the area during storms or hurricanes or developed during the early stages of a marine transgression. Irrespective, that carbonate sediment now forms the matrix of the phosphatic limestones.

Stage V: This terminal phase involved various diagenetic processes that led to lithification of the phosphatic sediments. Critical processes during this time were the precipitation of isopachous calcite cements, precipitation of P, Fe, and Al masses in the sediment, and precipitation of phosphate (including phosphate rafts) in some of the larger cavities.

10.4 Mangrove Deposits

Mangroves are found over wide areas of Grand Cayman (Figs. 10.11 and 10.12) and Little Cayman. On Grand Cayman, the area covered by mangroves has decreased in recent years due to the ongoing development of the island. This is especially true on the western peninsula where large areas of the mangroves have been replaced by housing and various other commercial developments. Extensive areas of mangroves, however, still exist in the central part of the island (Fig. 10.12). The mangrove swamps and their associated deposits were extensively documented by Woodroffe et al. (1980) and Woodroffe (1981, 1982, 1983) with the main focus being on those found on the Barkers Peninsula, the West Bay Peninsula, and the areas to the east and north of South Sound (Fig. 10.12).

10.4.1 Barkers Peninsula and West Bay Peninsula

The mangrove swamps on Barkers Peninsula (Fig. 10.13) and the area around Salt Creek on the West Bay Peninsula

Fig. 10.11 Mangrove swamps in area south of Morgan's Harbour, west coast of North Sound (Fig. 10.1a). **a** General view of mangrove swamp with drainage canal. **b** Mangrove swamp transected by drainage canals. Note exposed area with hard calcrete crust (CC) on top of the Ironshore Formation

(Fig. 10.14) are dominated by the mangroves *Rizophora mangle, Avicennia germanis*, and *Laguncularia racemose* that are found along with other types of vegetation, including log wood, *Conocarpus erectus* var *erectus*, the succulent herb *Batis marigtima*, and the fern *Arostichum aureum* (Woodroffe 1982). Although some areas are characterized by one species of mangrove, other areas have various combinations of species. Given that there are no natural exposures through the full thickness of the sediments that have accumulated in these swamps, Woodroffe (1981, 1982) relied on probing and cores to establish the thickness of the sediments and their constituent facies. Accordingly, he demonstrated that the sediment in the mangrove swamps on Barkers Peninsula is up to 258 cm thick, whereas on the West Bay Peninsula it is up to 4 m (and possibly 6 m) thick.

10.4.2 Barkers Peninsula and West Bay Peninsula

Cores through the sediments in the mangroves swamps on Barkers Peninsula and the West Bay Peninsula showed that the sediment succession that overlies the limestones of the Ironshore Formation, comprises (1) laminated and non-laminated crusts, (2) plastic mud, and (3) peat (Woodroffe 1981, 1982) (Fig. 10.15).

Laminated and Non-laminated Crusts

The hard laminated and non-laminated crusts that developed on top of the poorly lithified limestones of the Ironshore Formation (Fig. 10.11b) are the calcrete crusts that were described by Li and Jones (2014). These crusts,

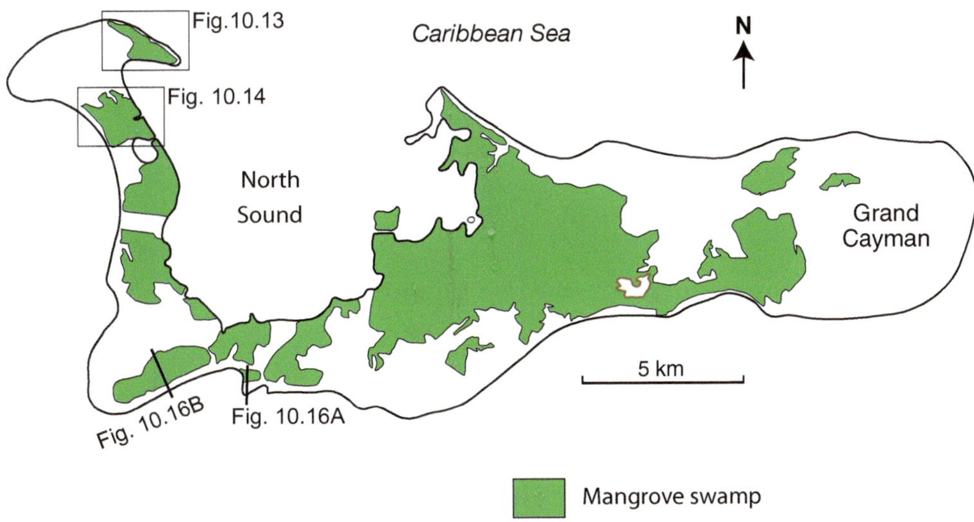

Fig. 10.12 Extent of mangrove swamps that once existed on Grand Cayman (modified from Woodroffe 1982, his Fig. 1). Large areas of mangrove swamps that existed on the western peninsula in 1982 have since been removed due to ongoing development

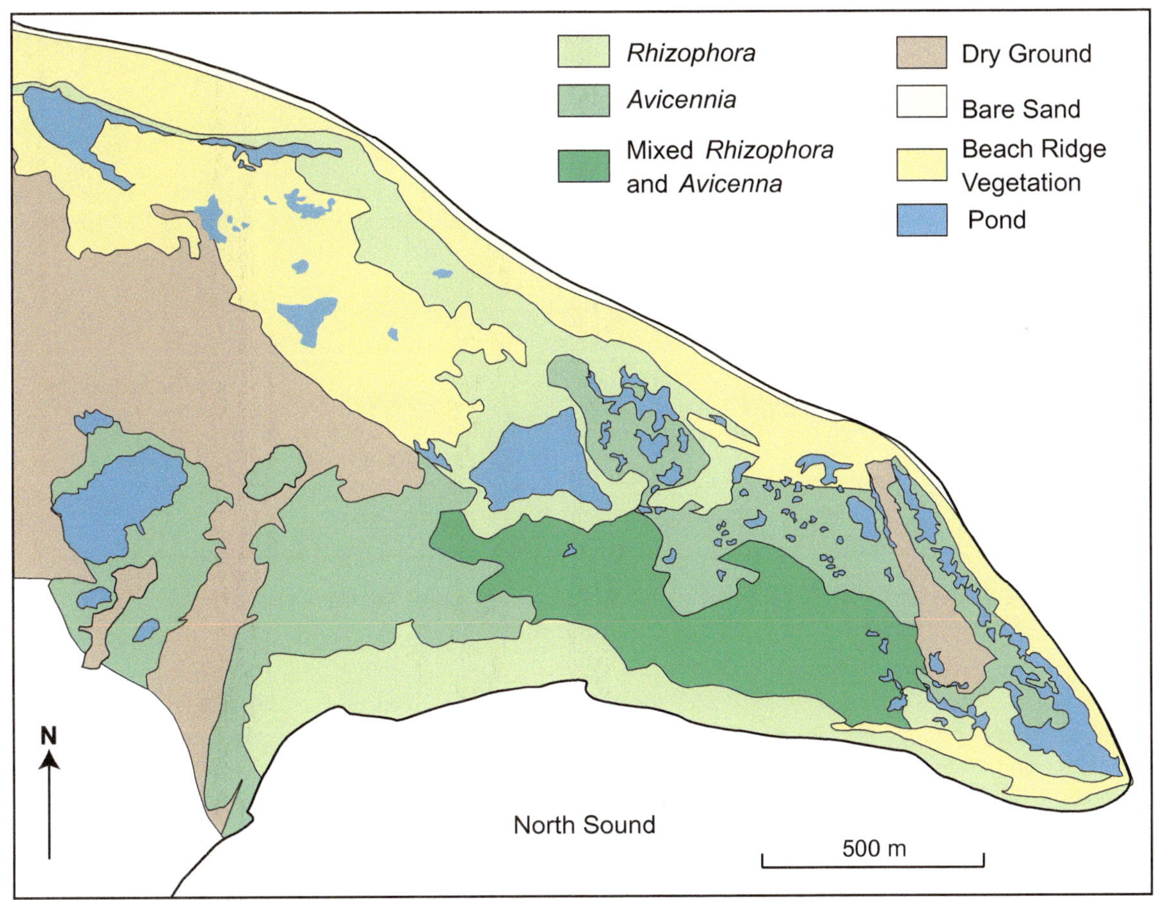

Fig. 10.13 Map showing distribution of different substrate types and plant species on Barkers Peninsula (modified from Woodroffe 1982, his Fig. 2)

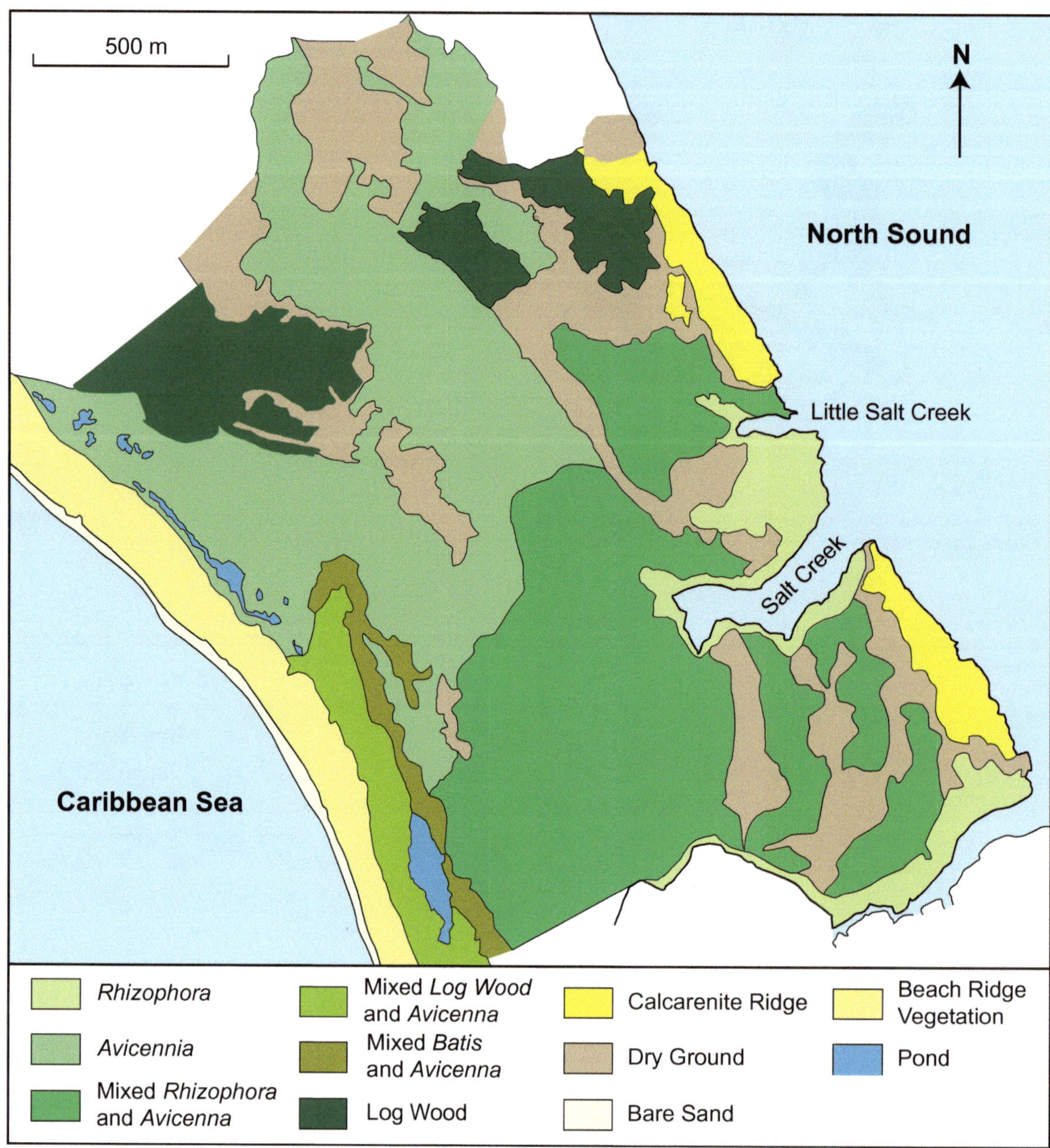

Fig. 10.14 Map showing distribution of different substrate types and plant species around Salt Creek on western peninsula of Grand Cayman (modified from Woodroffe et al. 1980, his Fig. 1)

characterized by their distinctive brown colour with localized black laminae that developed prior to the establishment of the mangrove swamps, have been attributed to a subaerial origin (Rigby and Roberts 1976; Li and Jones 2014).

Plastic Mud

Described as having a high plasticity, Woodroffe (1981) divided the mud into orange mud and green mud. The orange mud, up to 80 cm thick, is characterized by a low organic content, no fibres, a moisture content of 51–86%, and a pH of 6.3 to 7.1 when wet, and 4.8 to 6.6 when dry. In contrast, the green mud is non-fibrous, has a low content of organic matter (29–44%), a moisture content of 70–80%, and a pH of 6.4 to 6.7 when wet and 5.8 to 6.2 when dry.

Peat

Although most of the peat in the swamps is "mangrove peat", Woodroffe (1981) also found "sea grass peat" in one core. The mangrove peat, formed entirely of organic material derived from the mangroves, is generally <1 m thick but locally up to 4 m thick. Most of the organic matter is derived from *Rhizophora mangle* and *Avicennia germinas*. Roots

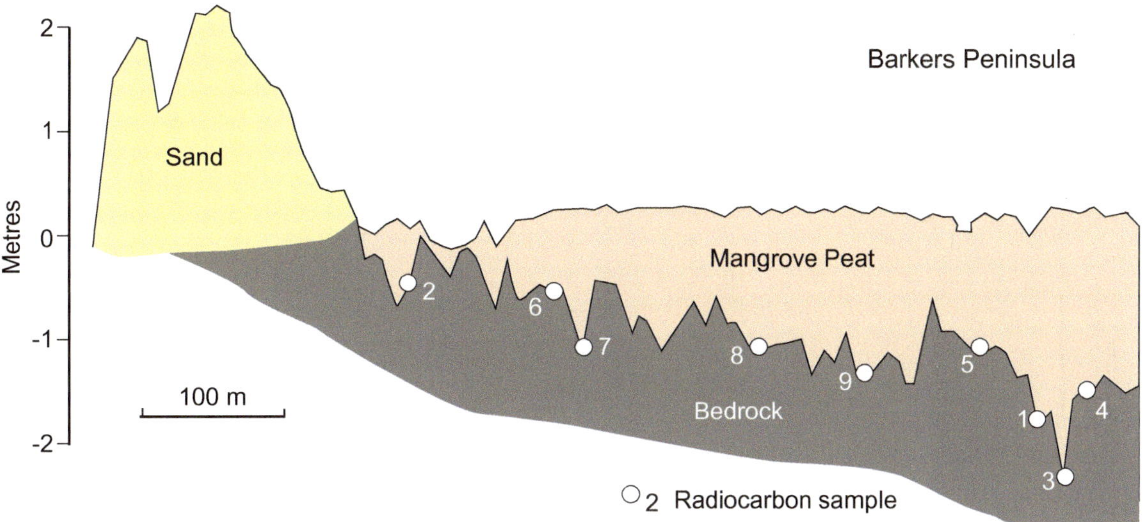

Fig. 10.15 Cross section across Barkers Peninsula showing location of samples that Woodroffe collected for radiocarbon dating (modified from Woodroffe 1981, his Fig. 4)

and rootlets are common in the peat. The swamps are invariably wet due to regular tidal flooding and periods of high rainfall. The organic content, determined by loss on ignition, is generally 50–80% but may be as high as 80–90% after flooding with tides. The pH of the peat is 6.0 to 7.0 when wet and 5.6 to 6.5 when dry.

The sea grass peat, found in only one core, is derived largely from *Thalassia testudinum* that is common in the shallow coastal areas of North Sound. The peat, formed of stacked blades of the sea grass has an organic content of 63–68%, and a pH of 6.6 to 7.2 when wet, and 6.0 to 7.0 when dry.

10.4.3 South Sound

Woodroffe (1983) examined the Prospect Point mangrove swamp that is located on the east side of South Sound and the larger mangrove swamp that existed north of the sound (Fig. 10.12). Both swamps were isolated from the ocean waters of South Sound by densely vegetated beach ridges. The Prospect Point swamp was dominated by *Rizophora mangle*, whereas the South Sound swamp was characterized by dense *Rhizophora* that was replaced landward by mixed Rhizophora and *Avicennia*. Both swamps are underlain by mangrove peat that is locally >2.0 m thick (Fig. 10.16). Flooding of these swamps is seasonal and dependent on rainfall. The swamps dry out during the dry season. Woodroffe (1983) suggested that the bedrock beneath the South Sound swamp belonged to the "Bluff Limestone". Although outcrops are rare in this part of Grand Cayman, mapping of areas to the north indicates that the bedrock probably belongs to the Cayman Formation.

10.4.4 Age and Development of Mangrove Swamps

Woodroffe (1981, his Table 1) provided radiocarbon ages of 567 ± 55 to 2160 ± 75 years that came from nine samples of mangrove peat samples obtained from Barkers Peninsula at depths from 47 to 209 cm below mean sea level (Fig. 10.15). Each sample came from peat just above the top of the underlying bedrock because he wanted to determine the age of the transgression that led to flooding of the area and establishment of the mangrove swamps. Woodroffe (1981), however, cautioned that these dates should be regarded as minimum ages because of the possibility that contamination by roots of younger mangroves may produce a younger age.

Woodroffe (1981) suggested that the plastic mud found in some areas of the swamp are localized deposits that formed in seasonally flooded habitats, whereas the mangroves became established during a period of marine transgression. As the transgression progressed, the mangroves expanded landward. He argued that the mangroves formed in response to a Holocene rise in sea level with the dates from the Cayman swamps suggesting that sea level has risen ~185 cm over the last 2,100 years. Woodroffe (1981) argued that this was consistent with the records determined from Florida and Belize.

10.5 Conclusions

Modern terrestrial sedimentological processes on the Cayman Islands have involved the accumulation of terra rossa, phosphates, and mangrove swamps. Depending on location, these deposits lie on top of the Cayman Formation, Pedro

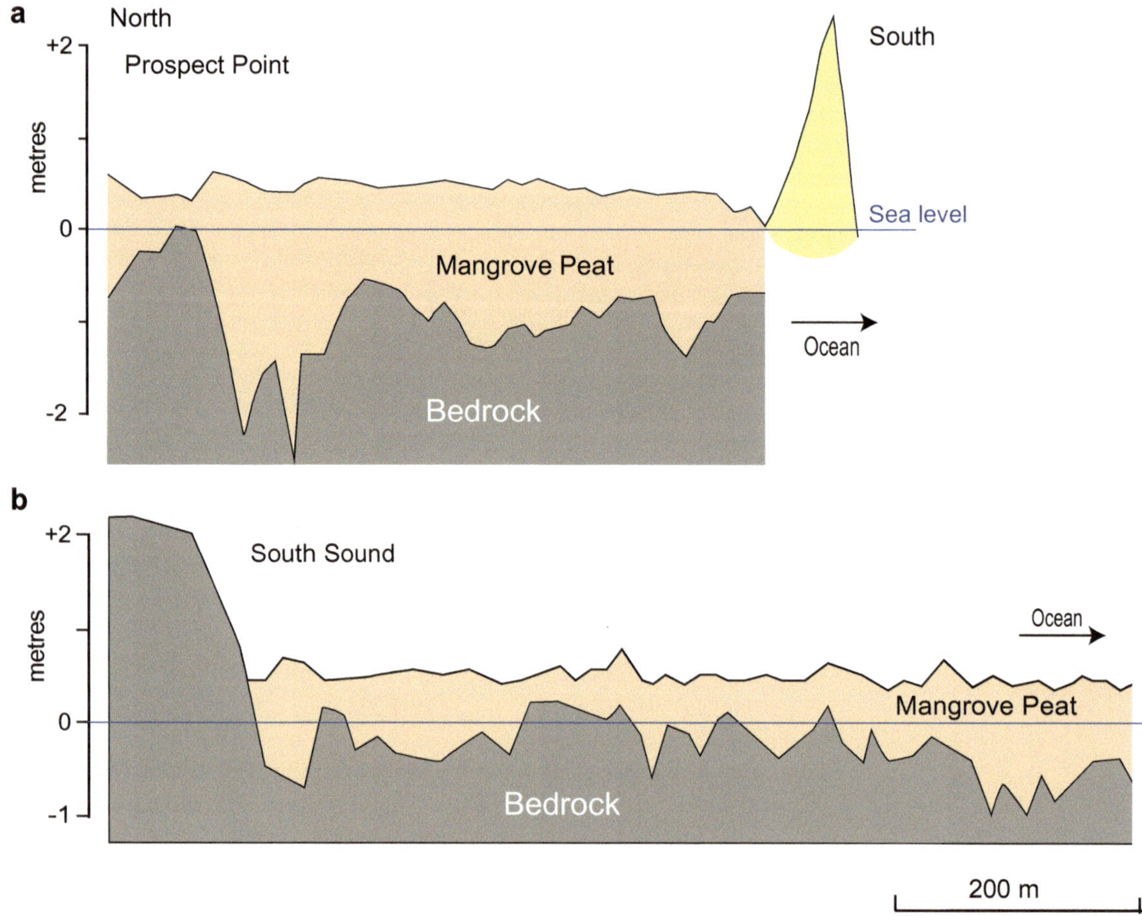

Fig. 10.16 Cross-sections showing relationship between the mangrove peat and bedrock at Prospect Point and South Sound, southwest corner of Grand Cayman (Fig. 10.12). Modified from Woodroffe (1983, his Fig. 5)

Castle Formation, or Ironshore Formation. Secondary processes have commonly transported these deposits into subsurface cavities. The following conclusions apply to these deposits.

- The terra rossa is formed largely of Al_2O_3 and CaO with lesser amounts of SiO_2, MnO, MgO, Na_2O, K_2O, and Ti_2O.
- Mineralogically, the terra rossa is formed of various combinations of X-ray amorphous material, anatase, hematite, goethite, various Mn precipitates, biofragments, glaebules, clay-coated grains, terra rossa ooids, and calcite cements.
- Available stratigraphic evidence indicates that terra rossa has been accumulating on the Cayman Islands for the last 5 million years.
- The composition and REE content of the terra rossa indicate that the Caymanian terra rossa is formed largely of wind-blown material that originated in the Sahara Desert.
- Between 1884 and 1890, phosphate was exported from Grand Cayman, Little Cayman, and Cayman Brac. Today, little remains of these deposits.
- The phosphate deposits found on Little Cayman are characterized by lithoclasts, detrital phosphate grains, glaebules, phosphatic rafts, and calcite cement.
- The phosphate deposits on Little Cayman developed in a low-lying coastal area, possibly near terra rossa. Guano, produced by a large nesting seabird community led to the formation of the phosphates. Storms and/or hurricanes resulted in the break-up of the phosphates and mixing with terra rossa. Subsequent marine incursions led to the deposition of calcareous muds.
- Mangrove swamps cover large areas of Grand Cayman and Little Cayman.
- Plastic mud and peat that developed in these swamps are up to 2 m thick.
- The mangrove swamps initially developed in associated with a transgression that started ~2,160 years ago.

Collectively, these terrestrial sediments are deposited on land surfaces that become exposed during periods of sea-level lowstands. Although they may be preserved, many of these deposits may be removed during the next transgressive phase.

References

Ahmad N (1996) Agricultural land capability of the Cayman Islands. The University of the West Indies, St. Augustine, Trinidad

Ahmad N, Jones RL (1969) Occurrence of aluminous lateritic soils (bauxites) in the Bahamas and Cayman Islands. Econ Geol 64:804–808

Baker RJ (1974) Soil and land-use survey No. 26 Cayman Islands. The Regional Research Centre, Department of Soil Science, University of the West Indies, Trinidad, p 42

Billymer JHS (1946) The Cayman Islands. Geogr Rev 36:29–43

Brewer R, Sleeman JR (1964) Glaebules: their definition, classification and interpretation. J Soil Sci Soc Am 15:66–80

Cerling TE, Quade J (1993) Stable carbon and oxygen isotopes in soild carbonates. Clim Change Cont Isotopic Rec Geophys Monogr 78:217–231

Comer JB (1974) Genesis of Jamaican bauxite. Econ Geol 69:1251–1264

Craton M (2003) Founded Upon the Seas: a History of the Cayman Islands and their People. Ian Randle Publishers, Kingston, Jamaica, p 532

Darwin C (1845) Journal of researches into the natural history and geology of countries visited during the voyage of H.M.S. Beagle round the world, under command of Capt. Fitz Roy, R.N. John Murray, London, UK

Darwin CR (1846) An account of the fine dust which often falls on vessels in the Atlantic Ocean. Q J Geol Soc Lond 11:26–30

Dawson JL, Smithers SG (2010) Shoreline and beach volume change between 1967 and 2007 at Raine Island, Great Berrier Reef, Australia. Global Planet Change 72:141–154

Fawcett W (1888) Cayman Islands. Bull Miscellaneous Inf (royal Botanic Gardens, Kew) 1888:160–163

Garrison VH, Foreman WT, Genualdi SA, Griffin DW, Kellogg CA, Majewski MS, Mohammed A, Rasmsubhag A, Shin EA, Simonich SL, Smith GW (2006) Saharan dust—a carrier of persistent organic pollutants, metals and microbes to the Caribbean? Rev Biol Trop 54:9–21

Goudie AS, Middleton NJ (2001) Saharan dust storms: nature and consequences. Earth Sci Rev 56:179–204

Hirst GSS (1910) Notes on the History of the Cayman Islands. P.A. Benjamin Manf. Co., Kingston, Jamaica, p 412

Hutchinson GE (1950) The biochemistry of vetebrate excretion. Bull Am Museum Nat Hist 96:1–554

Jones B (1992) Void-filling deposits in karst terrains of isolated oceanic islands: a case study from Tertiary carbonates of the Cayman Islands. Sedimentology 39:857–876

Jones B (1994) Geology of the Cayman Islands. In: Brunt MA, Davies JE (eds) The Cayman Islands: Natural History and Biogeography. Kluwer, Dordrecht, The Netherlands, pp 13–49

Jones B (2016) Cave fills in Miocene-Pliocene strata on Cayman Brac, British West Indies: implications for the geological evolution of an isolated oceanic island. Sed Geol 341:70–95

Jones B (2019a) Diagenetic processes associated with unconformities in carbonate successions on isolated oceanic islands: case study of the Pliocene to Pleistocene sequence, Little Cayman, British West Indies. Sed Geol 386:9–30

Jones B (2019b) Recycled insular phosphates and coated grains: case study from Little Cayman, British West Indies. Sediemntology 67:1844–1878

Jones B (2021) Formation, dispersal and accumulation of terra rossa on the Cayman Islands. Sedimentology 68:1–45

Jones B, Zheng E, Li L (2018) Growth and development of notch speleothems from Cayman Brac, British West Indies: response to variable climatic conditions over the last 125,000 years. Sed Geol 373:210–227

Li R, Jones B (2014) Calcareous crusts on exposed Pleistocene limestones: a case study from Grand Cayman, British West Indies. Sed Geol 299:88–105

Liang T, Jones B (2015) Petrographic and geochemical features of sinkhole-filling deposits associated with an erosional unconformity on Grand Cayman. Sed Geol 315:64–82

Matley CA (1926a) The geology of the Cayman Islands (British West Indies) and their relation to the Bartlett Trough. Q J Geol Soc Lond 82:352–387

Matley CA (1926b) Phosphate in the Cayman Islands. In: 14th International Geological Congress. Madrid, pp 777–779

McLennan SM (1989) Rare earth elements in sedimentary rocks: influence of provenance and sedimentary processes. Rev Mineral Geochem 21:169–200

Miller MW, Halley RB, Gleason ACR (2007) Coral Reefs of the World: Reef geology and biology of Navassa Island. In: Riegl BM, Diodge RE (eds) Coral Reefs of the USA. Springer, New York, USA, pp 407–434

Moskowitz BM, Reynolds RL, Goldstein HL, Berquó TS, Kokaly RF, Bristow CS (2016) Iron oxide minerals in dust-source sediments from the Bodélé Depression, Chad: implications for radative properties and Fe bioavailability of dust plumes from the Sahara. Aeol Res 22:93–106

Muhs DR, Budahn JR (2009) Geochemical evidence for African dust and volcanic ash inputs to terra rossa soils on carbonate reef terraces, northern Jamaica, West Indies. Quatern Int 196:13–35

Muhs DR, Budahn JR, Prospero JM, Carey SN (2007) Geochemical evidence for African dust inputs to soils of western Atlantic islands: Barbados, the Bahamas, and Florida. J Geophys Res 112:1–26

Muhs DR, Budahn JR, Prospero JM, Skipp G, Herwitz SR (2012) Soil genesis on the island of Bermuda in the Quaternary: the importance of African dust transport and deposition. J Geophys Res 117:1–26

Muhs DR, Bush CA, Stewart KC (1990) Geochemical evidence of Saharan Dust parent material for soils developed on Quaternary limestones of Caribbean and western Atlantic islands. Quatern Res 33:157–177

Ng KC (1990) Diagenesis of the Oligocene-Miocene Bluff Formation of the Cayman Islands—a petrographic and hydrogeochemical approach. Ph.D. thesis, University of Alberta, Edmonton

Piper DZ, Loebner B, Aharon P (1990) Physical and chemical properties of the phosphate deposit on Nauru, western equatorial Pacific Ocean. In: Cook PJ, Burnett WC, Shergold JH, Riggs SR (eds) Phosphate Deposits of the World, vol 3. Neogene to Modern. Cambridge University Press, Cambridge, pp 177–195

Ren M, Jones B (2018) Genesis of island dolostones. Sedimentology 65:2002–2033

Rigby JK, Roberts HH (1976) Grand Cayman Island: Geology, sediments and marine communities vol 4. Brigham Young University, Geology Studies, Special Publication no. 4, 97 p

Rogers KA (1989) Dahllite and whitlockite from Amatuku islet, Tuvalu. Mineral Mag 53:123–125
Sauer JD (1982) Cayman Islands seashore vegetation: a study in compartive biogeography. University of California Publications in Geography, vol 25. University of California Press, London, England, p 161
Scheuvens D, Schütz L, Kandler K, Ebert M, Weinbruch S (2013) Bulk composition of northern African dust and its source sediments—a compilation. Earth Sci Rev 116:170–194
Stoddart DR, Scoffin TP (1983) Phosphate rock on coral reef islands. In: Goudie AS (ed) Chemical sediments and geomorphology: precipitates and residue in the near-surface environment. Academic Press, San Diego, United States, pp 369–400
Stuut JB, Zabel M, Ratmeyer V, Helmke P, Schefuß E (2005) Provenance of present-day eolian dust collected off NW Africa. J Geophys Res 110:1–14
Tarhule-Lips RFA (1999) Karst processes on Cayman Brac, a small oceanic island. Ph.D. thesis, McMaster University, Hamilton.
Trichet J, Fikri A (1997) Organic matter in the genesis of high-island atoll peloidal phosphorites: the lagoonal link. J Sediment Res 67:891–897
Woodroffe CD (1981) Mangove swamp stratigraphy and Holocene transgression, Grand Cayman Island, West Indies. Mar Geol 41:271–294
Woodroffe CD (1982) Geomorphology and development of mangrove swamps, Grand Cayman, West Indies. Bull Mar Sci 32:381–398
Woodroffe CD (1983) Development of mangrove swamps behind beach ridges, Grand Cayman Island, West Indies. Bull Mar Sci 33:864–880
Woodroffe CD, Stoddart DR, Giglioli MEC (1980) Pleistocene patch reefs and Holocene swamp morphology, Grand Cayman, West Indies. J Biogeogr 7:103–113
Zhao H, Jones B (2012a) Genesis of fabric-destructive dolostones: a case study of the Brac Formation (Oligocene), Cayman Brac, British West Indies. Sed Geol 267–268:36–54
Zhao H, Jones B (2012b) Origin of "island dolostones": a case study from the Cayman Formation (Miocene), Cayman Brac, British West Indies. Sed Geol 243–244:191–206
Zhao H, Jones B (2013) Distribution and interpretation of rare earth elements and yttrium in Cenozoic dolostones and limestones on Cayman Brac, British West Indies. Sed Geol 284–285:26–38

11 Hydrogeology

Abstract

On Grand Cayman, the Lower Valley Lens, the East End Lens, and the North Side Lens are the major freshwater lenses, whereas the Tibbets Turn Lens is the major one on Cayman Brac. Other smaller lenses are found at various locations on both islands. No freshwater lenses have been identified on Little Cayman. There are no surface rivers or streams on the Cayman Islands because rainwater rapidly flows into the subsurface via fractures and joints that are common in the limestone and dolostone bedrock of the islands. Rare lakes and ponds, generally located near the coastline, are characterized by variable water levels and generally saline waters. The groundwater of the Cayman Islands is variably saturated with respect to calcite and dolomite. Evaluation of the major freshwater lenses on Grand Cayman, each floating on top of dense, saline water, is difficult because of aquifer heterogeneity, complex hydrogeology, and variable hydrological conditions. Water flow into and out of the lenses is largely controlled by the fractures in the karstic terrains that allow intrusion of seawater or saline water, enhance hydrodynamic dispersion, and define lens geometry. The porosity and permeability of the bedrock is difficult to accurately predict and model because it is variable at all scales. Over the last 30 years or so, reverse osmosis plants that produce freshwater from saline waters have become operational at various sites on Grand Cayman and Cayman Brac in order to meet the ever-increasing demand for potable water. The inland location of these plants has been based, in part, on a complete and thorough analysis of the porosity and permeability characteristics of the bedrock because of the large volumes of seawater that are needed for their successful operation.

11.1 Introduction

The availability of potable water (i.e. fresh drinking water) is a pre-requisite need for establishing populations on any isolated oceanic islands, like the Cayman Islands. Early visitors to the Cayman Islands in the 1500s and 1600s found little readily available water given that there are no surface rivers or freshwater lakes on the karstic terrains of the islands. In 1773, British Navy Surveyor George Gauld attached a note to his map of the islands that stated "*The only place of anchorage is at the west end, abreast of the Hogsties* [now known as George Town], *where you may come in to 6 or 7 fathoms, but great care must be taken to pick out a white Sandy Spot as there is a great deal of rocky ground, but it appears plainly of brown colour. This, in good weather is a very convenient place for wooding and watering and getting stock and other refreshments*". This appears to be the first reference to the fact that it was possible to obtain potable water from the George Town area of Grand Cayman. Later, Fawcett (1889) made note of a natural "cistern" at East End that was "50 ft across" and contained "clear, sweet water". Billymer (1946, p. 42) also made note of the freshwater from natural springs at East End on Grand Cayman. This natural Cistern may be the large, water-filled sinkhole that is commonly referred to as the "Big Wynters Land Cistern", which is located to the northeast of the East End Reservoir (Fig. 11.1)—this is the only known water source that would fit the description offered by Fawcett (1888). In that sinkhole, the change from freshwater to saline water is at a depth of $\sim$21 m.

In 2003, a "stepwell", discovered close to George Town may have been one of the four wells that Gauld referred to his notes of 1773. It is now preserved and covered by a framed piece of glass in the floor of one of the stores in the

B. Jones, *Geology of the Cayman Islands*,
https://doi.org/10.1007/978-3-031-08230-6_11

Fig. 11.1 Photographs of sinkhole filled with freshwater, located on "Big Wynters Land", east end of Grand Cayman. **a** Aerial photograph showing sinkhole surrounded by dense vegetation. **b** Ground view of sinkhole. Photographs courtesy of Water Authority of the Cayman Islands

Bayshore Mall. The well comprises eight steps that had been carved into the limestones of the Ironshore Formation to a depth of ~2.5 m (Fig. 11.2). The water in the well at the bottom of the stairs was about 30 cm deep. When the Water Authority tested the water from that well they found that it was still drinkable with an electrical conductivity of ~1.5 mS/cm (pers. comm., Hendrik van Genderen, Water Authority).

On the Cayman Islands, the issue of potable water became increasingly critical as the populations of the islands increased. When the population of the islands was low (<10,000), individual houses diverted the rainwater that fell on the house roof into a cistern that was located under or beside the house (Fig. 11.3). The availability of water was, therefore, linked directly to rainfall with the result that prolonged periods of dry weather commonly led to shortages of drinking water. As the population of Grand Cayman grew from <1,000 in 1802 to ~6,000 just before World War II to 8,400 in 1966, to 52,200 in 2006 (Hughes 2017), and to 69,656 according to the preliminary results of the 2021 Census, the need for a constant, reliable supply of potable water became critical. Accordingly, the provision of potable water to the ever-increasing population initially focused on development of the freshwater lenses on Grand Cayman and Cayman Brac, and later, on the development of desalination systems that converted saline groundwater into potable water.

With time, individual houses placed greater reliance on water that was trucked from the freshwater lenses. In many cases, such water was pumped directly from privately owned trench wells, commonly with high rates of abstraction with no precautions being taken to ensure the integrity of the groundwater system (Ng and Beswick 1994). Some of the smaller freshwater lenses that once existed on Grand Cayman, for example, were lost because of excessive extraction rates and/or contamination (Kreitler and Browning 1983). With the passing of the Water Authority Law in 1982, stricter controls were placed on water extraction from any of the freshwater lenses.

Today, the major freshwater lenses on Grand Cayman are the Lower Valley Lens, the North Side Lens, and the East End Lens (Fig. 11.4a). Freshwater has been supplied for public use from the Lower Valley and East End lenses, but the North Side lens has not been developed because of

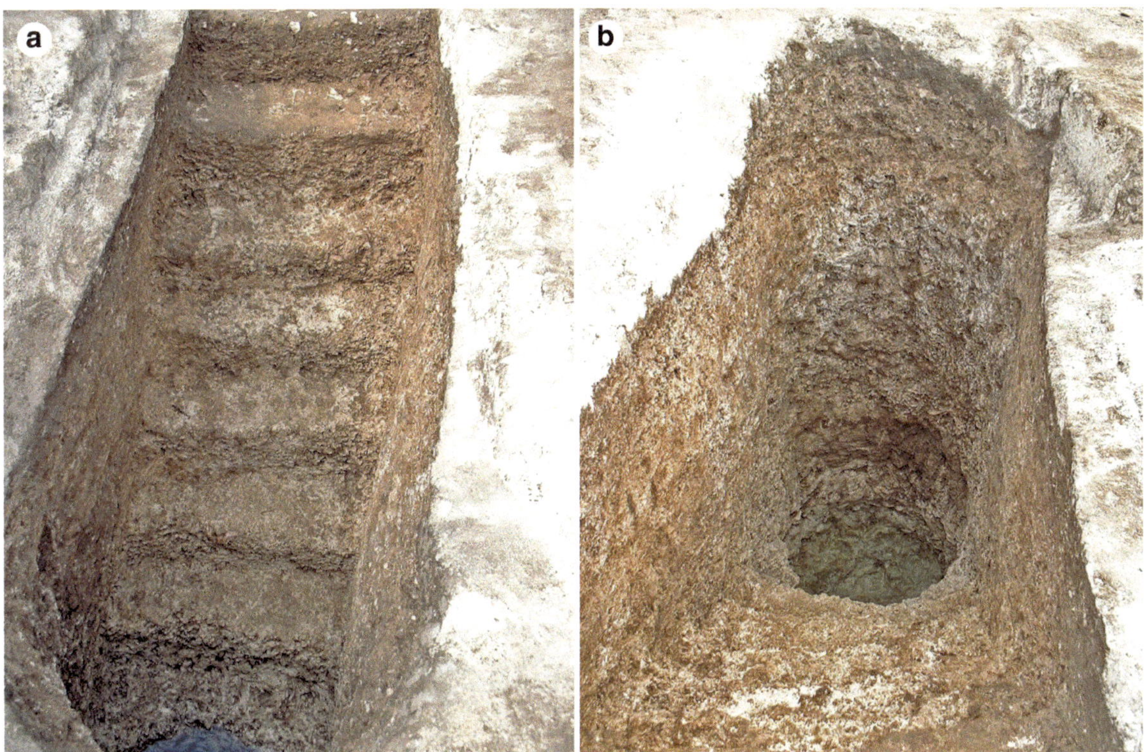

Fig. 11.2 Photographs of stepwell discovered in George Town in 2004. **a** Steps carved into limestones of the Ironshore Formation—water is at the bottom or the photograph. **b** View from top of stepwell showing pool of water at base of structure. Photographs courtesy of Hendrik van Genderen, Water Authority, Cayman Islands

Fig. 11.3 Photograph of traditional style house on Grand Cayman showing the main house with a pipe from its roof connected to the cistern (right side of image) where the rainwater was stored for usage. Photograph taken in 1971–72, courtesy of Daniel Doucet

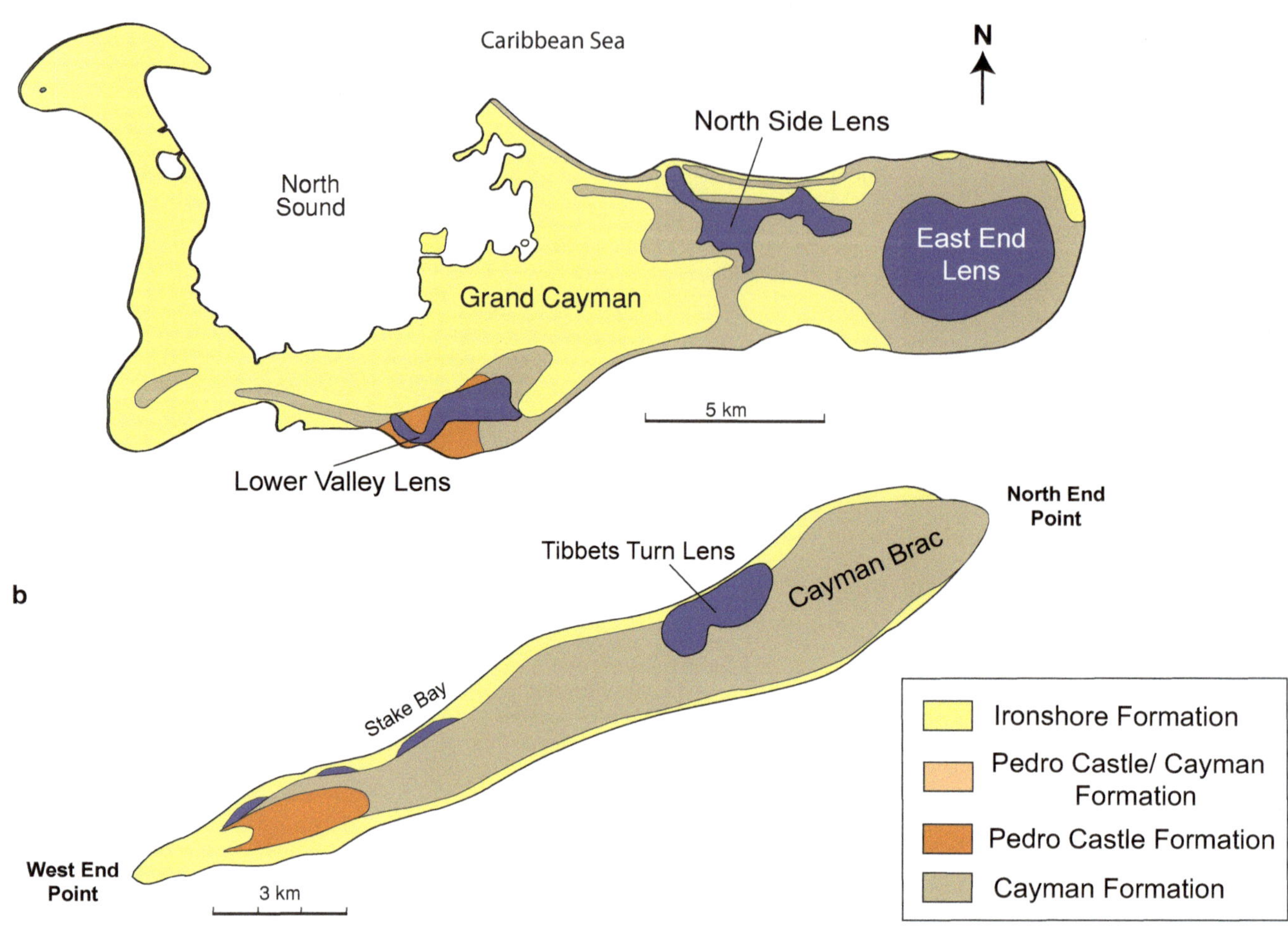

Fig. 11.4 Maps of **a** Grand Cayman and **b** Cayman Brac showing locations of main freshwater lenses on each island. Modified from Hydrological Survey Maps for Grand Cayman, Cayman Brac, and Little Cayman published in 1990 by the Water Authority, Cayman Islands

limited access and its location far from the western part of Grand Cayman where the demand for potable water is the highest. On Cayman Brac, the Tibbetts Turn water lens is the largest source of freshwater, whereas smaller lenses located along the northern coast on the west end of the island are only used locally (Fig. 11.4b). The Tibbetts Turn water lens was never developed as a source of potable water. As yet, no freshwater lenses have been found on Little Cayman.

Research on the water resources of the Cayman Islands that was directly related to the provision of potable water for all inhabitants can be divided into three loosely defined stages, as follows.

Stage I: Prior to enactment of the Water Authority Law in 1982

Stage II: After establishment of the Water Authority Law, focus was placed on the provision of potable water to the population of the islands while, at the same time, managing and protecting the freshwater lenses

Stage III: Increased installation and operation of reverse osmosis plants that started when the plant at Lower Valley first came on-line in 1998.

It is important to note that there is no exact date for the transition from stage 2 to 3 because extraction from the freshwater lenses has remained ongoing, given the cost and the challenges associated with the installation of the reverse osmosis plants and the underground piped water distribution system needed to deliver water to every house on Grand Cayman and Cayman Brac.

11.1.1 Stage I

This stage, which predated the enactment of the Water Authority Law, involved preliminary studies of the freshwater resources on Grand Cayman that resulted in publication of some scientific papers, including those by Mather (1972) and Bugg and Lloyd (1976), and various unpublished consultant reports such as those by Richards and Dumbleton International (1976, 1980a, b). Similarly, an unpublished report by Wallace Evans and Partners (1974), building on the work by Mather (1972), examined the feasibility of developing the Lower Valley Lens and other potential water sources on the islands. During this stage, however, relatively little attention was paid to Cayman Brac and Little Cayman. This research, which focused largely on the Lower Valley Lens, the North Side Lens, and the East End Lens on Grand Cayman, demonstrated that the freshwater lenses acted independently of each other and were complex in nature.

11.1.2 Stage II

The Cayman Water Authority was established in 1983. Peter Ravenscroft was appointed as the first hydrogeologist with the prime responsibility of undertaking research on the nature of the groundwaters on the islands. The Water Authority developed the Lower Valley wellfield and reservoir, which became operational in 1983. By 1985, a similar facility became operational for the East End Lens. At that time, the potable water from both wellfield operations was supplied to private trucking companies who transported the water to those houses that needed it. In order to maintain the condition of the lens, comprehensive monitoring of the Lower Valley water lens began in 1984 and for the East End lens in 1985.

Ravenscroft (1984), using data from the Cayman Islands, published a paper that dealt with the monitoring of fresh water lenses on small islands. Dr. Kwok-Choi Ng (University of Alberta) spent ~4 years with the Water Authority while he collected data and information regarding the freshwater lenses on the islands that he used for his Ph.D. thesis (Ng 1990). That research led to numerous papers that were published in various journals and conference proceedings (Ng and Jones 1990a, b, 1995; Ng et al. 1992; Ng and Beswick 1994, 1996).

Water abstraction from the three main freshwater lenses on Grand Cayman was closely monitored. Freshwater from Lower Valley and East End was trucked to individual houses under strict regulations that were designed to prevent damage to the lenses through over-extraction. Today, the East End reservoir and wellfield are still operated by the Water Authority so that truckers can deliver potable water to those houses that are not yet connected to the main distribution network on Grand Cayman.

11.1.3 Stage III

With continued population growth on the Cayman Islands, and especially on Grand Cayman, provision of freshwater became increasingly important as it became apparent that extraction from the freshwater lenses would soon exceed their replenishment and storage capacities. Thus, a transition to potable water production from reverse osmosis plants was initiated with the first plants being operative at the Governor's Harbour and the Red Gate Water Works in 1989, in Cayman Brac in 1991, and at the Lower Valley Water Works in 1998. Today, the Water Authority operates four reverse osmosis plants (one in Lower Valley, two at Red Gate, and one in North Side) on Grand Cayman that now collectively produce nearly 6 million cubic metres of water/year (1,500 million US gallons/year), and two reverse osmosis plants on Cayman Brac that have produced up to 225,000 cubic metres of water/year (60,000 million US gallons/year) (Fig. 11.5). On Grand Cayman, in recent years, ~60% of the water production came from Red Gate, ~7% from Lower Valley, and ~33% from the North Side plants. In addition, the Cayman Water Company operates several reverse osmosis plants that in 2020, for example, produced ~3.3 cubic metres of water (~860 million US gallons) of potable water for the Seven Mile Beach area and the West Bay district on Grand Cayman. Growth of these water supplies has been paralleled by ever–increasing networks of underground piping for the water distribution system. The Water Authority has also simultaneously developed a sewerage collection and treatment system that serves the Seven Miles Beach area of Grand Cayman.

Today, relatively few houses on Grand Cayman rely on cisterns for their water source and newly built houses are no longer required to have cisterns incorporated into their plans. On Cayman Brac, most residences still rely on cisterns, or large plastic storage tanks of 3.8 cubic metres (1,000 US gallons) for their water supply given that the Water Authority is still in the process of installing the necessary

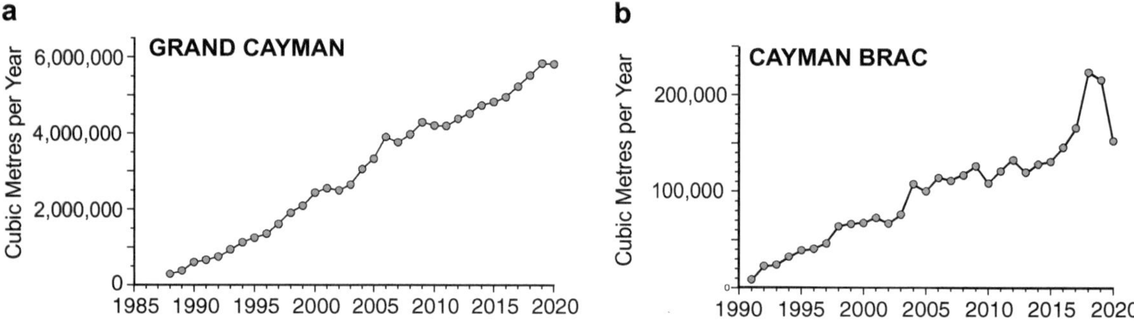

Fig. 11.5 Graphs showing continued increase in production of potable water by the Water Authority since 1988 from **a** the Red Gate, Lower Valley, and North Side sites on Grand Cayman, and **b** Cayman Brac

underground pipeline for supplying water from the reverse osmosis plant that is located on the west end of the island. On Little Cayman, virtually every house relies on water collected in cisterns and there are a few privately operated reverse osmosis plants that can supply the potable water.

On Grand Cayman, the reverse osmosis plants now provide large volumes of potable water to the ever-growing resident population of the islands and ever-increasing number of tourists. These plants are all located inland and away from coastal locations. As a result, the development and growth of these systems has relied heavily on knowledge of the bedrock geology of the islands because each plant needs a (1) continuous source of large volumes of saline water, and (2) location where the highly saline water that is a by-product of the desalination process can be disposed without it impacting the extraction zone or the local habitat. Each unit volume of salt water produces ~40% potable water and ~60% highly saline water. The highly saline water produced from the desalination process cannot, for example, be pumped directly back into the ocean because that would locally increase seawater salinity with the potential of causing the demise of the resident biota in those areas.

As the focus turned to the installation of reverse osmosis plants, research has largely focused on the geology of the subsurface carbonate successions with particular emphasis on the factors that control the porosity and permeability of the successions (Ng and Jones 1990b; Ng et al. 1992; Ng and Beswick 1994; Jones et al. 2001). This research has demonstrated that the subsurface carbonate successions are geographically variable and that considerable care has to be taken in determining where the feed water can be extracted and the processed highly saline water subsequently can be disposed.

11.2 Bedrock Geology

Across Grand Cayman, the Cayman Formation and Pedro Castle Formation, in which the freshwater lenses are located, vary from being formed almost entirely of limestone to being formed solely of dolostone (Fig. 11.6). Numerous cycles of karst development, as documented in Chapter 9, have greatly modified the porosity and permeability characteristics of the limestones and dolostones in these successions. The limestone successions in the Cayman Formation are confined to the east central part Grand Cayman, as is evident from wells GFN#1, NSC#1, BOG#1, and DTE#1 (Fig. 11.6). In contrast, the Cayman Formation in wells RWP#2, LV#2, SHT#4, and GTH#1 is formed almost entirely of dolostones with variable contents of LCD and HCD in each well (Fig. 11.6). Although the porosities of the limestones are generally higher than those of the dolostones, that is not universally true. Some of the dolostones in well LV#2, for example, have porosities that are similar to the porosities in the limestones in wells GFN#2 and NSC#1 (Fig. 11.6). There are no consistent patterns to the porosities from well to well. The dolostones in well RWP#2, located near the northeast coastline, are characterized by low porosities with most tested samples yielding values of <20% (Fig. 11.6). In contrast, the dolostones in the "porous unit" of the Cayman Formation in well LV#2 have porosities that are commonly >20%. Despite extensive research it has, to date, proven impossible to account for the variations from well to well.

Given the lack of core from the wells drilled on Cayman Brac, it has been impossible to examine the porosity and permeability of the strata of the Cayman Formation on that island. There, the formation is lithologically similar to that on Grand Cayman. For Little Cayman, knowledge of the subsurface geology is lacking because no deep sampled wells have been drilled there.

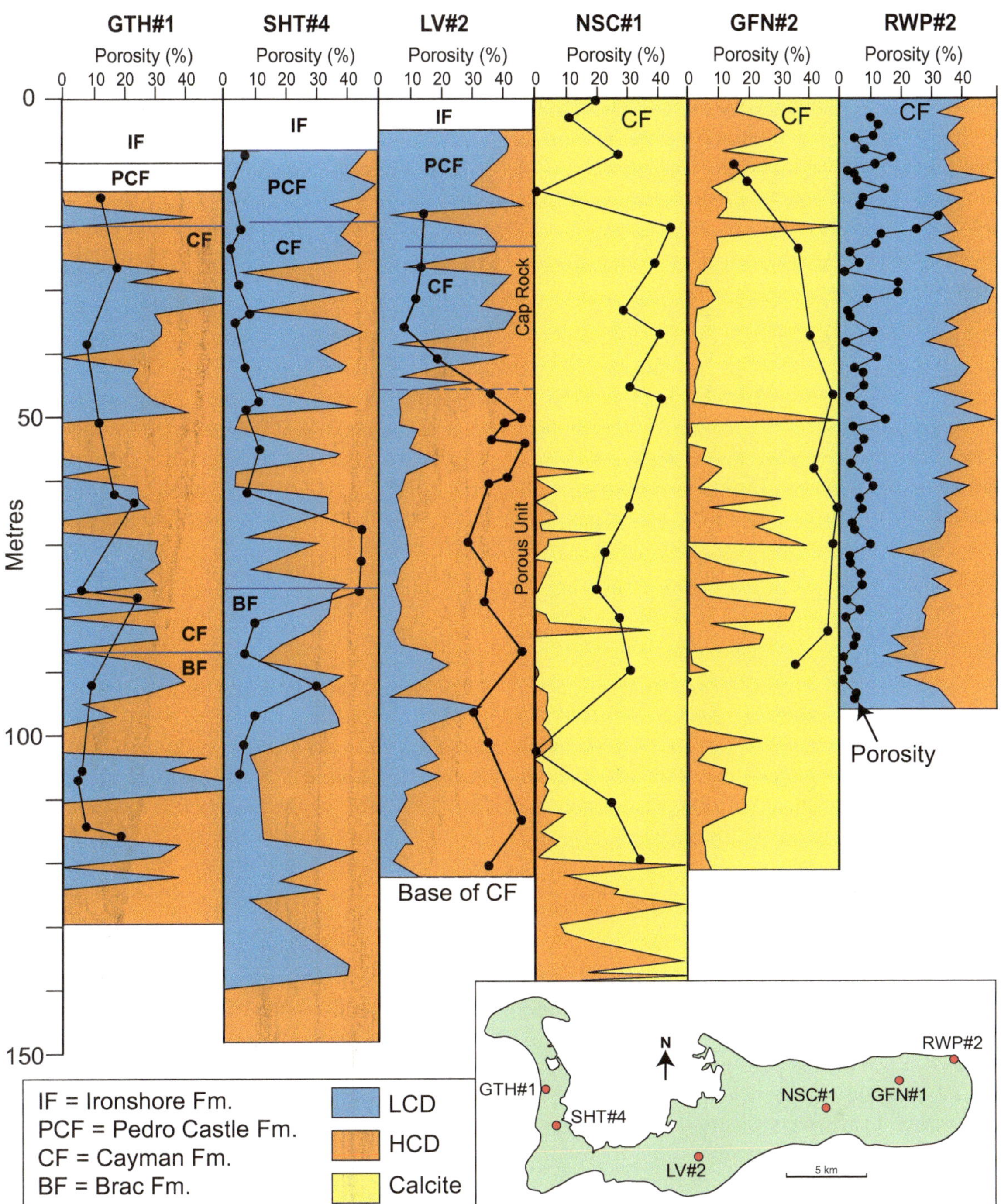

Fig. 11.6 Distribution of calcite, low calcian dolomite (LCD), high calcian dolomite (HCD), and tested porosity values for wells GTH#1, SHT#4, LV#2, NSC#1, GFN#1, and RWP#2 (locations shown on map in lower right corner). Note that variations in tested porosity values do not seem to correlate with the lithologies of the sequences in the different wells

11.3 Porosity and Permeability

On the Cayman Islands most of the groundwater resources reside in the dolostones and limestones of the Bluff Group with its storage and transmission through the bedrock being fundamentally controlled by the porosity and permeability of the constituent limestones and dolostones. The situation is never static because the groundwaters, as they pass through the dolostones and limestones of the Cayman Formation and Pedro Castle Formation, can either enhance the porosities and permeabilities of the rocks through progressive dissolution of the bedrock or reduce the porosity and permeability as calcite and dolomite cements are precipitated in any available cavity.

Large scale porosity features are commonly related to the development of cave systems with caverns of many different shapes and sizes developing (Fig. 11.7a). Dissolution of aragonitic skeletons of various organisms, including corals, bivalves, gastropods, and foraminifera has produced cavities and pores that mimic the size and shape of the original organism (Fig. 11.7b–d). Dolomitization of the limestones also produced inter-crystalline pores of various sizes and shapes (Fig. 11.7e). The permeability of limestones and dolostones in the Bluff Formation is fundamentally controlled by the degree to which the individual pores and cavities in the limestones and dolostones are linked together via fractures. In extreme cases, solution-widening of joints and fractures (Fig. 11.8) provide passageways for the transmission of large volumes of water in virtually any direction.

Cementation mediated by the groundwaters that flow through the bedrock can reduce the size of the pores and cavities (Fig. 11.7f) and ultimately reduce the porosity to zero. Such cementation may line walls of cavities so that remaining open cavity space is left open but isolated from other pores (Fig. 11.7f).

On the Cayman Islands, the porosity and permeability of limestones and dolostones in the Bluff Group is heterogeneous at all scales. In the succession in well GFN#1, for example, that is formed of limestones with variable percentages of dolomite, there is no obvious correlation between the composition of the rock and its porosity and permeability (Fig. 11.9). The porosity and permeability of the limestones in the Ironshore Formation are less complicated because they are younger and have not experienced as much diagenetic alteration as the those in the Bluff Group. The complexity of the porosity and permeability in the limestones and dolostones in the Bluff Group means that it is extremely difficult to make general predictions regarding water flow through the rocks, even though such information is critical for the development of desalinization plants and the provision of fresh drinking water

11.4 Rainfall

The long-term viability of the freshwater lenses on the Cayman Islands depends on the rainfall because there are no surface rivers or streams on any of the islands. Data provided in The Cayman Islands Compendium of Statistics 2020 (published in 2021) allows computation of the average temperatures and rainfall between 2011 and 2020 that clearly show the 'wet' and 'dry' seasons (Fig. 11.10a). The average values for the rainfall, however, mask the fact that some years are overall drier than other years (Fig. 11.10b). In the context of the freshwater lenses on Grand Cayman, the average rainfall amounts, based on the weather station located at the airport at George Town, must be treated with some caution because (1) the average monthly rainfall varies from year to year (Fig. 11.10b), and (2) annual rainfall totals vary across Grand Cayman with the highest amounts being in the southwest corner of the island and the lowest on the eastern part of the island (Fig. 11.10c). In contrast, Cayman Brac receives ~1,400 mm of rain per year with temperatures from 24 to 33 °C.

11.5 Freshwater Lenses

On Grand Cayman, the three main freshwater lenses (Lower Valley, East End, and North Side lenses—Fig. 11.4a) were first delineated by Mather (1972) and Bugg and Lloyd (1976). Chidley and Lloyd (1977) also examined some aspects of those lenses. Some small freshwater lenses that once existed near West Bay and George Town were irreversibly impacted by sewage contamination and over-extraction (Kreitler and Browning 1983). The aerial extent of these water lenses was originally mapped using the 600 ppm chloride contour to define the outer limit of each lens (e.g., Ng and Beswick 1994, their Figs. 4.10, 4.11). This value, which corresponds to an electrical conductivity (EC) of ~2 mS/cm, is used herein to separate the freshwater from the brackish water. The East End freshwater lens is ~20 m thick with a surface area of 18.6 km^2, whereas the Lower Valley freshwater lens is generally <12 m thick and covers an area of 3.9 km^2. The North Side freshwater lens is 13 m thick and covers an area of 6.2 km^2.

On Cayman Brac, the Tibbetts Turn freshwater lens is the main natural reservoir of freshwater (Fig. 11.4b). Other, smaller freshwater lenses are found along the northern rim at the west end of the island (Fig. 11.4b). Ng (1990) also documented the presence of perched water tables in the central part of the island. No freshwater lenses have been identified on Little Cayman.

On Grand Cayman, the top surfaces of the freshwater lens are generally ~0.2 m above mean sea level around their

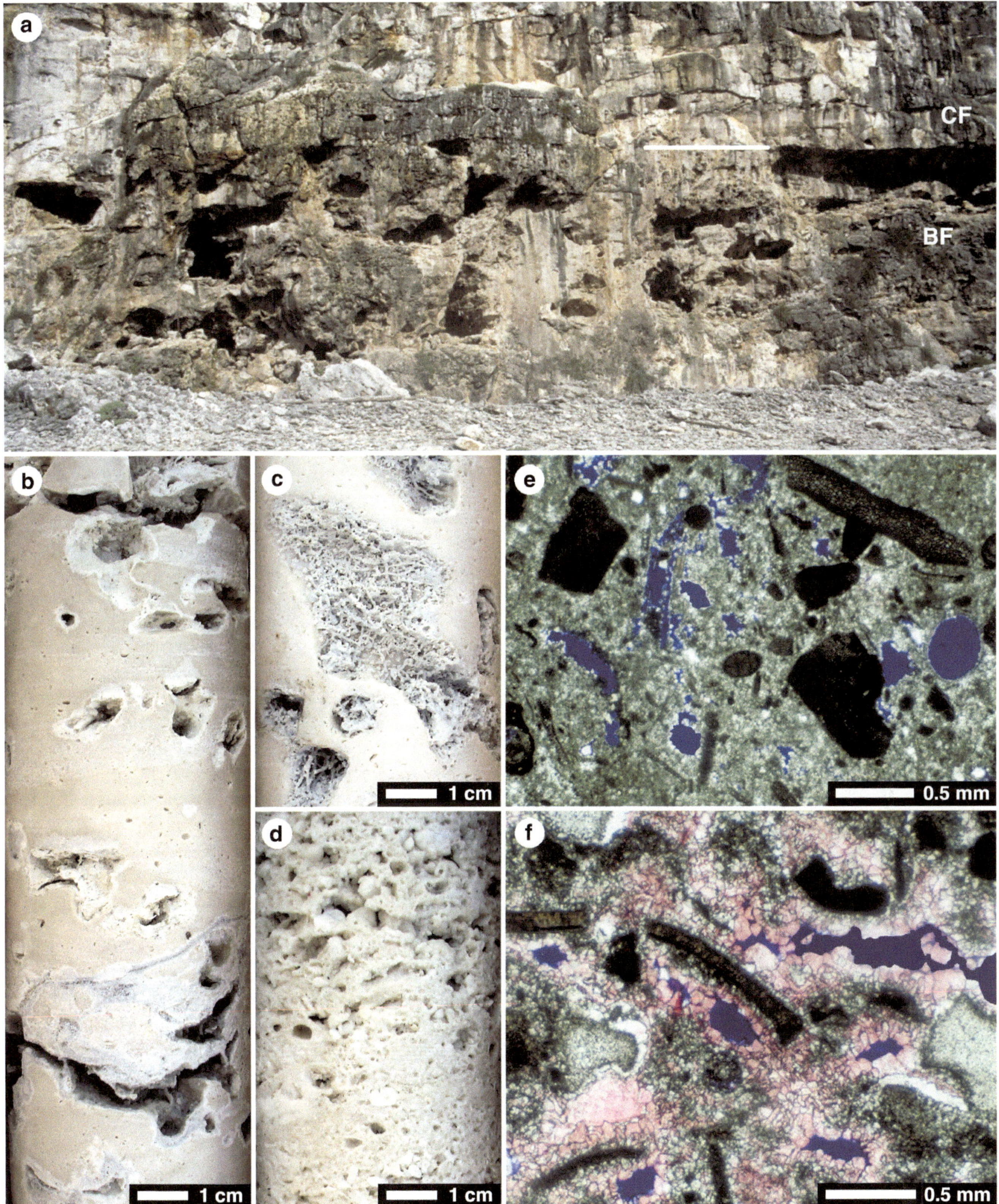
a
CF
BF
b
1 cm
c
1 cm
d
1 cm
e
0.5 mm
f
0.5 mm

◀ **Fig. 11.7** Examples of porosity features of all scales in the limestones and dolostones of the Bluff Group on the Cayman Islands. **a** Cliff face on south coast at east end of Cayman Brac showing cross-sections through caves of various sizes developed in the Brac Formation (BF). Note lack of caves in the overlying Cayman Formation (CF). White line indicates the unconformity between the Brac Formation and the Cayman Formation. **b–d** Examples of pores of different sizes, mostly developed by dissolution of aragonite skeletons of corals in the Cayman Formation, from well drilled just north of George Town, Grand Cayman. Samples from depths between 17 and 21 m below sea level. Pores in panels **b** and **c** formed by dissolution of coral skeletons, pores in **d** of unknown origin. **e** Thin section photomicrograph showing isolated pores (blue) in dolostone from lower part of Cayman Formation, west end of Cayman Brac. **f** Thin section photomicrograph of dolostone from the Cayman Formation, west end of Cayman Brac, showing pores filled with limpid dolomite cement and late calcite (stained red by Alizarin Red S solution) cements. Note isolated pores (blue) left in the middle of the cavities that were not completely filled with cement

Fig. 11.8 View of water -filled, solution-widened joint through the dolostones of the Cayman Formation. Scotts Quarry, west end of Cayman Brac. Joint is ~2 m wide at water level

edges and up to 0.7 m above mean sea level in the center of the lens. There is a thick brackish water zone transitioning from the freshwater to the saline water zones (Jones et al. 1997). The semi-diurnal tidal range around Grand Cayman is ~0.2 m with a seasonal variation of ~0.4 m. Given that the aquifers are hydraulically linked to the ocean, the water table oscillates in accord with the tide levels. For the East End lens the lag time between tides and water table levels is only one to two hours (Jones et al. 1997).

Overland runoff of rainwater is minimal because the water quickly drains into the subsurface via the open fractures, open joints, and other karst features that characterize the limestone and dolostone bedrock of the islands. Small surface ponds, however, may exist for short periods of time after prolonged periods of heavy rainfall. Evaporative losses of ~80% are due to the evapotranspiration that is underpinned by daytime temperatures of 30 °C in the summer and 25 °C in the winter (Ng et al. 1992). Using the method proposed by Vacher and Ayers (1980), Ng et al. (1992) estimated that there is a 75–85% evaporative loss from open water in sinkholes, large ponds and mangrove swamps, soil moisture, rainwater intercepted by vegetation cover, and groundwater. Where the water table is close to the surface, transpiration from the soil by plants and phreatophytes that extract water directly from the groundwater also causes a loss of fresh groundwater.

The low terrain of Grand Cayman means that the catchment area for each water lens is limited in size and depends almost entirely on water that remains after evapotranspiration and/or surface runoff. The annual recharge volume available to these lenses varies because of the combined effects of the fluctuating annual rainfall and evapotranspiration rates. The only exceptions occur in areas where water can flow down fissures and/or joints from the land surface directly to the subsurface bedrock that houses the water lens.

11.5.1 Joints, Factures, and Water Lenses

The joints and fractures that characterize the limestone and dolostone bedrock of the Cayman Islands are hydrologically critical because they provide avenues for rapid water flow in virtually any direction. Although the exact locations and extent of these joints and fractures are difficult to ascertain accurately, photolineaments evident on aerial photographs (Fig. 11.11) suggest that they form an extensive network throughout the bedrock over much of the islands (Rigby and Roberts 1976). Ng (1990) and Ng et al. (1992) argued their importance to the hydrology because they (1) provide a direct hydraulic connection between the ocean and the aquifer, (2) facilitate rapid recharge of the aquifers, (3) provide avenues for mixing of waters from different hydrochemical zones, (4) facilitate mixing dissolution, and (5) may define the geometry of the freshwater lenses. Ng

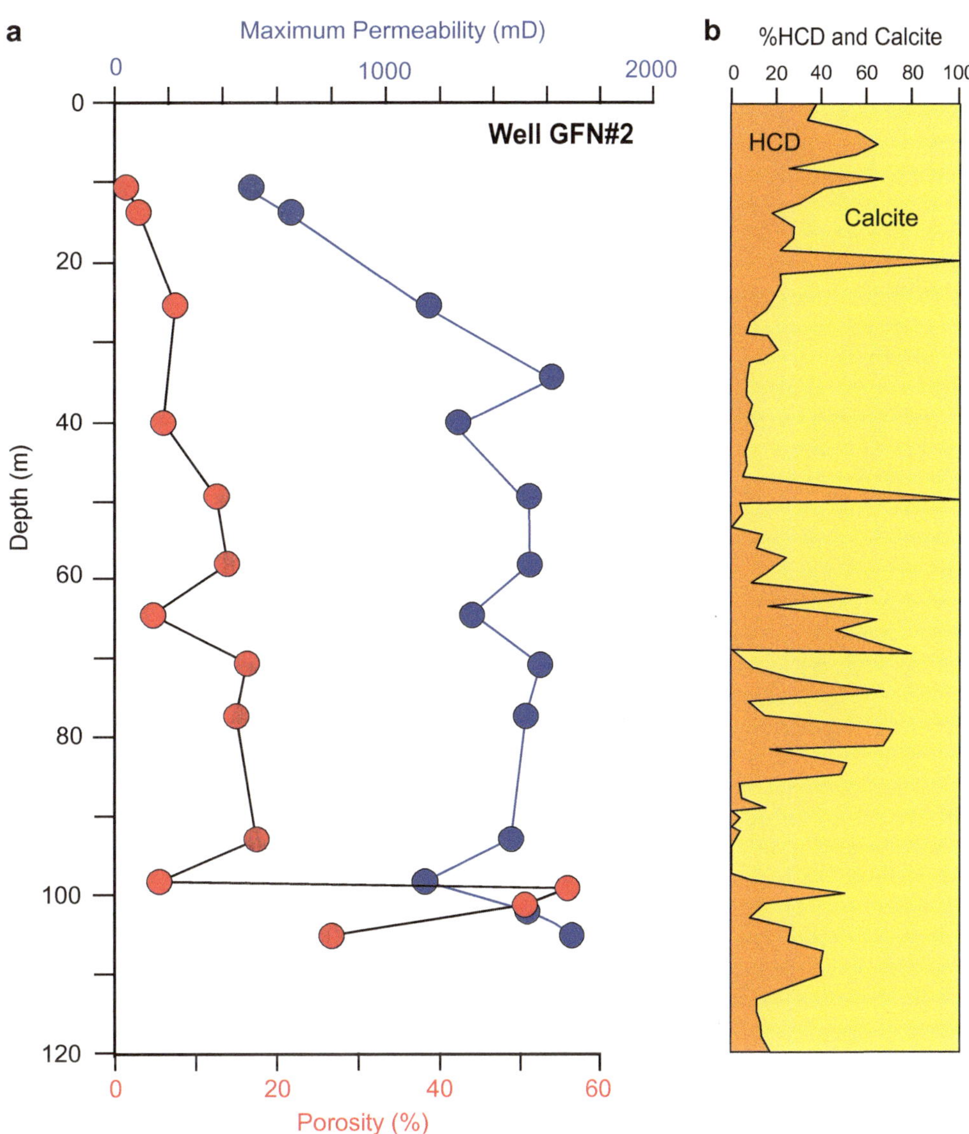

Fig. 11.9 Data for well GFN#2 (northwest corner of East End Lens, Grand Cayman) showing tested porosity and maximum permeability relative to the lithology of the bedrock in terms of the percentages of calcite and dolomite (all HCD)

et al. (1992) suggested that the V-cross-sectional shape of the East End freshwater lens was probably due to the profiles created by the two sets of joints and the fissure systems (Fig. 11.12). The fact that the "lenses" on the Cayman Islands do not conform to the standard model for freshwater lens on small, isolated islands is probably due to the presence of the extensive joint and fracture systems that transect the bedrock of the islands.

11.5.2 Groundwater on the Cayman Islands

Water found on the Cayman Islands includes seawater, rainwater, pond water, cave pond water, and groundwater with each type differing in terms of its pH, electrical conductivity (EC), chloride, and sodium content. The information for these different types of water comes from Ng (1990).

Seawater

Seawater around the Cayman Islands has a high chloride content, a high sodium content, an alkaline pH, a Mg/Ca ratio of ~5.2:1, and δ^{18}O and δ^{2}H values of +1.31‰ and +6.4‰, respectively. Globally, the isotopic variations in seawater are largely a reflection of local evaporation, precipitation, freezing, and/or dilution with freshwater. The relatively high δ^{18}O value of Cayman seawater indicates that it has probably been modified by evaporation. In some of the shallow water areas around the coast of Grand Cayman, the correlation between the electric conductivity (EC) and δ^{18}O values indicates that evaporation exceeds precipitation (Ng 1990).

Rainwater

The chemistry of rainwater, with a pH >7, electric conductivity (EC) between 29 and 133 μS/cm, and chloride content

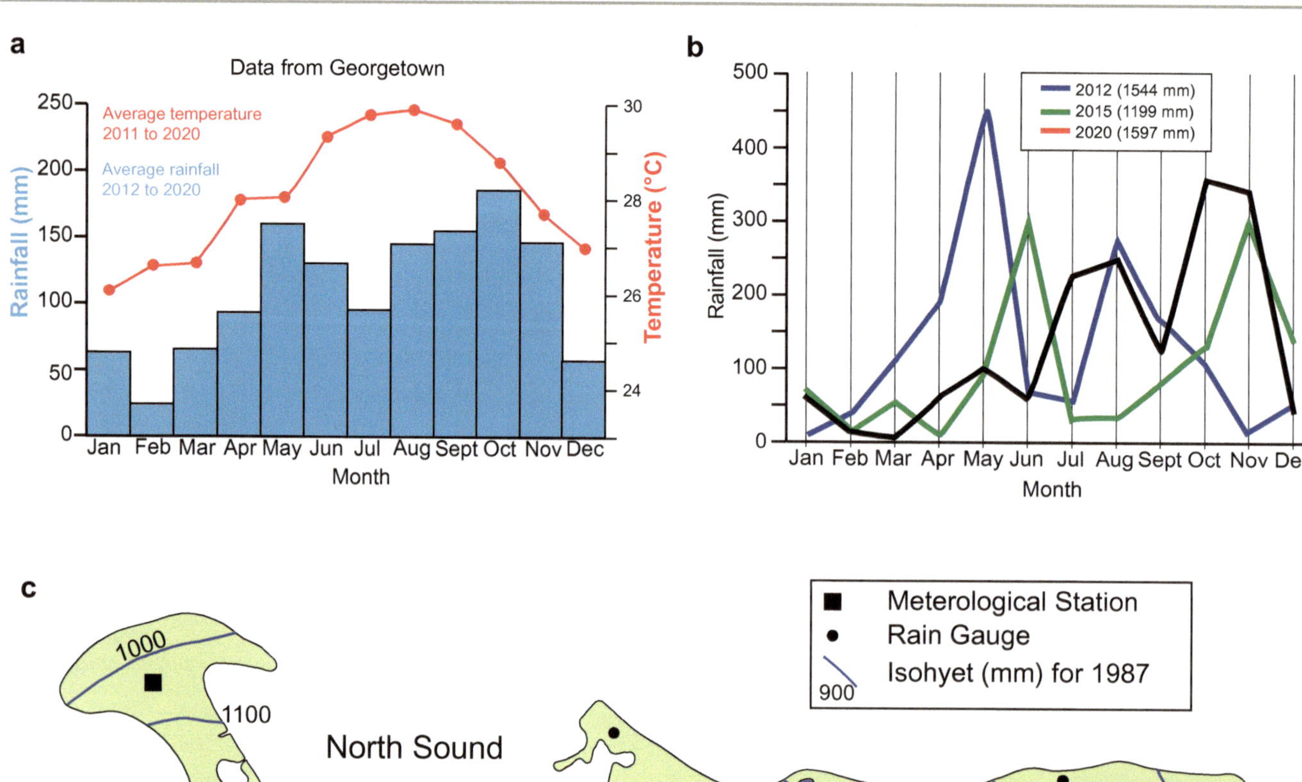

Fig. 11.10 Climate conditions on Grand Cayman. **a** Average rainfall (blue) (2012–2020) and temperature (red) (2011–2020) for Grand Cayman. Data from the National Weather Service, Cayman Islands Government (www.weather.gov.ky/portal/page/portal/nwshome/climate). **b** Variations in monthly rainfall totals for 2012, 2015, and 2020. **c** Isohyet (mm) map for Grand Cayman. Modified from Ng and Beswick (1994, their Fig. 4.8)

of 7–13.5 mg/l (possibly due to airborne sea spray), varies with geographic location and the season (Ng 1990). In general, rain nearer to the coast has a higher salinity than the rain further inland. Ng (1990) demonstrated that the EC can vary beyond the general range when he noted that the first major rainfall after a long dry spell in 1986 had an EC of 97 μS/cm but had fallen to 59 μS/cm by the next day, and that rainstorm Floyd in 1987 delivered rain with an EC of 29 μS/cm.

Rainfall on Grand Cayman has $\delta^{18}O$ that ranges from −1.56‰ to ~7.34‰ SMOW and $\delta^{2}H$ that varies from −8.4 to −53.8‰ SMOW. The relationship between the $\delta^{18}O$ and $\delta^{2}H$ of the rainwater, expressed by the equation $\delta^{2}H = 8.5^{*}\ \delta^{18}O + 9.1$, is very similar to the Meteoric Water Line, with its $\delta^{18}O$ and $\delta^{2}H$ related by the equation $\delta^{2}H = 8.0^{*}\ \delta^{18}O + 10$ (Ng 1990). On Grand Cayman, fresh groundwater has $\delta^{18}O$ values between −3.5 and −5.5‰, which indicate that most recharge occurs during heavy rainfalls when stable isotopes are low and exchange with the host rock is low (Ng 1990).

Recharge of the fresh groundwater by rainwater may involve (1) rapid recharge through the joints and fractures of

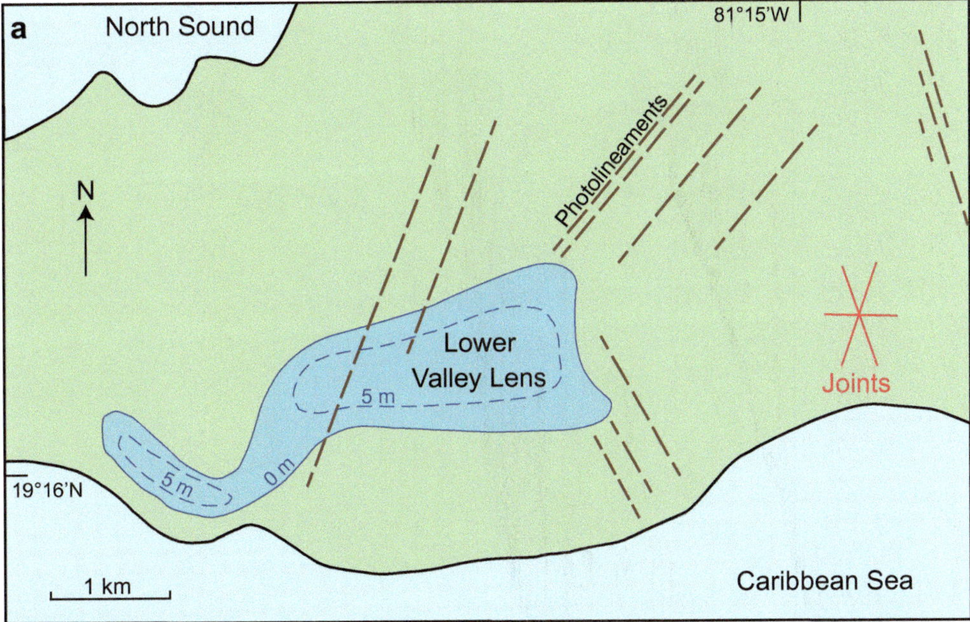

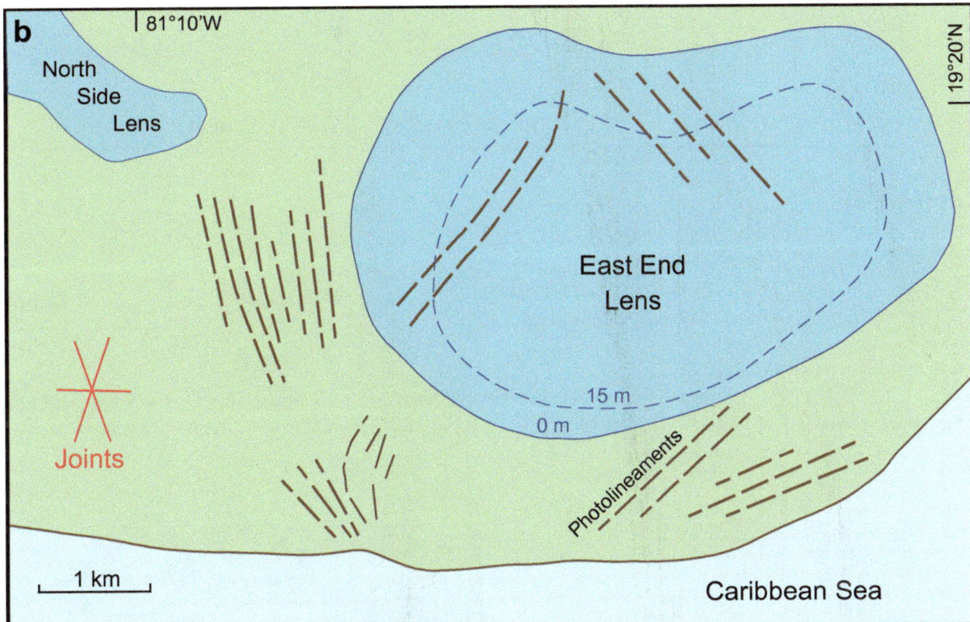

Fig. 11.11 Maps showing orientations of photolineaments relative to **a** Lower Valley Lens, and **b** East End Lens on Grand Cayman (Ng and Beswick 1994, modified from their Figs. 10 and 11, respectively). Insets show joint orientations (red) from Rigby and Roberts (1976, their Text-Fig. 30)

the bedrock (Figs. 11.11, 11.12) that will involve little or no interaction between the rainwater and the bedrock, or (2) slow infiltration through the vadose zone that may involve chemical modification of the water (e.g., decrease in Ca and Mg content, increase in HCO_3^-) as it passes through and interacts with the soils and bedrock. These two contrasting processes can have a significant impact on the chemistry of the subsurface waters.

Pond Water

Coastal ponds on Grand Cayman and Cayman Brac, attributed to beach-ridge damming during the Late Pleistocene, are the only surface water on the islands (Doran 1954; Rigby and Roberts 1976). Water levels in those ponds fluctuate with the tides. The water varies from lightly brackish after heavy rainfall to hypersaline at the end of the dry season. Some of the ponds, like Colliers Pond on the east coast of Grand Cayman (Fig. 11.13a) are characterized by complex chemical attributions that are, as yet, not fully understood. Water from Colliers Pond (Fig. 11.13a) commonly has high salinity and Ba and Cl concentrations higher than seawater but molar Mg/Ca ratios lower than seawater. The Mg /Ca ratios may reflect dissolution of $CaCO_3$ and/or influx of Ca-rich waters that flowed through the limestones of the

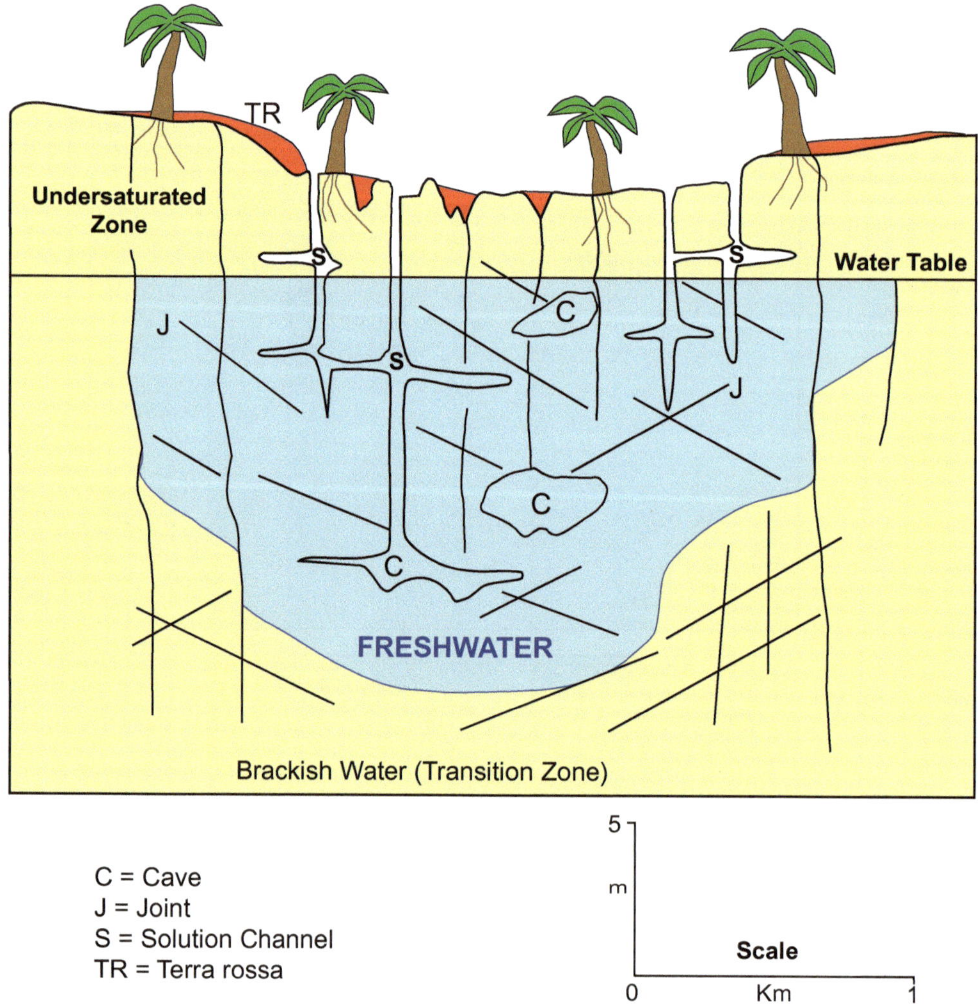

Fig. 11.12 Schematic diagram showing relationship between cross-sectional shape of a water lens and well-developed joint systems in the carbonate succession of Grand Cayman. Modified from Ng and Beswick (1994, their Fig. 4.5). Note vertical exaggeration

Fig. 11.13 Colliers Pond, northeast Grand Cayman showing comparison between conditions during **a** the wet season and **b** end of the dry season. Desiccated sediment at end of dry season is formed mostly of calcareous mud

Ironshore Formation that forms the shallow bedrock around Colliers Pond. By the end of the dry season Colliers Pond commonly dries out to reveal very fine-grained, desiccated calcareous sediments (Fig. 11.13b).

Water in Sinkholes

The fresh to slightly brackish water found in some sinkholes on the Cayman Islands has a higher salinity than rainwater. The stable isotope values of the surface waters are higher than those of the groundwater, but have isotopic concentrations that are similar to isotopically enriched rain and generally fall close to the rainwater line. Available evidence indicates that the water in the sinkholes probably came from groundwater that had been modified by evaporation and precipitation (Ng 1990).

Cave Water

Accessible caves on Grand Cayman and Cayman Brac are generally dry with water dripping from their ceilings only in the days immediately following periods of heavy rainfall. Free-standing water in the caves is a rarity. Ng (1990) obtained one sample of lightly brackish water (3650 mg l^{-1} TDS) from a pool in the Old Man Bay cave. One of the caves in the Crystal Cave complex (opened in 2016) included part of a large underground lake (Fig. 9.9a) that is probably part of the North Side Water Lens. A water sample from that lake, collected in August, 2017, yielded $\delta^{18}O$ of −4.52‰ and $\delta^{2}H$ of −25.0‰. A drip water sample from the end of a stalactite located near the lake yielded $\delta^{18}O$ of −3.01‰ and $\delta^{2}H$ of −25.0‰, and a sample from a well near the entrance to the complex gave a $\delta^{18}O$ of −4.34‰ and $\delta^{2}H$ of −25.0‰.

Groundwater—Perched Water Zone

Perched groundwaters from Cayman Brac are characterized by stable isotopes that plot around the global meteoric water line and the rainwater line of Grand Cayman. The stable isotopes are heavier than those from isotopically depleted rainwater and are the most isotopically depleted of any of the groundwaters (Ng 1990). The water chemistry of the perched groundwaters is controlled by evaporation that takes place on the land surface and as the water sinks or percolates into the perched lens, direct evaporation from the lens, and/or precipitation of calcite and/or dolomite.

Groundwater—Highly Brackish to Saline Water

On the Cayman Islands, the saturated zone is divided into the fresh (0–600 mg l^{-1} Cl^{-}), brackish 600–19,000 mg/l Cl^{-}), and saline zones (>19,000 mg l^{-1} Cl^{-}) zones (Ng and Beswick 1996). Na^{+}, K^{+}, Ca^{2+}, Mg^{2+}, Cl^{-}, and SO_4^{2-} are the major ions in all water on the Cayman Islands (Ng 1990). Water in the brackish and saline zones is of the sodium chloride type, whereas, the fresh groundwater is of the Ca-Mg bicarbonate type. Most of the chloride in the brackish water probably originates from the underlying saline water (Ng and Jones 1990a). The high HCO_3^{-} content and pH >7 is probably a reflection of carbonate dissolution (Ng and Jones 1990a).

Fresh groundwater on Grand Cayman, with temperatures ranging from 25 to 30 °C, contains >200 mg/l chloride even though the rainwater only contains 7–13.5 mg/l chloride. This may be due to evapotranspiration or the upward movement of water, caused by tidal oscillations from the underlying saline water that is typically ~20 m below the water table and has Cl and Sr ion concentrations akin to those in the surrounding seawater. Carbonate dissolution by the groundwater is indicated by the high bicarbonate ion concentrations and pH (>7). The molar Mg/Ca ratios of the fresh groundwater in the Lower Valley Lens and East End Lens on Grand Cayman, and Tibbetts Turn Lens on Cayman Brac are variable even though they are all housed in dolostones of the Cayman Formation.

The highly brackish and saline groundwaters have $\delta^{18}O$ values of −1.82 to +1.36‰ and $\delta^{2}H$ values of −8.9 to + 5.7‰, whereas the fresh and lightly brackish groundwaters have $\delta^{18}O$ values of −3.5 to −5.3‰ and $\delta^{2}H$ values of −22.0 to −35.0‰ (Jones et al. 1997). Chemical compositions indicate that the saline water originated from seawater whereas the brackish water formed as a result of mixing between the saline and fresh waters.

11.5.3 Hydrochemical Facies

The diagnostic chemical characteristics of groundwater can be considered in the context of hydrochemical facies that permit the classification and comparison of different types of water (Ng and Beswick 1994). Use of such a scheme shows that the water in the Lower Valley Lens and the East End Lens on Grand Cayman and the Tibbetts Turn Lens on Cayman Brac are hydrochemically different and also different from the saline groundwater and brackish groundwater found on the islands (Fig. 11.14).

The chemical components of the groundwater on the Cayman Islands are derived from various sources. As outlined by Ng and Beswick (1986, 1994) and Ng and Jones (1990a), the chemicals in the groundwater may come from (1) salts in the rainwater that may be concentrated by evaporation, (2) influx of carbonate minerals from limestones, (3) periodic incursions of water trapped in sinkholes, (4) precipitation and/or dissolution of carbonate minerals, and (5) salts that come from underlying saline and brackish waters. Variations in the sources and/or the amount of water from each source can greatly influence the composition of the groundwater.

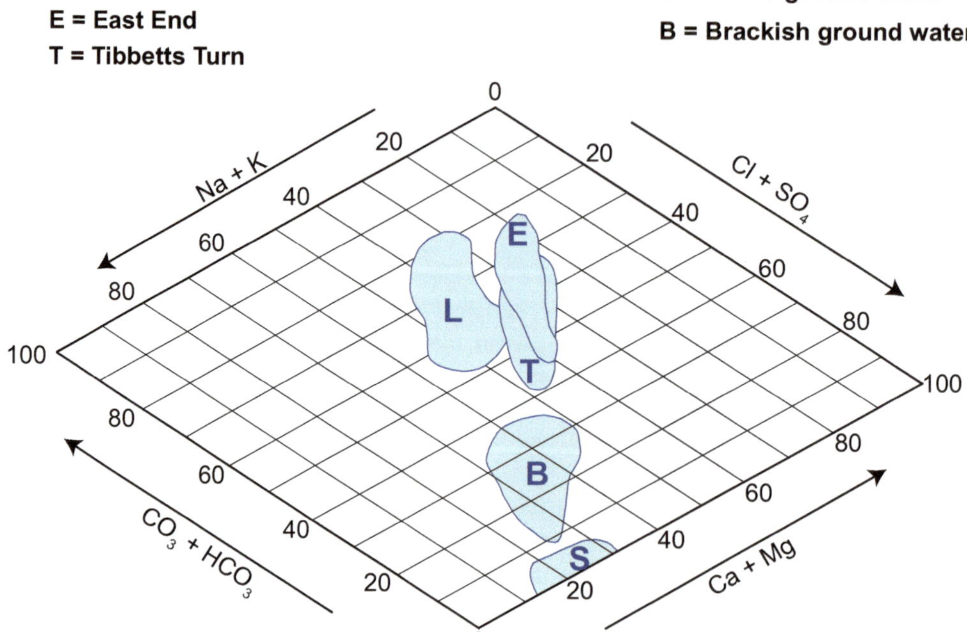

Fig. 11.14 Hydrochemical facies diagram showing comparisons between fresh groundwater from the Lower Valley lens, East End lens, and Tibbetts Turn lens, relative to brackish and saline waters. Modified from Ng and Beswick (1994, their Fig. 4.6)

Chloride Ions

On the Cayman Islands, rainwater typically contains 7–13.5 ppm chloride with the variation reflecting seasonal effects and distance from the coastline. This chloride source cannot, however, explain the >200 ppm concentrations found in the other potable water zones of the lenses. Ng and Beswick (1994) suggested that most of the chloride probably comes from the underlying saline water due to hydrodynamic dispersion associated with tidal oscillations.

Bicarbonate Ions

Dissolution of the limestone and/or dolostone bedrock of the islands produces the high bicarbonate ion content (300–400 ppm) and pH > 7 of the groundwater (Ng and Beswick 1994). Conversely, the significant amounts of calcite and/or dolomite cements found in the dolostones and limestones of the Cayman Formation and Pedro Castle Formation also indicate that subsurface precipitation of Ca and Mg is an active process.

Calcium and Magnesium Ions

The fresh groundwaters in the Lower Valley Lens, the East End Lens, and the Tibbetts Turn Lens have different Mg/Ca ratios even though they are all housed in dolostones of the Cayman Formation. The ratio in the East End groundwater is typically >0.5 whereas the groundwater from Lower Valley typically has a ratio of <0.5. These low values probably indicate that the water is being mixed with other water that may have come through the limestones of the Ironshore Formation (Ng and Beswick 1994).

11.6 Groundwater Calcite and Dolomite Saturation States

From a geological perspective, the saturation states of the various types of water on the Cayman Islands with respect to calcite and dolomite are critical because they determine the interactions that may occur between the groundwater and the limestones and dolostones in the bedrock of the islands. Ng (1990) examined this issue by applying speciation calculations under the assumption that the water–rock system operated under equilibrium conditions. These calculations, however, did not consider the kinetic factors that could impact these processes.

11.6.1 Rainwater

Rainwater on the Cayman Islands, with its low concentrations of calcium, magnesium, and bicarbonate, is undersaturated with respect to dolomite and calcite (Ng 1990). The fact that the P_{CO2} of the water is like that of the atmosphere, suggests that the CO_2 in the rainwater is in equilibrium with the atmosphere. Since the rainwater is undersaturated with respect to calcite and dolomite, it can mediate dissolution of the bedrock of the islands.

11.6.2 Perched Water

Ng (1990) showed that the water from the perched groundwaters on Cayman Brac are supersaturated with respect to calcite and dolomite, whereas the P_{CO2} is about three times that of the atmosphere. The high P_{CO2} probably resulted from CO_2 being added to the groundwater as it passed through the soil where plants were actively growing.

11.6.3 Groundwater—Fresh Water Zone

In the Lower Valley Lens, the freshwater is supersaturated with respect to calcium but is supersaturated with respect to dolomite in only 14 of the 31 wells that were sampled (Ng 1990). The latter may reflect a geographic variance because most of the water samples collected from the southernmost wells were saturated with respect to dolomite. Ng (1990) suggested that waters in the eastern and northern parts of the water lens may be under the influence of waters that come from the mangrove swamps (pH of 6.0–7.5) and calcium-rich waters that came from the Ironshore Formation. Along the south part of the lens, there is interaction with saline water that would provide the necessary Mg.

The freshwater in the East End Lens is supersaturated with respect to calcite and dolomite. Ng (1990) suggested that this may be due to its geographic location, the higher topography of the area, and/or the larger size of the lens. The P_{CO2} is about 6–12 times that of the atmosphere. Unlike the Lower Valley Lens, it is not affected by external factors. The freshwater in the Tibbetts Turn Lens on Cayman Brac is also supersaturated with respect to calcite and dolomite. The P_{CO2}, however, is lower than that in the East End Lens on Grand Cayman.

11.6.4 Groundwater—Lightly Brackish Water Zone

The lightly brackish groundwater zone on the Cayman Islands is supersaturated with respect to calcite and dolomite.

11.6.5 Groundwater—Types I and II

Ng (1990) identified two types of saline water on Grand Cayman. Although both types have a Cl^- content >19,000 mg/l, Type I has a variable chemical composition whereas Type II has a constant chemical composition.

The highly brackish to Type I saline water is generally saturated to supersaturated with respect to calcite and dolomite. In some locations, however, these waters have a pH <7 and are undersaturated with respect to calcite and dolomite, possibly because of an influx of water from nearby swamps.

Waters in the Type II saline water zone, which have a salinity like that of surrounding seawater, are supersaturated with respect to calcite and dolomite. The molar Mg/Ca ratios of 4.46–4.48, however, are lower than in seawater; possibly due to modification of the original seawater by calcite and/or dolomite dissolution in the subsurface environment.

11.7 Groundwater and Bedrock in and Around the East End Freshwater Lens

Since the original work on the East End Lens was undertaken by Mather (1972), Bugg and Lloyd (1976), various consulting companies, and Ng (1990), a considerable amount of research has been focused on the bedrock of the area in an effort to obtain a better understanding of the strata that house the East End Lens. Accordingly, numerous wells were drilled in the area so that a better understanding of the porosity and permeability characteristics of the bedrock could be developed. Sampling of the water at different depths in these wells has provided further information on the possible relationships between the groundwater and the bedrock.

Well HRQ#3, drilled close to the southwest corner of High Rock Quarry shows that the bedrock is formed primarily of low calcian dolomite (LCD) and high calcian dolomite (HCD) with calcite only being found in the uppermost part of the sequence (Fig. 11.15). Sampling of the groundwater from this well shows that the water temperature is relative constant at ~26 °C with only the near surface samples having a temperature of ~28 °C (Fig. 11.15). In this well the base of the freshwater zone is at a depth of ~18 m (2 mS/cm limit for EC), whereas the base of the brackish water is at a depth of ~54 m (50 mS/cm limit for EC) (Fig. 11.15). These boundary positions do not, however, coincide with any major lithological change in the bedrock succession (Fig. 11.15).

Water temperature profiles for wells LBL#1, GFN#1, and EEV#1, and are essentially the same as that for well HRQ#3 (Fig. 11.16). Conductivity profiles for the water from these four wells show the contrast between the different hydrological zones in LBL#1 and EEV#1 that are located outside of the East End Lens relative to those in wells HRQ#3 and GFN#1 that are within the lens (Fig. 11.16). Lithological cross-sections, SW to NE (Fig. 11.16) and SE to NW (Figs. 11.17, 11.18) show the variations in the distribution of

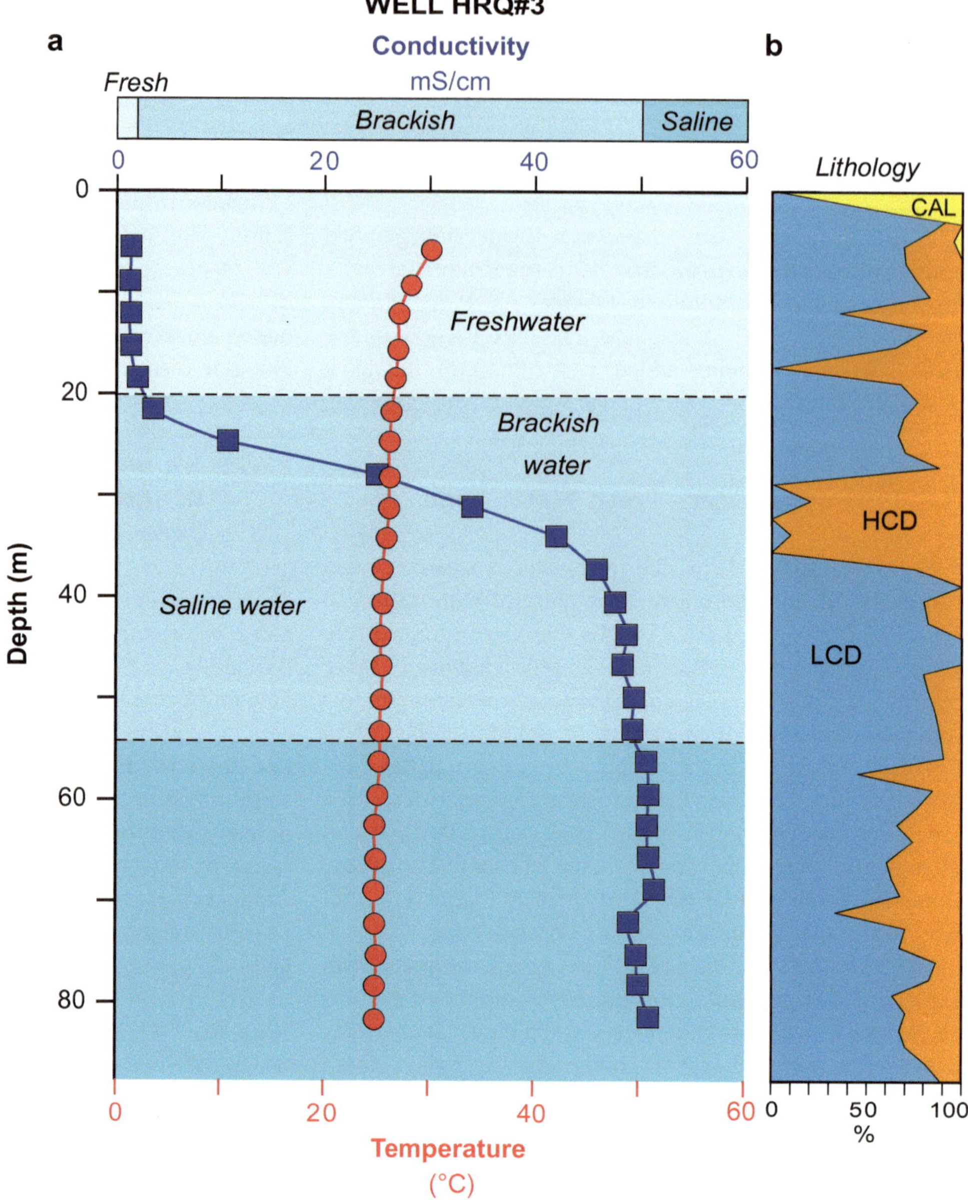

Fig. 11.15 Well HRQR#3 (near southwest corner of High Rock Quarry) showing water temperature and conductivity **a** relative to the bedrock that is formed of dolomite (LCD and HCD) with minor amounts of calcite (CAL). **b** The water zones are based on the electrical conductivity (EC) of the water

calcite, LCD, and HCD. Calcite is most common in the central area whereas the LCD and HCD are most common in the peripheral areas. There does not appear to be any direct correlation between the dominant lithologies and the distribution of the different types of water (Figs. 11.16, 11.17 and 11.18).

11.8 Development of Reverse Osmosis Plants

Installation of the reverse osmosis plants on the Cayman Islands fundamentally depends on the ability to (1) extract large volumes of saline groundwater over extended time periods, and (2) dispose of large volumes of highly saline water (~60% of the intake volume) that are produced by the desalination process. In addition to these issues, other local factors also have to be respected in the siting and development of these plants.

The reverse osmosis plant, developed between 1995 and 1998 by the Water Authority of the Cayman Islands, was located at the Lower Valley site where the Authority had previously abstracted fresh groundwater from the freshwater lens that is located there (Fig. 11.4a). From a geological perspective, this location is somewhat unique on Grand Cayman because it is situated over a large freshwater lens, and the subsurface succession includes the Ironshore Formation, Pedro Castle Formation, Cayman Formation and the

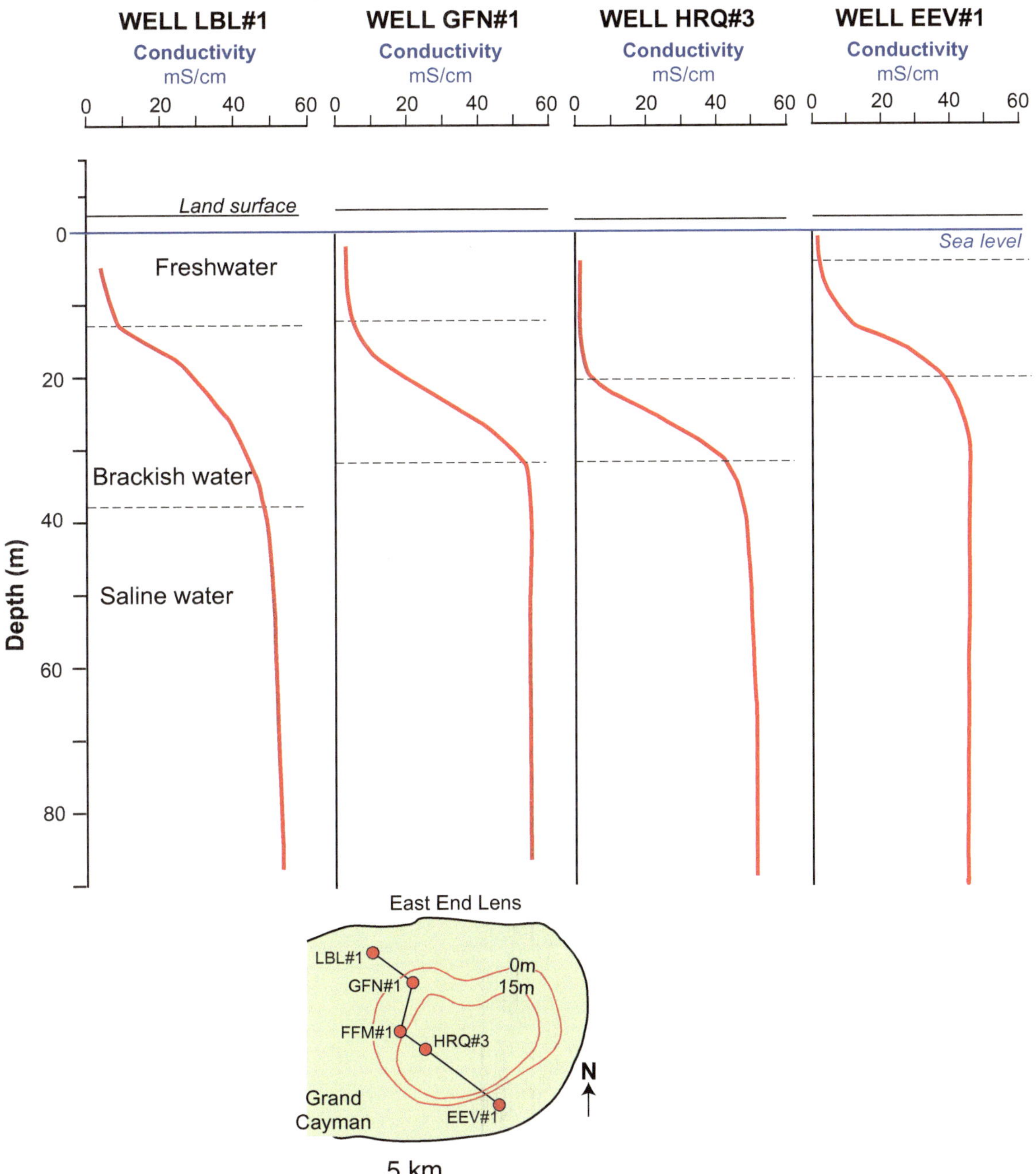

Fig. 11.16 Conductivity profiles for water in wells LBL#1, GFN#1, HRQ#4, and EEV#1 that are aligned along a southeast to northwest transect across the East End Lens. Note variations in thickness and distributions of the freshwater, brackish water, and saline water from well to well. Map shows location of wells relative to the East End water lens (thickness of freshwater water shown by red contour lines)

Brac Formation (Fig. 11.19). In contrast, the East End Lens is located in bedrock that is formed entirely of the Cayman Formation. Development of the wells for the Lower Valley reverse osmosis plant required preservation of the freshwater lens so that people in the neighbourhood could continue to use it for domestic and farming purposes. Thus, prior to development of the Lower Valley site, a detailed geological investigation was undertaken in order to identify the abstraction and disposal zones for the wells while assuring the integrity of the freshwater lens. This involved drilling a well to a depth of 150 m (the limit for drilling equipment on the island at that time) with 20 core samples (each one up to 3.3 m long, depending on drilling conditions) and well-cutting samples collected between (Fig. 11.19).

Analyses of all the core and well-cutting samples provided the depths of the formation boundaries, the thickness

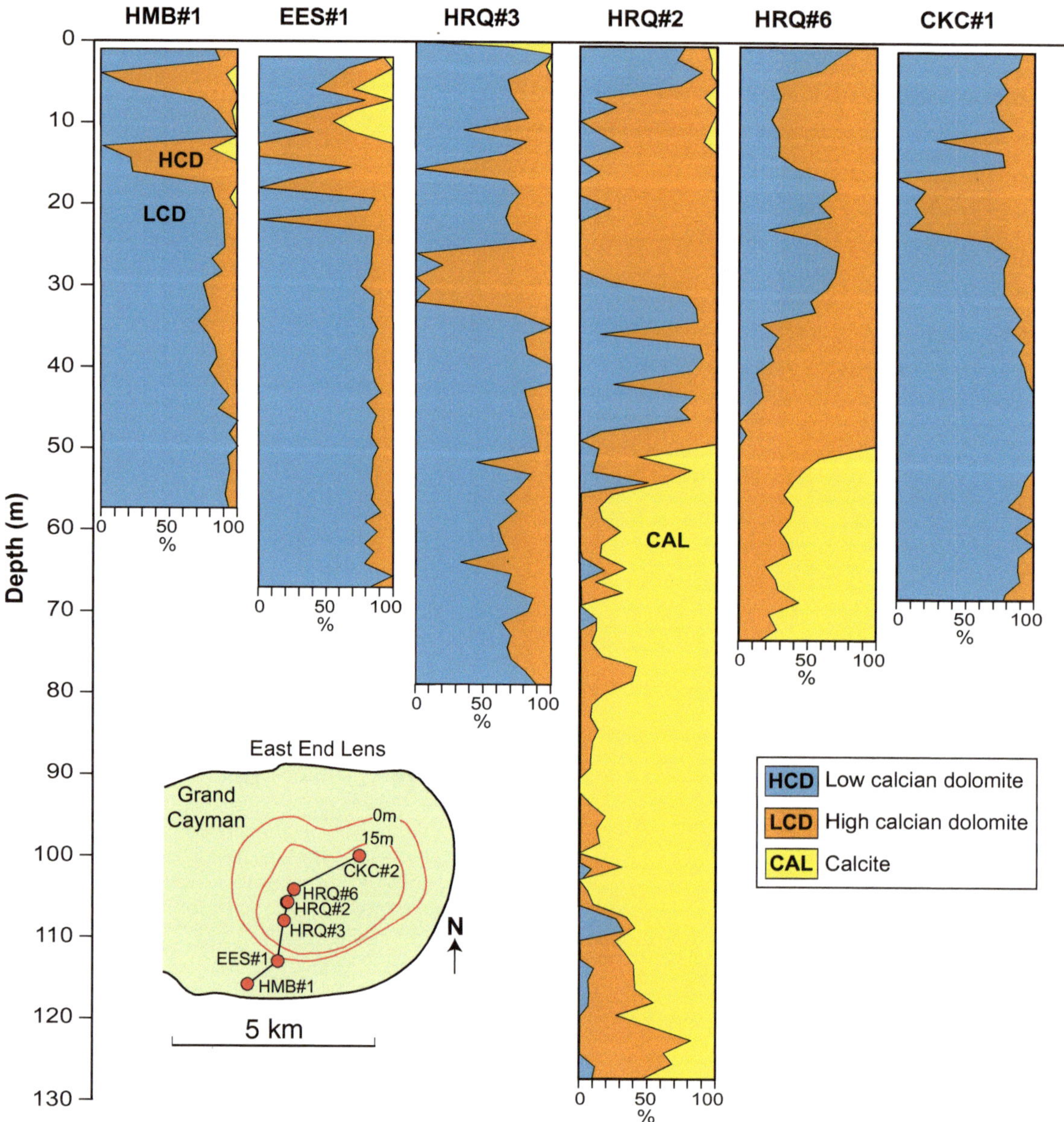

Fig. 11.17 Southwest to northeast cross-section across the East End Water Lens showing variable lithology of the bedrock. Inset map shows locations of sections, location of freshwater lens, and thickness (red contour lines) of freshwater lens

of each formation, and the constituent rock types (based on inspection of the samples, thin section analyses, and X-ray diffraction analyses (XRD)). The thin succession of limestones (<5 m thick) that forms the Ironshore Formation was not analyzed in detail because much of it had been disturbed, to varying degrees, by previous development of the site. With the exception of one sample at the top of the Pedro Castle Formation that contained minor amounts of calcite, the succession is formed entirely of LCD and HCD (Fig. 11.19). In general, the dolostones in the Pedro Castle Formation and the upper part of the Cayman Formation contained higher percentages of LCD than the dolostones from the lower part of the Cayman Formation and the underlying Brac Formation (Fig. 11.19).

The porosity and permeability (maximum, 90° to maximum, and vertical) were determined from each of the 20 cores (each tested sample being ~15 cm long, 10 cm in diameter, and devoid of any fractures or cavities that cut through the entire diameter of the core) in order to develop a clearer understanding of the subsurface strata (Fig. 11.19). Those analyses showed that the porosity ranges from 8.4 to 47.4% (average 32.5%), maximum permeability of 175–10,240 mD (average 2634.5 mD), permeability at 90° to maximum of 144–12,240 mD (average 2372.2 mD), and

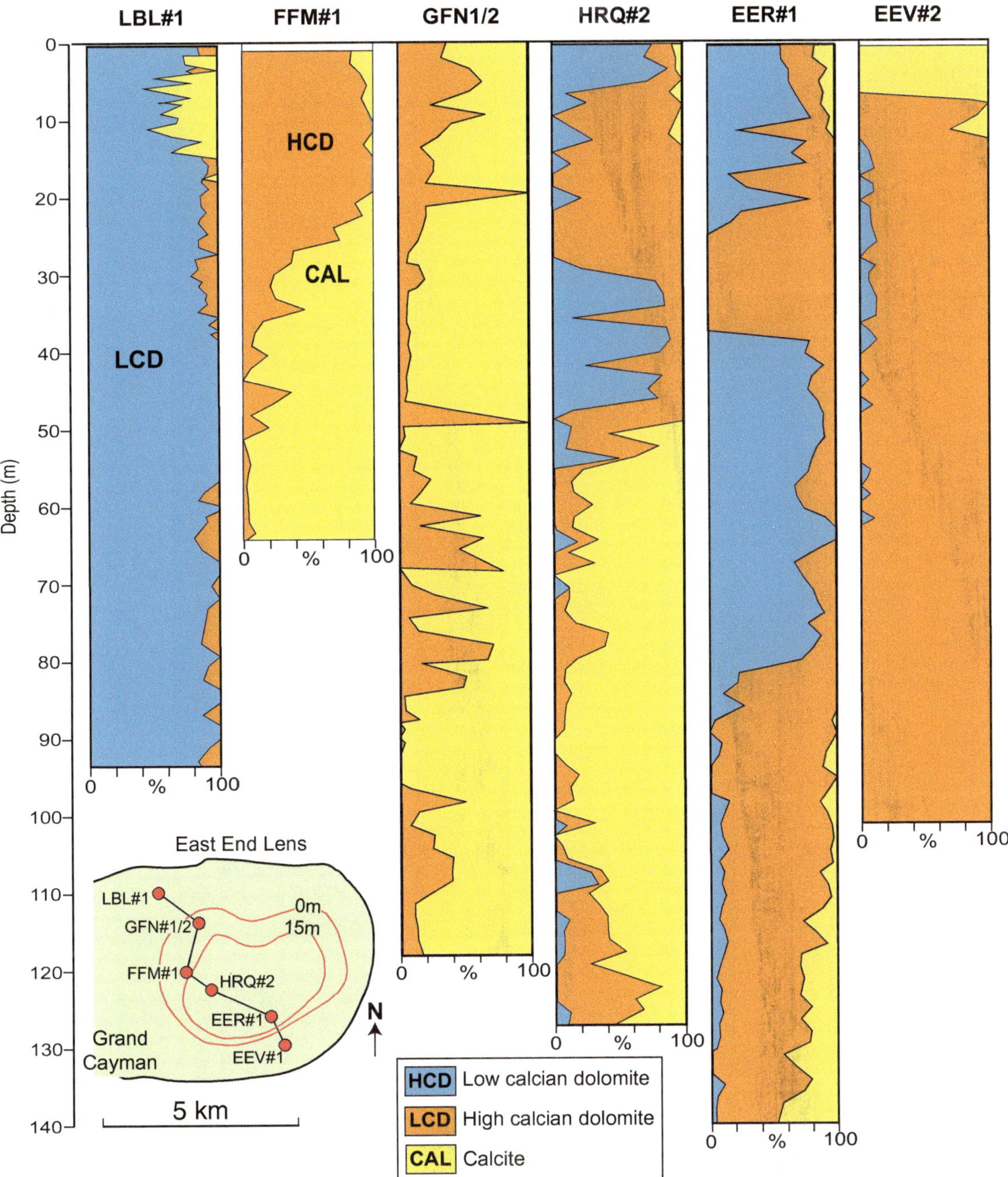

Fig. 11.18 Southeast to northwest cross-section across the East End Water Lens showing variable lithology of the bedrock. Inset map shows locations of sections, location of freshwater lens, and thickness (red contour lines) of freshwater water lens

vertical permeability of 0.09–10,240 mD (average 1570 mD). The porosity data allowed division of the Cayman Formation into the upper "cap rock" from 25 to 41 m that is characterized by low porosity (<15%) and low permeabilities and a lower "porous zone", from 41 to 148 m, that is characterized by high porosities (up to 47.5%, average 37.6%) and variable permeabilities (Fig. 11.19). These statistics, however, must be viewed relative to the manner in which they are determined. The analyses were determined from pieces of full diameter core that are devoid of any obvious cavities and/or fractures that cut through the core. As such the analyses provide porosity and permeability values that relate solely to the matrix of the rock. These measurements, for example, do not include any porosity/permeability that is related to large cavities/caves or fractures that exist in the subsurface. Thus, fractures and solution-widened joints like those known from surface exposures of the Cayman Formation (Figs. 11.8 and 11.11)

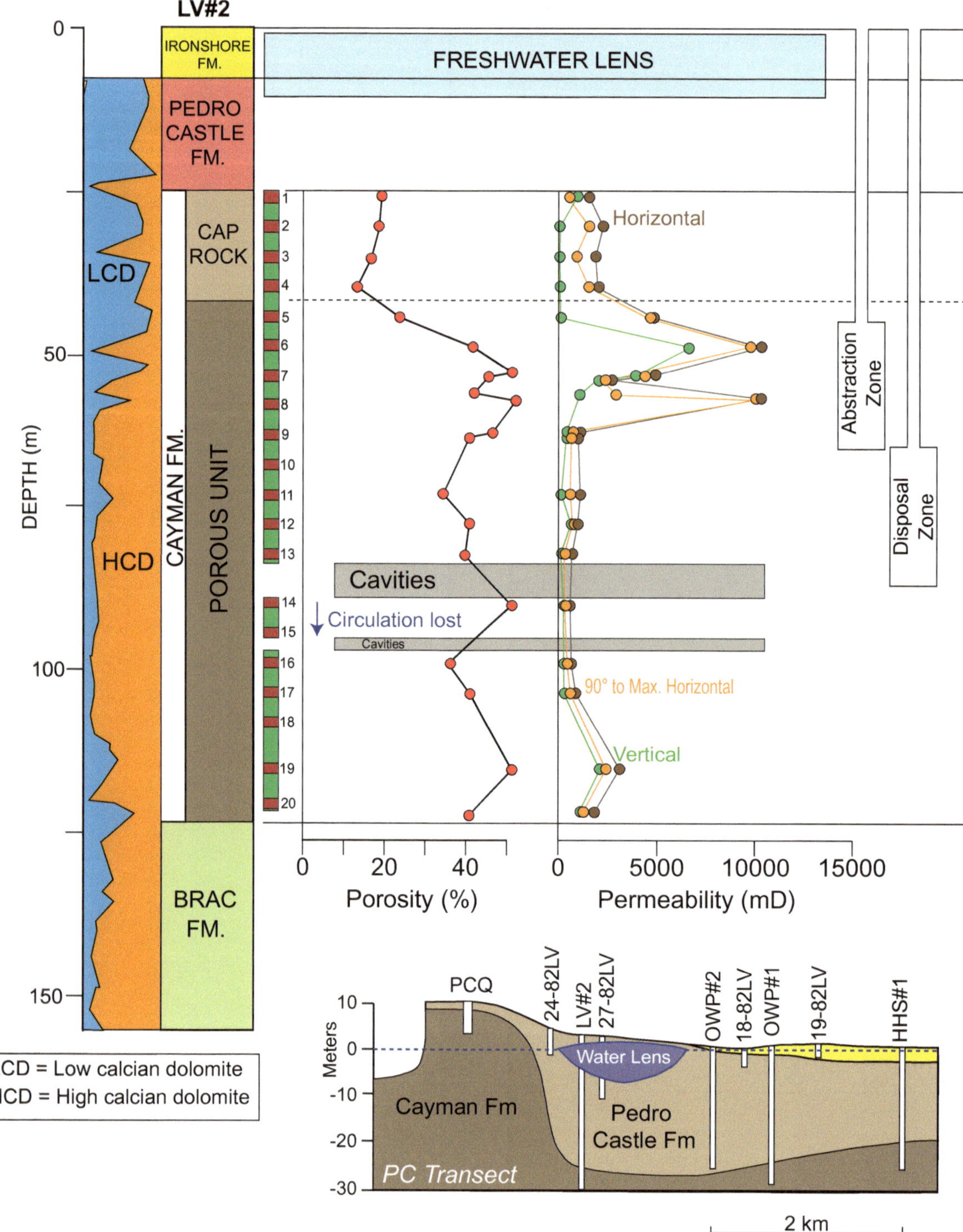

Fig. 11.19 Well LV#2, located in Lower Valley Lens, used for exploratory work to determine locations of abstraction and disposal zones for the salt water reverse osmosis plant. Diagram shows the stratigraphic succession, distribution of low calcian dolomite (LCD) and high calcian dolomite (HCD), location of cores used to determine porosity and permeability, measured porosity and permeability values, and positions of abstraction and disposal zones. Cross-section shows location of well LV#2 relative to Lower Valley water lens, other wells that were drilled in the area, and the stratigraphic succession of the area

cannot be factored into calculations of the overall porosity and permeability of the rocks.

At the Lower Valley site, the cap rock at the top of the Cayman Formation effectively segregates the porous dolostones of the lower part of the formation from the overlying Pedro Castle Formation and Ironshore Formation (Fig. 11.19). Given that the freshwater lens resides in the upper part of the Pedro Castle Formation and lower part of the Ironshore Formation, the cap rock provides an effective barrier against any upward movement of groundwater from the lower part of the Cayman Formation (Fig. 11.19). Based on these data, the abstraction zone was placed between 45 and 67 m below the surface, whereas the disposal zone was placed between 63 and 86 below the surface with the disposal well located ~100 m away from the abstraction well. That system has been maintained since the reverse osmosis plant first became operational in 1998.

As shown by Rigby and Roberts (1976, their Text-Fig. 30), Ng (1990), and Ng et al. (1992, their Fig. 8), numerous joints and fracture systems dissect the dolostones and limestones of the Cayman Formation and Pedro Castle Formation. Although some joints and fractures are filled with sediments and precipitates, many remain open and provide avenues along which large volumes of seawater can easily move inland. Over the last 10 years, the reverse osmosis plant at Lower Valley has abstracted between 2.7 and 0.14 million cubic metres of saline water per year. The presence of open fractures and joints also, at least in part, helps to explain the consistent flow of saline water into the abstraction zone over the last 10 years.

With the ever-increasing demand for potable water, the Water Authority developed the North Side desalinization plant, which is located in the central part of Grand Cayman approximately 625 m northwest of the Botanical Park. There is no freshwater lens in this area and relatively little was known about the subsurface geology because of the paucity of wells that had been drilled in that area. Given the challenges that were encountered with the well drilling and the nature of the succession, drilling was completed in two phases. In early 2005, two pilot wells (NSC#1/2) were drilled to a depth of ~150 m using the drilling rig that was available on the island—this produce well cuttings and some cores. Well NSC#2 was drilled close to NSC#1 after collapse of the bedrock prevented further drilling. Problems encountered with this well included poor cores because the friable carbonate bedrock was characterized by high porosities and it proved impossible to locate prospective abstraction and disposal zones. Thus, when the actual production and disposal wells were drilled in 2008, reverse circulation drilling was used for well NSC#3 so that the bedrock succession to a depth of ~245 m could be drilled. The overall succession, which is a composite based on wells NSC#1, 2, and 3, is different from that found in the wells around the East End Lens. The succession in NSC#1–3, can be divided into Units I to V based on the ratios of calcite to dolomite throughout the succession (Fig. 11.20).

- Unit I: surface to a depth of 61 m—entirely limestone.
- Unit II: 61–128 m—limestone with variable but generally minor amounts of dolomite
- Unit III: 128–171 m—mostly dolomite (HCD) but with variable amounts of calcite
- Unit IV: 171–218 m—dolostone (HCD)
- Unit V: 218–245 m—same as unit III

Porosity and permeability, which could only be obtained from 18 core samples in the upper part of the well from 0.9 to 120 m (Fig. 11.20), yielded tested porosities from 11 to 44% with an average of 29.6%, maximum permeability from 0.53 to 5105 mD (average 938 mD), permeability at 90° from 0.24 to 4500 mD (average 789.6 mD), and vertical permeability from 0.01 to 698 mD (average 83 mD). No analyses could be obtained from the deeper part of the well because reverse circulation drilling was used for depths below ~150 m. Since its inception in 2009, the North Side Water Works has been producing between 0.3 and 2 million cubic metres of potable water per year.

11.9 Conclusions

With the ever-increasing need for potable water, the need for a better understanding of the hydrogeology characteristics of Grand Cayman, Cayman Brac, and Little Cayman has become progressively more important. Detailed analyses of the hydrogeology of the islands has led to the following important conclusions.

- There are no surface rivers or streams on the any of the Cayman Islands because rainwater quickly flows into the subsurface via the complex arrays of fractures and joints that characterize the bedrock. Lakes and ponds, characterized by seasonally fluctuating water levels and compositions, are generally limited to coastal areas.
- On Grand Cayman, the Lower Valley, East End, and North Side are the largest water lenses. Other smaller lenses have commonly been lost due to over-extraction and/or contamination.
- On Cayman Brac, the Tibbets Turn Lens is the largest freshwater lens. Other smaller lenses are found along the north coast and perched water lenses are found in the elevated core of the island. On Little Cayman, little is known about the freshwater resources and further exploration is needed.
- Subterranean water flow on the each of the Cayman Islands involves a complex architecture of fractures,

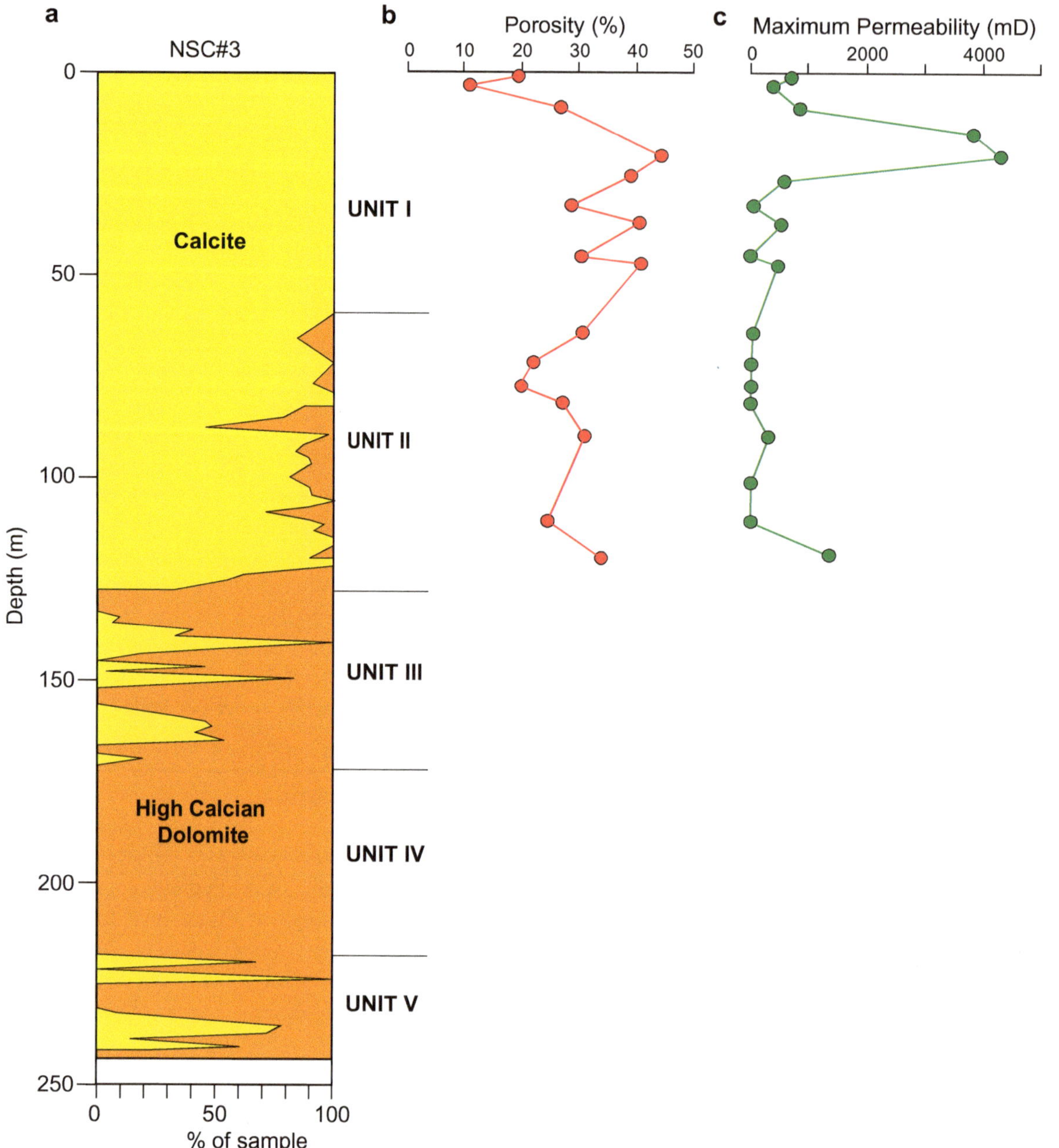

Fig. 11.20 **a** Lithological succession in well NSC1-3 showing Units I to V that are based on the relative proportions of calcite and dolomite (all HCD). **b**, **c** Porosity and maximum permeability profiles for the upper part of the well. Porosity and permeability could not be determined for the lower part of the succession because no cores were obtained from that part of the sequence

enlarged joints, and bedrock permeability with water levels and compositions being controlled by tidal activity and the periodic influx of rain.

- The porosity and permeability of the bedrock limestones and dolostones is highly variable at all scales and therefore difficult to accurately model for specific areas.
- The groundwater on the Cayman Islands is variably saturated with respect to calcite and dolomite.
- The Water Authority of the Cayman Islands has developed six reverse osmosits plants (four on Grand Cayman, two on Cayman Brac) in order to meet the ever-increasing demand for potable water. The Cayman Water Company also operates several reverse osmosis plants in order to supply potable to the western part of Grand Cayman.

References

Billymer JHS (1946) The Cayman Islands. Geogr Rev 36:29–43

Bugg SF, Lloyd JW (1976) A study of freshwater lens configuration in the Cayman Islands using resistivity methods. Q J Eng Geol 9: 291–302

Chidley RE, Lloyd JW (1977) A mathematical model study of freshwater lenses. Groundwater 15:215–222

Doran E (1954) Landforms of Grand Cayman Island, British West Indies. Tex J Sci 6:360–377

Fawcett W (1888) Cayman Islands. Bull Misc Inf (R Bot Gard Kew) 1888:160–163

Fawcett W (1889) Report by the director of public gardens and plantations on the Cayman Islands. Bull Bot Dep Jamaica 11:6–7

Hughes JD (2017) Grand Cayman environmental history: a case study of the Anthropocene. Ekonomska-I Ekoshistorija 8:88–95

International R.a.D. (1976) Report of feasibility study for water supply, sewerage and storm water drainage. Report to Government of the Cayman Islands

Jones B, Ng KC, Hunter IG (1997) Geology and hydrogeology of the Cayman Islands. In: Vacher HL, Quinn TM (eds) Geology and hydrogeology of carbonate Islands. Elsevier Science, The Netherlands, pp 299–326

Jones B, van Genderen HJ, van Zaten T (2001) Well-field design for a saltwater reverse osmosis plant located over a fresh water lens in Lower Valley, Grand Cayman, Cayman Islands. In: Frederick-van Genderen G, Tedd MB (eds) Innovative technologies in the water and waster industries for the 21st century. Proceedings of the 10th annual CWWA Conference and Exhibition, Grand Cayman, Cayman Islands. Water Authority – Cayman, pp 36–52

Kreitler CW, Browning LA (1983) Nitrogen isotope analysis of groundwater nitrate in carbonate aquifers: natural sources versus human pollution. J Hydrol 61:285–301

Mather JD (1972) The geology of Grand Cayman and its control over the development of lenses of potable groundwater. In: VI Conferencia Geologica Del Caribe- Margarita Venezuela Memorias, pp 154–157

Ng K-C, Beswick RGB (1986) Interpretation of hydrochemical facies of ground water in Grand Cayman. In: Qiuinones F, Sanchez AV, Smith HH (eds) Proceedings of the third Caribbean Islands water resources congress, St. Thomas, U.S. Virgin Islands, pp 51–55

Ng K-C, Beswick RGB (1994) Ground water of the Cayman Islands. In: Brunt ME, Davies JE (eds) The Cayman Islands: natural history and biogeography. Kluwer Academic Publishers, Holland, pp 61–74

Ng K-C, Beswick RGB (1996) Management of ground water resources on Grand Cayman: a methodology for developing small fresh water lenses. Apett Jl 30:51–59

Ng KC (1990) Diagenesis of the Oligocene-Miocene Bluff Formation of the Cayman Islands – a petrographic and hydrogeochemical approach. PhD thesis, University of Alberta, Edmonton

Ng KC, Jones B (1990a) Chemical and stable isotopic characteristics of ground water on Grand Cayman. Trop Hydrol Caribb Water Resour 411–420

Ng KC, Jones B (1990b) Porosity and permeability development in the Oligocene-Miocene Bluff Formation, Grand Cayman. In: Transactions of the 12th Caribbean geological conference, pp 115–124

Ng KC, Jones B (1995) Hydrogeochemistry of Grand Cayman, British West Indies: implications for carbonate diagenetic studies. J Hydrol 164:193–216

Ng KC, Jones B, Beswick RGB (1992) Hydrogeology of Grand Cayman, British West Indies: a karstic dolostone aquifer. J Hydrol 134:273–295

Ravenscroft P (1984) On the problems of monitoring fresh water lenses on small islands. In: 1st Caribbean Island Resource Congress of Conference, pp 169–190

Richards and Dumbleton International (1976) Report of feasibility study for water supply, sewerage and storm water drainage. Unpublished report

Richards and Dumbleton International (1980a) Further report on the ground water resources of the Cayman Islands. Report to Government of the Cayman Islands. Unpublished report

Richards and Dumbleton International (1980b) Report on the ground water resources of Cayman Brac. Report to Government of the Cayman Islands. Unpublished report

Rigby JK, Roberts HH (1976) Grand Cayman Island: geology, sediments and marine communities 4. Brigham Young University, Geology Studies, Special Publication no 4, 97pp

Vacher HL, Ayers JF (1980) Hydrology of small oceanic islands - utility of an estimate of recharge inferred from the chloride concentration of the freshwater lenses. J Hydrol 45:21–37

Wallace Evans & Partners (West Indies) Consulting Engineers (1974) Grand Cayman mains water supply feasibility report. Grand Cayman, Unpublished report, 64p

Epilogue

In the simplest sense, the geological development of the Cayman Islands at any given time involves the formation and accumulation of carbonate sediments below sea level and karst development and terrestrial sediment accumulations on areas exposed above sea level. This simplistic model, however, is in a constant state of flux because of changes in sea level that reflect the complex interplay between fluctuating eustatic sea level changes that are controlled by global factors and tectonic uplift/subsidence that are largely related to local tectonic factors. Grand Cayman, Little Cayman, and Cayman Brac are each located are separate fault blocks on the Cayman Ridge that experience independent vertical movements. Temporal climate changes are also important because they influence surface seawater temperatures and dictate the style and rates of subaerial weathering that affect the exposed limestones and dolostones on each island.

The diagenetic changes associated with the transformation of the carbonate sediments into lithified limestones is controlled by many different parameters that are largely controlled by the local climate and the local groundwaters. On the Cayman Islands, the transformation of thick successions of limestones to dolostones has further complicated the overall stratigraphic and sedimentological framework of the islands. The origin of these "island dolostones", which are similar to those found on other isolated islands throughout the oceans of the world, is still open to debate.

With the ever-increasing population of the Cayman Islands, the provision of potable water (i.e., drinking water) has become a major issue. Today, much of the needed water is produced by the reverse osmosis process with the production facilities being carefully located on inland sites where the geological framework allows extraction of the large quantities of saltwater that is needed for this process as well as disposal of the highly saline brines that are produced by this process. Knowledge of the geology of each island, as presented in this book, has played a key role in the development and provision of the potable water.

Although significant advances have been made with respect to our understanding of the geological development of the islands, many issues remain to be resolved. These include the following:

1. The geology of Little Cayman needs to be fully delineated so that its geological evolution can be compared to the temporal development of Grand Cayman and Cayman Brac. This goal can only be achieved once some deep wells have been drilled and sampled (ideally with core) – the very limited surface outcrops are insufficient for this purpose.
2. Many aspects of the "dolomite problem" remain unresolved. While much is known about the dolostones, the precise factors that trigger its formation remain elusive. Although true on a global scale, the dolostone successions on the Cayman Islands offer a natural laboratory that may provide some of the answers to this long-standing geological problem.

The Cayman Islands have proven to be ideal natural laboratories for the study of carbonate sediments, carbonate rocks, karst development, and the formation of phosphates, terra rossa, and swamp deposits.

B. Jones, *Geology of the Cayman Islands*,
https://doi.org/10.1007/978-3-031-08230-6